Lecture Notes in Physics

Volume 861

For further volumes:
www.springer.com/series/5304

The Lecture Notes in Physics

The series Lecture Notes in Physics (LNP), founded in 1969, reports new developments in physics research and teaching—quickly and informally, but with a high quality and the explicit aim to summarize and communicate current knowledge in an accessible way. Books published in this series are conceived as bridging material between advanced graduate textbooks and the forefront of research and to serve three purposes:

- to be a compact and modern up-to-date source of reference on a well-defined topic
- to serve as an accessible introduction to the field to postgraduate students and nonspecialist researchers from related areas
- to be a source of advanced teaching material for specialized seminars, courses and schools

Both monographs and multi-author volumes will be considered for publication. Edited volumes should, however, consist of a very limited number of contributions only. Proceedings will not be considered for LNP.

Volumes published in LNP are disseminated both in print and in electronic formats, the electronic archive being available at springerlink.com. The series content is indexed, abstracted and referenced by many abstracting and information services, bibliographic networks, subscription agencies, library networks, and consortia.

Proposals should be sent to a member of the Editorial Board, or directly to the managing editor at Springer:

Christian Caron
Springer Heidelberg
Physics Editorial Department I
Tiergartenstrasse 17
69121 Heidelberg/Germany
christian.caron@springer.com

Jean Souchay · Stéphane Mathis · Tadashi Tokieda

Editors

Tides in Astronomy and Astrophysics

 Springer

Editors
Jean Souchay
Dépt. Astronomie Fondamental
Observatoire de Paris
Paris, France

Tadashi Tokieda
Trinity Hall
University of Cambridge
Cambridge, UK

Stéphane Mathis
Laboratoire AIM Paris Saclay
Laboratoire AIM Paris Saclay Diderot
Gif-sur-Yvette, France

ISSN 0075-8450　　　　　　　　　　ISSN 1616-6361 (electronic)
Lecture Notes in Physics
ISBN 978-3-642-32960-9　　　　　　ISBN 978-3-642-32961-6 (eBook)
DOI 10.1007/978-3-642-32961-6
Springer Heidelberg New York Dordrecht London

Library of Congress Control Number: 2012953899

Preface

In the spring of 2009 in Cargèse, Corsica, we organized a school titled 'Tides in Astronomy and Astrophysics' with the support of CNRS (Centre National de la Recherche Scientifique), which brought researchers and students together and approached the theme of tides from as many directions as we could, offering lectures of a few hours each. This book is the outcome. We hope it will prove useful to a wide variety of readers.

The book focuses on the fundamental theories of tides at different scales of the universe—from tiny satellites to whole galaxies—and on the most recent developments. It also attempts to place the study of tides in a historical perspective.

We are grateful to Springer's Editorial Board for welcoming the book to the *Lecture Notes in Physics* series. We wish to express our thanks as well to the laboratory SYRTE (Systèmes de Référence Temps-Espace, Observatoire de Paris) and to CNRS and its staff, especially Victoria Terziyan who helped to coordinate the spring school.

J. Souchay
S. Mathis
T. Tokieda

Contents

Chapter 1
Tides: A Tutorial

Tadashi Tokieda

Abstract These notes, from a course I gave at a CNRS school in Cargèse in March 2009, have the aim of quickly letting non-experts pick up a physical intuition and a sense of orders of magnitude in the theory of tides. 'Tides' include ocean tides as well as tidal effects in astronomy. The theory is illustrated by a variety of back-of-the-envelope problems, some of them surprising, all of them simple.

1.1 What These Notes Do

The reader is asked to refer to, and sooner or later to memorize, the data listed in Section 1.2. These data allow performing order-of-magnitude estimates in all the illustrative

Problems, which are boxed against a grey background

... and whose solutions are proposed under the line.

Section 1.3 is a review of elementary material on gravitation. I tried to archive a sampling of neat factoids from the classical literature that are no longer always reproduced in the modern. The theory of tides proper is in 1.4 and 1.5, emphasizing ocean tides. 1.6 explores applications to astronomy.

However, the attitude adopted in these notes is an applied mathematician's, rather than an oceanographer's or an astronomer's: we want to form an *intuition* for the *principles* and to estimate *orders of magnitude* on *toy problems*. Predictions of day-to-day ocean tides subject to accidental features of sea floors and coastlines are outside our program: nor Laplace's tidal equations (1776), nor mapping of co-tidal lines (Whewell, –1836), nor harmonic analysis of tidal records (Kelvin, 1867–)

T. Tokieda (✉)
Trinity Hall, Cambridge CB2 1TJ, England
e-mail: tokieda@dpmms.cam.ac.uk

J. Souchay et al. (eds.), *Tides in Astronomy and Astrophysics*,
Lecture Notes in Physics 861, DOI 10.1007/978-3-642-32961-6_1,
© Springer-Verlag Berlin Heidelberg 2013

are touched upon,[1] let alone progress in the last half-century thanks to large-scale computing and satellite technology. Tout cela est prodigieusement conté dans les chapitres suivants de ce livre...

Technical terms are underlined on their first appearance.

Throughout the notes, an indicator (PIC1 ▶) means please look at the picture marked PIC1 on one of the plates on a later page, (◀ PIC∞) at PIC∞ on an earlier page.

1.2 Reference Data

	Earth ♁	Moon ☽	Sun ☉
radius	$R_{♁} = \dfrac{4 \times 10^7}{2\pi}$ m	$R_{☽} \approx \dfrac{1}{4} R_{♁}$	$R_{☉} \approx 100 R_{♁}$
mass	$M_{♁} \approx 6 \times 10^{24}$ kg	$M_{☽} \approx \dfrac{1}{80} M_{♁}$	$M_{☉} \approx \dfrac{1}{3} \times 10^6 M_{♁}$
density	$\rho_{♁} \approx 5.5\, \rho_{water}$	$\rho_{☽} \approx 3.3\, \rho_{water}$	$\rho_{☉} \approx 1.4\, \rho_{water}$

distance Earth-Moon	$D_{☽}$	$\approx 60\, R_{♁}$
distance Earth-Sun (1 A.U.)	$D_{☉}$	$\approx \dfrac{1}{4} \times 10^5 R_{♁}$
density of water	ρ_{water}	$= 10^3$ kg/m^3
gravitational constant		$\approx \dfrac{2}{3} \times 10^{-10}$ N m^2/kg^2
gravitational acceleration at sea level	$g = \dfrac{GM_{♁}}{R_{♁}^2}$	≈ 10 m/sec^2
weight of a small apple		≈ 1 N
speed of light in vacuo	c	$\approx 3 \times 10^8$ m/sec
1 year		$\approx \pi \times 10^7$ sec, with $|error| < 0.4\,\%$
		(alternatively $\approx 10^{7.5}$ sec, with $|error| < 0.25\,\%$)

From now on, we shall use the reference data all the time, everywhere.

---

[1] Except here.

Problem 1.21 Which looks wider to a terrestrial observer, the Sun or the full Moon?

We use the reference data to estimate their apparent angular diameters:

$$\frac{2R_{\odot}}{D_{\odot}} \approx \frac{2 \cdot 100}{\frac{1}{4} \times 10^5} = 8 \times 10^{-3} \text{ radian, or just under } \frac{1}{2} \text{ degree}$$

$$\frac{2R_{\mathbb{D}}}{D_{\mathbb{D}}} \approx \frac{2 \cdot \frac{1}{4}}{60} = \text{also just under } \frac{1}{2} \text{ degree}$$

which is a memorable round number (1 degree is the width of a finger at the end of an outstretched arm). This coincidence of apparent diameters is responsible for the occurrence of total eclipses.

Problem 1.22 How strong is the gravitational attraction between the Earth and the Moon?

We use the reference data to estimate

$$\frac{GM_{\oplus}M_{\mathbb{D}}}{D_{\mathbb{D}}^2} = \frac{GM_{\oplus}}{60^2 R_{\oplus}^2} \cdot \frac{1}{80}M_{\oplus} = g \cdot \frac{1}{60^2 \cdot 80} \cdot M_{\oplus}$$

$$\approx 10 \cdot \frac{1}{48 \cdot 6 \times 10^3} \cdot 6 \times 10^{24} \approx 2 \times 10^{20} \text{ N},$$

yet another memorable round number. Sometimes the trick of rewriting with the aid of g spares us parades of decimals.

1.3 Gravitation

1.3.1 Why $1/r^2$?

Why does the gravitational attraction $F(r)$ vary like $1/r^2$?

Imagine a point mass, which generates a vector field of gravitational force in the space surrounding it. Let us consider the flux of this field through a sphere of radius r centered at the mass (PIC1 ▶).

If we take any two concentric spheres of different radii, then the fluxes through these spheres must be the same, since we are assuming no other source/sink of gravitation in the vacuum between the spheres (PIC2 ▶). So

$$F(r) \cdot 4\pi r^2 = \text{const} \implies F(r) \propto r^{-2} \quad \text{in } \mathbb{R}^3.$$

The same argument works in any dimension.[2]

---

[2]Mathematically we have rediscovered the Green's function for the Laplacian in $\mathbb{R}^n$.

PIC1

PIC2

PIC3

PIC4

PIC5

$d\Omega$

PIC6

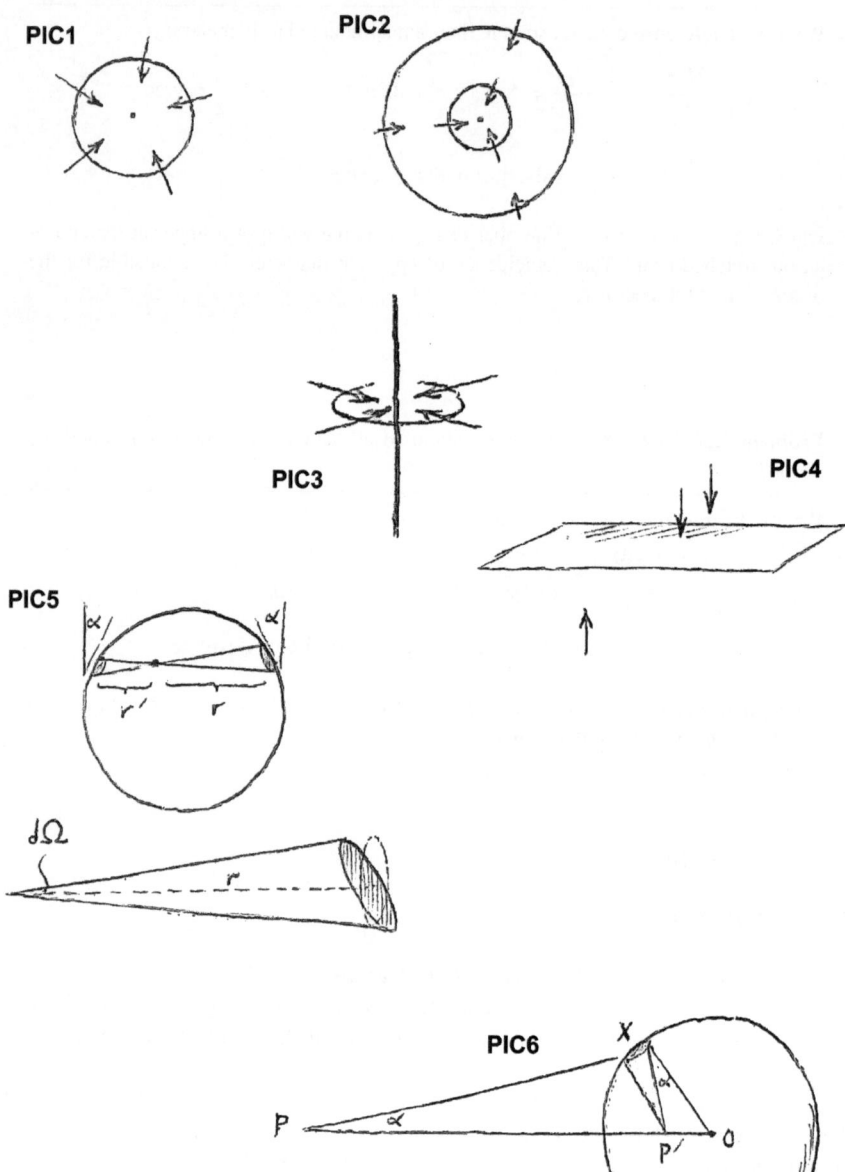

Theorem 1.3.11 *In $\mathbb{R}^n$, the force of gravitational attraction due to a point mass varies like $F(r) \propto r^{1-n}$. The potential varies like r^{2-n} in dimension $n \neq 2$, like $\log r$ in dimension $n = 2$.*

Problem 1.3.12 An infinite line on which mass is distributed uniformly ($\blacktriangleleft$ PIC3). How does the attraction F by the line depend on the distance r from the line? The same problem for an infinite uniform plane ($\blacktriangleleft$ PIC4).

For the line, we are solving the attraction problem effectively in dimension $n = 2$, so $F(r) \propto r^{-1}$. For the plane, effectively $n = 1$ and $F(r) \propto \pm$ const, i.e. the attraction does not depend on how distant we are from the plane, though of course it changes sign from one side of the plane to the other.

1.3.2 Attraction by a Spherical Shell

Theorem 1.3.21 *Inside a spherical shell on which mass is distributed uniformly, the force of gravitational attraction is zero.*[3]

Proof ($\blacktriangleleft$ PIC5) The attraction toward right is

$$\frac{r^{n-1}\,d\Omega}{\cos\alpha} \cdot r^{1-n} = \frac{d\Omega}{\cos\alpha}.$$

Likewise, the attraction toward left is $d\Omega/\cos\alpha$. These cancel each other, and such a cancelation occurs in every direction. $\square$

Theorem 1.3.22 *Outside a spherical shell, the attraction is as if the shell's entire mass were concentrated at its center.*[4]

Proof ($\blacktriangleleft$ PIC6) Take P' to be 'inverse' of P with respect to the sphere, such that $OP' \cdot OP = \text{radius}^2 = OX^2$. By similar triangles OPX and OXP', we have $P'X/PX = OX/OP$. Since by symmetry the overall attraction acts along OP only, we may consider

(mass element) $\cdot$ (attraction per unit mass) $\cdot$ (component along OP)

$$= \frac{P'X^{n-1}\,d\Omega}{\cos\alpha} \cdot \frac{1}{PX^{n-1}} \cdot \cos\alpha = OX^{n-1}\,d\Omega \cdot \frac{1}{OP^{n-1}}.$$

But $OX^{n-1}\int d\Omega$ is the mass of the sphere. $\square$

---

[3]The zero-gravity conclusion is equally valid for the inside of a uniform ellipsoidal shell; by 'shell' is meant a region bounded between similar concentric (not confocal) ellipsoids.

[4]Outside an ellipsoidal shell the result is more complicated.

Fig. 1.1
Newton (1643–1727)

These are propositions LXX, LXXI in book I of Newton's *Principia*.

Remark 1.3.23

(1) A similar idea proves that the attractions by the two shaded slices (PIC7 ▶) are equal.
(2) These are theorems in potential theory, about *any* field whose potential u is harmonic, $\nabla^2 u = 0$.

Digression 1.3.24 What do you think of the position of the Sun on this British pound note?

The Sun would be at the center if $F(r) \propto r$ (harmonic oscillator).

Problem 1.3.25 An infinitely long uniform cylindrical shell (PIC8 ▶). How does the attraction F by the cylinder depend on the distance r from the cylinder's axis?

The effective dimension is $n = 2$. $F(r) = 0$ inside, $F(r) \propto r^{-1}$ outside.

Beware: it is *not* the case that the attraction by a body is always directed toward its center of mass.

Problem 1.3.26 Along which direction does a uniform rod AB attract a given point P?

(PIC9 ▶) We have $x = h \tan\theta$, $dx = h \sec^2\theta \, d\theta$, while $h \sec\theta = r$. The attraction by dx is $\propto dx/r^2 = d\theta/h$, i.e. the contribution to the attraction is distributed uniformly in the angle θ. Hence the attraction on P is along the bisector of the angle APB subtended by the rod. This bisector does not pass through the center of mass unless P happens to lie on the perpendicular bisector of AB.

In the next problem, a nice property of conic sections allows us to describe the levels surfaces of the potential, equipotentials.

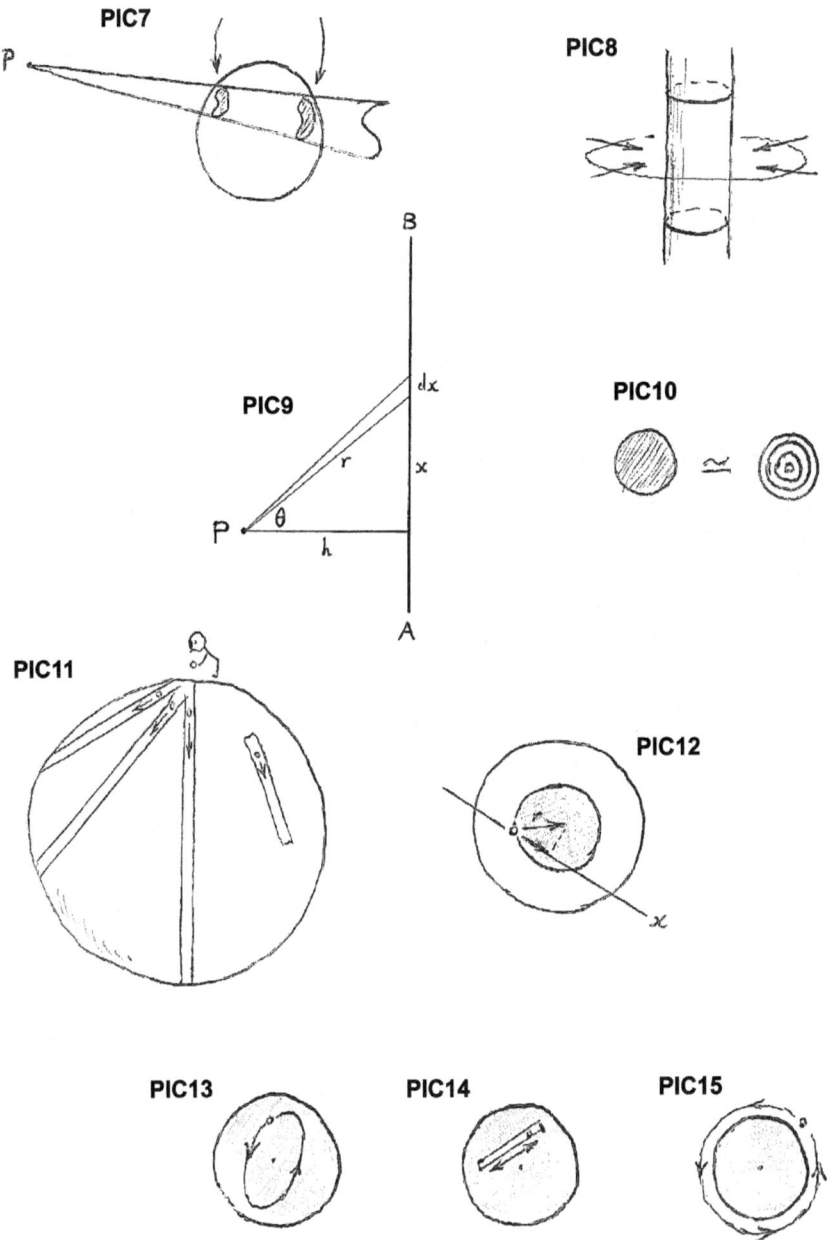

Problem 1.3.27 What are the equipotentials of the attraction by the rod?

Through P draw an ellipse of foci A, B. By the focal property of the ellipse, a light ray from A to P reflects and goes to B. Hence, at P, the angle bisector of APB is perpendicular to the ellipse. The equipotentials, which at every point are perpendicular to the attraction, are ellipsoids of revolution all having A, B as foci.

In case the attraction by a body *is* always directed toward some fixed point, we can prove the little-known converse to theorems above, using the expansion which will be introduced in Section 1.3.5:

Theorem 1.3.28 *Suppose there exists a point O such that, at every location outside the body, the body's attraction on that location is directed toward O.[5] Then O is the body's center of mass, and the moments of inertia about all axes through O are equal, i.e. the body is 'inertially spherical' around O.*

1.3.3 Attraction by a Solid Ball

This can be treated by 'onionifying' the solid ball as a layered assembly of spherical shells (◀ PIC10).[6]

In particular, as far as the gravitational field outside is concerned, solid balls with rotationally symmetric mass distribution can be replaced by points at their centers. This is why celestial mechanics started off as such a clean subject.

Problem 1.3.31 (◀ PIC11) Frictionless tunnels are dug in various directions through a planet of uniform density ρ. Drop stones in the tunnels. How does the stone's period of oscillation depend on the direction of the tunnel? What would the period be if the planet had the average density of the Earth?

Write V = volume of the unit ball. At the instant depicted in (◀ PIC12), only the inner ball attracts the stone. For every axis x,

$$x\text{-component of attraction} = \frac{G\rho V r^n}{r^{n-1}} \cdot \frac{x}{r} = G\rho V x,$$

so along this axis

$$\frac{d^2}{dt^2} x = -G\rho V x,$$

which represents a harmonic oscillator. Since $G\rho V$ is independent of the tunnel, the period $2\pi/\sqrt{G\rho V}$ ($= \sqrt{3\pi/G\rho} \approx 84$ min for the Earth, $n = 3$) is the same for all tunnels and all amplitudes (◀ PIC13,14,15).

---

[5]A weak hypothesis, only about the line of attraction passing through O, nothing about the *size* of attraction.

[6]*Principia* book I, proposition LXXIV.

Problem 1.3.32 Check directly that the caressing orbit (◄ PIC15) has the claimed period.

Balancing the centrifugal force and the attraction,

$$\frac{v^2}{r} = \frac{G\rho V r^n}{r^{n-1}} \implies \text{period} = \frac{2\pi r}{v} = 2\pi/\sqrt{G\rho V}.$$

1.3.4 Legendre Polynomials

In many problems in potential theory, the expression

$$\frac{1}{\sqrt{D^2 - 2Dd\cos\theta + d^2}} = \frac{1}{D}\left[1 - 2\frac{d}{D}\cos\theta + \left(\frac{d}{D}\right)^2\right]^{-1/2}$$

arises, cf. Section 1.3.5. The parameters d and D will be shown to have natural interpretations that make $d \ll D$, so we are led to expand the expression in powers of d/D. We define the coefficients by

$$[\cdots]^{-1/2} = \sum_{n\geq 0} P_n(\cos\theta)\left(\frac{d}{D}\right)^n$$

and call them Legendre polynomials.[7] The memorable, and the most important, low-degree Legendre polynomials are

$$P_0(z) = 1, \quad P_1(z) = z, \quad P_2(z) = \frac{3z^2 - 1}{2}, \quad P_3(z) = \frac{5z^3 - 3z}{2}, \quad \cdots.$$

(Alas, the memorable pattern does not continue.) It can be shown that in general

$$P_n(z) = \frac{1}{2^n n!}\frac{d^n}{dx^n}(z^2 - 1)^n \quad \forall n \geq 0,$$

$\deg P_n(z) = n$, $P_n(1) = 1$. Please familiarize yourself with their graphs (PIC16 ►).

Fig. 1.2
Legendre (1752–1833)

---

[7]Traditionally they are defined as solutions to a certain ODE that crops up when we try to separate $\nabla^2 u = 0$ in spherical polar coordinates. The definition chosen here is equivalent to the traditional one, but it is better motivated and easier to use for us.

The portrait of Adrien Legendre shown above is a famous one reproduced in many books. It recently came to light that this was a portrait of another, *Louis*, Legendre, cf. *Notices of the AMS* **56** (2009) 1440–1443.

1.3.5 Approximation Formulae for Bodies of Arbitrary Shape

As we saw in Sections 1.3.2 and 1.3.3, when the body is rotationally symmetric, its attraction is the same as that by a point mass. MacCullagh's formulae below give next-order corrections to the attraction when the body is no longer rotationally symmetric.

(i) The potential of a body in the far field.

   (PIC17 ▶) Notation: M mass of the body; I_1, I_2, I_3 its principal moments of inertia around the center of mass O; I its moment of inertia around the axis OP. Then *minus* the potential at P divided by G, acting on a unit mass, is

$$\int \frac{dM}{XP} = \int \frac{dM}{D}[\cdots]^{-1/2} = \frac{1}{D} \int dM \left\{ 1 + \cos\theta \frac{d}{D} + \frac{3\cos^2\theta - 1}{2}\left(\frac{d}{D}\right)^2 + \cdots \right\}$$

(cf. Section 1.3.4 for the expansion of $[\cdots]^{-1/2}$). The term $\cos\theta\, d/D$ gives 0 on being integrated. On the other hand,

$$\frac{3\cos^2\theta - 1}{2} = \frac{3(1 - \sin^2\theta) - 1}{2} = 1 - \frac{3}{2}\sin^2\theta,$$

so the above integral gives

$$\frac{M}{D} + \frac{I_1 + I_2 + I_3 - 3I}{2D^3} + \cdots.$$

In the 'inertially spherical' case $I_1 + I_2 + I_3 - 3I = 0$, and all the higher-order terms vanish, too.

(ii) The potential between two far bodies.

   Notation as in (PIC18 ▶).

$$-\frac{\text{potential}}{G} = \frac{MM'}{D} + \frac{M(I_1' + I_2' + I_3' - 3I')}{2D^3} + \frac{M'(I_1 + I_2 + I_3 - 3I)}{2D^3} + \cdots.$$

Fig. 1.3
MacCullagh (1809–1847)

PIC16

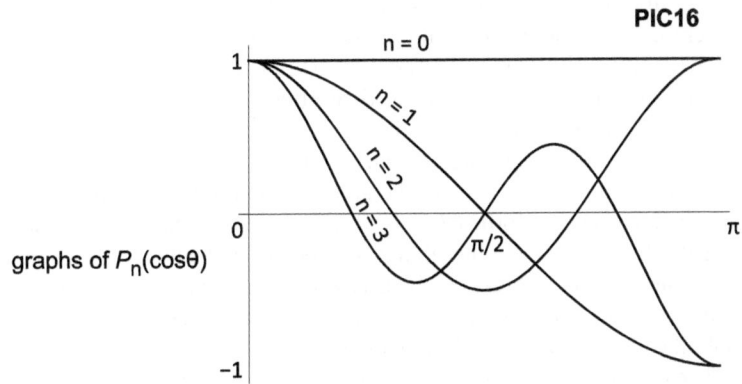

graphs of $P_n(\cos\theta)$

PIC17

PIC18

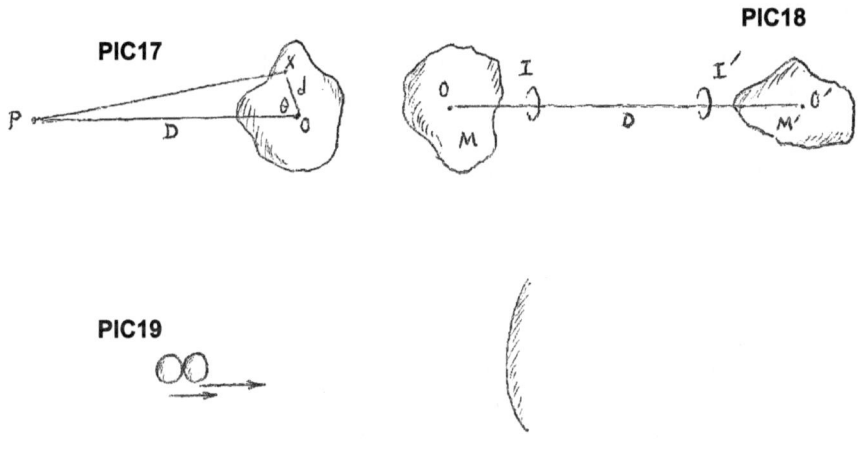

PIC19

PIC20

1.4 Tides—Static Picture

1.4.1 Plan

Why do tides exist?

Imagine two coconuts floating together near an attracting body (◀ PIC19). They feel a slight difference in the force of attraction, because one of them is slightly closer to, and the other one is slightly farther from, the attracting body. If we subtract the average attraction, then we see (◀ PIC20).

This, in a nutshell, is what the tidal effect is: difference, or *derivative*, in the attraction. The attraction varies like inverse square in the distance from the body, therefore the tidal effect, its derivative, varies like *inverse cube*. Qualitatively, the tidal effect is present as soon as the graph of the force as function of the distance is *concave*; the specific form $F(r) \propto -1/r^2$ is sufficient but not at all necessary.

Still, there are many other things we need to understand, for dynamic responses of a system to tidal effects can be tricky. For example: surely, as do the majority of textbooks since Newton (1687), we guess the (exaggerated) shape of the ocean to look like (PIC21 ▶)?... Well, that guess is *wrong*.

The right prediction looks rather like (PIC22 ▶).[8] This 'paradox' is one example among many of the characteristics about tides we try to understand in these notes.

Here is the plan we shall follow:

Theory of tides

- generating force for tides—static picture
- response of the ocean to this force—dynamic picture } neglecting dissipation
- effects of dissipation, astronomical applications, etc.

Pictorial Convention   In all the pictures, the body that is exerting the attraction, called

primary,

will be depicted on the right, while the body that is subjected to the attraction, called

secondary,

will be depicted on the left. In reality, everybody is attracting everybody all at once; 'primary' and 'secondary' are mere labels to clarify whose tidal influence on whom we are studying. Confusingly, primary and secondary can swap from one problem to the next, e.g. for ocean tides (1.4.3, 1.6.3) the Moon is primary and the Earth secondary, whereas for tidal locking (1.6.2) it is the other way around.

A good complementary reading is J. Lighthill, *Ocean Tides from Newton to Pekeris*, Israel Academy of Sciences and Humanities, 1995.

---

[8]It goes without saying that the observed tides of the real ocean are enormously complicated and do not resemble either of these pictures. But we are saying that, if we take the simplest model, of a spherical Earth covered by a sheet of ideal fluid, subjected to the dynamics of the Earth and the Moon, then the picture is (PIC22 ▶) rather than (PIC21 ▶).

1.4.2 Tidal Potential and Tidal Force

Let us figure out the tidal potential, then the tidal force, by making the picture of Section 1.4.1 quantitative.

We write the potential as

$$U_{\text{tide}} = U_{\text{cf}} + U_{\text{pr}}$$

where the three Us account for tidal force, centrifugal force, and attraction by the primary, in energy per unit mass. Since we are calculating the tidal effect of the primary at a given location in space like the black dot • of (PIC23 ▶), the attraction by the secondary is irrelevant to us. Let us erase the secondary (PIC24 ▶). Then our • is in orbit around the primary, so

$$\text{orbital centrifugal acceleration} \approx -\frac{GM}{D^2} \quad\Longrightarrow\quad U_{\text{cf}} = \frac{GM}{D^2} d \cos\theta.$$

Next,

$$U_{\text{pr}} = -\frac{GM}{D}[\cdots]^{-1/2} = -\frac{GM}{D}\left\{1 + \cos\theta\frac{d}{D} + \frac{3\cos^2\theta - 1}{2}\left(\frac{d}{D}\right)^2 + \cdots\right\}$$

(cf. 1.3.4 for the meaning of $[\cdots]^{-1/2}$ and its expansion in terms of Legendre polynomials). In the last sum { }, the constant 1 is immaterial for the potential and $\cos\theta d/D$ is canceled by U_{cf}. Altogether

$$U_{\text{tide}} \approx -\frac{GM}{D^3}d^2\frac{3\cos^2\theta - 1}{2},$$

whence a formula often quoted in the literature for the representative tidal force per unit mass

$$\boxed{\mathcal{F} = -\frac{\partial}{\partial d}U_{\text{tide}}\Big|_{d=r,\theta=0} \approx 2\frac{GM}{D^3}r.}$$

$\mathcal{F}$ varies like D^{-3} in the distance D from the primary. A quicker way to derive this formula is that the attraction varies like inverse square, while the tidal force is the small difference in the attraction over a displacement $\Delta D \approx -r$, i.e. it arises essentially as the *derivative* of D^{-2}:

$$\mathcal{F} \approx \frac{\partial}{\partial D}\frac{GM}{D^2} \cdot \Delta D \approx 2\frac{GM}{D^3}r.$$

(Recall the 'nutshell' comment in Section 1.4.1.)

Now

$$\frac{3\cos^2\theta - 1}{2} = \frac{1}{2}\left(3\frac{1 + \cos 2\theta}{2} - 1\right) = \frac{3}{4}\cos 2\theta + \frac{1}{4}.$$

The last term 1/4, independent of θ, cannot deform the sphere. Rewriting U_{tide} with the aid of $g = Gm/r^2$ (gravitational acceleration on the surface of the secondary), we finally obtain the part u_{tide} of the tidal potential responsible for deforming the sphere, as

$$u_{\text{tide}} = -\frac{3}{4}\frac{M}{m}\left(\frac{r}{D}\right)^3 \frac{d^2}{r} g \cos 2\theta$$

in energy per unit mass. Hence the tidal force f per unit mass that deforms the sphere has components (PIC25 ▶)

$$f_{\text{vert}} = -\frac{\partial}{\partial d} u_{\text{tide}}\Big|_{d=r} = \frac{3}{2}\frac{M}{m}\left(\frac{r}{D}\right)^3 g \cos 2\theta,$$

$$f_{\text{horiz}} = \frac{1}{d}\frac{\partial}{\partial \theta} u_{\text{tide}}\Big|_{d=r} = -\frac{3}{2}\frac{M}{m}\left(\frac{r}{D}\right)^3 g \sin 2\theta,$$

which over the surface of the secondary give the picture (PIC26 ▶). We see that the effect of the tidal force is to stretch the secondary in the primary's direction and to squeeze it in the transverse directions, in the shape of a rugby ball. The direction of f varies as a function of θ whereas its magnitude $\sqrt{f_{\text{vert}}^2 + f_{\text{horiz}}^2}$ does not. We name and retain for future use the key ratio

$$\boxed{\frac{f}{g} = \frac{3}{2}\frac{M}{m}\left(\frac{r}{D}\right)^3.}$$

Problem 1.4.21 Estimate f/g for the Earth (secondary) under the influence of the Moon (primary).

$$\frac{f}{g} = \frac{3}{2}\frac{M_{\mathbb{D}}}{M_{\oplus}}\left(\frac{R_{\oplus}}{D_{\mathbb{D}}}\right)^3 \approx \frac{3}{2}\cdot\frac{1}{80}\cdot\left(\frac{1}{60}\right)^3 \approx 8.6 \times 10^{-8},$$

which is tiny.

If the ellipticity of the Moon's orbit (PIC27 ▶) is taken into account, it turns out that f/g varies between 7.5×10^{-8} at the apogee and 10^{-7} at the perigee. Nevertheless, this tiny ratio produces the majestic ocean tides that wash the Earth.

1.4.3 Shape of the Ocean

Imagine an ocean that covers the secondary (PIC28 ▶).

$$-u_{\text{tide}}|_{d=r} = gh \quad \Longrightarrow \quad h(\theta) = \frac{3}{4}\frac{M}{m}\left(\frac{r}{D}\right)^3 r \cos 2\theta = \frac{1}{2}\frac{f}{g} r \cos 2\theta.$$

Spinning this about the line directed toward the primary (rightward in the pictures), we obtain the rugby-ball shape of the ocean as deformed by the primary's tidal force

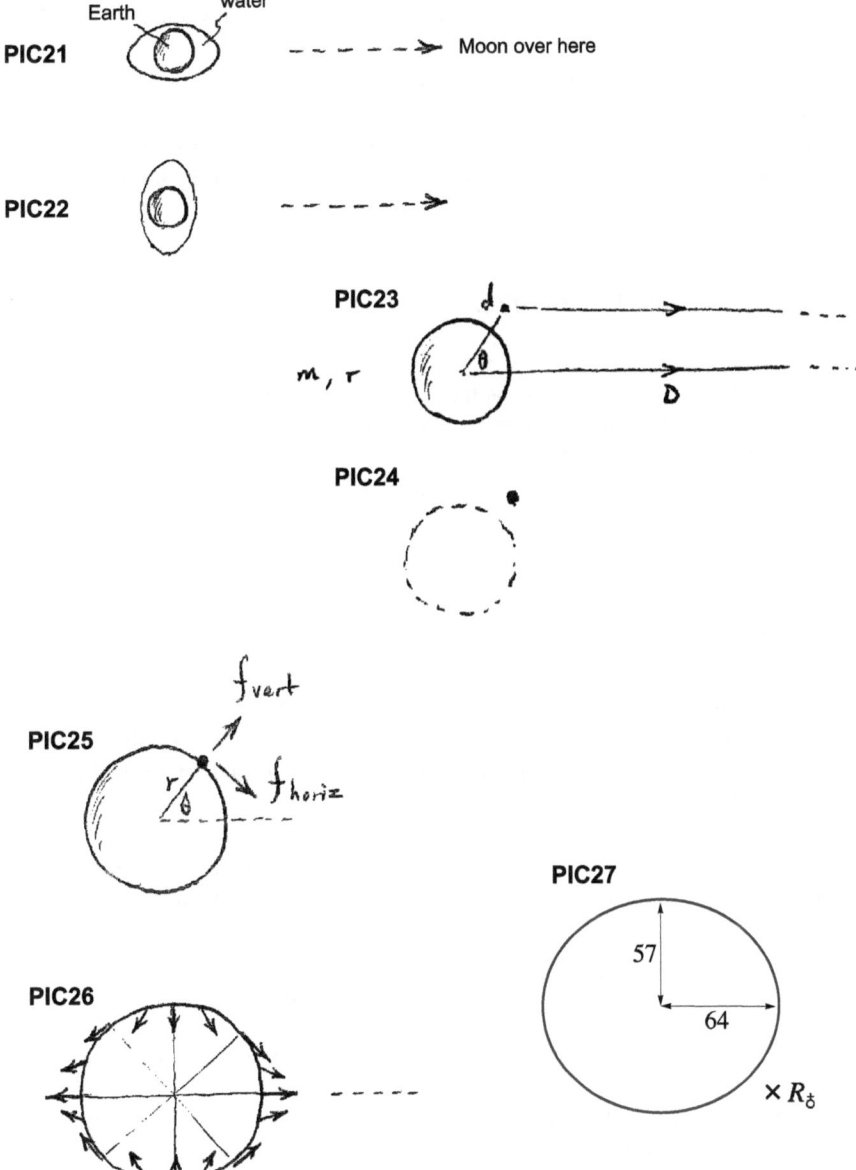

(PIC29 ▶). The reader is reminded that θ is the angle measured from this line, *not* colatitude from the North Pole. But on the side of the secondary facing the primary, θ happens to represent latitude. (PIC30 ▶) shows that, as the secondary rotates on its axis, a point on a given latitude θ traces a circle and attains h of the low tide at ×. This h also equals h at •. The upshot is that h has the common value of

$$\min h = \frac{1}{2}\frac{f}{g}r\cos\left(2\cdot\frac{\pi}{2}\right) = -\frac{1}{2}\frac{f}{g}r$$

for all latitudes θ. Clearly $\Delta h(\theta) = h(\theta) - \min h$, and we have[9] $\cos 2\theta - 1 = 2\cos^2\theta$. The daily amplitude of the tide $\Delta h(\theta)$ as a function of the latitude θ is

$$\Delta h(\theta) = \frac{f}{g}r\cos^2\theta.$$

It is proportional to the key ratio f/g.

Problem 1.4.31 At what latitude on the Earth does the daily amplitude of the tide attain its maximum? minimum? Estimate these amplitudes.

Using f/g from Problem 1.4.21,

$$\max_\theta \Delta h(\theta) = \frac{f}{g}R_{\oplus} \approx 8.6\times 10^{-8}\cdot\frac{4\times 10^7}{2\pi} \approx 0.5\text{ m}\quad\text{at the equator}\quad(\theta=0,\pi),$$

$$\min_\theta \Delta h(\theta) = 0 \qquad\qquad\qquad\qquad\qquad\qquad\text{at the poles}\quad(\theta=\pi/2).$$

So far we have pretended that the Earth's axis of rotation was perpendicular to the plane of the Moon's orbit. In reality the axis is *tilted* (PIC31 ▶): this produces two unequal high tides, 'small' high tide and 'big' high tide, and brings the low tides nearer the 'small' high tide (PIC32 ▶).

The axial tilt β varies between 17° and 29° owing to the precession of the Moon's orbit. Since the lunar revolution (period $\approx 27 + 1/3$ days) goes in the same direction as the terrestrial rotation (period = 24 hours), at a given location on the Earth a high tide arrives every

$$\frac{1}{2}\left(24 + \frac{24}{27+1/3}\right) \approx 12\text{ hours 26 minutes},$$

and this arrival gets delayed by 52 minutes (= twice the above number − 24 hours) per day.

The behavior of a real ocean tides depends sensitively on local geography. The largest Δh in the world is observed in the Bay of Fundy (Canada), where it attains 17 m.

---

[9]Undoing an earlier trigonometric transformation of Section 1.4.2.

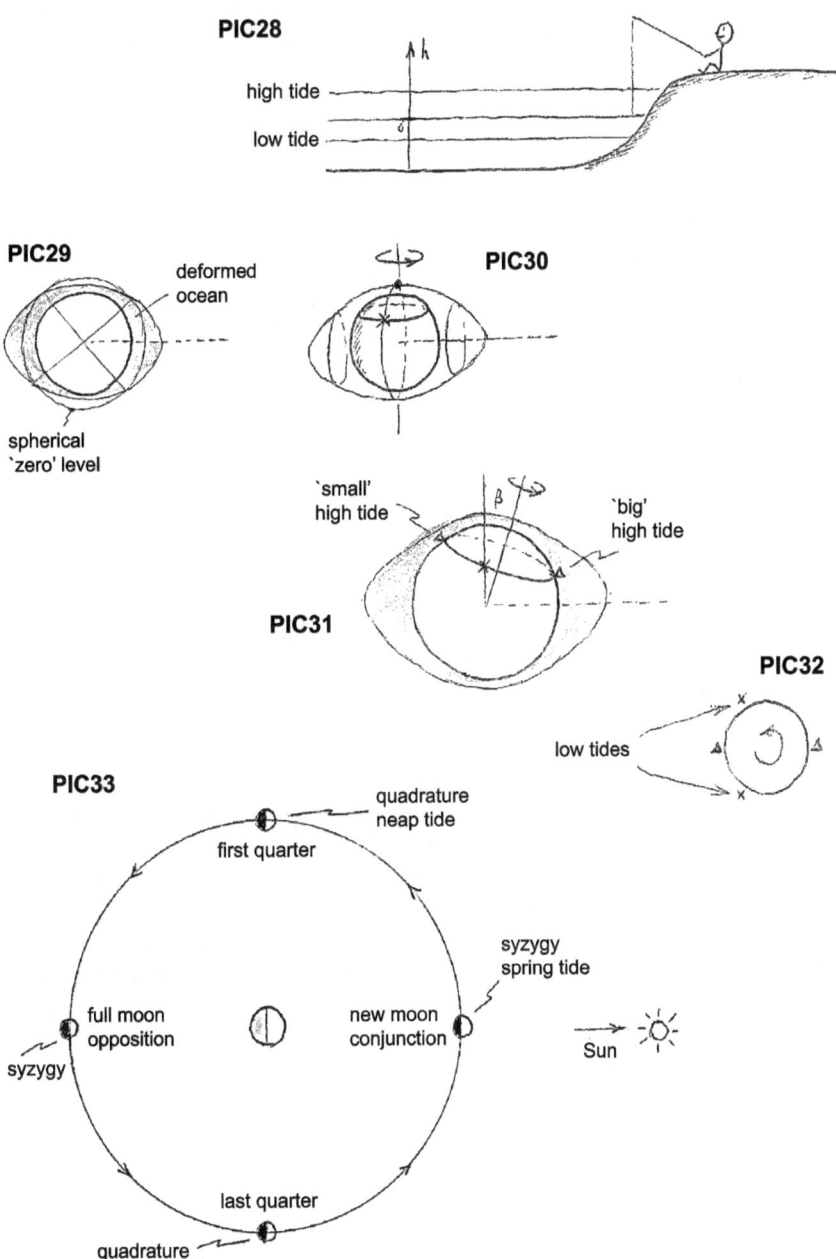

PIC28

high tide

low tide

PIC29

deformed ocean

spherical 'zero' level

PIC30

PIC31

'small' high tide

'big' high tide

PIC32

low tides

PIC33

quadrature neap tide

first quarter

syzygy spring tide

full moon opposition

new moon conjunction

Sun

syzygy

last quarter

quadrature

1.4.4 What About the Sun?

Problems 1.4.21 and 1.4.31 examined the Moon's tidal influence. Let us estimate how the Sun's influence compares with it. In view of the expression for the key ratio f/g in Section 1.4.2,

$$\frac{\Delta h_{\mathbb{D}}}{\Delta h_{\odot}} = \frac{f_{\mathbb{D}}}{f_{\odot}} = \frac{M_{\mathbb{D}}}{M_{\odot}}\left(\frac{D_{\odot}}{D_{\mathbb{D}}}\right)^3 \approx \frac{\frac{1}{80}}{\frac{1}{3}\times 10^6}\cdot 400^3 \approx 2.4.$$

Thus, the influence of the Sun is modest but far from negligible.

(◄ PIC33) When the three bodies (Sun, Earth, Moon) become aligned, the lunar and solar tides strengthen each other; this is spring tide, and the configuration is called syzygy.[10] In contrast, they weaken each other when the three bodies form a right angle; this is neap tide, and the configuration is called quadrature.

1.5 Tides—Dynamic Picture

1.5.1 Forced Oscillator

It is time to return to the 'paradox' of Section 1.4.1 (PIC34 ►).

A preparatory discussion on the relative motion between the Moon (primary) and the Earth (secondary). Normally, over a time-scale of a day, we think of the Moon as stationary and of the Earth as rotating on its axis. The ocean as a whole rotates with the Earth:[11] after all, if instead the ocean were stationary and the solid Earth rotated underneath it, the sea floor would be swept by the water at a mad speed of 4×10^7 m/day ≈ 1666 km/hour. In the (non-inertial) frame in which the Earth and the ocean are together stationary, it is the Moon that runs around them—once a day, and retrograde. Until the end of Section 1.5 we shall work in this frame.[12] With the preparation out of the way, back now to the 'paradox'.

What will happen if, in the absence of the revolving primary, the ocean on the stationary secondary is put in the state (PIC35 ►) and released? It will oscillate as in (PIC36 ►), with some period T_{free}. In the presence of the primary, the tidal effect exerts a *periodic external forcing*, with some period T_{ext}, and the oscillating ocean responds to it by modifying its behavior.

Theorem 1.5.11

$T_{\text{ext}} > T_{\text{free}} \Longrightarrow$ *oscillator's response*   *in phase*   *with external forcing.*
$T_{\text{ext}} < T_{\text{free}} \Longrightarrow$ $\cdots$ *out of phase* $\cdots$

---

[10]Etymology: *syzygy* < Greek σύζυγος (spouse) < ζυγός (yoke), cf. *conjugate* < Latin *jugum*.

[11]We are modeling a spherical Earth covered by a sheet of ideal fluid, cf. footnote 8 in 1.4.1.

[12]As does the human society, which insists that the Moon and the Sun rise in the east and set in the west. Actually most of this motion is caused by us spinning from west to east.

PIC34

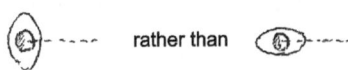 rather than

PIC35

PIC36

PIC37   **PIC38**

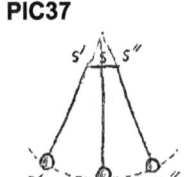 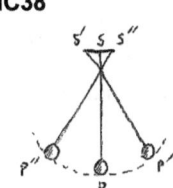

PIC39

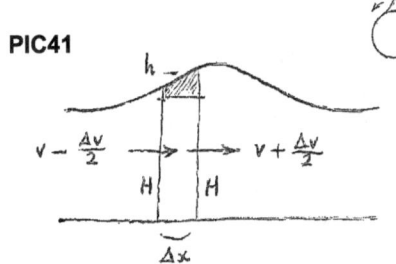

PIC40

PIC41

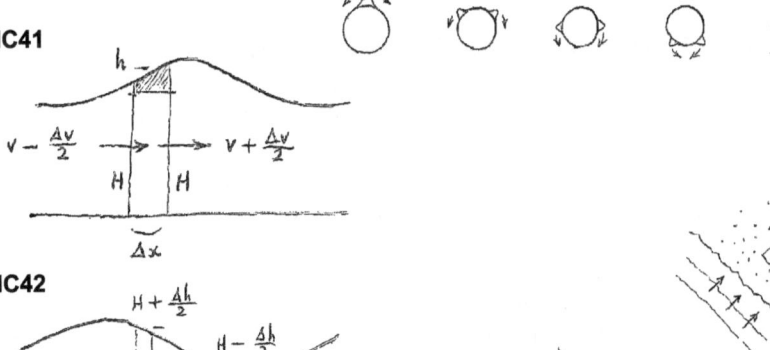

PIC42

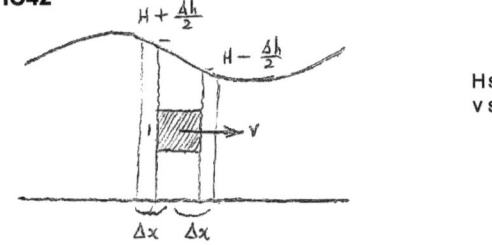

H shallow
v slow

H deep   **PIC43**
v fast

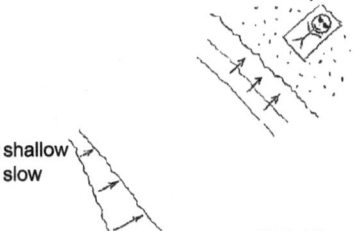

The conclusions can be checked by a home experiment. Hang a pendulum from a pivot, and wriggle the pivot horizontally. If this external wriggling is *slow* compared with the free oscillation of the pendulum ($T_{ext} > T_{free}$), then we observe the in-phase response (◄ PIC37). If the external wriggling is *fast* ($T_{ext} < T_{free}$), then we observe the out-of-phase response (◄ PIC38).

Proof The governing ODE

$$\frac{d^2}{dt^2}x + \left(\frac{2\pi}{T_{free}}\right)^2 x = f \exp\left(i\frac{2\pi}{T_{ext}}t\right)$$

is solved by

$$x(t) = \frac{f}{\left(\frac{2\pi}{T_{free}}\right)^2 - \left(\frac{2\pi}{T_{ext}}\right)^2} \exp\left(i\frac{2\pi}{T_{ext}}t\right).$$

The coefficient in front of exp has the same sign as f if $T_{ext} > T_{free}$, the opposite sign if $T_{ext} > T_{free}$.   □

1.5.2 Free Oscillation of the Ocean

To apply Theorem 1.5.11 to the tide, we estimate T_{free} for the free oscillation of the ocean, following Airy's canal theory (1845).

Imagine digging a canal of depth H all the way along the equator (◄ PIC39). Let a hump of water, collapsing under its own weight, propagate as a wave along this canal, as in (◄ PIC40).

Theorem 1.5.21 *The speed of propagation of this wave is $\sqrt{gH}$.*

Proof (◄ PIC41) Within a narrow slab of width Δx, the conservation of volume says

$$\frac{\partial}{\partial t}h\Delta x = H \cdot \left(v + \frac{\Delta v}{2}\right) - H \cdot \left(v - \frac{\Delta v}{2}\right)$$

$$\implies \quad \frac{\partial h}{\partial t} = H\frac{\partial v}{\partial x}.$$

(◄ PIC42) The equation of momentum per unit mass for a block of water of height 1 says

$$\frac{\partial}{\partial t}1 \cdot \Delta x \cdot v = \frac{g\left(H + \frac{\Delta h}{2}\right)\Delta x - g\left(H - \frac{\Delta h}{2}\right)\Delta x}{\Delta x}$$

$$\implies \quad \frac{\partial v}{\partial t} = g\frac{\partial h}{\partial x}.$$

Out drops $\partial^2 h/\partial t^2 = gH\partial^2 h/\partial x^2$, a wave equation with the propagation speed $\sqrt{gH}$.   □

Fig. 1.4
Airy (1801–1892)

Problem 1.5.22 Why do waves arrive with their fronts parallel to the beach?

Because the dependence: speed = $\sqrt{gH}$ causes the wave front to turn (◄ PIC43).

Problem 1.5.23 Why do waves get steeper as they approach a beach and eventually break (PIC44 ►)?

Because the profile of a wave strains as the depth of water varies. Say a wave passes over an underwater 'step'. In (PIC45 ►) the front of the wave moves faster than the rear $\implies$ the wave gets stretched and flatter. In (PIC46 ►) the rear of the wave moves faster than the front $\implies$ the wave gets squeezed and steeper.

For H = average depth of the ocean ≈ 4 km, we find

$$\sqrt{gH} \approx \sqrt{\frac{1}{100}} \cdot 4 \text{ km/sec} \approx 700 \text{ km/hour}$$

(only a bit slower than a jet plane). At this speed the wave tours half of the canal, i.e. half-circumference of the Earth, and the water humps due to the tide exchange their positions, in

$$\frac{\frac{1}{2} \cdot 40000}{700} \approx 30 \text{ hours} = T_{\text{free}}.$$

As regards T_{ext}, it is easy: the Moon runs from one side of the Earth to the other side, retrograde, in $T_{\text{ext}} \approx 12$ hours.

$T_{\text{ext}} < T_{\text{free}}$ implies, by Theorem 1.5.11, that the response of the ocean must be *out of phase* with the tidal force, which means (PIC47 ►).[13]

---

[13] In order to have the in-phase picture (PIC48 ►), we would require a deeper ocean $H > 20$ km.

Fig. 1.5
Roche (1820–1883)

Remark 1.5.24

(1) T_{free} of the oscillation of a *spherical sheet* of water of depth 4 km, which is more realistic than that in a canal, turns out to be $\approx$24 hours.[14] Thus the out-of-phase inequality $T_{\text{ext}} < T_{\text{free}}$ is satisfied with margin to spare.
(2) Our estimate of T_{free} is sensible: when a major earthquake strikes Chile, Japan receives a tsunami approximately 24 hours later (and vice versa), the tsunami having crossed the Pacific, which extends about half-way around the Earth.
(3) The oceanographers are interested in the ocean tides, but the astronomers are more interested in the tidal response of a solid crust. The latter response is (PIC49 ▶) rather than (PIC50 ▶) because, of the elastic waves in the crust, even the slowest[15] travels at $\approx$1 km/sec, which implies the in-phase inequality $T_{\text{free}} \approx 5.5$ hours $< T_{\text{ext}}$. The amplitude of such a tide is of the order of 0.5 m on the equator.

1.6 Astronomical Applications

1.6.1 Tidal Tearing

What holds us on the ground (PIC51 ▶)? It is g. If the tidal force f per unit mass of the primary becomes 'a few times' g of the secondary (PIC52 ▶), then the particles on the secondary can no longer hold together. Whereupon the secondary begins to be torn apart by the tidal force of the primary....

The <u>Roche limit</u> (1848) is the proximity within the primary at which this tearing begins. Roughly speaking, there exists a critical distance D_{Roche} such that

$$\frac{f}{g} = \frac{3}{2}\frac{M}{m}\left(\frac{r}{D_{\text{Roche}}}\right)^3 \approx \text{'a few times'}$$

$$\implies D_{\text{Roche}} \approx \text{'a few times'}\left(\frac{M}{m}\right)^{1/3} r = \text{'a few times'}\left(\frac{P}{\rho}\right)^{1/3} R.$$

In the last equality we used $m \sim \rho r^3$, $M \sim PR^3$.

---

[14] A hump spreads and propagates as a ring and meets as a new hump on the antipodes.
[15] Rayleigh wave (1885).

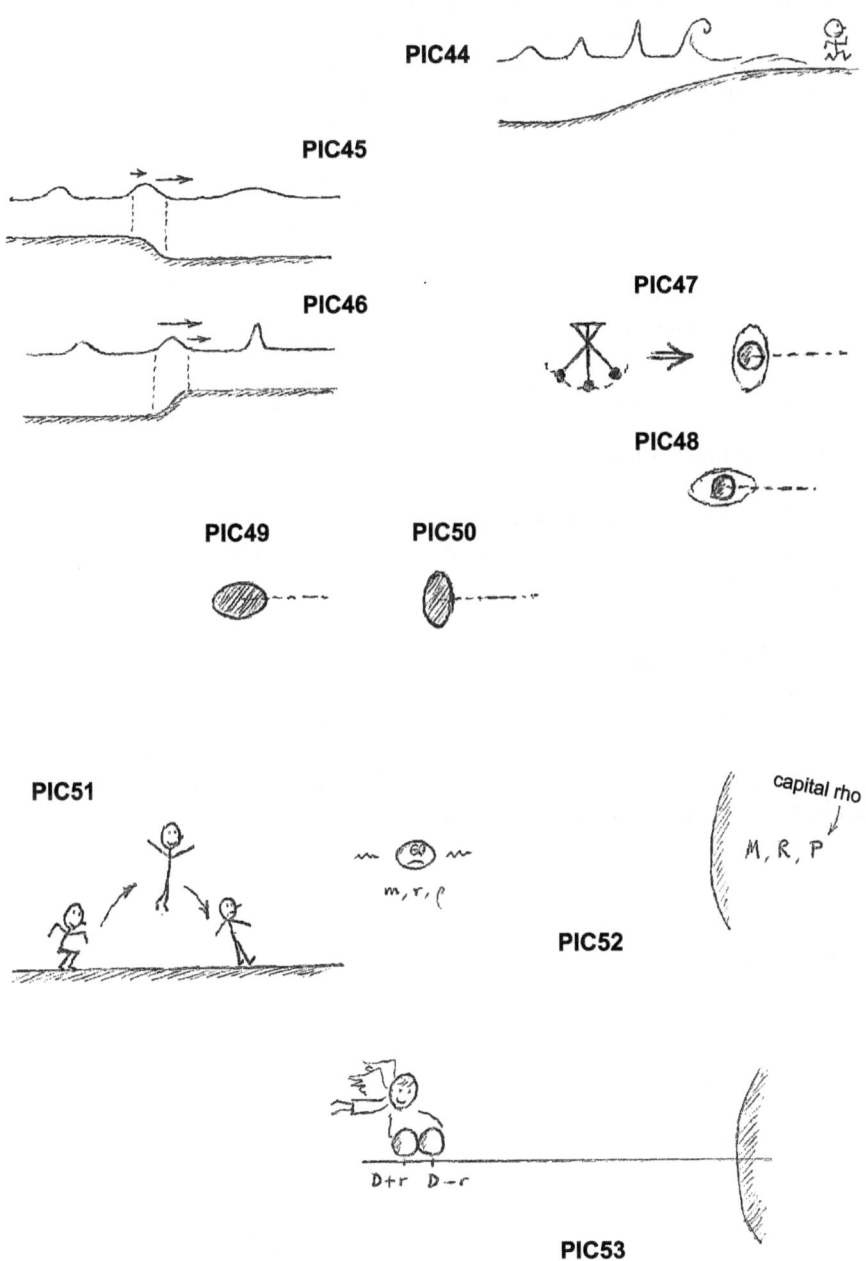

PIC44

PIC45

PIC46

PIC47

PIC48

PIC49

PIC50

PIC51

m, r, ρ

capital rho

M, R, P

PIC52

$D+r$ $D-r$

PIC53

Alternative interpretation: D_{Roche} is attained when the amplitude of the tide Δh of Section 1.4.3 exceeds the radius r of the secondary.

Let us estimate what 'a few times' should mean. (◄ PIC53) An angel deposits a pair of coconuts in contact (each with parameters m, r, ρ) at the distances $D \pm r$ from the primary. Will the coconuts detach themselves?

The gravitational cohesion between the coconuts is $Gm^2/(2r)^2$. Competing against this, the force of detachment due to the tide is

$$\text{attraction}|_{D-r} - \text{attraction}|_{D+r} \approx \frac{\partial}{\partial D}\left(\frac{GMm}{D^2}\right) \cdot (-2r)$$

$$= \frac{4GMm}{D^3}r,$$

which is $2m$ times the representative tidal force $\mathcal{F}$ per unit mass of Section 1.4.2. Therefore the detachment begins at

$$D_{\text{Roche}} = \left(16\frac{M}{m}\right)^{1/3} r \approx 2.5\left(\frac{M}{m}\right)^{1/3} r.$$

Thus, 'a few times' should be between $2\times$ and $3\times$.

Example 1.6.11 Instances of tidal tearing include: the formation of planetary rings and, in more recent history, the fragmentation of the comet Shoemaker-Levy 9 as it approached Jupiter (July 1992).

1.6.2 Tidal Locking

Why does the Moon always show the same face to us?

Imagine a barbell (secondary) in circular orbit around M (primary), at angular frequency $\omega = \dot{\varphi}$ (PIC54 ►). ω is determined by the condition that, in a circular orbit, the centrifugal force and the attraction balance:

$$\frac{(D\omega)^2}{D} = \frac{GM}{D^2} \quad \Longrightarrow \quad \omega = \sqrt{\frac{GM}{D^3}}.$$

The Lagrangian may be written down in terms of the parameters given in (PIC54 ►). We have

$$\text{kinetic energy} = \frac{1}{2}m\left(\dot{D}_+^2 + (D_+\dot{\varphi}_+)^2 + \dot{D}_-^2 + (D_-\dot{\varphi}_-)^2\right),$$

$$\text{potential energy} = -GMm\left(\frac{1}{D_+} + \frac{1}{D_-}\right)$$

and from geometry

$$D_\pm \approx D \pm \ell \cos\psi \qquad \dot{D}_\pm \approx \mp\ell\dot{\psi}\sin\psi$$

$$\varphi_\pm \approx \varphi \pm \frac{\ell}{D}\sin\psi \qquad \dot{\varphi}_\pm \approx \omega \pm \frac{\ell}{D}\dot{\psi}\cos\psi.$$

$(\ell/D)^2$ being neglected as high order, the approximate Lagrangian comes out to be

$$L = \text{kin} - \text{pot} \approx 3mD^2\omega^2 + m\ell^2\left(\dot{\psi}^2 - 4\omega\dot{\psi}\cos^2\psi + 3\omega^2\cos^2\psi\right).$$

The Euler-Lagrange equation

$$\frac{\partial L}{\partial \psi} - \frac{\mathrm{d}}{\mathrm{d}t} \frac{\partial L}{\partial \dot{\psi}} = 0$$

reads

$$(2\ddot{\psi}) + 3\omega^2 \sin(2\psi) = 0,$$

which is the equation of pendulum motion. The frequency of small oscillations of this pendulum is $\sqrt{3}$ times the frequency of the orbital revolution. With dissipation, any motion of the barbell asymptotes to the unique equilibrium $\psi = 0$.

Now suppose a secondary (e.g. Moon) revolves around a primary (e.g. Earth). Typically the secondary is *not* 'inertially spherical'. We can think of a barbell as its toy model. Then on the orbital revolution a tidal oscillation gets superposed (PIC55 ▶). With dissipation, the oscillation asymptotes to $\psi = 0$, i.e. settles in the radial direction to the primary, and so the secondary ends up getting *locked* with its face toward the primary.

This *is* a tidal effect. Indeed, in constant gravity (zero derivative in the attraction) there is no torque restoring the barbell to face the primary (PIC56 ▶).

Examples 1.6.21 Instances of tidal locking include: mutual locking of Pluto and Charon and, in more recent history, stabilization of artificial satellites (to keep them facing the Earth), notably Gemini 11 and 12 (1966).

1.6.3 Tidal Dissipation

As discussed in Section 1.5.1, in the absence of any tidal force, the ocean would rotate together with the solid Earth underneath as a single rigid body (PIC57 ▶) $\Longrightarrow$ no dissipation.

In the presence of the Moon and its tidal force (PIC58 ▶), the ocean is 'held in place' and *rubs* the solid Earth, which is rotating underneath $\Longrightarrow$ dissipation, another effect of the tide.

Let us estimate the rate of tidal dissipation.

We distinguish two kinds of angular momentum involved in this phenomenon: the one carried by the secondary (e.g. Earth)'s rotation about its own axis, which we call spin angular momentum, and the other carried by the revolution of the primary (e.g. Moon) around the secondary,[16] which we call orbital angular momentum. The spin a.m. and the orbital a.m. are respectively

$$I\dot{\psi} \quad \text{and} \quad L = D \times m_{\mathrm{red}} D\dot{\phi}$$

where we define $m_{\mathrm{red}} = Mm/(M + m)$, the so-called reduced mass.

The tidal torque τ due to the primary acts on the secondary's spin a.m. and does work at the rate

$$\dot{E} = \tau(\dot{\psi} - \dot{\phi}) = I\ddot{\psi}(\dot{\psi} - \dot{\phi}).$$

---

[16]More precisely, the revolution of the primary and the secondary around each other.

The details of the dissipation are so complex that it is impracticable to estimate $\dot{E}$ from first principles. But we *can* estimate $\dot{E}$ once we measure observationally the values of various parameters on the right-hand side of this equation. These stand for: $\dot{\psi}$ = how fast the secondary rotates about its axis, $\ddot{\psi}$ = how this rotation is being decelerated,[17] and $\dot{\phi}$ = how fast the primary revolves around the secondary. It will be helpful to note in addition that, since a month lasts approximately 30 days,[18] we have for the Earth-Moon pair

$$\frac{\dot{\phi}}{\dot{\psi}} \approx \frac{1}{30}.$$

Problem 1.6.31 Estimate the tidal dissipation on the Earth by the Moon.

$$I \approx \frac{2}{5} M_{\oplus} R_{\oplus}^2 \approx \frac{2}{5} \cdot 6 \times 10^{24} \cdot \left(\frac{4 \times 10^7}{2\pi}\right)^2 \approx 10^{38} \text{ kg m}^2$$
$$\dot{\psi} = \frac{2\pi}{24 \cdot 60 \cdot 60} \approx 7.3 \times 10^{-5} \text{ sec}^{-1} \quad \text{(childhood knowledge)}$$
$$\ddot{\psi} \approx -4.6 \times 10^{-22} \text{ sec}^{-2} \quad \text{(observational data)}.$$

Neglecting $\dot{\phi}/\dot{\psi} \ll 1$,

$$\dot{E} \approx I\ddot{\psi}\dot{\psi} \approx -3.7 \times 10^{12} \text{ J sec}^{-1} \approx -1.2 \times 10^{20} \text{ J year}^{-1}.$$

This is double the consumption of electricity in the world $\approx -6.1 \times 10^{19}$ J year^{-1}, according to the CIA data of 2005.[19]

I have been told that an astonishing 1/3, or some such fraction, of this dissipation takes place in the Bering Sea and the Sea of Okhotsk.

Digression 1.6.32 Sometimes we hear that the Coriolis force makes water spin one way or the other as it drains down a sink. Estimate how significant the Coriolis force is.

The Coriolis force per unit mass of water is $2\dot{\psi} \times v$. A natural standard for comparison is g, which the water also feels. From a sink of depth h water drains at speed $v = \sqrt{2gh}$ by Torricelli's law (1643). Hence, taking $h \approx 0.1$ m,

$$\frac{2\dot{\psi} \times v}{g} \approx \sqrt{\frac{8h}{g}}\,\dot{\psi} \approx 2 \times 10^{-5},$$

which is utterly invisible. The phenomenon is dominated by irregularities in the build of the sink and by random initial conditions of the water.

---

[17] $\ddot{\psi} < 0$ because the friction on the sea floor by the 'tidally held' ocean slows down the secondary's rotation.

[18] Lest the astronomers complain: anomalistic, draconic, sidereal, synodic.... For our approximate purposes here it does not matter which, they are all a little under 30 days.

[19] *Julius Caesar* IV. iii. 218–219 may come to some people's mind.

We take up another consequence of tidal dissipation that affects the fate of a system consisting of a primary revolving around a secondary.

Tidal dissipation brakes the secondary's spin a.m. $I\dot\psi$. Meanwhile, the total a.m. of the system $I\dot\psi + L$ is conserved. To compensate, the orbital a.m. L grows. As it orbits around faster,[20] the primary must *drift away* from the secondary (PIC59 ▶), along a spiral.

Let us estimate the rate at which the primary drifts away from the secondary. Refer again to (PIC58 ▶).

In quasi-circular orbit,

$$\frac{m_{\mathrm{red}}\left(D\dot\phi\right)^2}{D} \approx \frac{GMm}{D^2} \quad\Longrightarrow\quad D \approx \frac{L^2}{GMmm_{\mathrm{red}}}.$$

On account of the conservation of total a.m. we have $\dot L = -I\ddot\psi$, which implies

$$\dot D = -\frac{2D^2\dot\phi I\ddot\psi}{GMm} \approx -2\frac{\dot\phi}{\dot\psi}\frac{D^2}{GMm}\dot E = -2\frac{\dot\phi}{\dot\psi}\frac{\dot E}{\text{mutual attraction}}.$$

$\dot E < 0$ tells us that $\dot D > 0$.

Problem 1.6.33 Estimate $\dot D_{\mathrm{\tiny D}}$ for the Earth-Moon pair. What are we led to conclude if we extrapolate naively into the past?

Using $\dot E$ from Problem 1.6.31 and the size of the mutual attraction from Problem 1.22,

$$\dot D_{\mathrm{\tiny D}} \approx -2\cdot\frac{1}{30}\cdot\frac{-1.2\times 10^{20}}{2\times 10^{20}} \approx 0.04\ \mathrm{m\,year}^{-1} = 4\ \mathrm{cm\,year}^{-1}.$$

This is how fast the Moon is drifting away from the Earth. So

$$-\frac{D_{\mathrm{\tiny D}}}{\dot D_{\mathrm{\tiny D}}} \approx \frac{60\cdot\frac{4\times 10^7}{2\pi}}{0.04} \approx 10^{10}\ \text{years}$$

in the past, the Moon must have been in contact with the Earth.

Our estimate of $\dot D_{\mathrm{\tiny D}}$ is consistent with observational data, yet our naive extrapolation leads to double the geological estimate of the age of the Earth-Moon pair. The error is imputable to our linear extrapolation: the tidal dissipation was more efficient when the Moon was nearer the Earth.

For controversies surrounding other methods of estimating the age, I recommend T. W. Körner, *Fourier Analysis,* Cambridge UP, 1988, chapters 56, 57, 58.

Examples 1.6.34 Where does the energy dissipated by the tide go? It heats up the secondary. Tidal heating is dramatic when a tidal force periodically kneads a small

---

[20] And since simultaneously the Earth is spinning slower, we terrestrials have the impression that the Moon is orbiting all the faster. Halley was the first to notice this (1695).

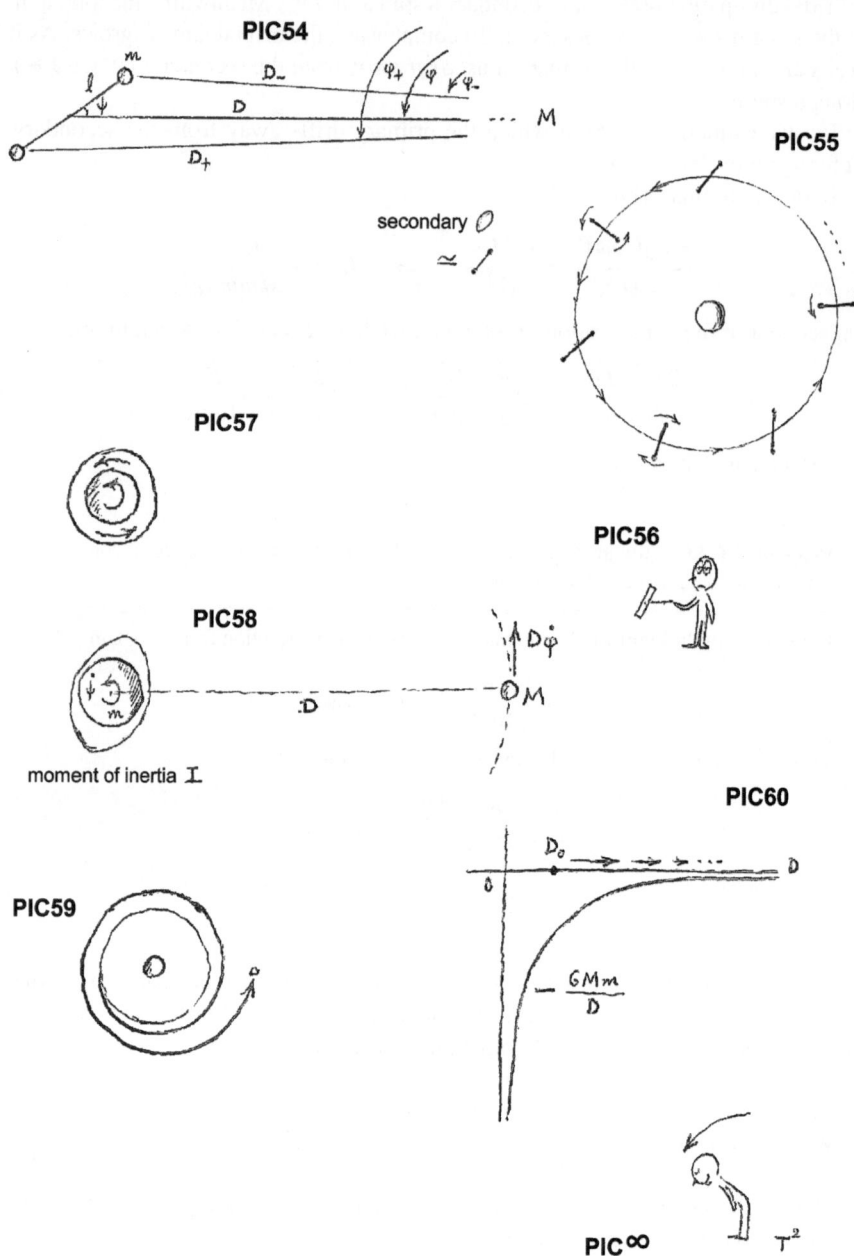

PIC54

PIC55

secondary $\mathcal{O}$
$\simeq$ /

PIC57

PIC56

PIC58

moment of inertia I

PIC60

PIC59

$-\dfrac{GMm}{D}$

PIC∞ T^2

secondary in resonance, e.g. Io : Europa : Ganymede (4 : 2 : 1 resonance) around Jupiter, driving on Io the most violent volcanism in the solar system.

1.6.4 Visit to the Horizon

Can we visit the horizon of a black hole without being torn apart by its tidal force? To answer this question, we prepare one concept.

(◄ PIC60) Imagine an astro-tourist in the gravitational field of a black hole. Initially she is at a distance D_0 from the black hole and launches herself with an outward speed v_0. Will she be able to escape to infinity? At the start, her energy was

$$\frac{1}{2}mv_0^2 - \frac{GMm}{D_0}.$$

At the end, if she manages to escape to infinity with no residual speed to spare, then her energy will be

$$\frac{1}{2}m0^2 - \frac{GMm}{\infty} = 0.$$

It follows that the escape requires the inequality

$$D_0 \geqslant \frac{2GM}{v_0^2}.$$

In other words, if the astro-tourist starts with $D_0 < 2GM/v_0^2$, she will exhaust her momentum before reaching infinity and will fall back toward the black hole. But v_0 available is at most the speed of light c. Therefore

$$D_{\text{Sch}} = \frac{2GM}{c^2} = \frac{2GM_\odot}{c^2}\frac{M}{M_\odot}$$

$$\approx \frac{2 \cdot \frac{2}{3} \times 10^{-10} \cdot 2 \times 10^{30}}{(3 \times 10^8)^2} \approx 3 \times 10^3 \frac{M}{M_\odot} \text{ in meters,}$$

called the <u>Schwarzschild radius</u> (1915),[21] represents the size of the <u>horizon</u>, from the interior of which nothing, not even light, can escape.[22]

Fig. 1.6
Schwarzschild (1873–1916)

---

[21] He wrote this paper while serving on the Russian front in WWI, a year before he died. The other paper he wrote in the same year supplied a quantum explanation of the Stark effect.

[22] A rigorous calculation using general relativity yields the same expression $2GM/c^2$ for D_{Sch}.

Problem 1.6.41 What are the Schwarzschild radii of the Earth, the Moon, the Sun?

Approximately 1 cm, 0.1 mm, 3 km.

It remains to relate the Schwarzschild radius to the Roche limit. To visit the horizon without being hurt, the astro-tourist would like the representative tidal force $\mathcal{F}$ per unit mass of Section 1.4.2 to be $Ng_{\oplus}$, with $N = 2$ or 3 at most. At the horizon,

$$\frac{\mathcal{F}}{Ng_{\oplus}} = \frac{2\frac{GM}{D_{Sch}^3}d}{Ng_{\oplus}} \quad \text{where } d = \text{your diameter}$$

$$= \frac{2GM_{\odot}}{(3 \times 10^3)^3 g_{\oplus}} \frac{d}{N}\left(\frac{M_{\odot}}{M}\right)^2 \approx \frac{2 \cdot \frac{2}{3} \times 10^{-10} \cdot 2 \times 10^{30}}{27 \times 10^9 \cdot 10} \frac{d}{N}\left(\frac{M_{\odot}}{M}\right)^2$$

$$\approx 10^9 \frac{d}{N}\left(\frac{M_{\odot}}{M}\right)^2.$$

This ratio is maintained within $\leqslant 1$ provided

$$M \geqslant \sqrt{10^9 \frac{d}{N}} M_{\odot} \approx 3 \times 10^4 M_{\odot}$$

for the choice $d \approx 2$ m, $N = 2$. A relatively painless visit to the horizon is possible provided the black hole is massive enough.

The center of our Galaxy is said to have mass $\approx 4 \times 10^6 M_{\odot}$. Its horizon, at $D_{Sch} \approx 0.01$ A.U., is a possible tourist attraction.

$(\blacktriangleleft \text{ PIC}\infty)$

Chapter 2
Investigations of Tides from the Antiquity to Laplace

Vincent Deparis, Hilaire Legros, and Jean Souchay

Abstract Tidal phenomena along the coasts were known since the prehistoric era, but a long journey of investigations through the centuries was necessary from the Greco-Roman Antiquity to the modern era to unravel in a quasi-definitive way many secrets of the ebb and flow. These investigations occupied the great scholars from Aristotle to Galileo, Newton, Euler, d'Alembert, Laplace, and the list could go on. We will review the historical steps which contributed to an increasing understanding of the tides.

2.1 Introduction

In the Western world, the first questionings about the ebb and flow date back to the 4th century B.C., when learned people of Greece began to acquire a precise knowledge of motions in the sea thanks to travels mainly driven by conquests. They raised basic questions such as: 'What causes this wide, periodic, breathing-like motion?' 'Why is it so small in the Mediterranean unlike in large oceans?' The phenomenon was disconcerting, for it is extremely regular in time and irregular in space. From that time to Newton and Laplace, explanations of the tidal phenomena were numerous, sometimes contradictory, often ingenious. They constitute an adventure of the human thought, which we will analyse in this chapter. In particular we will try to illustrate: when the origin of the tidal phenomena was discovered; how the mathematical and physical tools to describe the tides were developed; how

V. Deparis (✉)
77 Clos le Pastoral, 74250 Viuz en Sallaz, France
e-mail: vincent.deparis@neuf.fr

H. Legros
Ecole et Observatoire des Sciences de la Terre, 5 rue René Descartes, 67084 Strasbourg Cedex, France
e-mail: hilaire.legros@unistra.fr

J. Souchay
Observatoire de Paris, 61 av. de l'Observatoire, 75014 Paris, France
e-mail: Jean.Souchay@obspm.fr

J. Souchay et al. (eds.), *Tides in Astronomy and Astrophysics*,
Lecture Notes in Physics 861, DOI 10.1007/978-3-642-32961-6_2,
© Springer-Verlag Berlin Heidelberg 2013

the scientists succeeded in solving a problem which at the beginning had utterly bewildered them.

The concept of tide has two facets. First, we can consider the *tidal force*, which arises any time an extended body A is subject to the gravitation of another body B. The tidal force exists even in case A is a rigid body and not capable of deformation. Second, we can consider the *effects* of the tidal force exerted by B *on deformable parts* of A, for example ocean tides due to the combined gravitational tidal torque exerted by the Moon and the Sun on the bulk of the oceans. Another example is the terrestrial tides, due to the elasticity of the Earth's crust, ordinarily not perceptible but technically observable since the end of the 19th century. In our study, we will focus on the concept of tidal force rather than that of tidal deformation, even if it was via observations of the latter that the scientists have finally understood the real cause of the former.

We have divided our study chronologically. In the first section we discuss observations of tides from the Antiquity to the beginning of the 17th century. We show that learned people of the Antiquity and of the Middle Ages already had a good inkling of the nature and the behavior of ocean tides, and that hypotheses concerning their origin proliferated. In the 17th century several theories dominated the debates, which had little in common with one another. We describe in particular the theories by three great scientists: Kepler, Galileo, and Descartes. One section will be devoted entirely to Newton. We give details on his explanation of tides in his *Philosophiae Naturalis Principia Mathematica* (*Principa* for short), published in 1687. This work, based on a succession of geometric considerations, evaluates among other things the amplitude of the tidal force, and is regarded as the starting point of the true explanation of tides. Another section deals with the works of Daniel Bernoulli, Euler, and d'Alembert: in a relatively short span of time around 1740, these scientists improved the calculations of tidal effects exploiting then new, very efficient tools of calculus.

We conclude our history with the monumental work of Laplace, which was elaborated over a period of more than half a century. Laplace's work, supported by the tool of spherical harmonics, is the foundation of the modern theory of tides. A more complete study of the tides from the antiquity to modern times was done by D.E. Cartwright [2].

2.2 Study of Tides in the Antiquity[1]

The first precise recorded observations of tides go back to the Antiquity, outside the Mediterranean where the ebb and flow phenomena are negligible and cannot be easily detected. Greek mathematicians listened to travelers, often involved in military conquests, to form their description of oceanic tides. Among them, Nearchus,[2]

---

[1] We make an important use of the work of P. Duhem [8, 9].

[2] Born in Crete around 360 B.C., he participated in the expedition of Alexander the Great, being in charge of a fleet of 120 vessels, transporting 10 000 people. He was in charge of establishing a

Fig. 2.1 Poseidonios
(135–50 B.C.)

around 325 B.C., reported on tides of the Indian Ocean, whereas Pytheas,[3] an adventurer from Marseille, mentioned tides of the Atlantic during his trip from Gades in southern Spain to Brittany, 340–325 B.C. Pytheas noticed the fundamental correlation between ascending tides and the full Moon. The geometer Eratosthenes (276–194 B.C.)[4] carefully studied the flows inside the strait of Sicily. He pointed out that their frequency was nearly half a day, with a positive peak corresponding to the instant when the Moon is in the meridian or anti-meridian direction, and a negative peak when the Moon is close to the horizon. Around 150 B.C. Seleucos, a native of the Red Sea, pushed further the analysis of tides. He noticed that their amplitudes get all the greater as the declination of the Moon is larger.

Poseidonios[5] (Fig. 2.1) (135–50 B.C.), as a member of the Stoic school, was an adept of the Aristotelian conception of the universe, an imperfect sublunar world and a perfect supralunar one. For him the behavior of tides, governed by unexplained powers, confirmed the importance of the Moon in human destiny. This argument prevailed until the Middle Ages. In addition to the observation of the half-diurnal frequency of tides, Poseidonios recognized that their amplitudes (associated with what we call nowadays the tidal coefficient) are strongly linked with the phases of the Moon, being maximal during the syzygies (new or full Moon), minimal during the quadratures (when the Earth-Moon and Earth-Sun directions are perpendicular).

---

new maritime route between the Indus and the Persian Gulf. His achievements were described by Strabo in *Geography* (vol. XV).

[3] One of the oldest scientific explorers. His accounts and astronomical observations were used later by Eratosthenes and Hipparchus.

[4] Astronomer, geographer, philosopher, and mathematician, well known for his measurements of the Earth radius by studying shadows produced by the mid-day Sun at Cyrene and Alexandria.

[5] Geographer and historian, he was keen on measurements (meridian length, height of the atmosphere, distance to celestial bodies). He wrote treatises in physics and meteorology.

Fig. 2.2 Strabo
(58 B.C.–25 A.C.)

The geographer Strabo (58 B.C.–25 A.C.) (Fig. 2.2), who compiled the scientific knowledge of his times, emphasized the results obtained by Poseidonios. He mentioned that his predecessor recognized that oceanic tides undergo three kinds of motion, each related to an astronomical cycle: diurnal, monthly, and yearly. He also showed how Poseidonios understood that each time the elevation of the Moon reaches about 30°, the sea begins to rise progressively to reach a peak when the Moon crosses the meridian plane. Moreover Strabo reports that Poseidonios observed annual variations with peaks of amplitude around the equinoxes.

Supplementing these observations Pliny the Elder[6] (23–79 A.C.) made a remarkably precise discovery: he revealed a time lag between the instant when the Moon crosses the meridian/anti meridian and the instant when the tide reaches its maximum.

As the above enumeration shows, the Ancients knew the main characteristics of tides with remarkable accuracy and perspicacity. Nevertheless, a physical explanation remained to be found. Seleucos accepted the idea that the Earth rotates around its axis. He explained that this rotation creates a whirlwind which is modified by the presence of the Moon. The resulting effect is an activation of the oceanic motion. From his side, Poseidonios explained that the Moon had a larger influence than the Sun: the Sun, as a powerful fire, destroys all the vapor it creates at the surface of the ocean; the Moon, an attenuated fire, cannot vaporize the fluid masses and thus favors the ebb and flow. The Sun has no direct effect on the tides, but an indirect one as it lights the Moon, which in its turn acts on the oceans.

As far as we can gather from the surviving testimonies, Arab scholars did not add substantial knowledge or theory dealing with the tides. But as in the other fields of astronomy and mathematics, they played a key role in the transmission of scientific knowledge from the Greeks to the Western countries. The astronomer and astrologer

---

[6]Roman author, naturalist, and philosopher, he wrote *Naturalis Historia*, an encyclopedia of much of the knowledge of his time, the largest single work to have survived from the Roman empire to the present day, encompassing botany, zoology, astronomy, geology, and mineralogy.

Abu Maishar al-Bakhli (787–896), more often called Albumasar (787–886),[7] mentioned the three kinds of cycles accompanying the tides (semi-diurnal, fortnightly, and semi-annual) as well as the leading effect of the Moon as pointed out by Poseidonios. But in contrast with his Greek predecessor, he did not believe that moonlight was the cause of tides. Indeed two facts were difficult to explain according to the theory of Poseidonios: the existence of a peak of amplitude during a diurnal cycle when the Moon is located in the anti-meridian direction, and a maximum of the peak during a monthly cycle when the Moon is in conjunction with the Sun. Both cases correspond to a total absence of Moon light, which contradicts the theoretical foundations above.

Albumasar suggests an alternative explanation. As an astrologer ready to ascribe supernatural powers to celestial bodies, he says the Moon possesses a 'virtue' having the power of driving oceanic motions. The sea itself does not have the capacity to be disturbed under the influence of the moonlight, enhanced by the solar light. The cause of tides should be extrinsic to the sea and, after sieving various alternative explanations, he reaches the conclusion that the Moon is responsible of the uprising of oceanic masses, thanks to its own virtue. He supposes that the Sun too possesses a similar, though attenuated, virtue. Finally, Albumasar explains (correctly) that the lack of significant tidal phenomena in some basins comes from the their configuration, and not from a limitation of the lunar effect.

2.3 Variety of Theories in the Middle Ages

The medieval knowledge about tides came essentially from the writings of Pliny the Elder, until the Albumasar was translated into Latin in 1140. But it is worth mentioning the contribution of the Venerable Bede (about 672–735)[8] one whose interesting features is that it addresses the unexplored tides along the coasts of Great Britain. Bede made very valuable and accurate observations, remarking for instance that the maximum of the tide does not occur at the same time in various harbors along the coast, even when these harbors are located along the same meridian. This constitutes the first recognition of what is nowadays called the 'harbor establishment law'. It proves that Bede's perception of ebb and flow was particularly sharp, at an epoch when science generally stagnated. In contrast with these realistic observations, some original theories were proposed by other scholars. Paul Diacre (720–748), studying the maelströms in the North Sea, remarked that the direction of a whirlpool changed when the tide is reversed: therefore he attributed the tides to some abysses swallowing, then regurgitating, the oceanic masses.

---

[7] Persian astrologer, astronomer, and Islamic philosopher, he wrote a number of practical manuals on astrology that profoundly influenced the Muslim intellectual history.

[8] English monk at the Northumbrian monastery of St. Peter at Monkwearmouth. Well known as author and scholar, and for *The Ecclesiastical History of the English People*. In *The Reckoning of Time*, he deals with ancient and medieval views of cosmos, including explanations of astronomical phenomena.

In the 4th century, the philosopher and philologist Macrobe (about 370–430) imagined that the ocean had four arms crossed by big currents and that the tides originate from the conflict between these currents. This theory was popularized seven centuries later by the philosopher, mathematician, and naturalist Abelard de Bath (1080–1160), known for his interest in the Arab culture. None of these explanations involves the Moon or the Sun. A little later, we come back to a more convincing explanation of the nature of tides thanks to a professor of theology, Guillaume d'Auvergne (1190–1249), who reinstated the determining influence of the Moon at the center of the discussion. His explanation looks very close to the true one: willing to introduce some astrological principle involving the influence of the Moon, he proposed that the sea gets elevated toward the Moon which acts like a conductor, a disconcerting analogy with magnetism: when the Moon is ascending, it attracts the fluid as a magnet attracts iron when lifted. One of the interesting points in Guillaume d'Auvergne's conception of tides is the foreboding of gravitational attraction, another in the rejection of swallowing or regurgitating of fluid masses: the oceanic mass remains constant and the elevation is due to an agitation created by the Moon. Nevertheless, Guillaume d'Auvergne shows an ignorance of the semi-monthly cycle: he believes that a maximum of the tide occurs each month during the full Moon, attributing the lunar action to its lighting; he does not mention the symmetric case of the new Moon, when the satellite presents its dark face toward us.

Albert the Great (about 1200–1280) proposed similar explanations. For him, the Moon is doubly responsible for the tides. First it is a body of humid nature and so has the ability to attract the oceanic fluid as a magnet attracts iron. Second its brightness creates a heat which leads to the formation of a bulge—some kind of bubbling. He added that the water could be attracted only because of the salinity of the sea. St. Thomas Aquinas (1224–1274) still clung to the idea that the Moon possesses some virtue which gives it the capacity to stimulate motion inside fluids. Thus, the 12th and 13th centuries saw many theories dealing with the formation of tides, broadly based on two postulates. The first says that the Moon has some virtue; the second says that the Moon acts through its light. Either way, such theories face severe inconsistencies. For instance, how can the Moon cause the second semidiurnal tide when it is located in the anti meridian direction, at a position where its influence should be minimum in terms of power, brightness, or heat? According to some audacious theorists, such as Robert Grosseteste (1175–1253),[9] the power of the Moon, when it is below the horizon, is maintained through the reflection of its light on the celestial sphere. This theory, though highly hypothetical, was supported by many contemporary physicists, such as Roger Bacon (1214–1294). Also, inconsistencies with astrological principles arose: the idea that moonlight acts on oceans by a kind of bubbling is against the principle that the Moon is a humid body cooling down and condensing any vapor. Physicists of the Paris school such as Jean Buridan (1202–1363) hesitate to select one out of these many theories.

In summary, the main difficulties in the theory of tides during the Middle Ages were as follows.

---

[9]English scholar, bishop of Lincoln. He showed deep interest in geometry and optics.

- Precise observations of tidal mechanisms by Greek and Arab scholars had been forgotten, replaced by obscurantism or wrong beliefs. For instance some theories defend the idea of a monthly cycle for the maximum amplitude instead of the real semi-monthly one.
- The fallacious idea that tides are due to alternating swallowing and regurgitating of water competed the correct idea that oceanic masses remain constant and that a bulge is produced at some place, balanced by a mass deficiency at another place.
- The problem of the geographical variations of tides is unsolved. In particular the following questions do not find any answer. Why are the tides so strong at some coastal locations and nearly non-existent at others? Why is the diurnal inequality (difference of maximum amplitudes between successive semi-diurnal tides) clearly present at some locations like the Red Sea, while it does not appear at others like the Atlantic?
- Does the Moon have an influence? If yes, how to characterize this influence? In terms of its light? Or of some virtue? And how to explain the maximum of the tide when the Moon is lying on the anti-meridian, at its maximum angular distance below the horizon?
- What is the exact role of the Sun? Does it directly raise the water mass by heating? Or does it act indirectly by reflection off the Moon?
- How can we explain the various periods linked to the tides? If the interpretation of the semi-diurnal cycle can be found, what is the cause of the fortnightly and the semi-annual cycles?

2.4 Tides in the Renaissance and the 17th Century

In the Renaissance and especially in the 16th century, the developments of the theory of tides come largely from physicians and astrologers. Their main aim was to establish a link between celestial bodies and phenomena occurring on the Earth. For that aim they made a clear choice between the various explanations prevailing at the end of the Middle Ages and enumerated at the end of the last section. For them the water mass remains constant; tides are obviously caused by the action of the Moon and less predominantly by the Sun; these two bodies do not produce their action through their light but through a specific virtue, which is comparable to the attraction between a magnet and iron.

2.4.1 Renaissance

In the beginning of the 16th century, a physician of Sienna, Lucius Bellantius, explains that the rays with which the Moon attracts the oceanic masses are not light rays, as can be proved during the conjunctions (new Moon), when the Moon shows us its dark face. For him the Moon acts through virtual rays, in the same way as a magnet attracts iron. Another physician, Frederik Grisogono (1472–1538)[10] insists on the modulating influence of the Sun, which in some cases enhances, in others attenuates, the action of the Moon. His intuition is amazingly close to reality. The total tide can be divided into two components, one due to the Moon, the other due to the Sun. They both produce a swelling of the oceanic volume, maximum at the point of the oceanic surface closest to that body, and also at the antipodal point. Grisogono supposes that each of the Moon and the Sun distorts the sea to form an ellipsoid of revolution, whose major axis is oriented toward it. This helps to explain how twice a month, in the full and the new Moon (syzygies) when the two major axes coincide, the amplitude of the tide is maximum. Such ideas from physicians and astrologers spread rapidly, the majority of them supporting the 'magnetic model' of attraction. Jules César Scalinger (1484–1558) claimed that just as iron is moved by a magnet without any physical contact, so the sea can be moved by the presence of a 'noble body' such as the Moon. The English scholar and physician William Gilbert (1544–1603), who undertook pioneering studies in electrostatics and magnetism, discovered that the Earth acts like a giant magnet. He also adhered to the idea that the Moon does not act through its light but through forces analogue to the magnetic one.

2.4.2 Kepler's Views

Kepler (1571–1630) agreed with Scalinger's conception and with the magnetic analogy. Of gravitational phenomena he had a remarkable visions, which opened the path to Newton. First, though himself an occasional astrologer, he was a fierce opponent to the astrological principle according to which the Moon attracts the sea by their common humid nature. He defended the concept of a mutual gravity depending on the sizes of the bodies involved. He explicitly claims that if the Earth ceased its attraction of the oceanic masses, the latter would instantaneously rise toward the Moon. For him gravity is a mutual disposition to join between bodies sensitive to each other. For instance he presumes that if two stones were placed at a little mutual distance apart and far from any other body, they should undergo a mutual attraction, leading to a junction in some intermediary location. Notice that we cannot qualify this attraction as 'universal', because for Kepler the two bodies concerned must be of such nature as to favor attraction. The concept of gravity-driven tides were appreciated among some of Kepler's contemporaries, and by the beginning of the 17th century his ideas had spread quite a lot, even if they encountered opponents such as the mathematician, philosopher, theologian, and astronomer Pierre Gassendi (1592–1655) who rejected the idea that the Sun could have any action on tides, still arguing that the action of the Moon comes from its humid nature. Another strong opponent

---

[10]Also mathematician, physicist, astronomer, born at Zadan in Croatia and educated at Padova. In addition to *Commentaries on Euclid's Elements*, he developed an important theory of tides, published in Venice in 1528.

Fig. 2.3 Galileo Galilei
(1564–1642)

to Kepler's views was Galileo Galilei (Fig. 2.3) (1564–1642) who expressed his astonishment that such a 'free and subtle spirit' (Kepler's) could defend the idea of any power of the Moon on the water, thus betraying attachment to some occult and childish principles.

2.4.3 Galileo: An Original Concept

Galileo's opposition to Kepler's explanations was motivated by the fact that he himself developed a theory based on new principles of mechanics, very different from all those already described. The fourth day of his *Dialogues Concerning the Two Chief World Systems* [11] (published 1632) is devoted to the problem of tides, and gives a full account of his approach starting from the combination of the Earth's rotation around its axis and its orbital motion around the Sun. This theory could have been imported from the work of Celio Calcagnini[11] (1479–1541) published posthumously at Basel in 1544. Galileo's explanations relies on the analogy between the tides and the motion of water inside a vessel. When the vessel is accelerated or decelerated, the inertia inclines the surface of water toward the back or front side of the vessel. For him, the tidal motions of the oceans follow the same laws, governed by a varying acceleration of the water coming from the combination of the Earth's diurnal rotation plus its annual revolution around the Sun. According to this theory, if only one of these motions existed and not the other, the ocean would be in equilibrium.

These motions, when combined, produce the same kind of displacements as that of water in a vessel. For a given point at the circumference of the Earth, the two velocities due to the rotation and the revolution sometimes add together, sometimes subtract from each other (Fig. 2.4). Therefore the water masses are displaced alternately along the oriental and occidental coasts, causing a diurnal tide. Thus for

---

[11] Italian humanist and scientist from Ferrara, in his time a reputed astronomer.

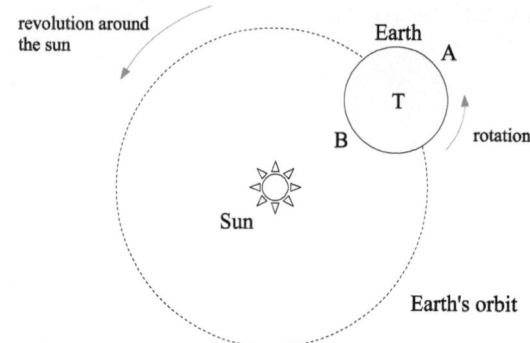

Fig. 2.4 Theory of tides according to Galileo: the ecliptic plane is supposed to coincide with the equator. Rotational and orbital motions are added in **A**, substracted in **B**

Galileo the origin of tides must be found exclusively from the combination of various terrestrial motions, and has no link at all with the influence of the Moon or of the Sun. In retrospect this theory does not look realistic. Nevertheless, the reality of the effect suggested by Galileo deserves some attention.

For him the two motions of rotation and revolution sometimes are added, sometimes substracted. Thus the points on the surface acquire a non-uniform velocity, implying an activation of water motion. Note that Galileo's idea can be associated with the concept which enabled the scholars of the Antiquity to explain the non-uniform motion of the Sun, the Moon, and the planets in the sky through a combination of motions with the help of a deferent and epicycles.

Souffrin [25] analysed this effect, illustrated in Fig. 2.5: the acceleration of a given point M at the surface of the Earth can be divided into two components: the first, γ_1, is the centripetal acceleration of the center of the Earth with respect to the Sun, due to the orbital motion, with angular velocity Ω; the second, γ_2, is the centripetal acceleration due to the rotational motion of the Earth around its center of

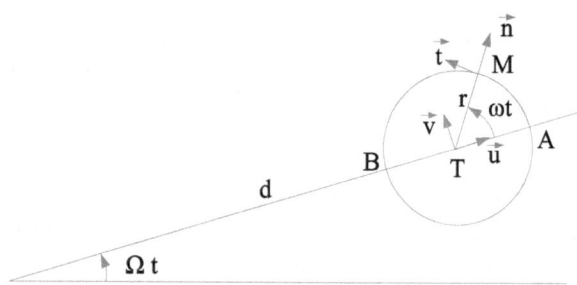

Fig. 2.5 Theory of tides according to Galileo: M is a point on the Earth surface, T the Earth center, S the Sun, r the Earth radius, d the radius of the Earth orbit, supposed circular. Ω is the angular velocity of the orbital motion, $\Omega + \omega$ the angular velocity of the rotational motion

mass, with angular velocity $\Omega + \omega$. Thus the total acceleration of M can be written as[12]

$$\gamma_M = \gamma_1 + \gamma_2 = -\Omega^2 d\mathbf{u} - (\omega + \Omega)^2 r\mathbf{n} \tag{2.1}$$

Since $\mathbf{u} = \cos \omega t \mathbf{n} - \sin \omega t \mathbf{t}$, the decomposition of γ_M along $\mathbf{n}, \mathbf{t}$ gives

$$\gamma_M = \gamma_n + \gamma_t = -\left(\Omega^2 d \cos \omega t + (\omega + \Omega)^2 r\right)\mathbf{n} + \Omega^2 d \sin \omega t \mathbf{t} \tag{2.2}$$

Thus the acceleration γ_M for a given point of the Earth has a normal component γ_n and a tangential component γ_t. The normal component has no significant effect, for it acts in the same direction as gravity and is negligible in comparison. The tangential component, though of very small size also, acts perpendicularly to gravity and can have a visible effect. This tangential component $\Omega^2 d \sin \omega t$ comes solely from the orbital motion. Because of the diurnal rotation, it is alternately directed eastward or westward.

Galileo's mistake comes from a misunderstanding of the orbital motion of the Earth. As Newton will show a little less than a century later, the Earth is kept in its orbit by the gravitational attraction of the Sun which acts on all terrestrial matters, the liquid part as well as the solid part. To a first approximation the water as well as the basins are subject to equal gravitational attractions $\Omega^2 d$ of the orbital motion. But the equality is exact only at the center of the Earth and not elsewhere, which constitutes the basis of the explanation of tides by Newton. The water and the basins, attracted in the same manner by the Sun, 'fall' toward it together: the relative motion between the water and the basins does not exist.

Of course Galileo did not know Newton's law of gravitation. No more than Kepler and other contemporaries was he able to understand the orbital motions of the planets. For him the revolution of the Earth is given naturally: it exists without any cause, guided by an imaginary physical principle. Consequently Galileo relied on purely kinematical principles and never adopted a dynamical one. Nothing and nobody could induce him to doubt his solution from the combination of the two Earth motions. Despite his misunderstanding, his will to develop a mechanical theory of tides was fundamentally new and his contribution was essential.

2.5 Descartes and His Theory of Vortices

In his *Principles of Philosophy* published in 1644 [7], Descartes (Fig. 2.6) (1596–1650) proposes an alternative theory of tides, relatively independent of the predecessors. He is convinced that everything in the universe is governed solely by the laws of motion, and that vacuum does not exist: as soon as vacuum arises, it gets filled with subtle matter organized in a system of vortices. This principle is applied to the solar system. The Sun occupies the center of the main vortex, and its proper

---

[12]Our method of proof, based on accelerations and vectors, is not that of Galileo who works with velocities only. But it is a faithful translation of his idea.

Fig. 2.6 René Descartes
(1596–1650)

rotation (discovered at the beginning of the 17th century) is transmitted to the vortex itself, which transports the planets on their orbit. Each planet is at the center of its own vortex. Their proper rotations lead to the rotation of these secondary vortices which transport the satellites in their revolutions (Fig. 2.7).

Thus, the Moon is transported by the Earth's vortex. Starting from this statement, Descartes built up an intricate theory where the Moon, despite being transported by the Earth's vortex, does not move at the same velocity. This creates an obstacle and perturbs the symmetric flow of subtle matter, causing a displacement of the Earth's center with respect to the Earth's vortex. Because of the presence of the Moon, the

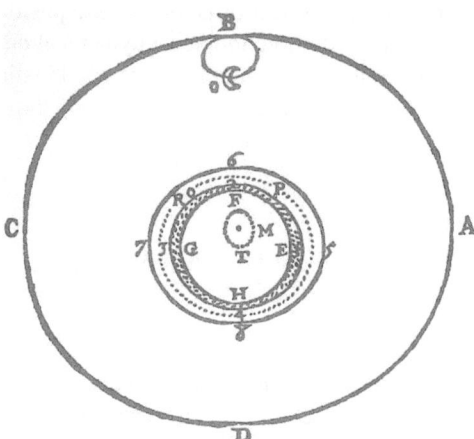

Fig. 2.7 The System of the World according to Descartes (from *Le Monde ou Traité de la lumière*, ed. Adam et Tannery, Paris, 1974). *ABCD* is the vortex of subtle material generated by the proper rotation of the Earth *EFGH*, with center *T*. *1, 2, 3, 4* represent the sea, and *5, 6, 7, 8* the atmosphere. Because of the presence of the Moon in *B*, the center *M* of the vortex does not coincide with *T*. The tides result from a differential pressure exerted by the vortex matter on the sea

Fig. 2.8 Issac Newton
(1642–1724)

vortex material surrounding the Earth cannot flow freely. Consequently it exerts a differential pressure at the surface of the oceans, giving rise to tides.

In addition Descartes explains the existence of two tides per day from the drift between the center of the Earth and the center of the main vortex. Astonishingly, according to his views, the low tides arise when the Moon is located in the meridian and anti-meridian directions, which is opposite what happens in reality. Moreover, Descartes attributes the geographical variations of the high tides along a given coast to the fact that the Earth is not entirely covered by oceans. For him time delays are caused by various factors, such as the form of the coasts, the varying depths of the oceans, the influx from the rivers, as well as the action of winds. These intuitions were valid, for we know today that all these elements have to be taken into account in order to construct accurate tide tables. For Descartes, the semi-monthly period as well as the alternation of large and low tidal amplitudes linked with it come from the non-circularity of the Earth's vortex. All these considerations let us think that his theory of tides, though unrealistic, attests to a remarkable view of mind. In any case it relies on statements which have never been explained or verified, as the existence of vortices, the drift between the center of the Earth and the center of the main vortices, etc. Despite these negative aspects Descartes's explanations became very popular during his life, in particular among his French disciples. In his *Geographia Generalis*, the German geographer Bernard Varenius (1622–1650) [26] adopts Descartes's theory as the preferred one among various others, and this will helped its popularization.

2.6 Newton and the Gravitational Attraction: A Giant Step

In 1687, Isaac Newton (Fig. 2.8) (1642–1727) published his *Philosophiae Naturalis Principia Mathematica* (*Principia* for short) [23], which revolutionized our perception of the Universe. In *Principia*, Newton set out the law of gravitation and the three fundamental laws of motion: the principle of inertia, the principle of the rate of change of momentum, and the law of action-reaction, showing that the behavior of celestial bodies is deducible from these laws. One of the tremendous results of his theory was an explanation of the tidal phenomena. For Newton, the oceanic tides are explained by mechanical principles, as Galileo wanted them to be. Moreover they are a consequence of an attraction at a distance, following Kepler's intuition.

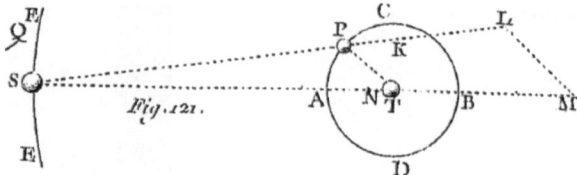

Fig. 2.9 The perturbing forces on the lunar motion (Fig. 121 of *Principia*, vol. I). *S* is the Sun, *T* the Earth, *P* the Moon with orbit *CABD*. The attraction of the Sun on the Earth is represented by the segment NS and the attraction of the Sun on the Moon by the segment LS, which can be divided into two parts: LM and MS

How did Newton reach the explanation of the tides, after explaining the orbital motion of the Earth around the Sun and that of the Moon around the Earth? In fact, his success came from a deep investigation of the orbital motion of the Moon, taking into consideration the departure of its trajectory from an exact ellipse, which was due to the gravitational perturbations of the Sun. In particular he observed the drift of the lunar nodes with respect to the ecliptic, with a 18.6 y period. Extrapolating these solar perturbations he guessed that they should also influence the oceanic masses.

2.6.1 The Solar Perturbation on the Orbital Motion of the Moon

As mentioned above, the tide was analysed by Newton when he was determining the perturbation by the Sun on the Moon orbiting around the Earth. We refer to Proposition 66 of *Principia*, vol. I. In order to solve the problem, Newton used the law of *parallelogram of forces*, well known since the 16th century when it was popularized by the Dutch engineer, mathematician, and philosopher Simon Stevin (1548–1620) and clearly set out by the mathematician Pierre Varignon (1654–1722) in his treatise *New Mechanics* published posthumously (1725). Newton's way of thinking is clearly shown in Fig. 121 of the Principia (see Fig. 2.9). It represents the Sun S, the Moon P, the Earth T, with their mutual distances SP, ST, PT. Newton represents the attraction by the Sun on the Earth (respectively on the Moon) by SN (respectively SL). The attraction SL itself is decomposed into two components: one, SM in the Sun to Earth direction, another one ML in the direction parallel to the Earth-Moon segment. One of the subtleties of Newton's proof is that he depicts the points N and T as coincident, although they have different statuses: T represents the position of the Earth and has a physical meaning, whereas N is used to measure the attraction SN exerted by the Sun, and has a mechanical meaning. From the law of gravitation and by calling F_{SP} (respectively F_{ST}) the forces exerted by the Sun on the Moon (respectively on the Sun) we have, using modern notations:

$$\text{SL} = F_{SP} = \frac{GM_S}{SP^2}, \qquad \text{SN} = F_{ST} = \frac{GM_S}{ST^2} \qquad (2.3)$$

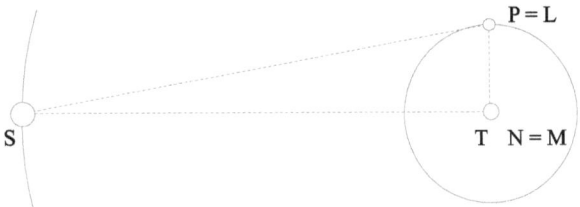

Fig. 2.10 The tidal force when the Moon is in quadrature. Because it is very far from S, P is considered as coinciding with L, and the attraction LS by the Sun on the Moon is practically equal to the attraction NS by the Sun on the Earth. The tidal force is represented by LN, equal to PT. It is added to the attraction of the Moon by the Earth

therefore

$$\frac{SL}{SN} = \frac{ST^2}{SP^2}, \qquad SL = \frac{ST^3}{SP^2} \tag{2.4}$$

Thus the Sun acts on the Earth with the force F_{ST} represented by SN and on the Moon with the force F_{SP} represented by SL. The length and the direction of the two segments are not the same, and these differences characterize the difference of attraction exerted by the Sun. To determine how the Sun perturbs the orbital motion of the Moon around the Earth, Newton searches which part of SL has a real effect. He remarks that if the accelerations SN and SM are equal, they will change nothing in the relative motion of the two bodies P and T, because they will bring the same attraction, both in amplitude and direction. Thus only the component NM plays a role in the difference of attraction and consequently in the perturbation of the orbital motion of the Moon around the Earth. We must also take into account the component ML, in such a way that finally the perturbing forces exerted by the Sun are reduced to the two segments NM and LM. In modern notation, this way of thinking should be equivalent to calculating the difference between the vectors LS and NS. In the proof above, Newton has just revealed the presence of a solar tidal force, i.e. a differential force which is not due to the total gravitational attraction by the Sun on the Moon, but rather to the difference of attractions by the Sun on the Moon and the on Earth. This is a fundamental discovery in the theory of tides.

2.6.1.1 Case of Quadrature

In quadrature (when ST and TP are perpendicular) the sketch is simplified (Fig. 2.10). M coincides with T, and the lengths of SP and ST can be treated as equal, given the large distance ST from the Sun to the Earth. Moreover we have LM = PT, and LM is oriented along the direction from the Moon to the Earth (*Principia*, vol. I, Proposition 66). We conclude that the tidal force F_{tidal} is reduced to its component PT, and that the ratio of its amplitude to the attraction of the Earth by the Sun is given by (*Principia*, Proposition 66, Corollary 14)

$$\frac{F_{tidal}}{F_{ST}} = \frac{PT}{ST}, \qquad F_{tidal} = \frac{F_{ST} \times PT}{ST} = \frac{GM_S \times PT}{ST^3} \tag{2.5}$$

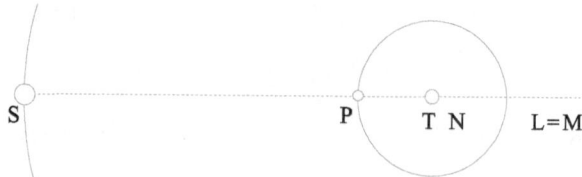

Fig. 2.11 The tidal force when the Moon is in opposition. It is represented by the segment LN, obtained by difference between the attraction LS by the Sun on the Moon and the attraction NS by the Sun on the Earth. Thus the tidal force is in opposition to the attraction of the Moon by the Earth

Thus in this particular case of quadrature we are in presence of a remarkable equivalence between the lengths of the segments representing the forces and the physical distances.

Moreover, applying Kepler's third law to the orbital motion of the Earth, we get

$$\omega_E^2 ST = \frac{GM_S}{ST^2}, \qquad \omega_E^2 ST^3 = GM_S \tag{2.6}$$

where ω_E is the angular velocity, or *mean motion*, of the Earth. In Newton's era, the distance ST from the Earth to the Sun was not known with accuracy. From the last two equations we get

$$F_{tidal} = \omega_E^2 PT \tag{2.7}$$

Now we can evaluate the ratio of the tidal force to the attraction F_{PT} exerted by the Earth on the Moon:

$$F_{PT} = \frac{GM_E}{PT^2} = \omega_M^2 PT \tag{2.8}$$

where ω_M is the angular velocity, or *mean motion*, of the Moon, and (*Principia*, vol. I, Corollary 17)

$$\frac{F_{tidal}}{F_{PT}} = \frac{\omega_E^2}{\omega_M^2} \tag{2.9}$$

2.6.1.2 Conjunction and Opposition

When the Moon P is in conjunction with the Sun (Fig. 2.11), L and M coincide. The Moon being closer than the Earth to the Sun, it is subject to a larger gravitational attraction. Newton shows that in this case the perturbing force by the Sun on the Moon is NM = 2 PT. This can be found from the equation

$$SM = SL = \frac{ST^3}{SP^2} \tag{2.10}$$

with SP = ST − PT. Expanding to the first order we have

$$SM = \frac{ST^3}{ST^2(1 - PT/ST)^2} \approx ST + 2\,PT = ST + NM \tag{2.11}$$

Therefore, to a first approximation NM = 2 PT. The tidal force NM is twice bigger than in the case of quadrature.

In a similar and symmetric way, when the Moon is in opposition we can easily prove that SM = ST − 2 PT ≈ ST − NM. We still have NM ≈ 2 PT. In conclusion, in syzygies (new Moon and full Moon) the perturbing force exerted by the Sun on the Moon has the same value (2 PT) and is directed in the direction opposite to that of the gravitational attraction exerted by the Earth on the Moon.

Thanks to this ingenious way of geometrical representation of the attraction, Newton could calculate the two components LM and NM of the perturbing force of the Sun on the orbital motion of the Moon, not only in the special cases of syzygies and quadratures but also at any position of the Moon on its orbit. This helped Newton to study in detail the characteristics of the lunar motion, showing in particular that the Moon is accelerated on its orbit from the quadratures towards the syzygies, and that the lunar nodal line[13] undergoes a linear retrogradation.

2.6.2 Ocean Tides

After studying the perturbations exerted by the Sun on the orbital motion of the Moon, Newton in vol. I, Proposition 66, Corollary 19 shows how to apply the same principle to the terrestrial phenomenon of the tides.

2.6.2.1 Analogy Between the Lunar Motion and the Ocean Tides

The fundamental idea consists in substituting for the Moon a set of fluid bodies, then to replace this set by a continuous fluid ring inserted in a canal surrounding the Earth. Under the gravitational attraction of the Sun, the fluid inside the canal undergoes the same kind of gravitational perturbation as the Moon in its orbit, that is to say an acceleration during the syzygies (for the part of fluid oriented in the direction of the Sun and in the opposite direction) and a deceleration during the quadratures (for the part of fluid in the directions perpendicular to the direction of the Sun). This alternating motion gives rise to the tidal phenomena. Thus, thanks to the analogy with the lunar orbital motion, Newton understands that the various parts of the terrestrial globe, located at different distances of the Sun, are subject to different attractions by the Sun, which in their turn cause the oceanic motions.

2.6.2.2 Agreement with the Observed Characteristics of Ocean Tides

In *Principia*, vol. III, Proposition 24, Newton returns to the problem of ocean tides. Here the Moon no longer plays the role of a test body whose irregularities of motion

---

[13]The intersection of the orbital plane of the Moon with the plane of the ecliptic.

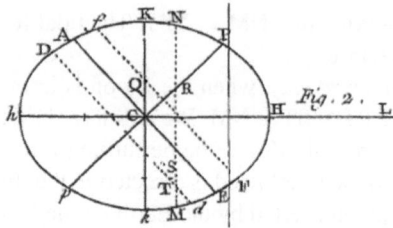

Fig. 2.12 The tidal ellipsoid and the diurnal inequality (Fig. 2 of *Principia*, vol. III). Under the effects of the tidal forces, the oceans change their global shape in an *ellipsoid* whose semi-major axis is directed toward a fictitious body delayed three hours with respect to the real body. *Pp* is the rotation axis, *AE* the equator, *Ff* a parallel. When the body is at a given declination, the two diurnal tides in *F* and *f* do not have the same amplitude: this characterizes the diurnal inequality

reveal the effects of the differential gravitational attraction exerted by the Sun, but as another celestial body providing the same kind of effects on the oceans. His clear objective goes beyond identifying the origin of tides, to explaining the actual tidal phenomena observed along the coasts. Attention is paid to their periodicities, amplitudes, and characteristics as a function of the relative positions of the Moon and the Sun.

Newton asserts that under the action of a celestial body (the Moon or the Sun), the sea at any instant takes the shape of an ellipsoid whose major axis is oriented toward the body. We have mentioned that 16th-century physicians and astrologers had already guessed at this phenomenon intuitively. Newton gives a full justification of the phenomenon, showing that it arises from a symmetry in the tidal force. As the Earth rotates, points on its surface pass alternately through the locations of maximum and minimum elevation of the water.

This explains the succession of low and high tides (Fig. 2.12). Moreover, Newton can explain the monthly periodicity of the tides: during the syzygies, the major axes of the ellipsoids due to the Moon and to the Sun are aligned, leading to the addition of the raising of the sea level, whereas during the quadratures, these axes are perpendicular and the effects at the sea level cancel, leading to an attenuation of the high tides. Newton also remarks that the maximum amplitude of the tides varies according to the distance of the perturbing body, which itself varies because of the ellipticity of its orbit around the Earth. In short, Newton fully explained the various periodicities of the tides, confirming the coherence and the validity of his theoretical assumptions.

By assuming that the seas are distorted into the figure of an ellipsoid, Newton accomplishes a significant step, that of adopting a simple figure of equilibrium, in the same way as he did to express the deformation of the Earth undergoing the effects of its rotation. Following this rotation, the ellipsoid moves in such a way that its major axis always points toward the perturbing body.

2.6.2.3 Remaining Questions

Nevertheless, this basic theoretical model faced several problems when confronted with the observations. First, there are significant time delays between the occurrence of the diurnal tides and the transit of the perturbing body at the meridian or at the anti-meridian. Second, the maxima of tides do not correspond exactly to the syzygies, as predicted by the theory. Third, the amplitudes of tides strongly change depending on the harbors where they are measured, even when the harbors are separated by short distances. Fourth, the diurnal inequality, i.e. the difference of amplitudes between successive high tides, does not show up significantly in observations, whereas theoretical calculations predict them to be large.

In fact the most important factor leading to these discrepancies between theoretical statements and observations lies on the fact that Newton's results are presented in the frame of a static theory: it is necessary to construct a dynamic theory in which the influence of the Coriolis force and the resonance phenomena are taken into account.

Nevertheless, even after recognizing the lack of perfect agreement between his theoretical investigations and the observations, Newton kept the explanations above, mentioning for example that the oceanic motions are delayed by the friction of the bottom of the basins. Notice that this phenomenon of inertia has been fully explored and validated in the 20th century to explain the secular deceleration of the proper rotation of the Earth. Thus in various places of his work, we see that Newton adheres to the idea of the inertia of oceans which necessitates further investigations. For instance he explains the absence of diurnal inequality as well as the presence of time delays between coastal points, by flow effects that conserve perturbing oscillations for some duration, in the same way as water moved in a vessel.

2.6.2.4 Calculation of the Solar Tide

After the periodicities of tides were explained and a hypothesis was made about their time delay, an important challenge remained to be undertaken by Newton: it consisted in starting from the tidal force exerted by the perturbing body (the Moon or the Sun) and deducing the amplitude of the ebb and flow. This could be done for the Sun, thanks to the previous calculations presented previously, of the tidal force exerted by the Sun on the Moon's orbital motion. But it could not be done for the Moon, because its mass was unknown. Thus in a first step, Newton attempted to calculate only the amplitude of the solar tides. Rather than trying to find a final formula, he proceeded by successive numerical approximations gathered in *Principia*, vol. III, Propositions 25 and 36. We saw in Eq. (2.9) that the ratio of the tidal force F_{tidal} exerted by the Sun on the Moon to the force of attraction F_{TP} exerted by the Earth on the Moon in quadrature could be expressed as

$$\frac{F_{tidal}}{F_{PT}} = \frac{\omega_E^2}{\omega_M^2} = \frac{T_M^2}{T_E^2} = 1/178.725 \qquad (2.12)$$

where $T_M = 27.32d$ and $T_E = 365.25d$ are the sidereal periods of revolution of the Moon and of the Earth. Once this result is obtained, it can be used to calculate the solar tide at the surface of the Earth, still in the case of quadrature. In particular it is possible to calculate the ratio of this solar tide to the gravitational acceleration g. Two facts are used for this purpose:

- Since the Moon being 60 times more distant from the center of the Earth than a point on the surface, $g/F_{PT} = 60^2 = 3600$.
- As the tidal force exerted by the Sun is directly proportional to the distance PT between the point considered and the center of the Earth, this tidal force F'_{tidal} on the surface of the Earth is 60 times smaller than the same tidal force F_{tidal} at the distance of the Moon: $F'_{tidal} = F_{tidal}/60$.

Hence the ratio F'_{tidal}/g is

$$\frac{F'_{tidal}}{g} = \frac{1}{60 \times 3600 \times 178.725} = \frac{1}{38\,604\,600} \tag{2.13}$$

This is the ratio for points on the surface of the Earth in quadrature, i.e. located at 90° with respect to the direction of the Sun. As seen previously, for points in conjunction with the Sun, i.e. for which the Sun is at zenith or at nadir, the ratio is twice bigger. Notice that in quadrature the tidal force is pushing the surface toward the bottom whereas in conjunction it raises the surface toward the attracting body, the Sun. Therefore, the amplitude of the total acting tidal force is 3 times bigger than the amplitude calculated above, that is to say $g/12\,868\,200$.

The next step consists in calculating the elevation of water under the sole action of the Sun. To simplify, Newton considers a fictitious Earth completely covered by oceans, and having the same density. Then he uses the same kind of trick as the one he used to calculate the bulging of the Earth under the centrifugal acceleration due to the rotation: he considers two channels filled with a homogeneous fluid extending radially from the center of the Earth to the surface, one in the direction of the Sun, the other in the direction perpendicular (Fig. 2.14). The first channel is longer, for the tidal force is substracted from the gravity, whereas in the second channel it is added. Assuming proportionality between the bulging of the surface and the perturbing force, he first remarks that the centrifugal force which is 289 times smaller than g at the equator leads to a difference of 27.7 km (in fact 85 472 Paris feet) between the equatorial radius (semi-major axis) and the polar radius (semi-minor axis) of the bulging Earth. By analogy, following the same proportionality, the solar tidal force being 12 868 200 times smaller than g will create a difference of level of 60 cm between a point in quadrature and another point in conjunction with the Sun.

2.6.2.5 Ratio of the Lunar Tide to the Solar Tide and the Mass of the Moon

The mass of the Moon being unknown, Newton cannot calculate directly the amplitude of the lunar tides. In *Principia*, Proposition 37, he considers the inverse problem: knowing the amplitude of the tides as a function of the relative positions of the Moon and the Sun, is it possible to calculate their respective attractions on the oceans and to deduce the mass of the Moon? The basic hypothesis is that the height of the tides caused by each body is proportional to the size of its tidal action. Close to the equinoxes, the two bodies are located on the equatorial plane and during an equinoctial syzygy, the height of the tide is maximal because the actions of the two bodies are maximal and added together. Thus the height of the tide $h_{syz.}$ can be written

$$h_{syz.} = A(M + S) \qquad (2.14)$$

where M and S are respectively the actions of the Moon and the Sun, and A is a coefficient of proportionality. About seven days after the syzygy, when the Moon is in equinoctial quadrature, the actions of the Moon and of the Sun must be substracted from each other. Moreover the Moon is no longer on the equator, its declination δ being roughly $23°$ (if we neglect the inclination of the lunar orbit with the ecliptic). This diminishes the strength of its action, the coefficient of diminution being $\cos^2 \delta$. This correction concerns only the semi-diurnal tides, as it will be shown by Laplace. Thus in this case the height of the tide is

$$h_{quad.} = A(M \cos^2 \delta - S) \qquad (2.15)$$

From the last two equations, we find

$$\frac{h_{syz.}}{h_{quad.}} = \frac{M/S + 1}{\cos^2 \delta (M/S) - 1} = \frac{\mu + 1}{\mu \cos^2 \delta - 1} \qquad (2.16)$$

where $\mu = M/S$ is the ratio of the action of the Moon to that of the Sun. With this formula it is theoretically possible to calculate μ from the observations of the tides at a given point to the surface of the Earth and in specific configurations (syzygy or quadrature). Equation (2.16) is not exactly that given by Newton, for he took into account the age of the tides, which led him to underestimate the action of the Sun. To apply this formula, he studied the observations made at Bristol harbor, during the days close to the equinoxes, in spring and autumn: he remarked that the tidal range, i.e. the difference of level between the high tide and the low tide, amounted to 45 feet during the syzygies and to 25 feet during the quadratures. From this observational data he concludes that the action of the Moon is 4.4815 times larger than that of the Sun. The value is off by a factor of 2: we know today that the true value is 2.18. The reasons for the discrepancy are first that the quality of the observations of tides is doubtful and second that dynamical effects are not taken into account, the calculations being made in the frame of a static model. Newton did not have at that time the mathematical tools that would have enabled him to tackle them. Laplace, at the end of the 18th century, will undertake Newton's calculations with substantial improvements leading to a ratio, much closer to the true value, of $\mu = 2.35$.

Nevertheless, using his value, Newton could give for the first time an estimate of the mass of the Moon. First proved that the tidal force exerted by the perturbing body (M or S) is proportional to its mass and to the inverse of the cube of its distance to the Earth. Thus by using the same notations as previously the ratio of the tidal force exerted by the Moon to that exerted by the Sun is given by

$$\frac{F_M}{F'_{tidal}} = \frac{M_{Moon}}{PT^3} \times \frac{ST^3}{M_{Sun}} = \frac{\rho_{Moon} R_{Moon}^3 ST^3}{\rho_{Sun} R_{Sun}^3 PT^3} \qquad (2.17)$$

where ρ_{Moon} and R_{Moon} (respectively ρ_{Sun} and R_{Sun}) stand for the density and the radius of the Moon (respectively of the Sun). Moreover, calling α_{Moon} and α_{Sun} the apparent diameters of the Moon and of the Sun,

$$\alpha_{Moon} = \frac{2R_{Moon}}{PT}, \qquad \alpha_{Sun} = \frac{2R_{Sun}}{ST} \tag{2.18}$$

These apparent diameters, varying as an inverse function of the distance (PT or ST), were already known with a very good accuracy in Newton's time, thanks to various astrometric measurements done for a little less than one century, since the first refractor by Galileo around 1610. On average we have $\alpha_{Moon} = 31'16''.5$ and $\alpha_{Sun} = 32'12''$, which immediately give $\alpha_{Sun}/\alpha_{Moon} = 1.0296$. From the equations above we get

$$\frac{F_M}{F'_{tidal}} = \frac{\alpha^3_{Moon}}{\alpha^3_{Sun}} \times \frac{\rho_{Moon}}{\rho_{Sun}}, \qquad \frac{\rho_{Moon}}{\rho_{Sun}} = \frac{\alpha^3_{Sun}}{\alpha^3_{Moon}} \times \frac{F_M}{F'_{tidal}} \tag{2.19}$$

Newton deduced the ratio $F_M/F'_{tidal} = 4.4815$ from the records of tides at Bristol. This led to $\rho_{Moon}/\rho_{Sun} = 4.891$. The next step is to determine ρ_{Sun}/ρ_{Earth}. Indeed, by using Kepler's third law and the value of g, we have

$$GM_{Sun} = \omega^2_{Earth}ST^3, \qquad g = \frac{GM_{Earth}}{R^2_{Earth}} \tag{2.20}$$

hence

$$\frac{\rho_{Sun}}{\rho_{Earth}} = \frac{\omega^2_{Earth}ST^3 R^3_{Earth}}{g R^2_{Earth} R^3_{Sun}} = \frac{\omega^2_{Earth} R_{Earth} ST^3}{g R^3_{Sun}} \tag{2.21}$$

Finally, with $\alpha_{Sun} = 2R_{Sun}/ST$ we find

$$\frac{\rho_{Sun}}{\rho_{Earth}} = \frac{8\omega^2_{Earth} R_{Earth}}{g\alpha^3_{Sun}} \tag{2.22}$$

In Newton's time all the quantities on the right-hand side of this equation were known with very good relative accuracy. The radius of the Earth had been set by Picard at $R_{Earth} = 6732$ km, while $g = 9.81$ ms^{-2}, $\omega_{Earth} = 2\pi/T_{Earth}$ with $T_{Earth} = 365.25d$, the value of α_{Sun} having been given previously. From the juxtaposition of the values found for ρ_{Moon}/ρ_{Sun} and ρ_{Sun}/ρ_{Earth}, a calculation gives $\rho_{Moon}/\rho_{Earth} = 11/9$ (*Principia*, vol. II, Proposition 37, Corollary 3). Thus, for Newton the Moon is slightly denser than the Earth. The ratio of the mass of the Moon to that of the Earth is obviously

$$\frac{M_{Moon}}{M_{Earth}} = \frac{\rho_{Moon} R^3_{Moon}}{\rho_{Earth} R^3_{earth}} \tag{2.23}$$

From determination both of the apparent diameter and of the parallaxes of the Moon, it is possible to provide the ratio R_{Moon}/R_{Earth} which, according to Newton is $1/3.65$. Finally he arrives at the mass ratio

$$\frac{M_{Moon}}{M_{Earth}} = \frac{11}{9} \times \frac{1}{3.65^3} = \frac{1}{39.79} \tag{2.24}$$

This is roughly twice bigger than the true value of 1/81. Newton over-estimated the mass of our satellite by a factor of 2, but recall that we are in presence of the first calculation of the mass of the Moon.

2.6.3 Assessment of Newton's Contribution

From his various calculations detailed in the previous sections, Newton demonstrated one of the most impressive consequences of his law of gravitation, a full explanation of the phenomenon of tides, through the differential gravitational action of the Sun and of the Moon on a particle on the surface of the Earth. These results were rapidly recognized throughout England as a real triumph of his theoretical investigations. This can be seen from the presentation of *Principia* by Edmund Halley (1656–1742) to King James in 1697, during which Halley singled out Newton's work on tides, explaining that his illustrious contemporary solved for the first time the mysterious problem of the ebb and flow.

Nevertheless, as it is well known, the diffusion of Newton's work and of his law of gravitation encountered strong opposition. Among opponents we find Huygens (1629–1695) who, while recognizing the unquestionable advances made by Newton, could not subscribe to his conception of gravity. The main trouble is with action at a distance: Huygens was not convinced that celestial bodies show a natural tendency for mutual attraction. This is confirmed by a letter to Leibnitz in 1690, in which he concedes that he cannot accept the reasons given by Newton on his theory of the ebb and flow as well as on other theories based on the principle of gravitational attraction [12]. Huygens's opinion is widely shared by scholars in France and other countries of the continent. Newton himself was disconcerted by the idea that a matter at rest can act on another matter without mutual contact. In *Principia* (vol. II, book III, scholium) he remarks that he explained celestial phenomena as well as terrestrial ones (tides) thanks to his law of gravitation without being able to assign a cause of this law.

Let us summarize the characteristics of Newton's theory of tides.

- Nowadays this theory is regarded as a by-product of his law of gravitation, but at that time it was taken by his supporters as an emblematic confirmation of his general theory of gravitation, including all the new fundamental tools of physics: law in $1/r^2$, calculus, geometrical combination of forces, etc.
- Newton did not tackle head-on the problem of ocean tides. This problem came gradually to his mind after he studied in detail the inequalities of the lunar orbital motion due to the perturbing gravitational action of the Sun, which in fact is based on the same dynamical principle as that which raises the ocean mass and causes the tides.
- Newton's tricky geometrical reasoning where he represents the attractions by segment lengths judiciously chosen allow him to quantify in a simple way the solar tidal force.

- Thanks to his theory of tides, and by fitting the results of his calculations to observational records of tidal range, Newton could make an estimate of the mass of the Moon relatively to that of the Earth. Moreover, his results explain such well-established characteristics as the presence of two tides per day and the variation of the tidal range according to the lunar phase with extrema during syzygies and quadratures.
- Newton understood perfectly the principle and the origin of tidal forces but his explanations about their consequences are imperfect, essentially because his conception of tides is static and in consequence he did not include the dynamical approach of the oceanic motions. Nevertheless his results mark a turning point and will be fully exploited by his successors like Daniel Bernoulli, Euler, and d'Alembert, and in a quasi-modern form by Laplace at the end of the 18th century.

2.7 Theory of Tides and Analytical Calculations Around 1740

For half a century after Newton, no substantial study on the subject appeared to improve or complete it. But during this same period, mathematics was progressing, notably thanks to the contributions of Leibniz, Jacob and Johann Bernoulli, l'Hôpital, and Varignon. All these mathematicians participated in the birth and the development of calculus. Whereas Newton showed a complex geometrical reasoning, these new tools allowed the development of analytical studies related to mechanics and more specifically to celestial mechanics. At the same time, the metaphysical opposition raised by the principle of Newton's action at a distance was gradually abandoned in the light of the obvious improvements it brought for the resolution of various problems. To illustrate this evolution, we can mention a testimony from Daniel Bernoulli (1700–1782) in 1740, who presents gravitation as an incomprehensible and essential principle that the famous Newton has so well established and that his contemporaries could no longer reject, without harming sublime knowledge and fortunate discoveries of the century. People spoke less of the 'absurdity' of gravitational attraction, and accepted the concept as it was, only preoccupied with investigating its consequences.

At the same time, a lot of systematic observations of tides were being carried out. In the period from 1700 to 1720, Jacques Cassini (1677–1756), a staunch supporter of Descartes's theory of vortices, gathered and discussed tidal observations in French harbors, Le Havre and Dunkerque (1701–1702), Lorient (1711–1712 and 1716–1719) and principally Brest (171–1716) [3, 24]. In the middle of the 18th century, the quasi-totality of scholars recognized the essential correctness of Newton's explanations, but they sometimes emphasized their insufficiencies. Marquise de Châtelet (1706–1749), in her commentaries on *Principia* [14] published posthumously in 1756, explained that people in her time knew that the tides are caused by the inequalities of the action of the Moon and of the Sun on the Earth; she added that Newton had established the mechanism of this cause so well that nobody could express any doubt on its validity. But she also pointed out that the famous scholar (Newton) did not investigate deeply enough the details of the important subject of tides.

In order to encourage scientists to investigate the problem more deeply, the French Académie des Sciences proposed in 1738 the precise elucidation of the tides as a prize to be awarded by the Académie in 1740. Four works received this prize. Three of them were based on the theory of gravitation. They were submitted by Daniel Bernoulli (Fig. 2.13) (1700–1782), Euler (1707–1785), and MacLaurin (1698–1746). The remaining one, by Cavalleri [4], was based on Descartes's theory of vortices, and must surely be, according to Laplace, the very last work dealing with this theory and considered by the Académie. MacLaurin's work entitled *De causa physica fluxus and refluxus maris* [22] is based on proofs of geometrical type. It presents remarkable theorems on the attraction of spheroids, but paradoxally offers few developments on the ocean tides. The two other works, entitled *Traité sur le flux et du reflux de la mer* by Bernoulli [1] and *Inquisto physica in causam fluxus ac refluxus maris* by Euler [10] both represent the real beginning of analytical studies on the subject of tides. These two works fully exploit Newton's calculations but in addition benefit from the drastic improvement accomplished at the beginning of the 18th century in the fields of calculus and of analytical mechanics. Thanks to these advantageous new tools, the two authors did not have to solve the problem of ocean tides by similitude with the problem of the lunar orbital motion perturbed by the tidal action of the Sun, as Newton did. They could directly tackle the resolution of the problem in the frame of terrestrial mechanics.

2.7.1 Prize of the Académie of 1740 for Bernoulli

Daniel Bernoulli's work honored by the prize of the Académie is entirely in the lineage of Newton. He deals with three major problems. The first, the most important according to Bernoulli himself, concerns the elevation of the ocean surface under the attraction of a perturbing body, the Sun. For that purpose he used exactly the same procedure as Newton. But his calculations were much clearer. The second concerns the exact time and amplitude of the high (or low) tides at any point on the surface of the Earth under the combined gravitational action of the two bodies (the Moon and the Sun). The calculations allow him to establish for the first time a tide table. The third concerns the estimation of the mass of the Moon. To this end, he did not follow the same procedure as Newton, based on the height of the tides, but an alternative one based on the interval of time separating successive high (low) tides one day after the other. Marquise de Châtelet, in her *Commentaires des Principes Mathématiques de la Philosophie Naturelle* [14] offers a very clear analysis of Bernoulli's treatise, following his arguments one by one, and often in a more understandable form. One of the fundamental principles of Bernoulli is that the attraction of the Earth by the Sun is rigorously equal to the centrifugal force coming from the revolution of the Earth, if we consider the Earth as a whole. If we consider locally a particle closer

Fig. 2.13 Daniel Bernoulli
(1700–1782)

to the Sun than the center of the Earth, the centrifugal acceleration will be the same whereas the attraction will be stronger. This leads to the characterization of the tidal force which can be regarded as the difference between the attraction of the perturbing body (the Sun) and the centrifugal force.

2.7.1.1 Calculation of the Elevation of Water

The most important question tackled by Bernoulli concerns the amplitude of the tides caused by the Sun. For that aim, he starts from the following hypotheses:

- The Earth at rest is spherical, completely covered by the sea, a thin fluid layer.
- Unlike Newton's hypothesis, the Earth is heterogeneous and made of concentric layers, each having its own density. A law of variation of density as a function of depth is given.
- At anytime the figure of equilibrium of the Earth undergoing the action of the perturbing body (here the Sun) is an ellipsoid, whose the major axis is directed toward the perturbing body.

Thus calculating the amplitude of the tides amounts to measuring the difference between the semi-minor and the semi-major axes of the ellipsoid. Following the same reasoning as Newton, Bernoulli imagined two channels, one directed toward the Sun and the other in a perpendicular direction (Fig. 2.14). In the first the tidal force is against the gravity, whereas in the second it increases it. The solution of the problem of the elevation of water is given by the equality of pressure at the bases of the channels. The problem is complicated by the fact that the ellipsoidal deformation alters the gravity of the Earth at any of point of the surface: in order to solve this additional difficulty, Bernoulli initiates subtle analytical calculations giving a model of self-gravity of the Earth. The height difference β between a high tide and the corresponding low tide is equal the difference above between the lengths of the two

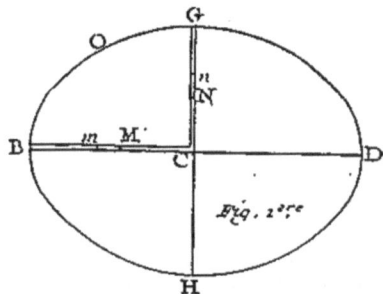

Fig. 2.14 Calculation of the amplitude of the solar tide with the equilibrium of two channels (Fig. 1 of Sect. V of the *Commentaires aux Principes Mathématiques* of la Marquise du Châtelet [14]). The two channels are directed one toward the Sun, the other in a direction perpendicular to the Sun, and join at the center of the Earth

channels. In the case of the simplified hypothesis of a homogeneous Earth with an ocean surrounding it and having the same density, Bernoulli gets

$$\beta = \frac{15}{4} \frac{M_{Sun}}{M_{Earth}} \frac{R_{Earth}^3}{d_S^3} R_{Earth} \tag{2.25}$$

where M_{Sun} and M_{Earth} are the mass of the Sun and of the Earth, d_S is the distance ST from the Earth to the Sun, and R_{Earth} is the Earth radius.

In the first half of the 18th century, d_S was not known with good accuracy. But by using Kepler's third law it is possible to substitute for it the angular velocity ω_E of the Earth around the Sun. Indeed we have the two relationships

$$GM_{Sun} = \omega_{Earth}^2 d_S^3, \qquad GM_{Earth} = \omega_{Moon}^2 d_M^3 \tag{2.26}$$

where ω_{Moon} and d_M are the angular velocity of the Moon around the Earth and the distance between the Moon and the Earth. Substituting these into Eq. (2.25), we get

$$\beta = \frac{15}{4} \frac{\omega_{Earth}^2}{\omega_{Moon}^2} \frac{R_{Earth}^3}{d_{Moon}^3} R_{Earth} \tag{2.27}$$

This formula enables Bernoulli to find the same numerical value as Newton, with $\beta \approx 60$ cm. But this value is obtained with the simplified model of oceans described above (a homogeneous Earth surrounded by an ocean with the same density). Bernoulli is convinced that the value is too small compared with what is expected from the observations of tides. Therefore he expounds various hypotheses about the interior of the Earth, considering for instance the case it is empty, or the case the density of an internal layer is proportional, or inversely proportional, to its radius. In some cases, with a density profile judiciously chosen, he succeeded in obtaining a value of β significantly larger than the value above. This profile corresponds to an increase of density with the depth of the layer, which is quite realistic.

Nevertheless, Bernoulli's calculations are something ambiguous and erroneous: first he believes that only a small part of the oceanic mass is moved by the attraction of the perturbing body, insisting that the Earth as a whole cannot be deformed;

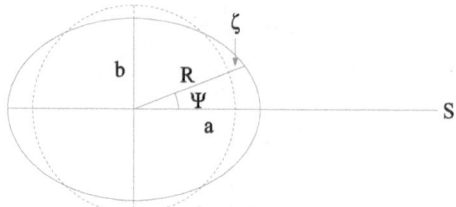

Fig. 2.15 The height of water ζ of a tidal ellipsoid is measured with respect to the initial spherical surface (when the water is at rest), with radius R. Ψ is the geocentric zenithal distance enabling one to define the position of the body S (Moon or Sun)

second, his calculations based on an equilibrium characterized by the ellipsoidal figure with equality of pressures at the bases of the two channels rely on the opposite hypothesis of a global deformation of our planet. There is a contradiction between the method of calculation and the model of the Earth chosen. This was pointed by d'Alembert in his *Reflexions sur la Cause Générale des Vents* in 1746 [5].

Nevertheless, though clumsy and erroneous, Bernoulli's approach deserves great interest. First it acknowledges the discrepancy between Newton's value for the amplitude of tides and the significantly larger amplitudes observed in various harbors. Second it inaugurates to some extent the concept of the internal structure of the Earth, based on a decomposition in layers with a gradual variation of density [6]. This new concept will be fully used a few years later by Clairaut in his study of the figure of the Earth, by Bouguer in 1749 in a study of the variation of amplitude and location-dependent direction of gravity on the surface of the Earth, and by d'Alembert and Euler in their works on the precession of equinoxes.

Bernoulli used geometrical arguments to show that when the perturbing body is at zenith, the elevation of water is twice the size of the depression when it is on the horizon, which is the result found by Newton. Moreover he states that each body (Moon or Sun) acts on the sea independently. In a first step, he makes the approximation that the two celestial bodies move on the celestial equatorial plane, with constant angular velocity. Under their combined action, the height of water with respect to the surface of the sea at rest (Fig. 2.15) is given by

$$\zeta = \beta_S \left(\cos^2 \psi_S - \frac{1}{3} \right) + \beta_M \left(\cos^2 \psi_M - \frac{1}{3} \right) \qquad (2.28)$$

where ψ_S and ψ_M are the zenithal angular distances of the Sun and the Moon, β_S and β_M are the differences between the semi-major and semi-minor axes of the ellipsoid representing the equilibrium tide for the Sun and the Moon (Fig. 2.16). Assuming β_S/β_M known, Bernoulli uses a formula derived from Eq. (2.28) to calculate the exact instant of the high tide during a lunar cycle. For that purpose he corrected his results by taking into account that the relative motions of the Moon and of the Sun are elliptical and inclined with respect to the equator. Finally he could construct the first theoretical tide tables, which proved satisfactory for any harbor where dominant tides have a semi-diurnal frequency. Finally, an important step in Bernoulli's calculations is the ratio of the lunar action to the solar one. We

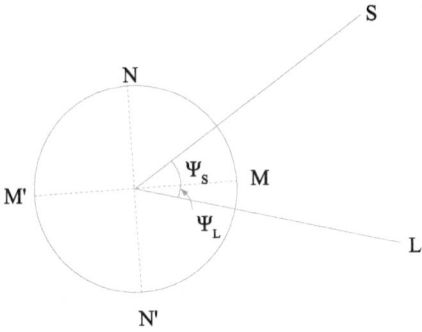

Fig. 2.16 Combined action of the Moon and the Sun. We are in the equatorial plane. S is the Sun, represented by the angle ψ_S. L is the Moon, represented by the angle ψ_L. $MNM'N'$ is the oceanic surface in the case of no deformation. Each body (Moon or Sun) deforms the water surface in an ellipsoid whose semi-major axis is directed toward it. The combination of the two ellipsoids generates a high tide in M and M' and a low tide in N and N'

saw that Newton's value of 4.48 was deduced from a comparison of the heights of tides during syzygies and quadratures. These values obtained in various harbors are not taken in a 'free sea': they show big variations and have consequently a large uncertainty. In contrast Bernoulli proposed an alternative method on the occurrence of tides, which seemed for him much easier to estimate.

Everyday the tides are delayed due to the fact the average time interval between two transits of the Moon at a given meridian is 24^h50^{mn}. But during syzygies, this interval is shorter (24^h35^{mn}) than during quadratures (25^h25^{mn}). These differences come from various combinations between lunar and solar tides and enable one to determine the ratio between the action of the Moon and of the Sun. By averaging over various tidal observations, Bernoulli gets a value of 2.5 for this ratio. From this new determination, he could get an updated value of the density and the mass of the Moon, respectively 5/7 (≈ 0.71) and 1/70 of those of the Earth, these two values being much closer to the true values (respectively 0.60 and 1/81) than Newton's ones.

2.7.2 Prize of Académie of 1740 for Euler

A second work honored by the prize of the Académie des Sciences in 1740 was by Euler (Fig. 2.17), who deepened several points of Newton's theory, from which he borrowed the definition of tidal force as the difference between the gravitational force exerted by an external body (the Moon or the Sun) on a point on the surface of the Earth, and this same gravitational force exerted at the center of the Earth. Euler established in a very modern way the analytical expression of the tidal force, which led him to deduce the formula of the radial and tangential components at the point considered. These formulas can be regarded as definitive.

Fig. 2.17 Leonhard Euler
(1707–1783)

Euler modeled the Earth as a spherical undeformable globe surrounded by an oceanic layer with limited thickness. Then, exploiting his formula, he defined the figure of equilibrium of the Earth subject to the effect of tides. He showed that at first order this figure is really an ellipsoid, as had been suggested without proof by Newton and Bernoulli. Moreover Euler did not have to rely on the artificial concept of two perpendicular channels joined at their bases. Instead, he found his inspiration in an idea already proposed by Huygens: the ocean surface is at rest on the condition that it is perpendicular to the direction of the vertical, as materialized by the plumbline.

2.7.2.1 Analytical Expressions for the Tidal Force

Euler establishes the expressions of the radial and tangential components of the tidal force in Chap. II, par. 24–27 of a work entitled 'On the lunisolar forces which put the oceans in motion'. Referring to Fig. 2.18, the gravitational force exerted by the Sun at the center C of the Earth and at any point M are given respectively by

$$\frac{GM_{Sun}}{d^2}\mathbf{u}_x, \qquad \frac{GM_{Sun}}{l^2}\mathbf{u}_l \tag{2.29}$$

The tidal force $\mathbf{F}$ is given by the difference between these two forces. Calling α the angle between SC and SM, the components of $\mathbf{F}$ are

$$F_x = \frac{GM_{Sun}}{l^2}\cos\alpha - \frac{GM_{Sun}}{d^2}, \qquad F_y = -\frac{GM_{Sun}}{l^2}\sin\alpha \tag{2.30}$$

$$F_x = GM_{Sun}\left(\frac{d-x}{l^3} - \frac{1}{d^2}\right), \qquad F_y = -GM_{Sun}\frac{y}{l^3} \tag{2.31}$$

Then Euler decomposes F_x and F_y into a radial and tangential components

$$F_r = -F_x\cos\psi + F_y\sin\psi \tag{2.32}$$

$$F_t = F_x\sin\psi + F_y\cos\psi \tag{2.33}$$

Fig. 2.18 Analytical expression of the tidal forces. S is the Sun, C the center of the Earth, M a point on the Earth for which the tidal force is determined. r, l, d stand respectively for the distances CM, MS, CS

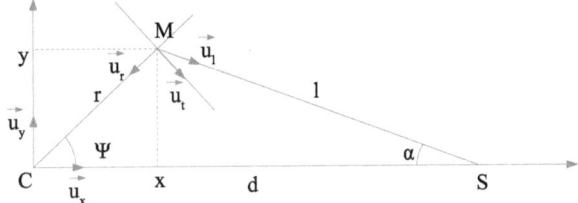

In view of

$$l = \sqrt{(d-x)^2 + y^2}, \qquad \sin\alpha = \frac{y}{l}, \qquad \cos\psi = \frac{x}{r}, \qquad \sin\psi = \frac{y}{r} \qquad (2.34)$$

these give

$$F_r = -\frac{GM_{Sun}}{(d^2 - 2dx + r^2)^{3/2}}\left(\frac{(d-x)x}{r} - \frac{y^2}{r}\right) + \frac{GM_{Sun}x}{d^2 r} \qquad (2.35)$$

and

$$F_t = \frac{GM_{Sun}}{(d^2 - 2dx + r^2)^{3/2}}\frac{dy}{r} - \frac{GM_{Sun}y}{d^2 r} \qquad (2.36)$$

Using the expansion

$$(d^2 - 2dx + r^2)^{-3/2} = \frac{1}{d^3}\left(1 + \frac{3x}{d} - \frac{3}{2}\frac{x^2 + y^2}{d^2} + \frac{15}{2}\frac{x^2}{d^2} + \cdots\right) \qquad (2.37)$$

it follows (Chap. II, art. 27, Fig. 15) that

$$F_r = \frac{GM_{Sun}}{d^3\sqrt{x^2 + y^2}}\left(y^2 - 2x^2 + \frac{3}{2}\frac{x}{d}(3y^2 - 2x^2)\right) \qquad (2.38)$$

$$F_t = \frac{GM_{Sun}}{d^3\sqrt{x^2 + y^2}}\left(3xy + \frac{3}{2}\frac{y}{d}(4x^2 - y^2)\right) \qquad (2.39)$$

These are the formulas established by Euler. To find the modern formula, we can introduce the angle ψ. Then

$$F_r = \frac{GM_{Sun}}{d^2}\left(\frac{r}{d}(3\cos^3\psi - 1) + \frac{3}{2}\frac{r^2}{d^2}(5\cos^3\psi - 3\cos\psi)\right) \qquad (2.40)$$

$$F_t = \frac{GM_{Sun}}{d^2}\left(\frac{r}{d}(3\cos\psi\sin\psi) + \frac{3}{2}\frac{r^2}{d^2}\sin\psi(5\cos^2\psi - 1)\right) \qquad (2.41)$$

Thus the analytical results obtained by Euler are very clear and give a quasi-definitive form to the expression of the tidal force.

2.7.2.2 Figure of Equilibrium and Tangential Component

A second part of Euler's investigations, particularly interesting, concerns the form of the surface of equilibrium of the oceanic mass under the combined gravitational

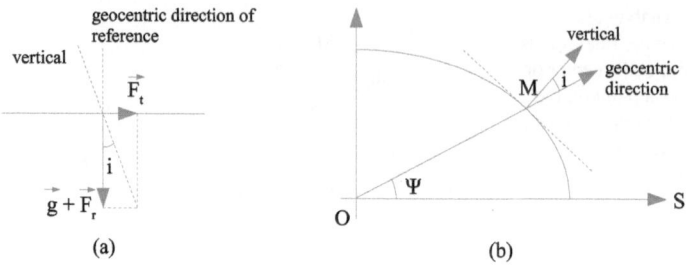

Fig. 2.19 Figure of equilibrium under the action of the tidal force: (**a**) deviation of the vertical under the influence of the tangential component of the tidal force, (**b**) the figure of equilibrium of the fluid layer at each point is perpendicular to the vertical. It is defined by the inclination i of the vertical with respect to the geocentric direction of M

action of the Moon and the Sun. In paragraphs 34 to 38 of Chap. 3, Euler explains that this form depends on the role of the tangential component of the tidal force. Making use of a statement first given by Huygens, Euler considers that the surface of the fluid is in equilibrium when at each point it is perpendicular to the direction of the gravity. The tidal forces modify slightly the direction of the vertical with respect to the geocentric direction of reference (Fig. 2.19): the tangential component F_t of the tidal force is responsible of the deviation i of the vertical line, which at first order can be calculated in a straightforward manner:

$$i \approx \tan i = \frac{F_t}{g + F_r} \approx \frac{F_t}{g} \qquad (2.42)$$

By taking into account the constraint that the total mass of water remains constant, Euler gets the general formula of the surface of equilibrium (Chap. II, art. 36)

$$r = R + \frac{GM_{Sun}}{dg}\left[\frac{R^2}{d^2}\left(\frac{3\cos^2\psi - 1}{2}\right) + \frac{R^3}{d^3}\left(\frac{5\cos^3\psi - 3\cos\psi}{2}\right)\right] \qquad (2.43)$$

where R is a reference radius and r is the radius form the center to the point considered at the surface of the ellipsoid, from which the angle ψ is measured. This expression by Euler constitutes a determining step toward a modern and accurate theory of the tides. The quantity $\zeta = r - R$ is the height of the equipotential represented by the surface of the ellipsoid. The right-hand side, when multiplied by g, is the tidal potential. The expressions $(3\cos^2\psi - 1)/2$ and $(5\cos^3\psi - 3\cos\psi)/2$ are called the Legendre polynomials of 2nd and 3rd degree respectively. They were introduced by Legendre (1752–1833) around 1780 and appear in Laplace's equations, studied below. By using the equality $g = GM_{Earth}/R^2$, Eq. (2.43) can be rewritten, at the first order in R/d,

$$r = R + \frac{M_{Sun}}{M_{Earth}}\frac{R^3}{d^3}R\frac{3\cos^2\psi - 1}{2} \qquad (2.44)$$

that is to say

$$r = b + \beta\cos^2\psi \qquad (2.45)$$

Fig. 2.20 Jean Le Rond
d'Alembert (1717–1783)

with

$$\beta = \frac{3}{2} \frac{M_{Sun}}{M_{Earth}} \frac{R^3}{d^3} R, \qquad b = R - \frac{\beta}{3} \qquad (2.46)$$

This is the equation of an ellipse with semi-major axis b and with difference β between the semi-major and semi-minor axes. From these calculations Euler proves what Newton and Bernoulli supposed without proof: at first order, the figure of equilibrium of the oceans under the action of the Sun or the Moon is an ellipsoid. Moreover the amplitude of β, which can be interpreted as the amplitude of the tides, is 2.5 times smaller than that found by Bernoulli, in the case of a homogeneous model of the Earth. The results by Euler are exact when neglecting the self-gravity of the oceans. The great advantage of Euler's method is that the solution is available for an oceanic surface layer surrounding a solid Earth, whereas in Bernoulli's method, relying on the equality of pressure at the bases of two perpendicular channels, the whole Earth (including the oceans) must be homogeneous and fluid.

2.8 D'Alembert and His 'Reflexions sur la Cause Générale des Vents'

D'Alembert (Fig. 2.20) (1717–1783) did not publish a specific work dealing with oceanic tides but his report entitled *Réflexions sur la Cause Générale des Vents*, submitted in 1746 to the Royal Academy of Sciences of Berlin, and published in 1747 [5], deals with several important points related to the tides. One of the subjects of great interest concerns the characteristics of regular winds in the tropical areas of the Earth. His aim is to study how the tidal forces exerted both by the Moon and the Sun on the atmosphere of the Earth can be regarded as the origin of winds on its surface. He tried to start from the calculation of the atmospheric tides and to infer in some detail the velocity distribution of the winds.

Some remarkable studies are included in the first part of the work. Two of them can be retained as emblematic of a totally new approach to the problem: the first is

Fig. 2.21 Determination of
the figure of equilibrium of
the fluid layer by d'Alembert
(Fig. 3 of *Réflexions sur la
Cause Générale des Vents*)

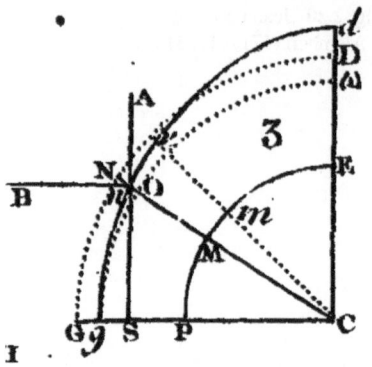

the determination of the surface of equilibrium of a given fluid surface layer when
it is subjected to a given external force. The consequence is the production of an
oscillation around a figure of equilibrium. D'Alembert calculates the eigenmodes
of this oscillation. The second study is oriented toward the question of self-gravity
of a deformed fluid layer. This can be naturally applied to the problem of ocean
tides. On this topic, d'Alembert completes the work of his predecessors, making the
link between Euler's and Bernoulli's results. In the following we present the main
features of these two studies.

2.8.1 Mechanics of Surface Layer of Fluid

D'Alembert does not show any interest on the global deformation of the Earth as
Newton and Bernoulli did. In contrast he analyses in detail the deformation of a thin
fluid layer surrounding the Earth considered as an undeformable spherical globe. In
that sense his method is similar to Euler's.

2.8.1.1 Figure of Equilibrium of the Fluid Layer

Here d'Alembert makes use of the same mathematical formalism both for the tidal
force and for the centrifugal force due to the rotation. For both, only the tangential component of the force has an effect on the figure of equilibrium (Fig. 2.21).
Moreover these tangential components have the same structure. They can both be
expressed by an equation of the type $F_t = \phi \cos \psi \sin \psi$. In the case of the tidal
force, ψ stands for the zenithal distance of the perturbing body and ϕ stands for
$\phi = 3GM_{Sun}R_{Earth}/d^3$, where M_{Sun} is the mass of the Sun, R_{Earth} is the Earth radius and d is the distance from the body to the Earth. In the case of the centrifugal
force ψ is the latitude and $\phi = \omega^2 R_{Earth}$ where ω is the rotational angular velocity
of the Earth. D'Alembert found the same results as Euler but in a more straight-
forward manner, thanks to the use of polar coordinates instead of rectangular ones.

The steps of his argument are similar: after determining the inclination of the gravity under the effect of the tangential component, he finds the difference β between the semi-major and semi-minor axes of the ellipsoid as

$$\beta = \frac{\phi R_{Earth}}{2g} \tag{2.47}$$

In the particular case of tides, this gives

$$\beta = \frac{3}{2} \frac{M_{Sun}}{M_{Earth}} \frac{R_{Earth}^3}{d^3} R_{Earth} \tag{2.48}$$

Then the mass conservation enables him to give the expression of the deformation (Fig. 2.21).

2.8.1.2 Oscillation of the Fluid Layer

D'Alembert's investigations are not restricted to the determination of the figure of equilibrium of the fluid layer, in the frame of a static theory. He wants to deal with a much more difficult problem: what is the law of displacement of the various parts of the fluid? For that purpose he determines the interval of time Δt necessary for a given particle to go from the initial spherical surface to the new ellipsoidal one. In a second step he compares Δt with the interval of time $\Delta\theta$ necessary for the same particle to fall from a height a above the surface when undergoing the acceleration of gravity g. By considering that the period of oscillation of the particle is $T = 4\Delta t$, and with $a = 1/2g\Delta\theta^2$, his calculations lead to the formula $T = 2\pi R_{Earth}/\sqrt{6gh}$, where h is the depth of the fluid layer. With respect to this equation d'Alembert makes the interesting remark: T does not depend on the gravity but only of the parameters R_{Earth} and h. This confirms for the first time the existence of a proper oscillation mode of the fluid, even when forcing is absent. About half a century later Laplace will reinforce this result, inaugurating a long series of similar works.

2.8.2 Self-gravity of the Fluid Surface Layer

A second remarkable study of d'Alembert's concerns the self-gravity of a fluid surface layer. When this layer is deformed, it creates gravitational changes, increasing or decreasing its own initial deformation. D'Alembert's argument relied on MacLaurin's and Daniel Bernoulli's calculations. In art. 49, he shows that to determine the figure of equilibrium of the fluid layer by taking into account its self-gravity, all we have to do is to multiply the perturbing force by a factor

$$\rho = \frac{1}{1 - \frac{3}{5}\frac{\delta}{\Delta}} \tag{2.49}$$

where δ is the density of the fluid layer and Δ is the mean density of the Earth considered as a solid body. Then the self-gravity of the fluid layer is expressed in a rather simple way, by inserting this law in the tidal force. The difference between the semi-major and semi-minor axes of the deformable ellipsoid becomes

$$\beta = \frac{3}{2(1 - \frac{3}{5}\frac{\delta}{\Delta})} \frac{M_{Sun}}{M_{Earth}} \frac{R^3}{d^3} R \qquad (2.50)$$

D'Alembert suggested that this formula, validated by Laplace half a century later, could enable one to deduce the unknown mean density δ if one takes β from observations of tides, for instance by measuring the difference of height between the low and high tides at a given point of the sea. Nevertheless this kind of measurement is particularly delicate because d'Alembert does not take into account dynamical phenomena which act on the figure of equilibrium. Laplace will make appropriate adjustments in 1790, remarking that the determination of δ/Δ will be more efficient when studying long periodic tidal components, less apt to affect the equilibrium tide.

Here, one of d'Alembert's important conclusions is that the self-gravity of a fluid layer accentuates the amplitude of the tides. But with the real ratio $\delta/\Delta = 1/5.5$, the calculations above lead to a small increase of 13 % of the amplitude of the tides due to the self-gravity. This value remains rather small, in comparison to the amplitudes of resonant phenomena in harbors that exhibit significantly larger effects. Finally, notice that when the auto-gravity of the sea is not taken into account we get Euler's equation (2.48). And when considering the fictitious case in which the sea has the same density as the Earth, we find

$$\beta = \frac{15}{4} \frac{M_{Sun}}{M_{Earth}} \frac{R^3}{d^3} d \qquad (2.51)$$

which fits with the expression found by Daniel Bernoulli. Therefore D'Alembert's expressions are completely in accordance with his two contemporaries.

2.9 Laplace's Masterpiece

Pierre Simon de Laplace (Fig. 2.22) (1749–1827) is just 25 years old when he begins his work on tides in 1774. In his introduction, when presenting the theories of his predecessors (Newton, Daniel Bernoulli, Euler, and d'Alembert) he points out their lack of validity when they are confronted with real observed tides. He proposes a complete renewal of the theoretical concepts, emphasizing the necessity to solve in a much more rigorous way what he considers as 'one of the most complex and interesting problems of the whole physical astronomy'. In 1825, after half a century of personal investigations on the subject of tides, he mentions that the motion of fluid covering a planet was a almost entirely new topic when he undertook its treatment in 1774. This terse comment seems excessive if we recall that Newton gave the definition of the tidal force, Euler found its precise formulation, and if we recall the calculations on the static tide by Bernoulli, Euler, and d'Alembert, as well as of the study by this last author of the oscillation and self-gravity of a fluid layer.

Fig. 2.22 Pierre-Simon de
Laplace (1749–1829)

However, Laplace's comment rings quite true if we remark that all these predecessors, though broadly explaining the phenomena related to tides, were unable
to propose an adequate model describing their effects. Laplace is in fact the first
scientist to construct a mathematical model of tides. Moreover for that purpose he
invented specific mathematical tools to handle the dynamical equations.

Laplace tackled the problem of tides in four successive memoirs, and gave a synthetic presentation of his calculations in his *Traité de Mécanique Céleste* (book IV,
vol. II; book XIII, vol. V) [19, 21]. The first two memoirs, written in 1775 and
1776 and published in 1778 and 1779 [15, 16], are entitled *Recherches sur plusieurs
points du Système du Monde*. They are devoted to theoretical aspects, accompanied
with general equations and a particular study of the influence of the bathymetry of
the sea on the oceanic tides. In the third memoir entitled *Traité du flux et du reflux*,
written in 1790 and published in 1797 [18], Laplace proposes a theoretical study of
the observations, in particular those recorded by Jacques Cassini at the beginning
of the 18th century and gathered in a treatise on tides by Lalande in 1781 [13]. The
contents of these memoirs are presented again by Laplace in a very clear and synthetic way in his book IV of *Traité de Mécanique Céleste*, published in 1799 [19]. In
particular he makes full use of results acquired in 1782 on the spherical harmonics
[17]. The fourth memoir, also entitled *Traité du flux et du reflux de la mer*, written
in 1818 and published in 1820 [20], is devoted to observations organized by himself
in harbor of Brest. At last in the book XIII of *Traité de Mécanique Céleste*, written
in 1824 and published in 1825 [21], Laplace analyse as set of observations carried
out at Brest from 1807 to 1822.

2.9.1 Development of Analytical Mechanics

Before taking a look at Laplace's capital contribution to the theory of tides, it is
worth making a brief summary of the then recent developments in the fields of analytical mechanics, from which Laplace could build his theoretical investigations.

First of all, we recall that in 1755 Euler published memoirs where he established the general equations of hydrostatics and hydrodynamics, whatever the compressibility of the fluid. By generalizing the ideas of Clairaut, developed in 1743 on the occasion of his researches on the figure of the Earth, Euler introduced the notion of pressure and gave the general condition of equilibrium of a fluid by showing that the pressure counterbalances at each point the effect of the acceleration. He established the general equation of motion of the fluid with respect to an absolute reference frame, introducing the internal force of pressure $-\nabla p$ and the external force $\mathbf{f}$ in such a way that

$$\rho\gamma = -\nabla p + \mathbf{f} \tag{2.52}$$

where ρ is the density and γ the acceleration. Euler also introduced the local equation of the conservation of mass, which characterizes the fact that the variation of mass inside a given volume of fluid is equal to the mass flux through the surface bounding the volume:

$$\frac{\partial \rho}{\partial t} + \operatorname{div} \rho \mathbf{v} = 0 \tag{2.53}$$

Together with the concept of pressure, Euler introduced also the concept of potential, although the evolution of this concept through his work is rather vague and progressive. As soon as 1736, he defines a function μ whose the differential is exact: $d\mu = -P\,dx - Q\,dy - R\,dz$, where P, Q and R are the rectangular components of the force per mass unit. In 1743, Clairaut showed the importance of such an expression in the equilibrium of a fluid mass and also proved that it must be an exact differential form. He added that the expression above represents the 'effort' of the gravity. This concept is close to the notion of work. Clairaut also showed that the surface of equilibrium of a fluid is given by setting the integral of $P\,dx + Q\,dy + R\,dz$ to a constant. The notion of potential, in a latent state, was finalized in 1774 and 1776 by Lagrange who showed that the gravitational attraction derives from a potential Ω and that the components of the force can be obtained by calculating the partial derivatives of this potential. He added that this way of representation of the forces can prove extremely advantageous by its simplicity and its generality. Finally we mention that also in 1774 Lagrange introduced the use of spherical coordinates, which Laplace used extensively later.

2.9.2 The Equations of 1775 and 1776

In 1775, Laplace uses the general equations of the hydrodynamics set up by Euler, to apply them to the Earth, in spherical coordinates. As his predecessors he models the ocean as a uniform fluid layer with variable depth, covering entirely a spherical, solid and undeformable Earth. The fluid is supposed incompressible. Laplace makes an essential hypothesis which simplifies noticeably his calculations: he remarks that the depth of the oceanic layer is small compared with the Earth radius. This implies

$$(6) \qquad y = -\frac{l}{\sin\vartheta}\frac{\partial . u\gamma \sin\vartheta}{\partial\vartheta} - l\gamma\frac{\partial v}{\partial\varpi},$$

$$(7) \qquad \frac{d^2 u}{dt^2} - 2n\frac{dv}{dt}\sin\vartheta\cos\theta = -g\frac{\partial y}{\partial\vartheta} + B\Delta + \frac{\partial R}{\partial\vartheta},$$

$$(9) \qquad \frac{d^2 v}{dt^2}\sin^2\vartheta + 2n\frac{du}{dt}\sin\vartheta\cos\theta = -g\frac{\partial y}{\partial\varpi} + C\Delta\sin\vartheta + \frac{\partial R}{\partial\varpi},$$

R étant égal à $K[\cos\vartheta\cos\nu + \sin\vartheta\sin\nu\cos(\varphi - nt - \varpi)]^2$.

Fig. 2.23 Laplace equations in 1776. *Equation (6)* characterize the mass conservation. *Equations (7) and (8)* are the dynamic equations. Laplace uses *l* for the depth of the layer, *y* for its deformation, *n* for the angular velocity of the Earth. The Earth radius is chosen to be the unit of length and *u* and *v* are horizontal displacements. $B\Delta$ and $C\Delta$ correspond to the self-gravity

Fig. 2.24 Reference frame and parameters of the Laplace dynamical equations. *M* is represented by its spherical coordinates: the colatitude θ, the longitude λ, counted positively eastward *l*. ω is the vector rotation of the Earth

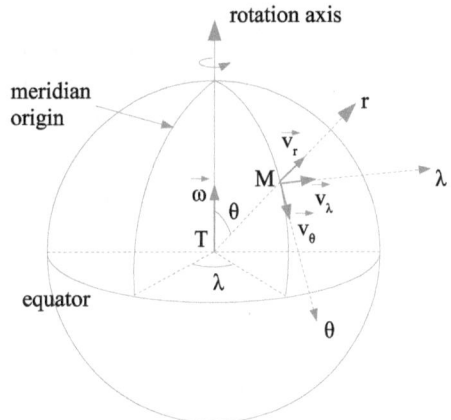

that the real large scale motions of the fluid are quasi-horizontal. In other words, the vertical velocity can be neglected, all the particles belonging to the same vertical line having a priori the same velocity. Then the general problem of tides can be treated by retaining only the tangential components, as a 2-dimensional problem. This fundamental simplification, known as the *long wave approximation*, will be fully used in later geophysical studies. Laplace kept on working on his equations inside his memoirs of 1776 (Fig. 2.23) and 1790, and his *Mécanique Céleste*, vol. IV, until he gave a precise and definitive formulation, which still stands nowadays.

Let θ and λ denote the colatitude and the longitude of a particle of the fluid layer at depth h, v_θ and v_λ the respective North-South and East-West components of the horizontal velocity $\mathbf{v}$ of the particle with respect to the Earth (Fig. 2.24). ζ is the radial deformation of the fluid layer, ω is the angular velocity of the Earth, g is the gravity, V is the tidal potential and Φ is the potential of self-gravity of the fluid layer.

With this notation, the dynamical equation and the equation of mass conservation are written with respect to a reference frame linked to the Earth

$$\frac{\partial \mathbf{v}}{\partial t} + 2\omega \wedge \mathbf{v} = -g\nabla\zeta + \nabla V + \nabla\Phi \tag{2.54}$$

$$\frac{\partial \zeta}{\partial t} + \operatorname{div} h\mathbf{v} = 0 \tag{2.55}$$

The pressure at a given point of the fluid is the sum of two terms

$$p = p_0 + p_\zeta \tag{2.56}$$

where p_0 is the hydrostatic pressure and p_ζ the additional pressure due to the deformation of the oceans. As the vertical accelerations due to the tides are very small in comparison to the gravity, the additional pressure p_ζ results only from the weight of the water column undergoing the deformation

$$p_\zeta = \rho g \zeta \tag{2.57}$$

ρ being the density of the fluid.

The hydrostatic pressure p_0 is counterbalanced by the gravity of the Earth combined to the force due to the rotation in such a way that these terms do not take part in the dynamical equations. In spherical coordinates, the preceding equations become

$$\frac{\partial v_\theta}{\partial t} - 2\omega\cos\theta v_\lambda = -\frac{1}{a}\frac{\partial}{\partial \theta}(g\zeta - V - \Phi) \tag{2.58}$$

$$\frac{\partial v_\lambda}{\partial t} + 2\omega\cos\theta v_\theta = -\frac{1}{a\sin\theta}\frac{\partial}{\partial \lambda}(g\zeta - V - \Phi) \tag{2.59}$$

$$\frac{\partial \zeta}{\partial t} + \frac{1}{a\sin\theta}\left(\frac{\partial}{\partial \theta}(hv_\theta\sin\theta) + \frac{\partial}{\partial \lambda}(hv_\lambda)\right) = 0 \tag{2.60}$$

2.9.3 Conservation of Mass

The way Laplace takes into account the conservation of mass differs significantly from his predecessors: it does not involve a global conservation of the ocean supposed to cover the whole Earth and whose surface shape changes from a sphere to an ellipsoid. Instead Laplace's calculations express conservation locally. The starting point is the equation given by Euler in 1755

$$\frac{\partial \rho}{\partial t} + \operatorname{div} \rho \mathbf{v} = 0 \tag{2.61}$$

Then Laplace substitutes a variable surface density $\rho(h + \zeta)$ for a constant volume density ρ, h being variable in space but constant in time:

$$\frac{\partial \rho(h + \zeta)}{\partial t} + \operatorname{div} \rho(h + \zeta)\mathbf{v} = 0 \tag{2.62}$$

As $\zeta \mathbf{v}$ has a 2nd-order amplitude, and taking into account that ρ is constant, this can be rewritten

$$\frac{\partial \zeta}{\partial t} + \text{div}(h\mathbf{v}) = 0 \tag{2.63}$$

This equation expresses the fact that the mass flux through the walls of a column of water is compensated for by variations in the height of the column.

2.9.4 Complementary Acceleration due to the Rotation of the Earth

One of the most essential improvements offered by the two Eqs. (2.58) and (2.59) is the presence of the components with the factor $2\omega \cos \theta$. They signal the existence of a complementary acceleration in a rotating frame, later called the Coriolis acceleration. They imply a deviation of the motions at the surface of the rotating Earth. Laplace's predecessors considered that the sole effect of the rotation of the Earth was to displace the ellipsoid of the equilibrium tides. In the introduction to his memoir of 1755, Laplace remarked the error of this simplified hypothesis, noticing that the change in the relative position of the Moon and the Sun at the surface of the seas is not the unique effect coming from the rotation of the Earth. This was already pointed out by MacLaurin in his treatise about the ebb and flow, but without any calculation. Laplace completed his remark with the following reasoning: the velocity of a particle of fluid remains the same when staying in the same parallel, its angular velocity increases or decrease according to its distance to the equator, and it drifts in meridian as it moves in parallel. The important fact is that the amplitude of the changes due to this effect is of the same order as the gravitational action of the two perturbing bodies. MacLaurin must not be considered as the only predecessor to mention the effect above. Galileo, in 1632, studied it in the problem of a bullet launched along a meridian, and Hadley, in 1735, when interpreting the deviation of trade winds westward. But Laplace was the first scientist to propose a quantitative analysis far before Coriolis (1792–1843).

2.9.5 A Decisive Innovation: Spherical Harmonics

In 1782, Laplace invented what turned out to be a decisive tool for tackling problems of tidal phenomena: spherical harmonics [17]. They occupy a fundamental place in his work dealing with terrestrial dynamics, notably by leading to a rewriting of his dynamical equations in a more elegant manner. In his memoir of 1790 (art. 2 and 3) he already amends his notation, by introducing the spherical harmonics of order 2 in the expression of the tidal potential and by taking into account in a simple manner the self-gravity of the fluid layer. But the mathematical expressions of the potential as well as of the dynamical equations reach their full maturity in the *Mécanique Céleste* (vols. III and IV) of 1799.

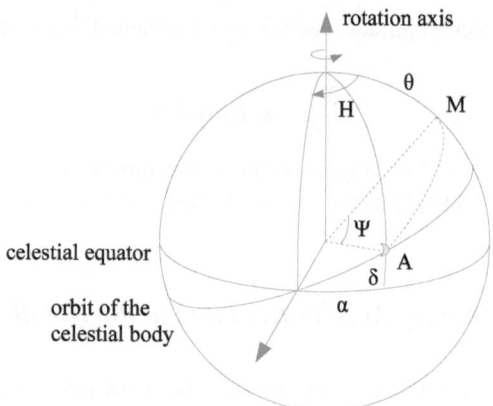

Fig. 2.25 Coordinates of a celestial body in the sky. M is the zenith of a surface point, with colatitude θ and longitude λ (counted positively eastward). A is a point corresponding to the direction of the celestial body (Moon or Sun), with declination δ and right ascension α. H is the hour angle of the intersection of the orbit with the equator. The angle between the 'meridian' of the body and the meridian of the surface point is given by $H - \alpha = \omega t + \lambda - \alpha$

2.9.6 Tidal Potential

In vol. III, art. 23, Laplace establishes the definitive expression for the tidal potential V exerted by a perturbing body (Fig. 2.25):

$$V = \frac{Gm_p}{d}\left(\frac{r^2}{d^2}P_2(\cos\psi) + \frac{r^3}{d^3}P_3(\cos\psi) + \frac{r^4}{d^4}P_4(\cos\psi)\right) \qquad (2.64)$$

where d is the distance of the perturbing body from the center of the Earth, m_p its mass, ψ the geocentric angle of zenithal distance, r the radius of the Earth and P_n are the Legendre polynomials defined by

$$P_n(x) = \frac{1}{2^n n!}\frac{d^n[(x^2-1)^n]}{dx^n} \qquad (2.65)$$

Thus the first Legendre polynomials are

$$P_0(x) = 1, \qquad P_1(x) = x, \qquad P_2(x) = \frac{3x^2-1}{2}, \qquad P_3(x) = \frac{5x^3-3x}{2}$$

$$P_4(x) = \frac{35x^4 - 30x^2 + 3}{8} \qquad (2.66)$$

This expression of the tidal potential is very close to the expression of the deformation of an oceanic layer as given by Euler in 1740. Thanks to classic relationships in a spherical triangle, Laplace can replace $\cos\psi$ by a function of the colatitude θ and the longitude λ of the point considered on the surface of the Earth, and of the equatorial coordinates of the perturbing body, i.e. its right ascension α and its declination δ:

$$\cos\psi = \cos\theta\sin\delta + \sin\theta\cos\delta\cos(\omega t + \lambda - \alpha) \qquad (2.67)$$

By inserting this expression in the tidal potential, Laplace shows that it can be written naturally as a combination of functions called *spherical harmonics* $Y_n(\theta, \lambda)$:

$$V(\theta, \lambda) = V_2 Y_2(\theta, \lambda) + V_3 Y_3(\theta, \lambda) + \cdots = \sum_{n=2}^{\infty} V_n Y_n(\theta, \lambda) \qquad (2.68)$$

Here the coefficients V_n themselves can be written as functions of the spherical harmonics $Y_n(\delta, \omega t - \alpha)$

2.9.7 Potential for Self-gravity

Thanks to their various properties, especially of orthogonality, spherical harmonics form a basis in which any surface function can be expanded. Thus Laplace expands the deformation of the fluid layer $\zeta(\theta, \lambda)$ in spherical harmonics

$$\zeta(\theta, \lambda) = \sum_{n=0}^{\infty} \zeta_n Y_n(\theta, \lambda) \qquad (2.69)$$

With the help of this expansion Laplace finds a simple and subtle expansion of the potential for self-gravity of the fluid layer (vol. III, art. 11; vol. IV, art. 2)

$$\Phi(\theta, \lambda) = \sum_{n=0}^{\infty} \Phi_n Y_n(\theta, \lambda) \qquad (2.70)$$

with

$$\Phi_n = \frac{3 g \rho_w}{\rho_e} \frac{\zeta_n}{2n + 1} \qquad (2.71)$$

where ρ_w and ρ_e are the density of water and the mean density of the Earth. A remarkable point of the formula above is that each component of degree n of this potential is expressed as a function only of the corresponding degree of deformation ζ_n. Therefore each spherical harmonics can be treated separately, according to their degree.

2.9.8 Dynamical Equations with Spherical Harmonics

In his *Mécanique Céleste* (book III, art. 3) Laplace rewrites the equations of motion of a fluid particle, which are similar to those given in 1776, by relying on a completely new approach which expands in spherical harmonics all the parameters concerned. This approach is extremely general and valid at each degree of the harmonics, which can be treated independently. They are shown in Fig. 2.26.

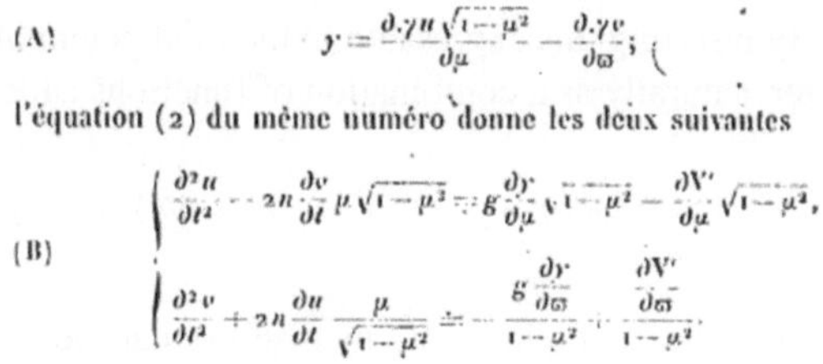

Fig. 2.26 Laplace equations in 1799. μ corresponds to $\cos\theta$. The potential V' is the combination of the tide potential V and the potential of self-gravity Φ of the fluid layer

(A)
$$y = \frac{\partial . \gamma u \sqrt{1-\mu^2}}{\partial \mu} - \frac{\partial . \gamma v}{\partial \varpi};$$

l'équation (2) du même numéro donne les deux suivantes

(B)
$$\begin{cases} \dfrac{\partial^2 u}{\partial t^2} - 2n\dfrac{\partial v}{\partial t}\mu\sqrt{1-\mu^2} = g\dfrac{\partial y}{\partial \mu}\sqrt{1-\mu^2} - \dfrac{\partial V'}{\partial \mu}\sqrt{1-\mu^2}, \\[2ex] \dfrac{\partial^2 v}{\partial t^2} + 2n\dfrac{\partial u}{\partial t}\dfrac{\mu}{\sqrt{1-\mu^2}} = g\dfrac{\partial y}{\partial \varpi}\dfrac{1}{1-\mu^2} - \dfrac{\partial V'}{\partial \varpi}\dfrac{1}{1-\mu^2}. \end{cases}$$

2.9.9 Oscillation of the Fluid Layer in Case of a Static Earth

In 1799, after much effort, Laplace succeeded in making progress on the difficult problem already raised by d'Alembert, that of determining the oscillation of the fluid layer in the case of a static Earth. In that specific case, the three Eqs. (2.58), (2.59), (2.60) can combine to give one equation in ζ. By differentiating the equation of mass conservation (2.63) and using the fact that h is constant, Laplace gets

$$\frac{\partial^2 \zeta}{\partial t^2} + \frac{h}{a}\left[\frac{1}{\sin\theta}\frac{\partial}{\partial\theta}\left(\sin\theta\frac{\partial v_\theta}{\partial t}\right) + \frac{1}{\sin\theta}\frac{\partial}{\partial\lambda}\left(\frac{\partial v_\lambda}{\partial t}\right)\right] = 0 \qquad (2.72)$$

and, by replacing the partial derivatives of the velocities by their expressions from (2.58) and (2.59),

$$\frac{\partial^2 \zeta}{\partial t^2} = \frac{h}{a^2}\Delta_t(g\zeta - V - \Phi) \qquad (2.73)$$

where Δ_t denotes the tangential Laplacian given by

$$\Delta_t = \frac{1}{\sin\theta}\frac{\partial}{\partial\theta}\left(\sin\theta\frac{\partial}{\partial\theta}\right) + \frac{1}{\sin^2\theta}\frac{\partial^2}{\partial\lambda^2} \qquad (2.74)$$

Equation (2.73) was found by Laplace as early as 1776. But it took further 20 years for Laplace to find way of solving this complex equation. Once more thanks to the spherical harmonics, he found a way out, as it shown in *Mécanique Céleste* (vol. IV, art. 2). Each term in the equation is expressed with the help of spherical harmonics, and by use of their remarkable property

$$\Delta_t Y_n = -n(n+1)Y_n \qquad (2.75)$$

Equation (2.73) becomes equivalent to a set of equations for different degrees n:

$$\frac{\partial^2 \zeta_n}{\partial t^2} = -n(n+1)\frac{h}{a^2}(g\zeta_n - V_n - \Phi_n) \qquad (2.76)$$

with

$$\Phi_n = \frac{3g\rho_w}{\rho_e}\frac{\zeta_n}{2n+1} \qquad (2.77)$$

Laplace gets

$$\frac{\partial^2 \zeta_n}{\partial t^2} + n(n+1)\left(1 - \frac{3\rho_w}{(2n+1)\rho_e}\right)\frac{gh}{a^2}\zeta_n = n(n+1)\frac{h}{a^2}V_n \qquad (2.78)$$

This represents the equation of an oscillation forced by the tidal potential V_n. For $n = 2$, and by setting the tidal potential to zero, we have

$$\frac{\partial^2 \zeta_2}{\partial t^2} + 6\left(1 - \frac{3\rho_w}{5\rho_e}\right)\frac{gh}{a^2}\zeta_2 = 0 \tag{2.79}$$

This represents an oscillation of the fluid layer, with characteristic period

$$T = \frac{2\pi a}{\sqrt{6gh(1 - \frac{3\rho_w}{5\rho_e})}} \tag{2.80}$$

This corresponds exactly to the eigenmode of oscillation found by d'Alembert. But Laplace generalized the work for all degrees of the spherical harmonics.

2.9.10 Hydrostatic Equilibrium

In his *Mécanique Céleste* (vol. IV, art. 12), Laplace remarks that his equations recover in a simpler manner his predecessors' results on another fundamental topic: the equilibrium tide. A simple hypothesis is adopted that the surface of the sea takes the form induced by the instantaneous forces acting on it, in other words the velocities and their derivatives are ignored. With this hypothesis, the value of the deformation ζ of the fluid layer can be determined immediately for arbitrary depth and density of the sea. Indeed the equation becomes

$$g\nabla\zeta = \nabla V + \nabla\Phi \tag{2.81}$$

or after integration

$$\zeta = \frac{V + \Phi}{g} \tag{2.82}$$

This formula says that equipotentials are equivalent to equipressures. It shows the advantageous of working with potential, which quickly yields the static deformation. Expanding in spherical harmonics and recalling

$$\Phi_n = \frac{3g\rho_w}{\rho_e}\frac{\zeta_n}{2n+1} \tag{2.83}$$

we get, for each harmonic of degree n,

$$\zeta_n = \frac{1}{(1 - \frac{3}{2n+1}\frac{\rho_w}{\rho_e})}\frac{V_n}{g} \tag{2.84}$$

When self-gravity is neglected, this formula is very close to that given by Euler in 1740. Truncating the expression at the second order, we get

$$V = \frac{3Gm_p a^2}{2d^3}\left(\cos^2\psi - \frac{1}{3}\right) \tag{2.85}$$

and

$$\zeta = \frac{3}{2(1 - \frac{3}{5}\frac{\rho_w}{\rho_e})}\frac{m_p a^4}{m_e d^3}\left(\cos^2 \psi - \frac{1}{3}\right) \qquad (2.86)$$

which now corresponds to the formula given by d'Alembert. Laplace is interested in the hypothesis of hydrostatic equilibrium in order to show that it conflicts with the observations. He showed that when the Moon and the Sun are in conjunction in the summer solstice, when their declination is maximal, the hypothesis implies that the excess of water at midday high tide over the following low tide should be roughly 8 times bigger than the excess of midnight high tide over the following low tide, whereas the observations show these excesses to be of the same size.

2.9.11 Three Species of Oscillation

An extremely important contribution of Laplace's study concerns the special structure of the tidal potential: he showed that this potential generates three different kinds of oscillations, for which he studied the influence of bathymetry of the oceans. His research on this topic began in the memoirs of 1775 and 1776 (art. 25–28), then in *Mécanique Céleste* (art. 4–10). He shows that when restricting to the degree 2 of spherical harmonics, the leading term, the tidal potential exerted by an external body is

$$V = \frac{3Gm_p a^2}{2d^3}\left((\cos\theta \sin\delta + \sin\theta \cos\delta \cos(\omega t + \lambda - \alpha))^2 - \frac{1}{3}\right) \qquad (2.87)$$

where the parameters are either local, like colatitude θ and longitude λ, or related to the ephemerids of the perturbing body, as celestial coordinates α and δ, ωt being the sidereal angle of rotation of the Earth. By expanding (2.87) and combining the terms, we find

$$V = \frac{Gm_p a^2}{d^3}\left(\frac{3\sin^2\delta - 1}{2}\right)\left(\frac{3\cos^2\theta - 1}{2}\right)$$
$$+ \frac{3Gma^2}{d^3}\sin\theta \cos\theta \sin\delta \cos\delta \cos(\omega t + \lambda - \alpha)$$
$$+ \frac{3Gma^2}{4d^3}\sin^2\theta \cos^2\delta \cos 2(\omega t + \lambda - \alpha) \qquad (2.88)$$

This modern way of writing the expressions is not exactly the same as Laplace's one, but strictly equivalent. It shows the symmetry between θ, δ, and between λ, $\omega t - \alpha$. The distance d of the external body (Moon or Sun) as well as its equatorial coordinates α and δ vary relatively very slowly with respect to the diurnal variable ωt. This led to the conclusion by Laplace that the three terms of the potential V in Eq. (2.88) give rise to three different species of oscillation. He mentions that the three species mix without interacting and can be studied separately.

2.9.11.1 Oscillations of the First Species

The oscillations of the first species do not depend on the longitude of the surface point, but vary as a function of the orbital parameters of the perturbing body: they have a zonal structure. Among these oscillations figure monthly and fortnightly components for the Moon, annual and semi-annual ones for the Sun. The amplitudes of these oscillations do not depend on the bathymetry of the oceans.

2.9.11.2 Oscillations of the Second Species

These have a quasi-diurnal period (close to 24^h50^{mn} for the Moon, 24^h00^{mn} for the Sun) due to the presence of the argument ωt with diurnal frequency. Their amplitudes are modulated by the orbital motion of the body. The amplitude is zero when this body is on the celestial equator and gets all larger as the declination gets high. This occurs during the solstices for the Sun and twice a month for the Moon.

In theory, when the declination is maximal, the oscillations should generate a large difference of amplitude between successive high tides occurring on the same day. This is in fact contrary to what is observed in various harbors of the Atlantic, where these two tides show approximatively the same amplitudes. Laplace discovered that these oscillations depend on the depth of the seas and vanish if the depth is constant. As early as 1775 he expressed his satisfaction in observing his predictions, mentioning that this agreement constituted one of the main accomplishments of his research.

2.9.11.3 Oscillations of the Third Species

They are the most prominent in harbors of the Atlantic. Their period is semi-diurnal (close to 12^h25^{mn} for the Moon, 12^h00^{mn} for the Sun) and their amplitudes are also modulated by the relative celestial motion of the celestial body. The amplitude is maximum when the body lies on the celestial equator. Laplace sought the condition in which these oscillations vanish and found that this requires an ocean of infinite depth.

2.10 Methodology, Organization, and Analysis of Observations

The numerous calculations by Laplace on the influence of bathymetry and his research on the necessary conditions for the oscillations of second and third species to vanish reach some limits. The impossibility of explaining the variety of tides by a direct deterministic calculation led him, in his memoir of 1790, then in *Mécanique Céleste* (vols. IV and XIII), to fully exploit observational data and to develop semi-empirical methods.

Thus, if Laplace must be considered as the founder of the dynamical theory of tides, his activities in this field were not restricted to theoretical studies. He was also concerned by more practical aspects and was at the origin of the development of systematic observations of the tides. When in 1790 he sought to determine the local laws of the ebb and flow using a semi-empirical method, the only observations available were those carried out between 1711 and 1716 at Brest harbor and those on Lalande's initiative in 1777. Laplace remarked that they were too vague and incomplete to enable a fruitful analysis. Then he exhorted the scientific community to undertake tidal measurements with 'the same care as astronomical observations'.

In 1803, with the help of Pierre Lévêque and Alexis de Rochon, he participated in a commission in charge of the planning of tidal observations. *Memoir on the observations it is important to carry out on the tides in different harbors of the Republic* was written on this occasion. It establishes an extremely precise protocol of observations, underlying the fact that if earlier the observations guided the theory, now the theory guides the observations. In 1806, following this memoir, a long series of observations was undertaken at Brest. Laplace in *Mécanique Céleste* (vol. XII) analysed the data from 16 years of observations (1807–1822) and in 1843 the Bureau des Longitudes published the observations from 1807 to 1835.

2.10.1 Semi-empirical Methods Based on Partial Flows

Whereas the purely theoretical expressions of the tidal potential established by Laplace reached a quasi-definitive status and incurred few modifications until now, his semi-empirical methods were rather complex and merely gave a starting point for the development in the 19th and 20th centuries. Laplace was aware of why his calculations are not satisfactory: tides are modified by the distribution of continents and oceans, irregularities of the ocean depths, the positions and the slopes of shores, currents, the drag of water. It is true that for these reasons tides have no direct and simple relationship with the tidal potential. But they should obey some laws. Laplace established a principle that should give access to local tides laws and is still used up to the present. It relies on two basic ideas:

- The tidal potential can be decomposed as a series of sinusoidal terms with various periodicities. This decomposition explains the modulation of tidal waves according to the characteristics of the lunar and solar orbital motions (variations of declination, of distance, of longitude). In his decomposition, Laplace introduced a significant number of waves, which are found later in the decompositions used by Lord Kelvin in 1867, Darwin in 1883, and Doodson in 1921.
- Despite numerous perturbations listed above, tides conserve something of their periodicities. In other words, the sea is subject to the same periods as those of the forcing tidal potential: each wave of this potential generates a partial sea flow itself expressed by a sinusoidal function with the same period. The coefficients and phases of the partial flows are modified differently for each harbor and for each wave. The total sea flow at a given point is reconstructed as the sum of the individual partial flows, using the principle of superposition.

2.10.2 Determination of the Amplitudes and Phases of the Partial Flows

The determination of the parameters acting on the water height is possible only from observations. Laplace's method tries to employ a shrewd combination of observations to disentangle the phenomenon being studied. For instance, for the characterization of the semi-diurnal tides, high and low tides were recorded in the vicinity of solstitial and equinoctial syzygies or quadratures. In vol. IV, Laplace uses Cassini's observations from the beginning of the 18th century, refined by observations that he himself organized at Brest between 1807 and 1822. He gathered more than 6000 observations for the purpose. Thus he could determine the fundamental parameters (coefficients and phases) which take part in the diurnal and semi-diurnal oscillations, and find a very good agreement between his semi-empirical formula and observations. In particular he could show that, under the effect of the terrestrial rotation and of various perturbations listed above, the amplitude of the diurnal flow in Brest harbor is reduced by a factor of 1/3 compared with the value predicted by the theoretical equilibrium tide, whereas the semi-diurnal flows is multiplied by a factor of 16. Pushing further the treatment of observations, Laplace sought to put in evidence the flow depending on the lunar potential of degree 3, that is to say involving $1/d_M^4$. His semi-empirical method was powerful indeed.

2.10.3 Determination of the Ratio of Lunar/Solar Tides

Finally, Laplace could attempt a fresh estimate of the ratio μ of the amplitude of the lunar tide to that of the solar tide. We saw that in 1687 Newton had set the value $\mu = 4.5$ by using the height of tides in Bristol harbor, and that later Bernoulli lowered this value to $\mu = 2.5$ by using the precise times of high tides instead of their height. Then in 1749 d'Alembert and Euler lowered the value further to 2.33 thanks to the study of the precession-nutation of the Earth. Finally Laplace found the value 2.35 and concluded that 'the agreement of values found by various means is remarkable'. This ratio also enabled him to calculate the mass of the Moon, for which he found 1/75 of the mass of the Earth, very close to the real fraction of 1/81.

2.10.4 Laplace and Atmospheric Tides

On the margin of his research about the oceanic tides, it is worth mentioning that Laplace was also interested in atmospheric tides. The subject had already been initiated by Daniel Bernoulli and d'Alembert: Laplace explained that the gravitational influence of the Moon and the Sun generates in the atmosphere periodic motions similar to the oceanic ones but extremely weak. The barometer variation he calculated theoretically should be of the order of 0.6 mm of mercury (80 Pa). These variations are too small to explain the strange variations of the barometer, with a 12-hour period and 1.5 mm amplitude observed in the 18th century in tropical areas, in particular by Lamanon in 1785 during La Perouse expedition (1785–1788). Laplace concluded that these variations should be due to thermal forcing. In 1825, using the analysis of 8 years of pressure measurements at the Observatoire de Paris, he tried to show the existence of atmospheric tides with a lunar origin: he found an oscillation of 0.055 mm of mercury, but emphasized that his results are not statistically convincing. The existence of such a tide was demonstrated for the first time in 1842, from observations on the island of St. Helens.

2.11 Conclusion on Laplace's Work

Laplace's work is a landmark in the study of tides. We can condense our discursive text above on his contributions into three bullet points.

- The origin of the tides and the ultimate outcome of Newton's ideas. Instead of the generating force of tides, Laplace used the fruitful tidal generating potential, and pioneered the use of the spherical harmonics.
- Even more fundamental, the establishment of a dynamical theory of tides. By neglecting the vertical velocity in the fluid layer, Laplace establishes the general equation of the dynamics of water in the oceans, which to this day remains the basis of tidal theory. He highlighted the Coriolis force and the fact that each oscillation of the tidal potential generates a partial flow with the same period which, mixed with various local perturbations, gives a great variety of geography-dependent tidal behaviors.
- The organization of an observational network, with a very precise protocol. In parallel he developed a method of analysis of observations.

From these various point of view, Laplace can be considered as the true founder of the modern science of ocean tides.

2.12 Overall Conclusion

Since the first ideas on the influence of the Moon put forward by the Ancients, until the mathematical work of Laplace, the improvements of the theory of tides have been considerable. But why did it take such a long time to solve the problem? Three reasons can be found.

- First, solving the problem necessitated the discovery of the universal law of gravitation, so had to wait Newton.
- Second, tides mingle two causes, a deterministic and precise law of gravitation and perturbations by local environments. To understand the tides, we had to separate the two causes. Newton took the first step by giving the tidal force. Laplace

took the second step, by showing that the sea flows in each harbor have the same periodicities as the tidal potential but with phases and amplitudes depending on the local characteristics of each site.

- Third, mathematical tools had to be invented, not only calculus but also spherical harmonics and the equations of fluid mechanics. By the end Laplace's career, all the tools are ready and provide the theoretical basis for the future.

After Laplace, improvements continued. Several of them are:

- The increase of the observations, in particular thanks to the floating tide gauges invented in 1843 by Rémi Chazallon (1802–1872).
- Understanding that tides result largely from resonances of basins to astronomical excitations: pioneering work by John William Lubbock and William Whewell in 1830–1840, then by Rollin Harris in 1897.
- Refining the harmonic expansion of the tidal potential: 91 terms for George Darwin in 1883, 378 for Doodson in 1921, and 12 935 for Hartmann and Wenzel in 1995.
- Understanding that tides concern only oceanic masses but also the solid part of the Earth: elastic deformations.
- Last but not least, fantastic computing tools that integrate the dynamical equations, replacing the partial integrations done by George Biddell Airy, Lord Kelvin, Henri Poincaré, Carl Gustav Rossby, and others.

In addition to the numerical modeling of tides, the problem today consists in dealing with the tides in a global way to determine the motions of a deformable Earth, partially covered by oceans, containing a fluid core and subject to the action of external bodies.

References

1. Bernoulli, D.: Traité sur le flux et le reflux de la mer. In: Pièces qui ont remporté le prix de l'Académie Royale des Sciences en 1740. Martin, Coignard et Guérin, Paris, pp. 55–191
2. Cartwright, D.E.: Tides. A Scientific History. Cambridge University Press, Cambridge (1999)
3. Cassini, J.: Sur le flux et le reflux, Mémoires de l'Académie Royale des Sciences (1710) (1712, 1713, 1714 et 1720)
4. Cavalleri, A.: Dissertation sur la cause physique du flux et du reflux de la mer. In: Pièces qui ont remporté le prix de l'Académie Royale des Sciences en 1740, pp. 1–51. Martin, Coignard et Guérin, Paris (1741)
5. D'Alembert Jean le Rond: Réflexions sur la cause générale des vents. David l'Aîné, Paris (1747)
6. Deparis, V., Legros, H.: Voyage à l'intérieur de la Terre. CNRS Editions, Paris (2000)
7. Descartes, R.: Les principes de la philosophie (1644). Réédition in Cousin V. (ed.) Oeuvres de Descartes, tome III, Levrault, Paris (1824)
8. Duhem, P.: La théorie physique, son objet, sa structure. Rivière & Cie, Paris (1906). Réédition Vrin (2007)
9. Duhem, P.: Le système du monde. Hermann, Paris (1958), tome II, pp. 267–390; tome III, pp. 112–125; tome IX, pp. 7–78

10. Euler, L.: Inquisitio physica in causam fluxus ac refluxus maris. In: Pièces qui ont remporté le prix de l'Académie Royale des Sciences en 1740. Martin, Coignard et Guérin, Paris, pp. 235–350

11. Galileo, G.: Dialogue sur les deux grands systèmes du monde. Points Sciences. Seuil, Paris (1992)

12. Huygens, Ch.: Oeuvres complètes, tome IX, p. 538. Nijhoo, La Haye (1888–1950)

13. Lalande, J.J.L. (de): Traité du flux et du reflux de la mer. In: Astronomie, vol. 4, pp. 1–348. Desaint J. C., Paris (1781)

14. La marquise du Châtelet: Commentaires des principes mathématiques de la philosophie naturelle, Publiés à la Fin de Sa Traduction de L'Oeuvre de Newton. Réédition Jacques Gabay, Sceaux, Sect. V, art. 1, p. 260 (1990)

15. Laplace, P.S. (de): Recherches sur plusieurs points du système du monde, 1775. Mémoire de l'Académie Royale des Sciences de Paris (1778). Réédition in Oeuvres complètes de Laplace, tome IX. Gauthier-Villars, Paris, pp. 71–183 (1893)

16. Laplace, P.S. (de): Recherches sur plusieurs points du système du monde, 1776. Mémoire de l'Académie Royale des Sciences de Paris (1779). Réédition in Oeuvres complètes de Laplace, tome IX. Gauthier-Villars, Paris, pp. 187–280 (1893)

17. Laplace, P.S. (de): Théorie des attractions des sphéroïdes et de la figure de la terre, 1782. Mémoire de l'Académie Royale des Sciences de Paris (1785). Réédition in Oeuvres complètes de Laplace, tome X. Gauthier-Villars, Paris, pp. 341–419 (1894)

18. Laplace, P.S. (de): Sur le flux et le reflux de la mer, 1790. Mémoire de l'Académie Royale des Sciences de Paris (1797). Réédition in Oeuvres complètes de Laplace, tome XII. Gauthier-Villars, Paris, pp. 3–126 (1898)

19. Laplace, P.S. (de): Traité de mécanique céleste, tome II, livre IV. Duprat, Paris (1799). Réédition in Oeuvres complètes de Laplace, tome II. Gauthier-Villars, Paris, pp. 183–314 (1878)

20. Laplace, P.S. (de): Sur le flux et le reflux de la mer, 1818. Mémoire de l'Académie Royale des Sciences de Paris (1820) Réédition in Oeuvres complètes de Laplace, tome XII. Gauthier-Villars, Paris, pp. 473–546 (1898)

21. Laplace, P.S. (de): Traité de mécanique céleste, tome V, livre XIII, Bachelier, Paris (1825). Réédition in Oeuvres complètes de Laplace, vol. V. Chelsea Publishing Company, New York, pp. 163–269 (1969)

22. Maclaurin, C.: De causa physica fluxus et refluxus maris. In: Pièces qui ont remporté le prix de l'Académie Royale des Sciences en 1740. Martin, Coignard et Guérin, Paris, pp. 193–234

23. Newton, I.: Principes mathématiques de la philosophie naturelle (Traduction de la marquise du Châtelet) (1756). Réédition Jacques Gabay, Sceaux: tome I (1990)

24. Pouvreau, N.: Trois Cents ans de mesures marégraphiques en France: outils, méthodes et tendances des composantes du niveau de la mer au Port de Brest, pp. 57–82. Université de la Rochelle, Rochelle (2008)

25. Souffrin, P.: La théorie des marées de Galilée n'est pas une théorie fausse. Epistémologiques **1–2**, 113–139 (2000)

26. Varenius, B.: Géographie générale, tome II. Vincent et Lottin, Paris (1755)

Chapter 3
Oceanic Tides

Bernard Simon, Anne Lemaitre, and Jean Souchay

Abstract The phenomena of tides are a matter of common experience: ocean tides under the influence of the Moon and the Sun, differences of the surface level of the oceans reaching several meters, following well-established cycles. In the present chapter we propose a first step in the general and classical mathematical formulations of the tidal potential and tidal force. Then we apply this formulation to the concrete case of the lunisolar ocean tides at a given point of the surface of the sea. At the end we give a review of various tidal manifestations all around the world.

3.1 Introduction

It is a well-established fact that the origin of the tides is the gravitational action of the Moon and the Sun on objects bound to the Earth [21], but the tidal generating force should not be confused with the gravitational attraction exerted by each of these bodies on the water particles. The tidal generating force is actually the difference between this attraction and what the attraction would be if the particle were located at the center of the Earth. Indeed, the centrifugal force resulting from the orbital motion of the Earth (around the center of mass of the Earth-Moon system or of the Earth-Sun system) is the same at every point of the Earth, while the gravitational attraction varies with the proximity of the celestial bodies according to where the particle is positioned on the surface of the Earth. At the center of mass of the Earth, these two forces balance exactly. Since the Earth radius is small compared with the distance to the Moon (and *a fortiori* to the Sun), to a first approximation for

B. Simon (✉)
19, rue du port, Portsall, 29830 Ploudalmézeau, France
e-mail: bfsimon@neuf.fr

A. Lemaitre
naXys, University of Namur, 8, Rempart de la Vierge, 5000 Namur, Belgium
e-mail: anne.lemaitre@fundp.ac.be

J. Souchay
SYRTE, UMR8630 CNRS, Observatoire de Paris, 61, av. de l'Observatoire, 75014 Paris, France
e-mail: Jean.Souchay@obspm.fr

J. Souchay et al. (eds.), *Tides in Astronomy and Astrophysics*,
Lecture Notes in Physics 861, DOI 10.1007/978-3-642-32961-6_3,
© Springer-Verlag Berlin Heidelberg 2013

someone on the Earth surface the magnitude of the force is the same as if the body (Sun or Moon) were at the zenith or the nadir. This explains the semi-diurnal tide (two high tides and two low tides per day). During the day, a maximum force occurs when the Moon crosses the upper semi-meridian and another maximum when it crosses the lower semi-meridian, the minimum occurring when it crosses the horizon (attractive and centrifugal forces are then nearly opposite). However, owing to the inclination of the axis of the Earth's rotation relative to the axis of the Earth's orbit, the two extrema are generally not equally pronounced, and sometimes, at higher latitudes, the Sun or the Moon never set or rise (polar night). In this case, a maximum disappears and the resulting type is diurnal (one maximum and one minimum per day). This qualitatively explains some aspects of the tide as described in the preceding paragraph: the generative force of the tide entails diurnal and semi-diurnal components.

In his dynamic theory presented to the Académie Royale des Sciences in 1790, Laplace introduced the concept of tidal generating potential [12]. He was the first to treat the tide as a problem of dynamics of water masses and not as a static problem. According to his dynamic theory, the sea response to the tidal generating force takes the form of extensive waves crossing the oceans with a velocity depending essentially on depths. Moreover, like any wave phenomenon, these waves are reflected, refracted, and diffused according to the nature of the propagation medium and the shape of ocean basins. It follows that the observed tide at any point is the result of the superposition of elementary waves which come from all parts of the ocean, each of them being subject, during its travel, to different propagation conditions. All these components can obviously interfere with one another, resulting in strengthened or attenuated amplitudes according to frequencies.

The hydrodynamic equations of this phenomenon, first formulated by Laplace [13], cannot be easily solved even with modern computing tools available, but they remain the basis of all subsequent developments. Above all, they allow establishing a formula, known as 'Laplace equation', applicable to tidal predictions and based on two principles. The first one is that a water mass undergoing a periodic force is subject to a periodic oscillation with the same frequency. The second one is that the total motion of a system subject to small forces is equal to the sum of the elementary motions created by each force.

These two principles express the assumption of the oceans' linear response to the action of the tidal generating force. It turns out that this assumption is well verified in the case of Brest harbor, where tidal observations were used by Laplace to test his theory. The tidal generating force being divided as a sum of elementary periodic forces, the Laplace equation implies that the tide may itself be decomposed into oscillations of similar periods. The assumption of linearity is not inconsistent with the fact that two parameters, the proportionality factor and the phase shift between the tidal component and corresponding power generator, may depend on the frequency. These parameters also depend on hydraulic conditions of wave propagation, different from one point to another, and in practice must be determined experimentally by analysis of available observations.

The main interest of the Laplace theory lies in its ability to provide a practical method for prediction of high and low tides, known as 'Laplace method'. In 1839, the hydrographer Chazallon [4] published the first precise scientific timetable of tides based on this method. In this timetable hours and heights of high and low tides at Brest were calculated. For other harbors, they were obtained using time differences and amplitude factors. The Laplace equation has remained the basis of calculation of tides in France for over 150 years. Before the advent of computers, no competing method could indeed claim to provide better accuracy for the calculation of the tide at Brest. However, because of the assumption of linearity, the Laplace equation could not claim to be universally applicable. In fact, they have never been used to calculate the tides at other places than Brest.

Subsequently, we must note the works of two Englishmen, William Whewell [24] (1794–1866) and George B. Airy [1] (1801–1892) [2], who were particularly interested in the propagation of the tidal wave, the first in oceans, the second in canals and rivers, taking friction into account. But we must wait until the late 19th century, with the contribution of Sir William Thomson, better known as Lord Kelvin (1824–1907), to note a real progress in the calculation of tidal predictions [23]. In 1867, the British Association for the Advancement of Science (BAAS) set up a committee to promote the improvement and widespread implementation of the harmonic analysis of tides. The report of this committee was written by Kelvin himself. Some other reports appeared on this subject, but the major contribution was the paper published in 1883 by George H. Darwin (1845–1912) [5]. This paper presents the precise harmonic expansion of the tidal potential, which has been universally used up to the present day as the basis of most studies on tides. Today, the tidal harmonic components are designated by the names assigned by Darwin. In addition, methods of calculation, developed and adapted to the means of that era, were often transposed without changes, even with the technological evolution of computers. However, this development, based on an ancient lunar theory in which all elements are referred to the orbit, was not entirely satisfactory because it is not purely harmonic: it was necessary to introduce correction factors to account for slow changes in the components, mainly due to the slow retrograde motion of the orbit of the Moon. The long-term variations associated with these correction factors can be regarded as constant over periods of the order of one year. Calculated over many years, these factors are available as published tables [22]. The use of these tables is not quite satisfactory for modern computing, but was very useful for manual calculation. That is probably why Darwin remained popular, while as early as 1921 more satisfying purely harmonic expansions such as those proposed by Arthur T. Doodson (1890–1968), were available. Doodson [9] published in the Proceedings of the Royal Society an expansion based on the lunar theory proposed by Brown in 1919 [3]. This new expansion, digital and purely harmonic, provides many more terms than those presented by Darwin and does not require correction factors. Thus tables for these factors were no longer necessary and automatic processing could be greatly improved to come into practical use in the late 1950s. Other expansions, more complete or more accurate, have been proposed since. However, for practical applications in tidal calculations, they do not bring significant progress with respect to the Doodson expansions which remain the reference.

3.2 Basic Mathematical Tidal Theory

In this section we consider the general case of a celestial body orbiting a non rigid planet P. This will give rise to a deformation of this planet. The hypothesis is that this deformation is proportional to the force, to the stress itself. This is why our fundamental aim is to calculate the force exerted on each point P of the planet surface, due to the presence of the celestial body.

Let M be the mass of the non-rigid planet, R its mean radius. O is the position of its center of mass, chosen as origin of the coordinates x, y, z. The celestial orbiting body is regarded as a point mass m, with position Q. While orbiting around O it deforms the planet, and a surface element of the planet is denoted by a position P, at a distance $r \simeq R$ from the center O.

We introduce the following vector and scalar notation:

$\mathbf{r} = \mathbf{OP} = (x, y, z)$, a point on the surface of the planet of mass M

$r = \|\mathbf{r}\|$, its norm

$\mathbf{d} = \mathbf{OQ} = (u, v, w)$, the position of the perturbing body of mass m

$d = \|\mathbf{d}\|$, its norm

$\Delta = \|\mathbf{QP}\|$, the distance between Q et P

3.2.1 Tidal Potential

The potential V calculated at the point P due to the presence of the orbiting body with mass m is given by

$$V(x, y, z) = -\mathscr{G}\frac{m}{\Delta} \quad \text{where } \Delta^2 = d^2 + r^2 - 2dr\cos\psi \tag{3.1}$$

with $\mathbf{r} \cdot \mathbf{d} = rd\cos\psi$ and $\mathscr{G}$ the gravitational constant.

Let us expand this expression:

$$V(x, y, z) = -\mathscr{G}\frac{m}{d}\left(1 + \left(\frac{r}{d}\right)^2 - 2\left(\frac{r}{d}\right)\cos\psi\right)^{-\frac{1}{2}}$$

$$= -\mathscr{G}\frac{m}{d}\sum_{n\geq 0}\left(\frac{r}{d}\right)^n P_n(\cos\psi). \tag{3.2}$$

Then, if we assume that $R \ll d$,

$$V(x, y, z) = V_0 + V_1 + V_2 + \cdots \tag{3.3}$$

$$\simeq -\mathscr{G}\frac{m}{d}\left(1 + \left(\frac{r}{d}\right)P_1(\cos\psi) + \left(\frac{r}{d}\right)^2 P_2(\cos\psi)\right) \tag{3.4}$$

where P_n represents the Legendre polynomial of degree n.

- V_0 is a constant term with respect to (x, y, z) and can be dropped,

$$V_0 = -\mathscr{G}\frac{m}{d}, \tag{3.5}$$

- V_1 is the potential corresponding to a system of two masses orbiting around their center of mass,

$$V_1 = -\mathscr{G}\frac{m}{d}\left(\frac{r}{d}\right)P_1(\cos\psi) = -\mathscr{G}\frac{mr}{d^2}\cos\psi, \tag{3.6}$$

- V_2 is the first part corresponding to the tidal deformation, which is studied in detail in this chapter,

$$V_2 = -\mathscr{G}\frac{m}{d}\left(\frac{r}{d}\right)^2 P_2(\cos\psi) = -\mathscr{G}\frac{mr^2}{d^3}\frac{1}{2}(3\cos^2\psi - 1). \tag{3.7}$$

The next term in the expansion would be:

$$V_3 = -\mathscr{G}\frac{m}{d}\left(\frac{r}{d}\right)^3 P_3(\cos\psi) = -\mathscr{G}\frac{mr^3}{d^4}\frac{1}{2}(5\cos^3\psi - 3\cos\psi). \tag{3.8}$$

Let us rewrite the amplitude factor A_2 in V_2:

$$A_2 = \mathscr{G}\frac{mr^2}{d^3} = \frac{\mathscr{G}M}{r^2}\frac{m}{M}\frac{r^3}{d^3}r = g\xi, \tag{3.9}$$

where $g = \frac{\mathscr{G}M}{r^2}$ is the gravity at the surface of the planet of mass M, and $\xi = \frac{m}{M}\frac{r^3}{d^3}r$ depends on the position and mass of the perturbing body.

We recall that the planet is rotating. Consequently it makes sense to speak of its equator and **P** can be positioned with its latitude φ measured from this equator, and an angle of longitude λ varying with the rotation of the planet. In a similar way the orbiting point mass can be positioned through variable coordinates which are its latitude δ and its longitude λ'.

Now we can introduce the spherical coordinates:

$$\begin{aligned} x &= r\cos\varphi\cos\lambda, & y &= r\cos\varphi\sin\lambda, & z &= r\sin\varphi, \\ u &= d\cos\delta\cos\lambda', & v &= d\cos\delta\sin\lambda', & w &= d\sin\delta. \end{aligned} \tag{3.10}$$

They allow us to write:

$$\begin{aligned} \cos\psi &= \cos\varphi\cos\lambda\cos\delta\cos\lambda' + \cos\varphi\sin\lambda\cos\delta\sin\lambda' + \sin\varphi\sin\delta \\ &= \cos\varphi\cos\delta(\cos\lambda\cos\lambda' + \sin\lambda\sin\lambda') + \sin\varphi\sin\delta \\ &= \cos\varphi\cos\delta\cos(\lambda - \lambda') + \sin\varphi\sin\delta, \end{aligned} \tag{3.11}$$

and consequently,

$$\begin{aligned} P_2(\cos\psi) &= \frac{1}{2}(3\cos^2\psi - 1) \\ &= \frac{1}{2}(3\sin^2\varphi - 1)\frac{1}{2}(3\sin^2\delta - 1) + \frac{3}{4}\cos^2\varphi\cos^2\delta\cos 2(\lambda - \lambda') \\ &\quad + \frac{3}{4}\cos 2\varphi\cos 2\delta\cos(\lambda - \lambda'). \end{aligned} \tag{3.12}$$

3.2.2 Tidal Force

From the tide potential V_2 we can extract the expression for the corresponding tidal force per unit of mass, $\mathbf{F}_2$, due to the presence of the external body situated at (u, v, w) and acting on the surface point $\mathbf{P}$ with coordinates (x, y, z):

$$\mathbf{F}_2(x, y, z) = -\nabla V_2(x, y, z) = \nabla W_2(x, y, z) = \left(\frac{\partial W_2}{\partial x}, \frac{\partial W_2}{\partial y}, \frac{\partial W_2}{\partial z} \right). \quad (3.13)$$

For that purpose, let us rewrite W_2 in terms of x, y and z:

$$\begin{aligned}
W_2(x, y, z) &= \mathscr{G} \frac{mr^2}{d^3} \frac{1}{2} \left(3 \cos^2 \psi - 1 \right) \\
&= \mathscr{G} \frac{m}{d^5} \frac{1}{2} \left(3r^2 d^2 \cos^2 \psi - r^2 d^2 \right) \\
&= \mathscr{G} \frac{m}{d^5} \frac{1}{2} \left(3(xu + yv + zw)^2 - d^2 (x^2 + y^2 + z^2) \right). \quad (3.14)
\end{aligned}$$

It is now easy to calculate the three partial derivatives

$$\begin{aligned}
\frac{\partial W_2}{\partial x} &= \mathscr{G} \frac{m}{d^5} \left(3(xu + yv + zw)u - d^2 x \right) \\
&= \mathscr{G} \frac{m}{d^5} \left(3rd \cos \psi u - d^2 x \right), \\
\frac{\partial W_2}{\partial y} &= \mathscr{G} \frac{m}{d^5} \left(3rd \cos \psi v - d^2 y \right), \\
\frac{\partial W_2}{\partial z} &= \mathscr{G} \frac{m}{d^5} \left(3rd \cos \psi w - d^2 z \right),
\end{aligned}$$

and the corresponding force per unit of mass, $\mathbf{F}_2$, acting on $\mathbf{P}$ due to $\mathbf{Q}$

$$\mathbf{F}_2 = \mathscr{G} \frac{m}{d^3} \left(3r \cos \psi \frac{\mathbf{d}}{d} - \mathbf{r} \right). \quad (3.15)$$

We can already analyse the first term, depending on ψ.

- $\psi = 0$ when $\mathbf{r}$ and $\mathbf{d}$ are aligned: in this configuration, $\mathbf{F}$ is maximal and points toward the perturbing body.
- $\psi = \pi$ when $\mathbf{r}$ and $\mathbf{d}$ are anti-aligned: in this configuration, $\mathbf{F}$ is again maximal but points away from the perturbing body.

Extrapolating these remarks to any point on the surface, we can say that the deformation at any point $\mathbf{P}$ of the surface of the planet is instantaneous and directly proportional to the tidal force which creates it. In the case of a deformable planet (like the Earth), the external surface is deformed in such a way that it corresponds to an equipotential surface.

3.3 Expression of the Tidal Potential for the Earth

Here we apply the mathematical principles of the previous section to express the tidal potential $V_2(\mathbf{P})$ or the associated force function $W_2(\mathbf{P}) = -V_2(\mathbf{P})$, exerted by an external body (Moon, Sun, planet) with mass m on a point $\mathbf{P}$ of the Earth [c??]. $\mathbf{P}$ is classically labeled by its spherical coordinates (r, φ, λ), with respect to the terrestrial equator, where φ is the latitude and λ the terrestrial longitude, r being the distance from the center $\mathbf{O}$ of the Earth to $\mathbf{P}$.

The position of the perturbing body is given, as before, by its declination δ with respect to the celestial equator and by its longitude λ', which we replace by its hour angle H defined by $H = \lambda' - \lambda$.

Then the potential $W_2(\mathbf{P})$ is given by:

$$W_2(\mathbf{P}) = \mathscr{G}m\frac{r^2}{d^3}\frac{9}{4}\left(\sin^2\varphi - \frac{1}{3}\right)\left(\sin^2\delta - \frac{1}{3}\right)$$

$$+ \mathscr{G}m\frac{r^2}{d^3}\frac{3}{4}\sin 2\varphi \sin 2\delta \cos H$$

$$+ \mathscr{G}m\frac{r^2}{d^3}\frac{3}{4}\cos^2\varphi \cos^2\delta \cos 2H \qquad (3.16)$$

$$= W_2^{zonal} + W_2^{tesseral} + W_2^{sectorial}. \qquad (3.17)$$

We can decompose W_2 into three terms, named *zonal*, *tesseral* and *sectorial*, which are characterized in the next section, where we make a large use of [c??].

3.3.1 Zonal Part of the Tidal Potential

Let us start with the zonal term, W_2^{zonal}:

$$W_2^{zonal} = \mathscr{G}m\frac{r^2}{d^3}\frac{9}{4}\left(\sin^2\varphi - \frac{1}{3}\right)\left(\sin^2\delta - \frac{1}{3}\right). \qquad (3.18)$$

This term is called *long period* or *low frequency* because it does not contain the hour angle H, which is by far the highest-frequency variable. Its variations come from the squares of the sines of the declination $(\sin^2\delta)$ of the perturbing body (Moon, Sun, planet) around the Earth, which in reality vary slowly. It introduces a period which is half the time of the relative revolution of the perturbing body, i.e. roughly 14 days in the case of the Moon and 6 months in the case of the Sun. Given the extremal values reached by the declinations, $28°30'$ for the Moon and $23°27'$ for the Sun, the last factor is always negative.

The factor $(\sin^2\varphi - \frac{1}{3})$ vanishes at latitudes such that $\sin\varphi = \pm 1/\sqrt{3}$, i.e. at latitudes $35°16'$N and $35°16'$S. The locus where this term vanishes are parallels (lines of equal latitude). Taking into account that one factor is always negative, it follows that the long-period term of potential is always positive for latitudes between $35°16'$N and $35°16'$S and negative elsewhere.

The partition in zones of latitude of this part of the potential, as it is shown in Fig. 3.1, justifies the terminology of *zonal potential*.

Fig. 3.1 Zonal distribution
of the long period component
of the tidal potential

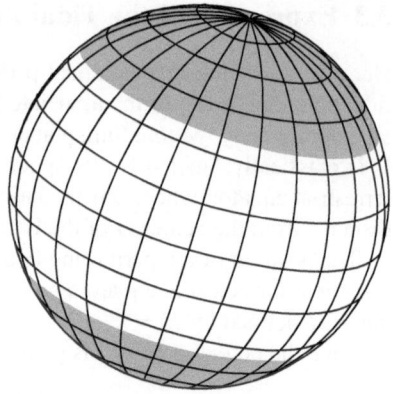

Fig. 3.2 Tesseral distribution
of the diurnal component of
the tidal potential

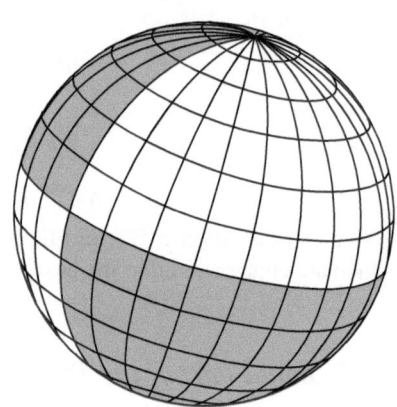

3.3.2 Tesseral Part of the Tidal Potential

The second term $W_2^{tesseral}$ is given by:

$$W_2^{tesseral} = \mathscr{G}m\frac{r^2}{d^3}\frac{3}{4}\sin 2\varphi \sin 2\delta \cos H. \tag{3.19}$$

It is called *diurnal*, for it contains H with period of roughly one day, regardless of the celestial body m being considered. Its nodes are the meridians normal to the direction of the perturbing body, and the equator (Fig. 3.2). It gives a *tesseral* structure of equipotential lines, whose sign changes with the declination. The period in the hour angle is approximately 24 hours for the Sun and 24 h 50 min for the Moon. The declination δ and parallax $1/r$ vary very slowly in comparison to this diurnal frequency.

Thus they act as modulations on the diurnal term. The diurnal local maximum is reached when the perturbing body crosses the upper or lower meridian of the observer. The maximal extrema on the Earth are reached at latitudes 45°N and 45°S when δ is itself at its maximum value (23°27′ for the Sun and 28°30′ for the Moon).

Fig. 3.3 Sectorial
distribution of the
semi-diurnal component of
the tidal potential

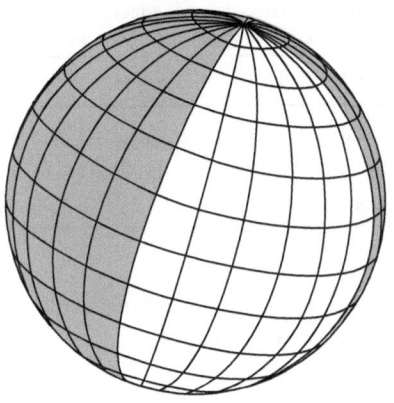

The potential is zero for the points at the equator ($\varphi = 0°$) and at the poles ($\varphi = 90°$)
or when the declination δ of the orbiting body (Moon, Sun or planet) is zero.

3.3.3 Sectorial Part of the Tidal Potential

Finally the third term $W_2^{sectorial}$ is given by:

$$W_2^{sectorial} = \mathscr{G}m\frac{r^2}{d^3}\frac{3}{4}\cos^2\varphi\cos^2\delta\cos 2H. \qquad (3.20)$$

It is called *semi-diurnal*, for it contains $2H$, with period of roughly 12 h for the Sun
and 12 h 25 min for the Moon. Its nodes are the meridians located at 45° of longitude
eastward or westward of the meridian containing the perturbing body. These nodes
divide the Earth into four sectors. The sectorial potential is positive in the section
containing the great circle of the perturbing body and its opposite, and negative in
the other two sections.

This is the reason for which it presents a sectoral distribution (Fig. 3.3) over
the Earth. This component has two maxima and two minima per day, due to the
periodicity of $\cos 2H$. The maximal extrema are reached at the equator ($\varphi = 0°$),
when the declination of the body (δ) is zero. The semi-diurnal part of the potential
is zero at the poles ($\varphi = 90°$).

3.3.4 Components of the Local Tidal Force

The local tidal force $\mathbf{F} = (F_r, F_\varphi, F_\lambda)$ at a given position $\mathbf{P}$ can be deduced by
simple differentiation of the tidal potential along the three local coordinate axes:
the first direction is zenital, the other two directions are horizontal, respectively in
the North-South and East-West directions. Let us recall that $H = \lambda' - \lambda$, and then

Table 3.1 Trigonometric functions appearing in the expansions of the local tidal force according to the three rectangular directions and the three tidal families

Family	Factor	Vertical	North-South	East-West
Zonal	$\frac{9}{4}\mathscr{G}m\frac{r}{d^3}$	$2(\sin^2\varphi - \frac{1}{3})(\sin^2\delta - \frac{1}{3})$	$\sin 2\varphi(\sin^2\delta - \frac{1}{3})$	0
Tesseral	$\frac{3}{4}\mathscr{G}m\frac{r}{d^3}$	$2\sin 2\varphi \sin 2\delta \cos H$	$\cos 2\varphi \sin 2\delta \cos H$	$\sin\varphi \sin 2\delta \sin H$
Sectorial	$\frac{3}{4}\mathscr{G}m\frac{r}{d^3}$	$2\cos^2\varphi\cos^2\delta\cos 2H$	$-\sin 2\varphi\cos^2\delta\cos 2H$	$2\cos\varphi\cos^2\delta\sin 2H$

$\frac{\partial W_2}{\partial\lambda} = -\frac{\partial W_2}{\partial H}$. Thus we have, with a relationship similar to Eq. (3.15) but adapted to spherical coordinates:

$$\mathbf{F} = \nabla W_2 = \left(\frac{\partial W_2}{\partial r}, \frac{1}{r}\frac{\partial W_2}{\partial\varphi}, \frac{1}{r\cos\varphi}\frac{\partial W_2}{\partial\lambda}\right). \tag{3.21}$$

Consequently we have three trigonometric functions: vertical or $\frac{\partial W_2}{\partial r}$, North-South or $\frac{1}{r}\frac{\partial W_2}{\partial\varphi}$, East-West or $\frac{1}{r\cos\varphi}\frac{\partial W_2}{\partial\lambda}$, for each tidal family (zonal, tesseral, sectorial). The nine resulting trigonometric functions characterizing the tidal force are summarized in Table 3.1.

Notice that the deviations of the vertical n_1 and n_2 along the two horizontal axes (North-South and East-West) are immediately derived from

$$n_1 \approx \tan n_1 = \frac{1}{rg}\frac{\partial W_2}{\partial\varphi} \tag{3.22}$$

and

$$n_2 \approx \tan n_2 = \frac{1}{rg\cos\varphi}\frac{\partial W_2}{\partial\lambda}. \tag{3.23}$$

3.4 Doodson Expansion of the Tidal Potential

The Doodson expansion is based on the theory of the orbital motion of the Moon proposed by Brown (1919) [3], which describes the motion of the Moon in ecliptic coordinates. Brown provided harmonic expansions of the mean longitude, latitude, and the average horizontal parallax of the Moon in a series of trigonometric functions whose arguments are linear in the mean time.

3.4.1 Previous Expansions of the Lunisolar Potential

The expressions of the lunisolar potential given by Laplace [13] in the form of Eqs. (3.18), (3.19) and (3.20) and their derivatives are not directly suitable for the analysis of tidal phenomena because the term $1/d^3$ as well as the trigonometric functions containing δ and H exhibit very complicated time-variations due the complexity of the orbital motions of the Earth around the Sun and of the Moon around the Earth.

Laplace already had the idea of expanding the Moon and Sun potentials in sinusoidal functions whose arguments are linear in time. Each term of such an expansion can be understood as the potential of a fictitious body describing a uniform circular motion in the equatorial plane, generating an elementary tidal component with the period of revolution of the fictitious body, and with amplitude and phase depending on the harbor considered. Assuming that the ocean's response is a linear function of the period of revolution of the Sun and the Moon for diurnal components on one hand and semi-diurnal on the other (i.e. in narrow frequency ranges), Laplace was able to avoid resorting to a purely harmonic expansion. He also showed how to make the potential expansion in a form of purely harmonic components, taking into account the main inequalities of the Moon. Then he deduced the corresponding expressions, independently of any assumption on each component amplitude and phase.

Kelvin [23] and Darwin [6] in 1883 [5] continued Laplace's work by improving the harmonic expansion of the tidal potential. Darwin's expansions were the starting point for the harmonic method of calculating tides which have then been used universally [8]. However, the Moon orbital theory available at that time did not allow Darwin to find a comprehensive expansion of the potential. In particular, defects induced by the motion of the lunar nodes were regarded as disturbances requiring the use of correction factors called nodal factors. In 1921, Doodson remedied this situation and published a purely harmonic expansion containing those 386 components whose amplitude coefficient exceeds 10^{-4} of the leading one. All the expansions of the tidal potential published after Doodson's have shown an excellent agreement with them. In the following we explain in detail the principles of construction for Doodson's series of the tidal potential.

3.4.2 Doodson's Constant

In the expression of the potential W_2 in Eq. (3.16) or in Table 3.1, the trigonometric functions are multiplied by the factor $\frac{3}{4}\mathscr{G}m(r^2/d^3)$, where r and d are respectively the distances of $\mathbf{P}$ and of the perturbing body (Moon, Sun, planet) to the center $\mathbf{O}$ of the Earth. Therefore it seems judicious to introduce a constant scaling factor close to it [c??].

A natural way of doing this is to replace d by its mean distance c, that is to say its averaged value during a revolution, and to replace r by a radius $\dot{a}$ such that the volume of the sphere of radius $\dot{a}$ is the same as the volume of the Earth. Consequently

$$\bar{a} = \sqrt[3]{a^2 b} \tag{3.24}$$

where a and b are respectively the semi-major and semi-minor axis of the Earth (considered as an ellipsoid with a circular equator).

Then we can define the general Doodson's constant D:

$$D = \frac{3}{4}\mathcal{G}m\left(\frac{\bar{a}}{c}\right)^3,$$ (3.25)

and for the Moon and for the Sun, also called the Doodson's constants:

$$D_M = \frac{3}{4}\mathcal{G}m_M\left(\frac{\bar{a}}{c_M}\right)^3, \qquad D_S = \frac{3}{4}\mathcal{G}m_S\left(\frac{\bar{a}}{c_S}\right)^3.$$ (3.26)

Thus the ratio of these two constants is

$$\frac{D_S}{D_M} = \frac{m_S}{m_M} \times \frac{c_S^3}{c_M^3} \approx 0.4590.$$ (3.27)

This shows why the influence of the Moon on the tides is roughly twice bigger than that of the Sun.

With the help of the Doodson's constant given by Eq. (3.25), Eqs. (3.18), (3.19) and (3.20) for the potential can be rewritten as

$$W_2^{zonal} = 3D\frac{c^3}{d^3}\left(\sin^2\varphi - \frac{1}{3}\right)\left(\sin^2\delta - \frac{1}{3}\right),$$ (3.28)

$$W_2^{tesseral} = D\frac{c^3}{d^3}\sin 2\varphi \sin 2\delta \cos H,$$ (3.29)

$$W_2^{sectorial} = D\frac{c^2}{d^3}\cos^2\varphi \cos^2\delta \cos 2H.$$ (3.30)

3.4.3 Basic Principle

The three tidal potentials in Eqs. (3.28), (3.29), (3.30) depend on the latitude φ which is constant for a given point **P**, and on astronomical expansions involving the position of the perturbing body through the variables H, δ, and d. The principle is to truncate the complete expansion to get linear functions of time to approximate the motions, on suitable timescales. Doodson performed this calculations for specific variables which are defined in the next section.

3.4.4 Six Fundamental Variables

The choice of six variables, which over a span of a century may be regarded as linear functions of time, was made by Doodson on the basis of the accumulated results in fundamental astronomy. These variables are:

- τ is the *hour angle of the mean Moon shifted by* 180°: $\tau = H_M + 180°$.
- s is the *mean tropic longitude of the Moon* ('selene' is Greek for the Moon).

Table 3.2 Main variables involved in the tidal potential with definitions, rates, periods

Variable	Period definition	Hourly angle	Period
$t = t' - h$	mean solar day	$15°.0000000$	1 d
$t' = t + h$	sidereal day	$15°.0410686$	0.997270 d
$\tau = t - s$	mean lunar day	$14°.4920521$	1.035050 d
s	tropic month	$0°.5490165$	27.321582 d
h	tropic year	$0°.0410686$	365.242199 d
p	rev. mean perigee of Moon	$0°.0046418$	8.847 y
N'	rev. of lunar nodes	$0°.0022064$	18.613 y
p_s	rev. of perihelion of Earth	$0°.0000020$	20.940 y
$s - N$	mean draconitic month	$0°.5512229$	27.21222 d
$s - p$	mean anomalistic month	$0°.5443747$	27.55455 d
$s - h$	mean synodic month	$0°.5079479$	29.53059 d
$s - 2h + p$	evection	$0°.4715211$	31.812 d
$h - p_s$	mean anomalistic year	$0°.0410667$	365.25964 d
$h - p$		$0°.0364268$	411.78471 d
$2(s - h)$		$1°.0158958$	14.76530 d

- h is the *mean tropic longitude of the Sun* ('helios' is Greek for the Sun).
- p is the *mean tropic longitude of the lunar perigee*.
- $N' = -N$ is the *mean tropic longitude of the ascending lunar node with respect to the ecliptic*. The sign is changed because N is the only variable decreasing with time.
- p_s is the *mean tropic longitude of the Earth perihelion*.

All these variables as well as elementary combinations of them are presented together with their period definition (referring to some well-defined astronomical cycle), their hourly angle, and their period, in Table 3.2.

3.4.5 Preliminary Expansions of Astronomical Trigonometric Functions

3.4.5.1 Lunar Motion

To a first approximation, the orbit of the Moon is quasi-elliptic. However this is too rough when we require more accuracy on its orbital motion, which is especially the case when computing the lunar tidal potential. In fact two main irregularities must be taken into account, respectively called *evection* and *variation*.

The evection arises because the Sun crosses twice a year the projection of the semi-major axis of the Moon on the ecliptic (if we neglect the slow motion of the lunar perigee). This results in a gravitational excitation of the eccentricity, its frequency being $(\dot{s} - \dot{p}) - 2(\dot{h} - \dot{p}) = \dot{s} - 2\dot{h} + \dot{p}$. The variation arises because the eccentricity of the Moon is modified during the syzygies, when the three bodies (Sun, Earth, Moon) are in conjunction. These two irregularities, due to the perturbing gravitational action of the Sun in the framework of a three body problem, well known as the *main problem*, affect the ratio c/d and the longitude λ_M, as well as the classical formula of the elliptic motion. We thus have [18]

$$\frac{c_M}{d_M} = 1 + C_{ell.}\cos(s - p) + C_{ev.}\cos(s - 2h + p) + C_{var.}\cos(2s - 2h)$$
$$= 1 + 0.0549\cos(s - p) + 0.010\cos(s - 2h + p) + 0.008\cos(2s - 2h)$$

which gives

$$\left(\frac{c_M}{d_M}\right)^3 = 1 + 0.1647\cos(s - p) + 0.030\cos(s - 2h + p) + 0.024\cos(2s - 2h)$$
$$\tag{3.31}$$

and

$$\lambda_M = \dot{s}_0 t + C'_{ell.}\sin(s - p) + C'_{ev.}\sin(s - 2h + p) + C'_{var.}\sin(2s - 2h)$$
$$= \dot{s}_0 t + 0.110\sin(s - p) + 0.023\sin(s - 2h + p) + 0.011\sin(2s - 2h).$$

In order to use these expansions inside the tidal potential, a last step consists in expressing the declination of the Moon as a function of λ_M:

$$\sin\delta_M = \sin\varepsilon\sin\lambda_M = 0.398\sin\lambda_M \tag{3.32}$$

where ε is the obliquity of the Earth ($\varepsilon = 23°27'$). It follows that:

$$\sin^2\delta_M = 0.0792(1 - \cos 2\lambda_M) \tag{3.33}$$
$$\cos^2\delta_M = 0.921 + 0.0792\cos 2s - 0.036\cos N$$
$$+ 0.036\cos(2s - N) + \cdots \tag{3.34}$$
$$\sin 2\delta_M = 2\sin\delta_M\cos\delta_M = 0.764\sin s + \cdots \tag{3.35}$$

3.4.5.2 Solar Motion

In the case of the Sun, the unperturbed elliptic motion is quite acceptable:

$$\frac{c_S}{d_S} = 1 + e_{Earth}\cos(h - p_s) = 1 + 0.0167\cos(h - p_s) \tag{3.36}$$

which gives

$$\left(\frac{c_S}{d_S}\right)^3 = 1 + 0.0502\cos(h - p_s) + \cdots \tag{3.37}$$

and, for the longitude,

$$\lambda_S = \dot{h}_0 t + 0.0335\sin(h - p)t. \tag{3.38}$$

The expression for $\cos^2 \delta_S$ involves the same coefficients as for the Moon in Eq. (3.34):

$$\cos^2 \delta_S = 0.921 + 0.0792 \cos 2h + \cdots \tag{3.39}$$

3.5 Tidal Spectrum

Now that we have obtained the necessary expansions of the trigonometric functions of astronomical angles involved in the tidal potential W_2, it is possible to express each part of W_2 (sectorial, tesseral, zonal) as a combination of sinusoidal functions whose arguments are expressed themselves as combinations of the Doodson's variables.

3.5.1 Characterization of the Semi-diurnal Waves

The semi-diurnal (or sectorial) waves come from the sectorial part of the potential given by Eq. (3.30).

3.5.1.1 Lunar Sectorial Part

In the case of the Moon, $H_M = \tau - 180°$, and by using the expansion of $(c_M/d_M)^3$ and $\cos^2 \delta_M$ respectively given by Eqs. (3.31) and (3.34), we get:

$$\left(W_2^{sectorial} \right)_M = D \left(\frac{c_M}{d_M} \right)^3 \cos^2 \varphi \cos^2 \delta_M \cos 2H_M, \tag{3.40}$$

$$= D_M \cos^2 \varphi \left[1 + 0.165 \cos(s - p) + 0.030 \cos(s - 2h + p) \right.$$
$$\left. + 0.024 \cos(2s - 2h) + \cdots \right]$$
$$\times [0.921 + 0.0792 \cos 2s + \cdots] \cos 2\tau. \tag{3.41}$$

The result is an infinite number of terms which show frequencies symmetrically distributed on both sides of the half lunar-day frequency. The leading oscillation is obviously $0.921 D \cos^2 \varphi \cos 2\tau$. It is classically called the M_2 wave and its period is the mean lunar day, that is to say $12^h 25^{min} 14^s$. The following biggest wave is called N_2 with argument $2\tau + (s - p)$ associated with a symmetrical wave with much smaller amplitude and argument $2\tau - (s - p)$. Another big wave is named K_{2M} with argument $2\tau + s$ which corresponds to the sidereal day. The index M stands for 'Moon part', as this wave is also present in the case of the solar part.

3.5.1.2 Solar Sectorial Part

In a way similar to what was done above for the lunar part, we take into account that the hour angle of the Sun is $H_S = \tau + s - h$ and we use the expansions of $(c_S/d_S)^3$ and $\cos^2 \delta_S$ respectively given by Eqs. (3.37) and (3.39). We get:

$$\left(W_2^{sectorial}\right)_S = D_S \left(\frac{c_S}{d_S}\right)^3 \cos^2 \varphi \cos^2 \delta_S \cos 2H_S \tag{3.42}$$

$$= D_S \cos^2 \varphi \left[1 + 0.0502 \cos(h - p_S) + \cdots\right]$$
$$\times \left[0.921 + 0.0792 \cos 2h + \cdots\right] \cos(2\tau + 2s - 2h). \tag{3.43}$$

The leading wave with argument $2\tau + 2s - 2h$ has a half mean solar-day period, that is to say exactly $12^h 00^{min} 00^s$ and is called S_2. Other main oscillations are called *elliptic* or *declinational* because they come either from the ellipticity of the Earth orbit or from the declination of the Sun. The leading waves of the first category are named R_2 and T_2 with respective symmetrical arguments $2\tau + 2s - 2h - (h - p_S)$ and $2\tau + 2s - 2h + (h - p_S)$, those of the second category have arguments $2\tau + 2s - 2h + 2h$ (named K_{2S}) and $2\tau + 2s - 2h - 2h$.

3.5.2 Characterization of the Diurnal Waves

The diurnal (or tesseral) waves come from the sectorial part of the potential given by Eq. (3.29).

3.5.2.1 Lunar Tesseral Part

Still by taking $H_M = \tau - 180°$, and by using the expansion of $(c_M/d_M)^3$ and $\sin 2\delta_M$ respectively given by Eqs. (3.31) and (3.33), we get

$$\left(W_2^{tesseral}\right)_M = D_M \left(\frac{c_M}{d_M}\right)^3 \sin 2\varphi \sin 2\delta_M \cos H_M \tag{3.44}$$

$$= -D_M \sin 2\varphi \left[1 + 0.165 \cos(s - p) + 0.030 \cos(s - 2h + p)\right.$$
$$\left. + 0.024 \cos(2s - 2h) + \cdots\right]$$
$$\times \left[0.0764 \sin s + \cdots\right] \cos \tau. \tag{3.45}$$

In contrast with the semi-diurnal part, a leading oscillation with no symmetrical counterpart does not exist because the mean value of $\sin 2\delta_M$ is zero (there is no constant part in the second term of the right hand side). Thus the leading oscillations are the two symmetrical declinational waves K_{1M} which corresponds to the sidereal day with argument $\tau + s$ and period $23^h 56^{min} 04^s$, and O_1 with argument $\tau + s$ and period $23^h 49^{min} 10^s$.

3.5.2.2 Solar Tesseral Part

By analogy with the lunar part and taking into account that $H_S = \tau + s - h$ we have:

$$\left(W_2^{tesseral}\right)_S = D_S \left(\frac{c_S}{d_S}\right)^3 \sin 2\varphi \sin 2\delta_S \cos H_S$$

$$= D_S \sin 2\varphi \left[1 + 0.052 \cos(h - p_S) + \cdots\right]$$

$$\times \left[0.0764 \sin h + \cdots\right] \cos(\tau + s - h). \qquad (3.46)$$

As for the Moon we find the term K_1 (called here K_{1M}) with a sidereal day period, and argument $\tau + s$ (or $t + h$), coming from the combinations of the terms with argument h and $\tau + s - h$ and its symmetric counterpart, with argument $\tau + s - 2h$ (or $t - h$).

3.5.3 Characterization of the Long Periodic Waves

The long periodic (or zonal) waves come from the zonal part of the potential given by Eq. (3.28).

3.5.3.1 Zonal Lunar Part

The lunar zonal part is written

$$\left(W_2^{zonal}\right)_M = 3 D_M \frac{c_M^3}{d_M^3} \left(\sin^2 \varphi - \frac{1}{3}\right) \left(\sin^2 \delta_M - \frac{1}{3}\right). \qquad (3.47)$$

By using the expansions of $(c_M / d_M)^3$ and $\cos^2 \delta_M$ from Eqs. (3.31) and (3.34) and after combining the trigonometric functions,

$$\frac{c_M^3}{d_M^3} \left(\sin^2 \delta_M - \frac{1}{3}\right) = \frac{c_M^3}{d_M^3} \left(\frac{2}{3} - \cos^2 \delta_M\right)$$

$$= -0.254 - 0.0792 \cos 2s + 0.036 \cos N + 0.036 \cos(2s - N)$$

$$- 0.049 \cos(s - p) + \cdots \qquad (3.48)$$

Thus the main zonal lunar oscillation has an argument $2s$ and a semi-monthly (fortnightly) period 13.66 d. It is called M_f (Moon, fortnightly). The second most important term is named M_m (Moon, monthly) with argument $s - p$ which corresponds to the anomalistic month.

3.5.3.2 Zonal Solar Part

The solar zonal part is written

$$\left(W_2^{zonal}\right)_{Sun} = 3D_{Sun}\frac{c_S^3}{d_S^3}\left(\sin^2\varphi - \frac{1}{3}\right)\left(\sin^2\delta_S - \frac{1}{3}\right). \tag{3.49}$$

We use the expansion

$$\frac{c_S^3}{S_M^3}\left(\sin^2\delta_S - \frac{1}{3}\right) = \frac{c_S^3}{d_S^3}\left(\frac{2}{3} - \cos^S\delta_S\right)$$
$$= -0.254 - 0.0792\cos 2h - 0.0128\cos(h - p) + \cdots \tag{3.50}$$

The dominant term here is a semi-annual wave with argument $2s$ and period 182.62 d and an annual one with argument $h - p_S$ corresponding to the anomalistic year with period 365.26 d.

3.5.4 Catalogue for the Lunisolar Potential

Of course it is possible to expand the lunisolar potential into an infinite series of sinusoidal terms. The number of terms taken into account expresses the level of truncation. George Darwin (1883) kept 91 terms, Doodson [9] kept 378 terms, and Hartmann and Wenzel [11] kept 12 935 waves in their catalogue, called HW95, including 1 483 waves due to the direct planetary effects. These last authors did their calculations with DE200 numerical ephemerids of the planets and the Moon, between the years 1850 and 2150.

In Table 3.3 we present the principal tidal waves. We have separated the lunar waves form the solar ones. The coefficients are those coming from Doodson's expansion, very close to those calculated by Darwin. In practice, only their relative magnitudes are considered.

This table requires some comments:

- The coefficients of Sa and S1 are very weak: these components should not be included because there exist other more important components which are not mentioned. They are introduced to take into account the annual and diurnal height variations of tidal observations, of meteorological origin.
- As already mentioned, the components K1 and K2, sometimes called 'sidereal components' since their periods equal respectively the sidereal day and the half-sidereal day, are present in both the solar potential and the lunar potential. For all studies concerning these components, the coefficients to consider are the sum of the coefficients coming from both sources.
- The constant terms obviously do not intervene in the tide. The long period components are usually very weak. They are often masked by noise of meteorological origin and are not easily detected in tidal observations. Only components Sa and Ssa, reflecting seasonal variations in the sea level can generally be identified.

Table 3.3   Main components of the lunisolar tides

Symbol	Name	Angular speed/hour	Periodicity	Coefficient $\times 10^5$
MOON				
M0	constant term	0.00000000		50458
Mm	monthly	0.54437468	27.55455017 days	8253
Msf	variational	1.01589576	14.76529408 days	1367
Mf	bimonthly	1.09803304	13.66079044 days	15640
2Q1	elliptic 2d order	12.85428623	28.00622177 hours	952
Q1	main elliptic	13.39866092	26.86835670 hours	7206
$\rho 1$	evectional	13.47151452	26.72305298 hours	1368
01	lunar principal	13.94303560	25.81934166 hours	37689
M1	elliptic minor	14.49669396	24.83324814 hours	2961
K1	declinational	15.04106864	23.93446922 hours	36232
J1	elliptic 2d order	15.58544332	23.09847641 hours	2959
OO1	lunar 2d order	16.13910168	22.30607414 hours	1615
2N2	elliptic 2d order	27.89535487	12.90537453 hours	2300
$\mu 2$	variational	27.96820848	12.87175751 hours	2777
N2	main elliptic	28.43972956	12.65834808 hours	17391
NU2	evectional maj.	28.51258316	12.62600422 hours	3302
M2	lunar mean	28.98410424	12.42060089 hours	90812
$\lambda 2$	evectional min.	29.45562532	12.22177410 hours	669
L2	elliptic min.	29.52847892	12.19161987 hours	2567
K2	declinational	30.08213728	11.96723461 hours	7852
M3		43.47615636	8.28040123 hours	1188
SUN				
	constant term	0.000000000		23411
Sa	annual	0.041068640	365.24218966 days	1176
Ssa	semi-annual	0.082137280	182.62109375 days	7245
P1	solar principal	14.95893136	24.06588936 hours	16817
S1	radiational	15.00000000	24.00000000 hours	−423
K1	declinational	15.04106864	23.93446922 hours	16124
T2	elliptic major	29.95893332	12.01644897 hours	2472
S2	solar mean	30.00000000	12.00000000 hours	42286
R2	elliptic minor	30.04106668	11.98359585 hours	437
K2	declinational	30.08213728	11.96723461 hours	3643

- The diurnal main components are K1, O1, P1, Q1, and the main semi-diurnal ones are M2, S2, K2, N2. They contain the main part of the tidal signal energy and arc the only waves generally taken into account in the first approximation for quick studies.

3.6 Tidal Behavior and Predictions Around the World

In terms of tidal prediction, through Doodson's work in particular, the harmonic method has provided a practical, precise and potentially universal tool. It is not fundamentally different from Laplace's method for it too relies on a theoretical formulation including a number of fixed parameters which must be determined experimentally by analyzing the available observations. For a good accuracy, these observations must extend over a sufficiently long time. Generally, a year of hourly measurements is necessary to achieve the accuracy required for the purposes of navigation. Moreover the results are useful only for the site where observations have been made.

A more ambitious approach, based on the hydrodynamics of ocean basins, had been proposed since a long time ago, by such pioneers as Bernoulli, Whewell, Poincaré, and Harris. However given the complexity of bathymetry and coastlines, it was not possible to obtain an accurate solution to the problem of tide modeling until powerful computing resources came into existence. Analytical solutions are nevertheless capable of explaining qualitatively the main features of tide propagation, for example the existence of amphidromic points. However it was the development of numerical methods, becoming possible with the ever improving computers, that really allowed progress in this direction. In particular, the work of the German specialist Hansen (1949) has been the source of new attempts to solve the Laplace equation for the real ocean [10].

It should be noted that altimetry from satellite tracking and geodesy have created new needs for an elaborate knowledge of tides and have led to a renewed interest in world ocean modeling. In particular, satellite altimetry, which measures the sea-level with a quasi-centimeter accuracy, has enabled the development of much more realistic tidal models by assimilating always more abundant data.

3.6.1 Global Characteristics

Laplace described the tides as 'the most difficult problem of all celestial mechanics'. The complexity of this phenomenon lies primarily in its description. The more we want to refine it, the more we realize that some empirical rules can be established from partial observations, which can only be coarsely generalized. It is very difficult indeed to detect a temporal 'rhythm' in tidal phenomena. It is even theoretically impossible because, in contrast to common belief, the tides are not periodic: there is no period after which the height variations repeat exactly the same way. Indeed, there are periods after which the same conditions are almost fulfilled, the best known being the Saros, equal to 223 lunar months, or 6585.32 days. After this time interval, the Moon, the Sun are nearly in the same relative positions and their orbital elements are also nearly the same. It follows that the tidal generating force takes nearly the same value. This does not mean that the Saros is a period of tides: after several Saros, the resemblance with the initial tide diminishes further and further.

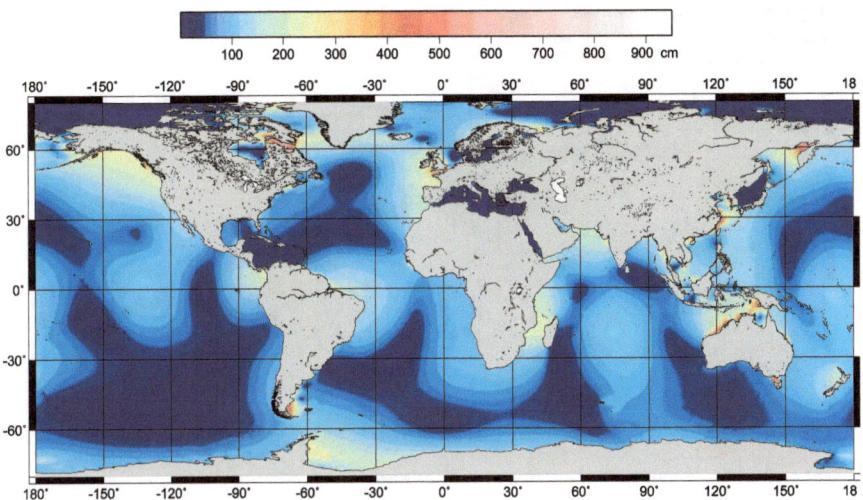

Fig. 3.4 Tidal amplitudes of oceans worldwide

3.6.2 Tide Amplitudes in the Oceans

Besides the difficulties of temporal description of the tides, a spatial description presents another set of difficulties. In terms of height first, the geographical distribution of amplitudes in the oceans (Fig. 3.4) seems to follow no obvious a priori pattern. However, we can note that the highest amplitudes are mainly located on continental shelves around the continents, or in shallow seas such as the English Channel. These amplitudes are very weak in semi-enclosed seas of small size (Sea of Japan, Caribbean, Baltic, Mediterranean). Apart from these qualitative observations implying the effect of depth and size of oceanic basins, no general rule can be established.

3.6.3 Tide Characterization

As we have shown in the previous section, the tides are mainly due to the superposition of a diurnal component (daily maximum and minimum height) and a semi-diurnal component (two maxima and two minima per day). Nevertheless, the relative importance of these two components varies geographically, defining types, according to a more or less conventional classification:

- a semi-diurnal type characterized by a negligible component of the diurnal tide,
- a semi-diurnal type with diurnal inequality: the semi-diurnal component is dominant but is modified by the diurnal one,
- a mixed type: the diurnal component dominates, but is modified by the semi-diurnal one,
- a diurnal type: the semi-diurnal component is negligible.

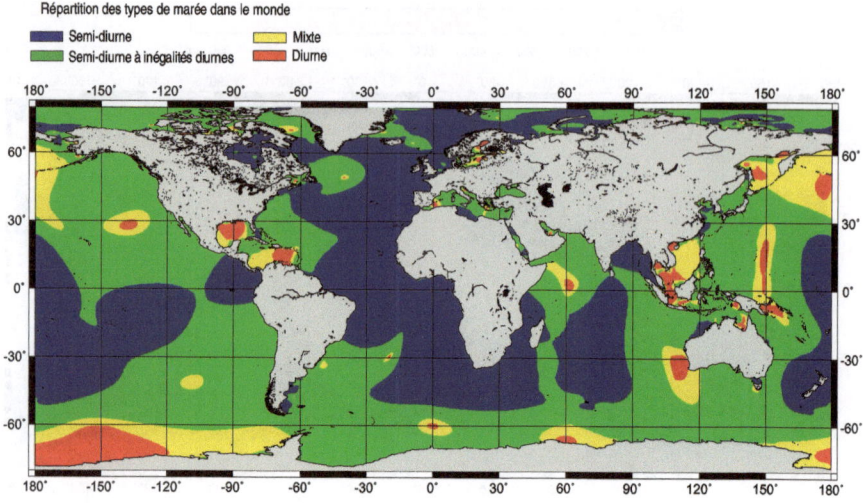

Fig. 3.5 Types of the dominating tides in the oceans all around the world

The distribution of these 4 types of tide in the ocean worldwide (Fig. 3.5) shows that no general rule can be established, apart the observation that the semi-diurnal type is dominant in the Atlantic, the other types appearing only when the semi-diurnal amplitude is low.

3.6.4 Amphidromic Points

Another feature of the tide is its mode of spreading. The crests of each wave component propagate around points called *amphidromic points*.[1] These points occur because of the combined action of the Coriolis force and the interference with oceanic basins, seas, and bays. Each tidal component is at the origin of a different amphidromic system. Amphidromic points for a tidal constituent (diurnal, semi-diurnal, etc.) is characterized by the property that there is almost no vertical motion of the oceanic mass from tidal action. Nevertheless tidal currents can appear when water levels on two sides of the amphidromic point are not the same. This leads to a well-defined wave pattern called an *amphidromic system*.

In the example of a semi-diurnal pattern spread in the Atlantic Ocean shown in Fig. 3.6, each line, called *co-tidal line*, indicates the position of the crest of the wave at a given hour, referred to the transit of the Moon at the Greenwich meridian. We can note for example that the wave progresses from south to north along the coasts of Europe, but from north to south along the North American coast. The rotation

---

[1] 'Amphidromic' derives from the Greek words *amphi* (around) and *dromos* (running).

Fig. 3.6 Co-tidal lines in Atlantic Ocean

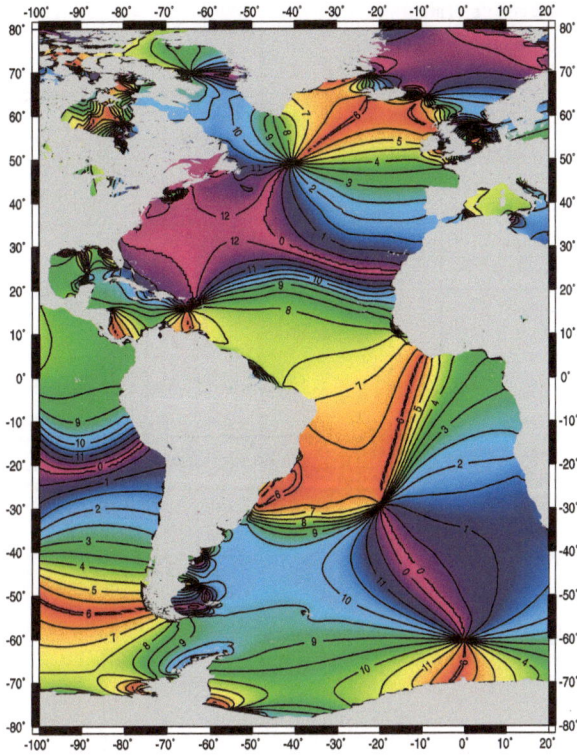

around an amphidromic point does not seem to follow any general rule: for example, the two major networks of the South Atlantic rotate in opposite directions. These co-tidal lines, representing the average semi-diurnal tide, does not exactly match the actual tide. Amphidromic points are not absolutely fixed, and it would be wiser to speak of amphidromic area. In addition, the diurnal component propagates very differently: the corresponding number of amphidromic points is approximately half in the case of the semi-diurnal component. All these tidal characteristics, with gradually more precise and abundant data, have long been subject to questions, hypotheses (often fallacious), theoretical developments, and scientific studies conducted with the help of technologies becoming more sophisticated and more adequate in particular thanks to the innovation of artificial satellites and powerful computers.

3.6.5 Tidal Curves

The graph versus time of sea level measurements or predictions at a given surface point of the ocean is called a *tidal curve*. As an example, we show in Fig. 3.7 the tidal curve obtained from observations at Brest of the semi-diurnal tide for one day time span. Each minimum of the curve is called low tide and each maximum high-tide. From the low tide to the high tide, the sea level rises during the flow phase,

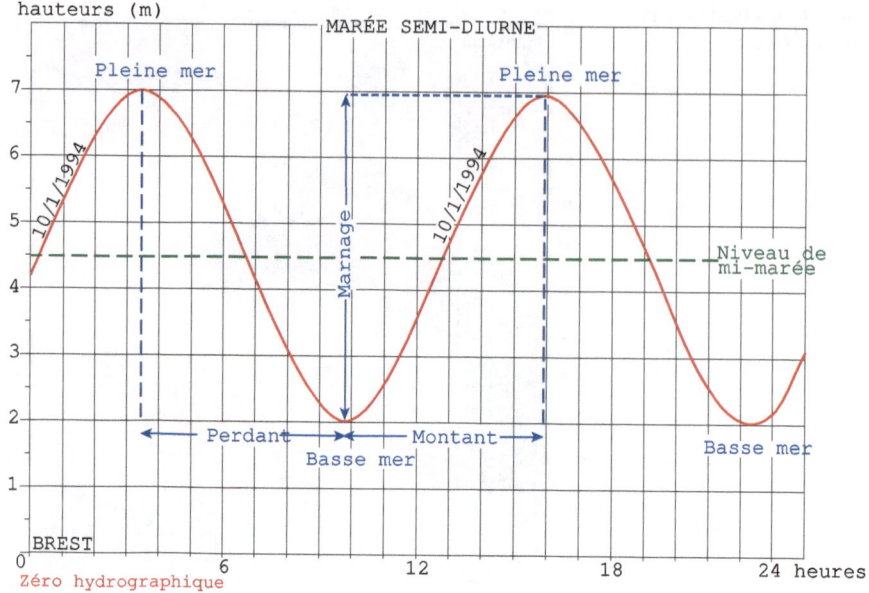

Fig. 3.7 Tidal curve observations at Brest

and decreases from high to low during the ebb phase. The difference between the high tide level and the low tide one is called the *tidal range*, not to be confused with the amplitude, which is the norm of a sinusoidal function. Nevertheless the word 'amplitude' is sometimes used for the tide, for which it means 'half the tidal range'. The heights are referred to a reference level which often comes from a nautical chart.

Figure 3.8 shows another example of the semi-diurnal tide curve deduced from a prediction for roughly thirty days. At the times of new and full Moon, the lunar- and solar-induced ocean bulges line up (and add up) to produce tides having the highest monthly tidal range (i.e. the highest high tide and the lowest low tide): they are called the *spring tides*. In the opposite case, at the first and third quarter phases of the Moon, the Sun's pull on the Earth is at right angles to the Moon's pull. At this time tides have their minimum monthly tidal range (i.e. unusually low high tide and unusually high low tide). These are called the *neap tides* or *fortnightly tides*.

Changes in tidal range are generally recorded from a minimum (neap tide) to a maximum (spring tide). The alternative phases of increasing and decreasing tidal ranges are called respectively *revival* and *waste*. The time interval between one phase as a full Moon or a new Moon, and the tidal extremum which follows immediately, is called the *age of the tide*.

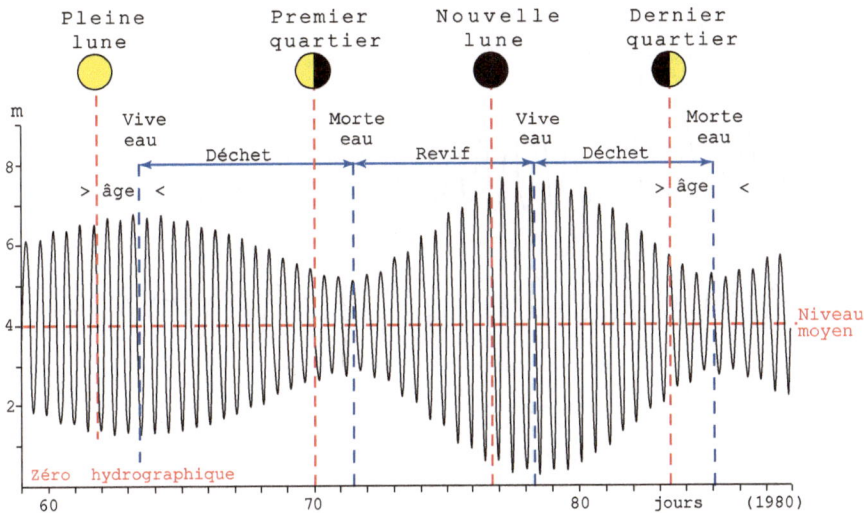

Fig. 3.8 Semi-diurnal tides with range variations during one month at Brest

3.6.6 Tidal Curves According to Tidal Types

As mentioned before, the distinction between types of tides is somewhat conventional. A classification into three types is often suggested, but the classification in four types corresponding already defined in Sect. 3.6.3 (see Fig. 3.5) is proposed hereafter, where we show in Fig. 3.9 four different tidal behaviors on the Earth.

3.6.6.1 Semi-diurnal Tide (Casablanca, Morocco)

This type of tide has been presented before (Fig. 3.8). It exhibits every day two high tides and two low tides of nearly the same level, with nearly equal tidal ranges throughout the daytime. This type of tide dominates in the Atlantic, especially in Europe and Africa. However, as has been noted above, other types of tide are likely to be encountered.

3.6.6.2 Semi-diurnal Tide with Diurnal Inequality (Vung Tau, Vietnam)

During a lunar day, two relatively small tidal ranges are followed by two larger tidal ranges, or vice versa. The difference between large and small tidal ranges, called the *diurnal inequality*, is maximized when the declinations of the Moon and the Sun are themselves close to their maximum. The diurnal inequality is also observed on European coasts, although the tide is characterized as semi-diurnal, for the diurnal inequality is small. However, it may be very important in many ports in the Pacific and Indian Oceans.

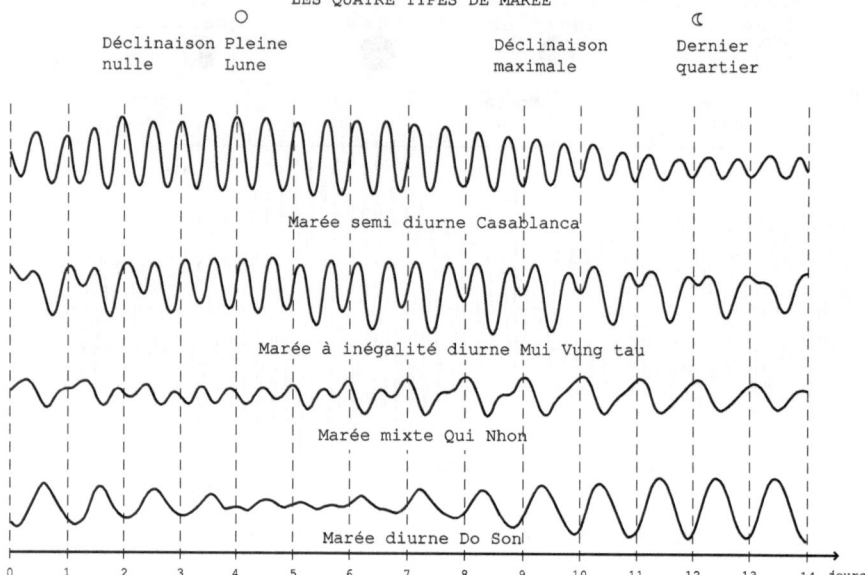

Fig. 3.9 The four types of tides: examples of Casablanca (semi-diurnal), Vung Tau (semi-diurnal with diurnal inequality), Qui Nhon (mixed) and Do Son (diurnal)

3.6.6.3 Mixed Tide (Qui Nhon, Vietnam)

Mixed tide is characterized by the succession of a semi-diurnal type and a diurnal type during a lunar month. This type of tide is common in Indonesia, Indochina, on the coasts of Siberia, and Alaska. It is also found in the Atlantic Ocean and the Caribbean Sea.

3.6.6.4 Diurnal Tide (Do Son, Vietnam)

Diurnal tide presents only one high tide and one low tide per lunar day, with a tidal range varying with the declinations of the Moon and the Sun. This type of tide, rather uncommon, is observed mainly in the Pacific Ocean: in Siberia (with very large ranges), in Alaska and also in Southeastern Asia.

3.6.7 Tides in Shallow Water

When propagating through shallow water, almost all primitively sinusoidal deep-water offshore tides are deformed. The periodic components, issued from the generating force, combine themselves through nonlinear processes, creating harmonics

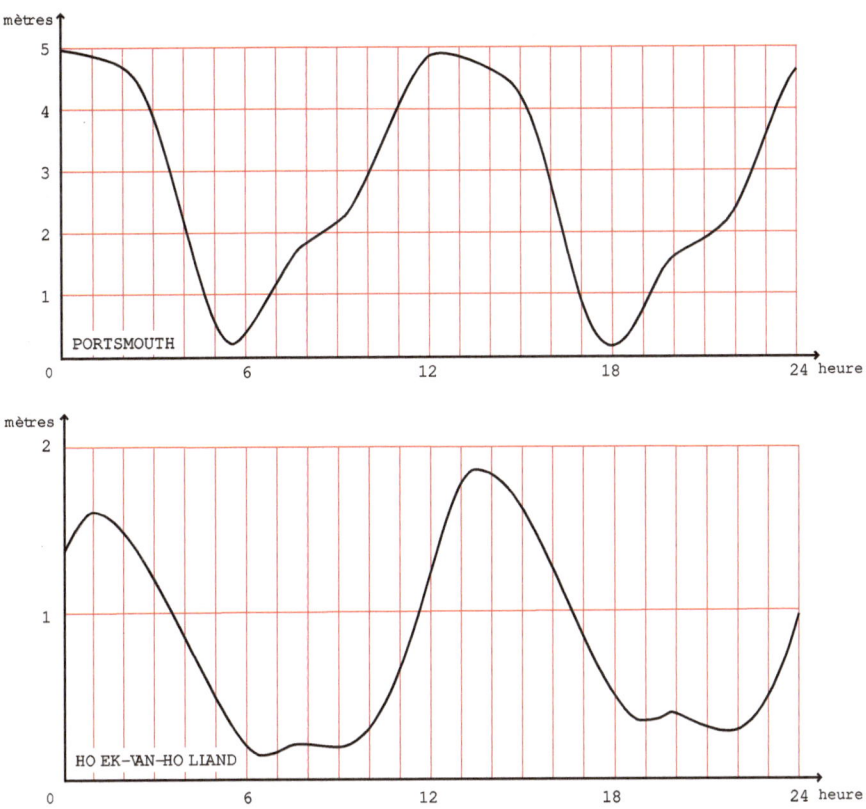

Fig. 3.10 Shallow water tidal waves in the Channel at Portsmouth (*above*) and in the North Sea at Hoek Van Holland (*below*)

which can propagate independently. The tidal curves observed in the English Channel and the North Sea coasts (Fig. 3.10) are typical examples of tidal curves after a long progression of the tidal wave over a shallow shelf.

The propagation in estuaries exhibits other examples of distortion of the tidal wave in shallow water. This kind of deformation is related to the laws of hydrodynamics, which states that the speed of a hydraulic wave is proportional to the square root of the depth. In deep water, the difference of magnitude does not alter the speed of propagation. On the contrary in shallow water, the peak of the wave moves faster than the trough, so that a wave crest tends to overtake the preceding wave. The example of the Gironde estuary in Fig. 3.11 is telling of such a behavior.

In extreme cases it forms a bore, a water bar moving upstream along a river. This phenomenon is present in many estuaries of major rivers. The height of the bar can reach several meters, especially in the estuary of the Amazon, the Hoogly and Indus rivers in India and the Tsien Tang in China.

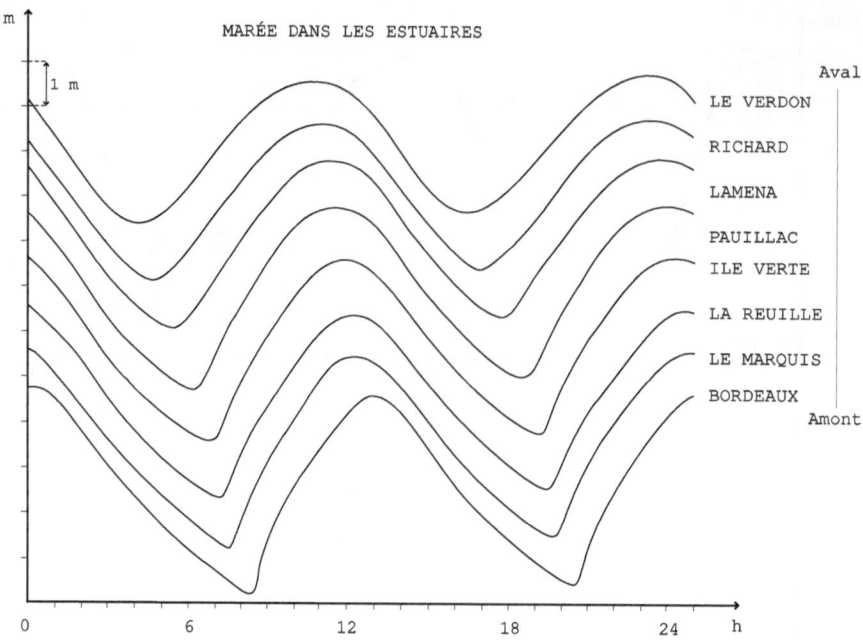

Fig. 3.11 Tidal curves in the Gironde estuary

3.6.8 Spectral Characteristics of Tides

The tidal spectrum, despite being the result of a calculation, is really an objective mode of representation of tidal phenomena, independent of any theory. It is particularly suited to tidal studies. It is not necessary to give an exact definition of the spectrum. It only matters that it represents the amplitude, or energy, as a function of frequency.

In Fig. 3.12 we present the spectra at two points of the Loire estuary. They are characterized by a low resolution, which means that there is an imperfect separation of adjacent frequencies. The comparison between the two spectra shows the evolution of the structure from the mouth of the river to upstream. These examples show that the main characteristic of the tidal spectrum is split into separate, regularly spaced clusters. The main group is the semi-diurnal one (two cycles per lunar day).

It is worth noting that the energy increases as the frequencies decreases. A significant noise originates from atmospheric influences. Figure 3.12 shows also the increase of the number of harmonics when the tidal wave progresses from the mouth of the Loire to Nantes, located one hundred miles upstream. The upstream spectrum shows presence of energy in high frequencies. Indeed, only the first 3 groups (diurnal, semi-diurnal, and third-diurnal) represent the bulk of the astronomical tide issued directly from the actions of the Moon and the Sun. Other groups appear during the progress of the tidal waves in shallow waters.

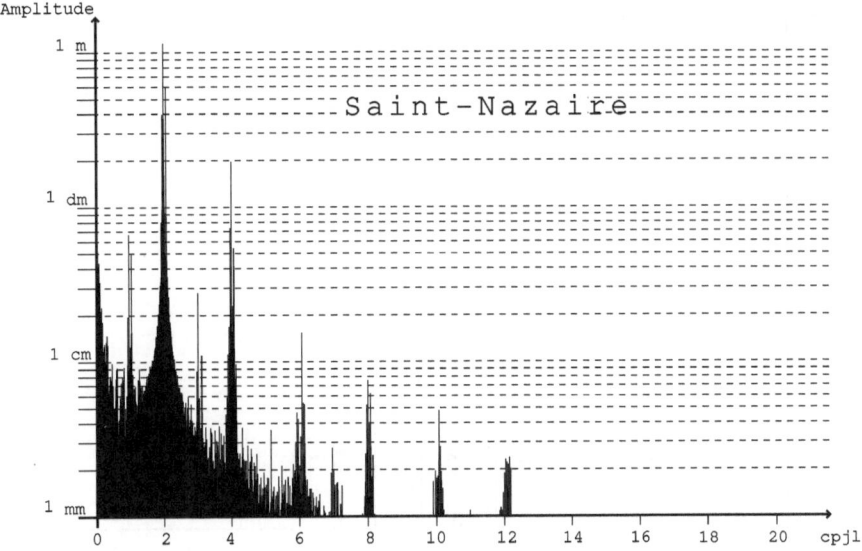

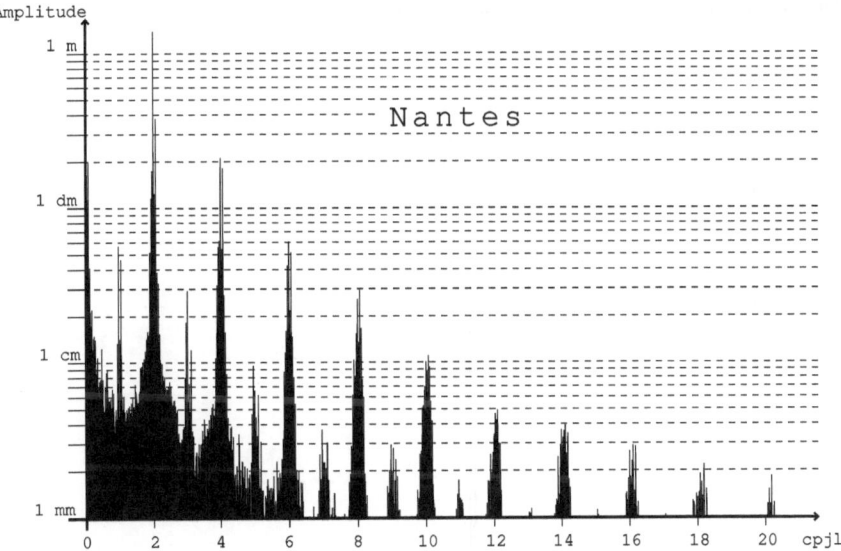

Fig. 3.12 Semi-diurnal tidal spectrum at Saint-Nazaire at the mouth of the Loire estuary and at Nantes, about 100 km upstream

After analyzing more than 120 years of almost continual observations, the spectral signature at high resolution of the semi-diurnal group in Brest (Fig. 3.13) exhibits thin, well separated components, justifying (retrospectively) the representation of tide as harmonic series. An even better illustration is given in Fig. 3.14,

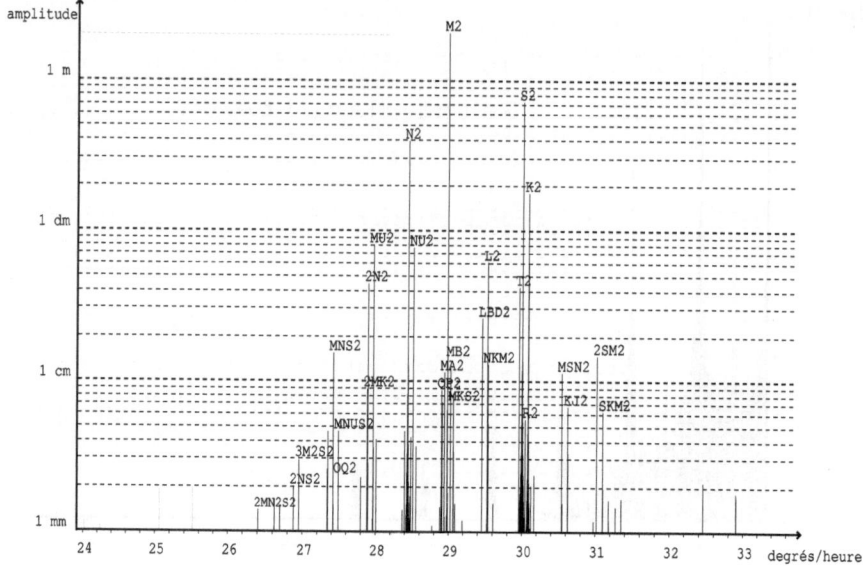

Fig. 3.13 High resolution semi-diurnal spectrum at Brest

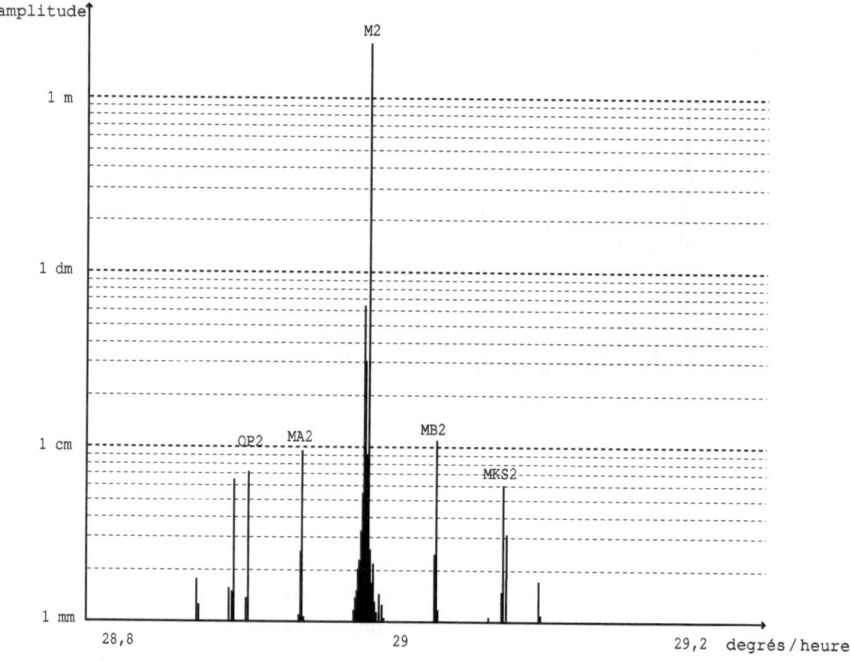

Fig. 3.14 High resolution M2 group

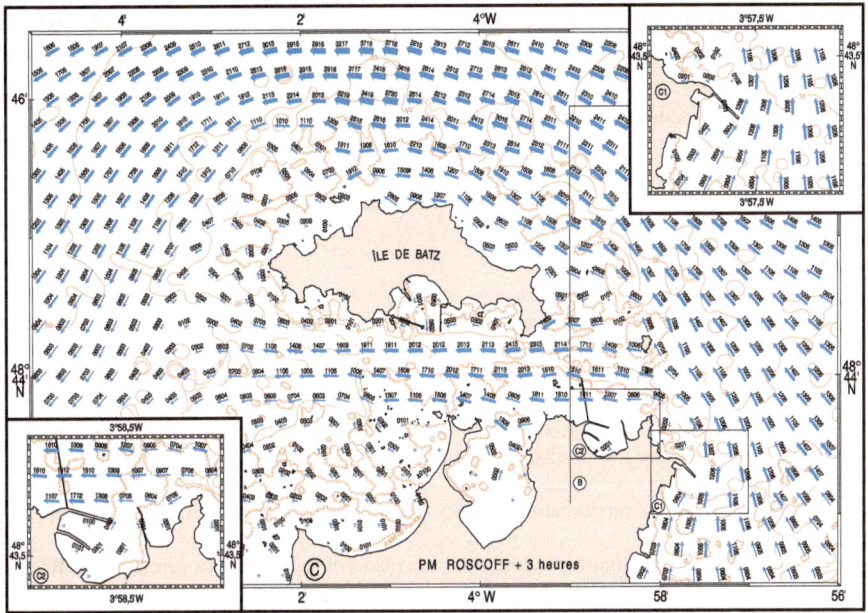

Fig. 3.15 Model of tidal currents around the island of Batz (France)

which presents the expansion of the spectrum close to the M2 component. The main components, clearly identified, are named after Darwin (or Doodson).

3.6.9 Tidal Currents

We can consider that any tide is an oscillation similar to a swell. In both cases, water molecules approximately describe closed trajectories in a vertical plane. However, unlike swell, the tide wavelength is always greater than the depth. In a homogeneous and deep ocean, tidal motions affect the whole depth of water. All molecules of a given vertical plane describe extremely flattened orbits. The vertical motion is the tide, whereas the horizontal motions, incomparably more prominent, constitute the tidal currents.

In a density stratified ocean, internal tidal waves are created, especially near continental slopes. They change the vertical structure of currents. In extreme cases, as in the Strait of Gibraltar, for instance, the currents caused by these internal waves, mainly semi-diurnal, may be opposite in direction between the surface and the bottom. Moreover the energy dissipation of tides is mainly due to current friction at the bottom. The study of currents may be conducted with the same tools as the study of tides, but it is more difficult at least for two reasons: first, because of the large spatial variability of their characteristics from one point to another and, second, because of the much more important influence of atmospheric factors. Strong tidal currents in some areas justify their study in order to provide valuable help to sailors, which constitutes an important activity of hydrographic offices. Figure 3.15 shows results of the modeling of tidal currents around the island of Batz (France). It comes from a navigation aid document, particularly useful in some areas where tidal currents are sometimes violent.

References

1. Airy, G.B.: Tides and waves. In: Rose, H.J. et al. (eds.) Encyclopaedia Metropolitana. Mixed Sciences, vol. 3, pp. 1817–1845 (1841)
2. Airy, G.B.: Tides and waves. In: Encyclopedia Metropolitana, vol. 5, pp. 241–396. London (1845)
3. Brown, E.W.: Tables of the Morion of the Moon. Yale University Press, New Haven (1919)
4. Chazallon, M.: Annuaire des marées des cotes de France. Dépot des Cartes et Plans, Paris (1839)
5. Darwin, G.H.: Report on the harmonic analysis of tidal observations. Brit. Assoc. Rep, pp. 49–117 (1883)
6. Darwin, G.W.: On the figure of equilibrium of a planet of heterogeneous density. Proc. R. Soc. Lond. 1(36), 158–166 (1883)
7. Darwin, G.: On tidal prediction. Proc. R. Soc. Lond. 49, 130–133 (1890)
8. Darwin, G.H.: The harmonic analysis of tidal observations. In: Scientific Papers, vol. 1, pp. 1–70. Cambridge University Press, Cambridge (1907)
9. Doodson, A.T.: Harmonic development of the tide-generating potential. Proc. R. Soc. Lond. Ser. A 100, 305–329 (1921)
10. Hansen, W.: The Reproduction of the Motion in the Sea by Means of Hydrodynamical Numerical Methods. Pub. N°5, pp. 1–57. Mitteilung Inst. Meereskunde, Univ. Hamburg, Hamburg (1966)
11. Hartmann, T., Wenzel, H.G.: The HW95 tidal potential catalogue. Geophys. Res. Lett. 22(24), 3553–3556 (1995)
12. Laplace, P.S.: Mémoire sur le flux et le reflux de la mer. Mém. Acad. Sci., pp. 45–181 (1790)
13. Laplace, P.S.: Traité de mécanique céleste, vol. 2, livre 4 (1799); vol. 5, livre 13 (1825)
14. Le Provost, C., Genco, M.L., Lyard, F., Canceil, P.: Spectroscopy of the world ocean tides from a finite element hydrodynamic model. J. Geophys. Res. 99(C12), 24777–24797 (1994)
15. Lubbock, J.W.: On the tides. Philos. Trans. R. Soc. Lond. 127, 97–104 (1837)
16. Marchuk, G.I., Kagan, B.A.: Dynamics of Ocean Tides. Kluwer, Dordrecht (1989), 327 pp
17. Mazzega, P.: The M2 ocean tide recovered from seasat altimetry in the Indian ocean. Nature 302, 514–516 (1983)
18. Melchior, P.: The Tides of the Planet Earth, 2nd edn. Pergamon, Elmsford (1982), 641 pp
19. Munk, W.H., Cartwright, D.E.: Tidal spectroscopy and prediction. Philos. Trans. R. Soc. Lond. A 259, 533–581 (1966)
20. Munk, W.H., MacDonald, G.J.F.: The Rotation of the Earth—A Geophysical Discussion. Cambridge University Press, Cambridge (1960), 323 pp
21. Newton, S.I.: Principia. University of California Press, Berkeley (1687)
22. Schuremann, P.: Manual of harmonic analysis and prediction of tides (1971); U.S. Coast and Geodetic Survey (1971)
23. Thomson, W.: On gravitational oscillations of rotating water. Proc. R. Soc. Edinb. 10, 92–100 (1879)
24. Whewell, W.: Researches on the Tides, 14th series. On the Results of Continued Tide Observations at Several Places on the British Coasts, pp. 227–233

Chapter 4
Precession and Nutation of the Earth

Jean Souchay and Nicole Capitaine

Abstract Precession and nutation of the Earth originate in the tidal forces exerted by the Moon, the Sun, and the planets on the equatorial bulge of the Earth. Discovered respectively in the 2nd century B.C. by Hipparcus and in the 18th century by Bradley, their existence and characteristics were deduced theoretically by Newton for the precession and by d'Alembert for the nutation. After a historical review we explain, both in an intuitive manner and by simple calculations, the gravitational origin and the main characteristics of the precession-nutation. Then we describe in detail two fundamental theories, one using the Lagrangian formalism, the other the Hamiltonian one. A large final part is devoted to successive improvements of the precession-nutation theory in the last decades, both when considering the Earth as a rigid body and when taking into account the small effects of non-rigidity.

4.1 Introduction

Among various astronomical phenomena that have their origin in the lunar and solar tides, the precession of the equinoxes exhibit a very small effect on the time-scale of a human life. Yet it was discovered as early as the 2nd century B.C. by Hipparcus, who was comparing the positions of the stars in his era to those recorded by his predecessor Timocharis, about 150 years earlier. In Chap. 2 of volume III of *Almagest*, Claudius Ptolemy reports the work of Hipparchus on the length of the year[1]: the most surprising fact for him was that when the return of the Sun at an equinox is measured, 1 year amounted to a little less than 365 + 1/4 days, whereas when this

---

[1] No writing by Hipparcus survives. According to O. Neugbauer in *A History of Ancient Mathematical Astronomy*, he lived between 190 and 120 B.C., whereas Ptolemy lived between 100 and 170. The observations attributed to Timocharis seem to date back to a period between 300 and 270 B.C.

J. Souchay (✉) · N. Capitaine
SYRTE, UMR8630 CNRS & UPMC, Observatoire de Paris, 61, av. de l'Observatoire, 75014 Paris, France
e-mail: Jean.Souchay@obspm.fr

N. Capitaine
e-mail: Nicole.Capitaine@obspm.fr

J. Souchay et al. (eds.), *Tides in Astronomy and Astrophysics*,
Lecture Notes in Physics 861, DOI 10.1007/978-3-642-32961-6_4,
© Springer-Verlag Berlin Heidelberg 2013

return is compared to the fixed stars, he found it a little longer than this value. From this observation, Hipparcus deduced that the celestial sphere itself (with the stars fixed on it) was undergoing a slow motion with respect to the equinoxes, and vice versa.

A physical explanation of the precession had to wait for more than eighteen centuries, until Newton (1642–1727). In *Philosophiae Naturalis Principia Mathematica* published in 1687, he understood for the first time that the Earth was flattened oblately and that the precession was caused by the gravitational torque exerted by an external body (the Moon or the Sun), owing to this flattened asymmetry with respect to the direction of the external body. Newton tackles the problem of the motion of the axis of rotation of a spheroid in Corallaries 18, 20, 21, 22 of Proposition 66 of *Principia*, as an application of the three-body problem. First, he studies the motion of a satellite revolving around a planet and perturbed by the Sun. Second, he replaces the satellite by a fluid ring, composed of infinitely many independent particles. Third, he replaces the fluid ring by a rigid one, fixed to a homogeneous sphere. The rigid ring, which represents the equatorial bulge, imparts to the sphere its own motion. Newton shows that the nodal line of the equatorial plane of the sphere (containing the ring) with respect to the orbital plane of the planet around the Sun undergoes a retrograde motion, due to the gravitational perturbation of the Sun. Thus precession was explained for the first time. Later on, Newton observed that, for the Earth, the precessional motion contains two components, due to the Moon and to the Sun, and that the ratio of the amplitudes of the two components is the same as the ratio of the forces they exert in the phenomena of tides. He evaluates this ratio to be 4.5.

More than half a century later, in 1747, James Bradley (1693–1762), after observing for a period of roughly 20 years the transit of zenithal stars at Kew and Wansted, remarked that the polar axis traces, in addition to the by then well-known precession, a small loop of 18.6-year period, with an amplitude close to $9''$, that is to say far beyond the capacity of detection with the naked eye.[2]

Just after this discovery, d'Alembert (1717–1783) realized that the 18.6-year period corresponds exactly to the period of retrogradation of the nodes of the lunar orbit with respect to the ecliptic. Then he elaborated in barely a year a complete theory of the Earth's rotation, improving Newton's calculations of precession by correcting a substantial number of errors and approximations, and by making full use of the new mathematical tools of calculus. He succeeded in proving the existence and the nature of the nutation loop, showing that it originates from exactly the same cause as the precession. His book, entitled *Recherches sur la Précession des Equinoxes et sur le Nutation de l'Axe de la Terre dans le Système Newtonien* and published in 1749, must be considered as the first treatise dealing with a complete theory of the precession-nutation of the Earth. It opens the path to a new era of

---

[2]The discovery of nutation, following that of aberration, by Bradley, is officially recorded in a memoir as a letter to Lord Macclesfield, his protector and friend, later President of the Royal Society 1752–1764. Bradley's memoir, dated 31 December 1747, was read at the Royal Society on 14 February 1748, and published later in the Philosophical Transactions.

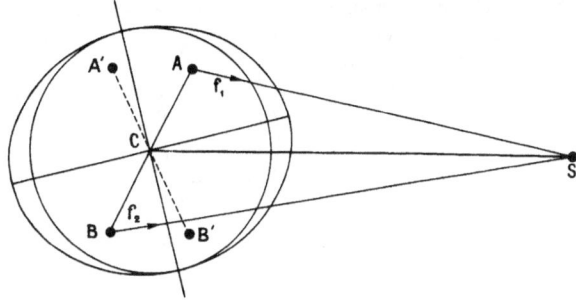

Fig. 4.1 Geometric proof of the cancellation of the torque exerted by an external body S over the sphere tangential to the ellipsoid (from [16])

increasingly refined theories of the Earth's rotation, which we will describe in the following.[3]

4.2 A Simple Geometric Explanation

In this section we show how the precession mechanism can be explained in a simple geometric way, following the exposition in Danjon [16], from Newton's calculations. We suppose here (Fig. 4.1) that the Earth is an ellipsoid of revolution, homogeneous and flattened at the poles. What is the effect of the solar attraction on the orientation of its axis of rotation in space? It is easy to show that the effect involves both a force, responsible for the orbital motion of the Earth around the Sun, and also a torque, which tends to make the equator coincide with the ecliptic.

4.2.1 Precession-Nutation due to the Sun

First we can show that the attraction of the Sun on the mass included in the sphere centered in C, and internally tangential to the ellipsoid is reduced to a force. For this let us figure out that we cut the sphere by a plan containing the center S of the Sun and the axis of the Earth. A and B represent two ranges of material, identical, normal to the plan of figure, and symmetrical with respect to the center C (Fig. 4.1). Of course the two gravitational forces $\overrightarrow{f_1}$ and $\overrightarrow{f_2}$ exerted by the Sun on these ranges are not equivalent, neither in amplitude, nor in direction. Their effects can be replaced by a force positioned at the center C and a torque perpendicular to the plan of figure. Now we can consider two other ranges of material A' and B', symmetrical respectively to the ranges A and B with respect to the plan crossing C and perpendicular to the direction CS. If we reduce in the same manner as above the gravitational attraction exerted by the Sun on A' and B', we observe that the resulting force is equal to that of the ranges AB, whereas the resulting torque is exactly the opposite. As the

---

[3]Notice that Newton in his *Principia* predicted the semi-annual and semi-monthly nutations which their small amplitude rendered undetectable at his time.

Fig. 4.2 Geometric proof of
the existence of a torque
exerted by an external body S
on an ellipsoid (from [16])

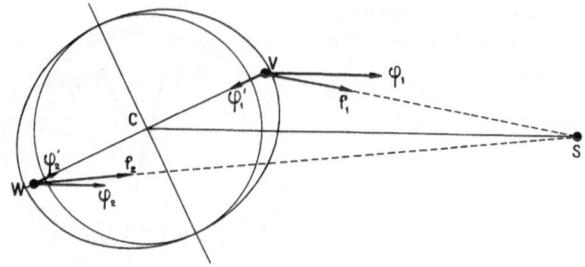

total mass of the sphere can be completely decomposed into such groups as A, B, A′, B′, we conclude that the resulting torque on the sphere is zero.

In contrast, this is not the case of the torque exerted on the part of the ellipsoid exterior to the sphere, generally called the *equatorial bulge*. Here the symmetry which ensured the annihilation of the torques does not generally exist. A simple geometric argument proves that fact. Let us consider (Fig. 4.2) two point like elements of material V and W located at the surface of the Earth on the equator itself, symmetrical with respect to C. Then it is possible to decompose the gravitational force $\mathbf{f}_1$ exerted by the Sun on V into two other ones: $\boldsymbol{\varphi}_1$, parallel to CS and $\boldsymbol{\varphi}'_1$ towards the direction of the center CV. In the same way it is possible to decompose the force $\mathbf{f}_2$ into two sub-components $\boldsymbol{\varphi}_2$ and $\boldsymbol{\varphi}'_2$. The forces $\mathbf{f}_1$ and $\mathbf{f}_2$ are obviously inverse to the square of the distances CV and CW to the Sun. As a consequence it is also easy to prove that the forces $\boldsymbol{\varphi}_1$ and $\boldsymbol{\varphi}_2$, parallel one to each other, are proportional to the inverse of the cube of these distances. Therefore the torque due to the action of the Sun on V has an amplitude larger than the torque due to this same action on W. As a conclusion the resulting torque is represented by a vector perpendicular to the plane of figure and oriented backward. It tends to make a rotation of the segment VW clockwise, in such a way that the obliquity, i.e. the angle between the ecliptic and the equator planes, decreases.

We can remark that for two particular cases, the resulting torque exerted in the elements of material V and W is equal to zero: when the line CS is along the equatorial plane, or when it is perpendicular to that plane. The first case occurs during the equinoxes, and the second one never occurs. Indeed these two cases correspond respectively to a declination of the Sun $\delta_{Sun} = 0°$ and $\delta_{Sun} = 90°$, whereas the declination of the Sun varies in the range $\pm 23°27'$. Following the fact that the resulting torque vanishes for the two values of the declination above, it follows that the torque reaches its maximum for a value included between these two extrema. In summary we can conclude that the Sun exerts on the Earth considered as an homogeneous ellipsoid a torque varying with its declination. This torque is zero during the equinoxes, maximum during the solstices. The moment of this torque is located along the equator, and tends to decrease the obliquity.

It is easy to construct a vector which verifies those properties: let us consider (Fig. 4.3) a constant vector $\mathbf{M}_1$ in the equatorial plane directed towards the equinox γ, and a second one $\mathbf{M}_2$ with the same amplitude and symmetric of $\mathbf{M}_1$ with respect to the line CU, itself perpendicular to the solar hour circle (the great circle

Fig. 4.3 Decomposition of the external torque M into two components: a fixed $\mathbf{M}_1$ and a moving $\mathbf{M}_2$ with the same amplitude and symmetric with respect to the CU line (from [16])

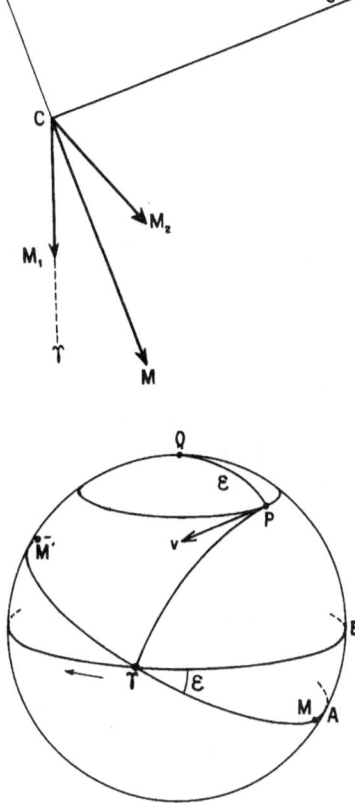

Fig. 4.4 Motion of the celestial pole P under the effect of the solar torque (from [16])

containing the poles and passing through the line CS). Then the resultant vector $\mathbf{M} = \mathbf{M}_1 + \mathbf{M}_2$ satisfies the properties above: its amplitude is zero during the equinoxes, maximum during the solstices and it corresponds to the moment of a torque leading to a decrease of the obliquity. As a consequence it is interesting to study the action of the solar torque by analyzing independently the effects of $\mathbf{M}_1$ and $\mathbf{M}_2$.

4.2.1.1   Effect of M_1: Precessional Motion

The first moment $\mathbf{M}_1$ tends to impart a rotation around $C\gamma$. In the same time, the Earth undergoes a rotation around its instantaneous axis of rotation (which at first approximation can be considered as coinciding with the axis of figure). This angular velocity is represented by a vector $\mathbf{v}$ oriented towards the pole P (Fig. 4.4). Thus the perturbation caused by the solar attraction due to the first component $\mathbf{M}_1$ tends to move P closer to the equinox γ. In other words; the instantaneous pole of rotation is moving in such a way that its velocity $\mathbf{v}$ remains tangent to the hour circle of the equinox γ. At the same time, this hour circle remains tangent at P to a small circle centered in Q, the ecliptic pole, in such a way that the angle $\widehat{CQP} = \varepsilon$, where ε is the obliquity. In summary $\mathbf{v}$ is oriented towards the tangent common to the two

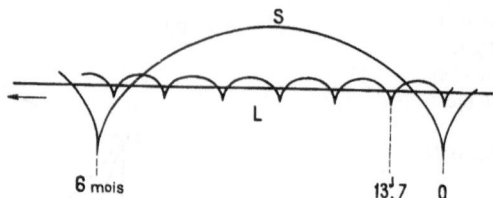

Fig. 4.5 Loops traced by the celestial pole under the effect of the precession added to the semi-annual and fortnightly nutation components. The leading component with period of 18.6 years is not included in this simplified model (from [16])

circles. As a result P describes the small circle above, the obliquity ε remaining constant. The motion of both P and γ is uniform and retrograde. In the same time, the celestial equator undergoes a rotation around the diameter MM' perpendicular to the hour circle Pγ, and the point γ retrogrades along the ecliptic with an angular velocity of 15''.8/year. In parallel P undergoes its circular motion with an angular velocity of $15''.5 \times \sin\varepsilon = 6''.3$/year.

4.2.1.2 Effect of M_2: Semi-annual Nutational Motion

The instantaneous effect of the torque with moment M_2 is the following one: if the torque acts alone, the velocity of P on the celestial sphere should be included in the hour circle of the fictitious point towards the moment M_2. This point moves on the equator and it accomplishes two complete tropical revolutions during one tropical year. As can be verified when referring to Fig. 4.3, it is opposed to γ during the equinoxes and coincides with it during the solstices. As a result, the velocity vector of P undergoing the effect of M_2 moves around P in the plane tangent to the celestial sphere in P, accomplishing two revolutions per year. In conclusion:

– the displacement of the pole P under the effect of M_2 is characterized by a periodic orbit, very close to a circle, with radius $0''.55$, with a 6-month period;
– this nutation of P is naturally accompanied both by a periodic displacement of γ along the ecliptic, with amplitude $1''.3$, and a periodic variation of the obliquity, with amplitude $0''.55$. This obliquity is maximum during the equinoxes and minimum during the solstices.

4.2.1.3 Combined Precession-Nutation Motion

Now that we have characterized individually the solar precession caused by M_1 and the solar nutation caused by M_2 we can combine the two effects: they result in a cycloidal motion of the pole P in the celestial sphere, represented in Fig. 4.5. The turning back points correspond to the equinoxes, when the amplitude of the two torques M_1 and M_2 added together is zero.

4.2.2 Precession-Nutation due to the Moon

All that has been demonstrated above for the Sun can be repeated by analogy for the Moon. We know that the perturbing forces involved are proportional both to the mass of the perturbing body and to the inverse of the cube of its distance. Knowing the ratios between the masses and the distances of the Sun and the Moon, we can conclude that the amplitude of the lunar torque is roughly 2.2 times that of the Sun. As a consequence, the lunar precession in longitude, i.e. the linear retrogradation of the point γ along the ecliptic due to the sole action of the Moon, reaches roughly 2.2 times that due to the Sun, that is to say $2.2 \times 15''.8 = 34''.6$ per year.

Concerning the nutation, by analogy with the semi-annual nutation coming from the moment $\mathbf{M_2}$ due to the Sun, the corresponding moment due to the Moon gives birth to a nutation with period half that of the tropical revolution of the Moon (27.5 days). The amplitude of this semi-monthly (also called *fortnightly*) nutation is much smaller than the semi-annual one due to the Sun. The point γ oscillates around its mean (precession) motion with a $0''.2$ amplitude (instead of $1''.3$) and in the same time its obliquity can be increased or decreased by $0''.09$ (instead of $0''.55$). The corresponding displacement of the pole P in the celestial sphere is a small circle with amplitude $0''.09$.

Therefore, taking into account the effect of the Moon alone on the combined precession and nutation leads, as it is the case for the Sun, to a cycloidal motion of the pole (Fig. 4.5). In that case, this motion is characterized by 27 arches per year (27 is the number of lunar half-period cycles during one year). In fact this basic representation of the lunar nutation has been established with the implicit and simplified idea that the Moon is moving along the ecliptic, which is not the case, for its orbit presents an inclination of roughly $5°$ with respect to this last plane. Moreover the Moon's orbit is not fixed: its ascending node with respect to the ecliptic is precessing in the retrograde direction, with a 18.6 years period. The effect of this retrogradation is to create another nutation component, much larger than the semi-monthly one explained above. This component is often called the *principal nutation*. It is characterized by an elliptical loop (close to a circle) with $9''$ amplitude described by the pole of rotation with respect to the celestial sphere, in this same 18.6 years period. As a direct consequence, the obliquity is varying with this same amplitude of $9''$ and the γ point is undergoing an oscillation with an amplitude of roughly $18''$ alternatively in the prograde and retrograde direction along the ecliptic.

4.2.3 Global Motion of the Pole of Rotation in Space

Following all the leading effects described above, the main components of the combined gravitational torque exerted by the Moon and the Sun on the equatorial bulge of the Earth are:

- a linear retrograde displacement of the γ point (the ascending node of the ecliptic on the celestial equator) which was traditionally called the *luni-solar precession in longitude* with amplitude $34''.6 + 15''.8 = 50''.4$ per year

Fig. 4.6 Parametrization for the study of the precession-nutation motion of the Earth. $\Re_0 = (O, x_0, y_0, z_0)$ is a fixed inertial coordinate system; $\Re' = (O, x', y', z')$ is fixed to the Earth. $\Re = (O, x, y, z)$ is an intermediate coordinate system (from [16])

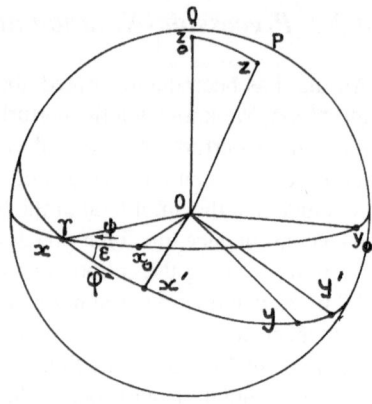

- an elliptical loop of lunar origin with amplitude roughly $9''$ described in a prograde sense in 18.6 years.
- a quasi-circular loop of solar origin with amplitude $0''.55$ described in a prograde sense in a half year period.

To these motions we must add the semi-monthly nutation loop already described above and a series of smaller nutational oscillations coming from secondary components in the perturbing function due in particular to the irregularities in the relative orbital motion of the Moon and the Sun around the Earth; the number of these components depends of course on the truncature limit for their amplitudes.

4.3 A Basic Mathematical Proof of the Precession-Nutation Phenomena

As we saw in the previous section, the lunisolar precession and nutation are due to the action, on the oblate Earth, of the torque exerted both by the Moon and the Sun on the Earth assimilated to an ellipsoid flattened at the equator. Since Euler's pioneering work, we know how to deal with the motion of a body around its center of gravity subject to a torque. Following the exposition in Danjon [16] we give the expression for the lunisolar torque acting on the Earth as a function of the celestial coordinates of the Sun and the Moon. Then we present the classic Euler equations for the rotational motion of our planet. The integration of the equations will furnish both the lunisolar precession and the main nutation components.

4.3.1 Reference Frames and Parametrization

In this subsection we define the rectangular reference systems which enable one to characterize the rotation of our planet, O being its center (Fig. 4.6):

Fig. 4.7 Parameters involved
in the calculations. O is the
Earth barycenter. A is a point
in the Earth. r is the distance
OA. S is the Sun barycenter.
ρ and a are respectively the
distances AS and OS
(from [16])

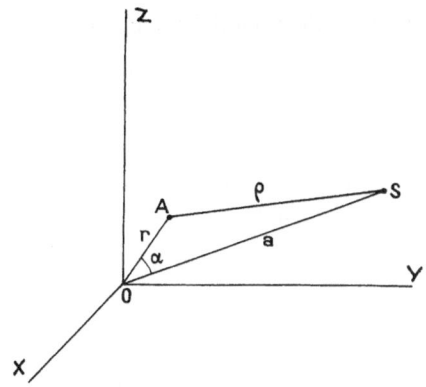

- $\Re_0 = (O, x_0, y_0, z_0)$ is constructed in such a way that (O, x_0) and (O, y_0) are two
  fixed directions in the ecliptic, considered as fixed in first approximation, (O, z_0)
  being directed towards the ecliptic pole.
- $\Re' = (O, x', y', z')$ is fixed with respect to the Earth, in such a way that the axes
  (O, x') and (O, y') are located in the equator, and (O, z') is oriented towards the
  pole of figure. We choose these axes in such a way that they coincide with the
  principal axes of inertia of the Earth.
- $\Re = (O, x, y, z)$ is a third rectangular coordinate system insuring the link between the two previous reference frames $\Re_0$ and $\Re'$: the (O, x) axis is oriented toward the ascending node of the ecliptic with respect to the equator,
  $(O, z) = (O, z')$ has been defined previously, and (O, y) completes the triad. We
  call A, B, C respectively the moments of inertia of the Earth with respect to these
  three rectangular axes.

Thus the rotation of the Earth is defined by a set of 3 angles, called Euler's angles
(Fig. 4.6):

$$\psi = \widehat{x_0 x}$$
$$\varphi = \widehat{x x'}$$
$$\varepsilon = \widehat{z_0 z}$$

4.3.2 Expression of the Tidal Torque

Now let us name x, y, z the rectangular coordinates in $\Re$ of an element A of the
Earth with infinitesimal mass dm, and X_S, Y_S, Z_S the coordinates of the Sun in $\Re$.

The distances OS, OA, AS are respectively noted a, r, ρ (Fig. 4.7). Let us call
$M_\odot$ the mass of the Sun, and G the constant of gravitation. Therefore, the gravitational force exerted by the Sun per mass is

$$f = \frac{GM_\odot}{\rho^2} \tag{4.1}$$

By using the vectorial notation this force is expressed as:

$$\frac{GM_\odot}{\rho^3}\overrightarrow{AS} = \frac{GM_\odot}{\rho^3}\overrightarrow{AO} + \frac{GM_\odot}{\rho^3}\overrightarrow{OS} \qquad (4.2)$$

Here we can already notice that the torque with respect to O due to the first component at the right-hand side of this equation is zero. Therefore we can let it aside. The amplitude of the second component can be rewritten

$$\frac{GM_\odot a}{\rho^3} = \frac{GM_\odot}{a^2} + GM_\odot a\left[\frac{1}{\rho^3} - \frac{1}{a^3}\right] \qquad (4.3)$$

The first component at the right hand side of Eq. (4.3) corresponds to an acceleration independent on the position of A on the Earth. Integrated over the whole Earth, it is the acceleration of the orbital motion of the Earth. The second component has a small amplitude, because of the relatively very close values of a and ρ. We can write

$$\frac{1}{\rho^3} - \frac{1}{a^3} = \frac{a^3 - \rho^3}{a^3\rho^3} = \frac{(a-\rho)(a^2+a\rho+\rho^2)}{\rho^3 a^3} \approx \frac{3(a-\rho)}{a^4} \qquad (4.4)$$

and

$$GM_\odot a\left[\frac{1}{\rho^3} - \frac{1}{a^3}\right] \approx -\frac{3GM_\odot}{a^3}(\rho - a) \qquad (4.5)$$

In first approximation, we can regard a as constant, which means that the orbital motion of the Earth is a circle with radius a. Moreover, according to Kepler's third law,

$$GM_\odot = \frac{4\pi^2 a^3}{T^2} = n^2 a^3 \qquad (4.6)$$

Then the combination of (4.5) and (4.6) gives

$$GM_\odot\left[\frac{1}{\rho^3} - \frac{1}{a^3}\right] \approx -3n^2(\rho - a) \qquad (4.7)$$

Still in first approximation, we can consider that ρ and a are parallel (Fig. 4.7). Thus we have

$$\rho - a = -r\cos\alpha = -\frac{xX_S + yY_S + zZ_S}{a} \qquad (4.8)$$

α standing for the angle $\widehat{AOS}$, x, y, z and X_S, Y_S, Z_S representing respectively rectangular coordinates of A and S in $\mathfrak{R}$ Finally, the elementary perturbing force applied to the element of mass dm is given by

$$dF = 3\frac{n^2}{a}(xX_S + yY_S + zZ_S)\,dm \qquad (4.9)$$

This force is parallel to OS. Its components along the three equatorial axes are respectively $(X_S/a)\,dF$, $(Y_S/a)\,dF$ and $(Z_S/a)\,dF$.

4.3.3 Expression of the Solar Torque

Now we can express the moment of the torque coming from the elementary force dF on A:

$$d\vec{M} = \begin{pmatrix} dL_S \\ dM_S \\ dN_S \end{pmatrix} = \frac{dF}{a} \begin{pmatrix} x \\ y \\ z \end{pmatrix} \wedge \begin{pmatrix} X_S \\ Y_S \\ Z_S \end{pmatrix} \tag{4.10}$$

which gives, after projection on the axes (O, x), (O, y) and (O, z):

$$dL_S = \frac{dF}{a}(yZ_S - zY_S), \qquad dM_S = \frac{dF}{a}(zX_S - xZ_S)$$

$$dN_S = \frac{dF}{a}(xY_S - yX_S) \tag{4.11}$$

By combining (4.9) and (4.11), and after integration, we deduce the total torque exerted on the whole Earth:

$$L_S = \frac{3n^2}{a^2} \int (xX_S + yY_S + zZ_S)(yZ_S - zY_S)\,dm \tag{4.12.1}$$

$$M_S = \frac{3n^2}{a^2} \int (xX_S + yY_S + zZ_S)(zX_S - xZ_S)\,dm \tag{4.12.2}$$

$$N_S = \frac{3n^2}{a^2} \int (xX_S + yY_S + zZ_S)(xY_S - yX_S)\,dm \tag{4.12.3}$$

The expansion of the expressions at the right hand side of these three equations is considerably simplified when using the following properties:

$$\int x\,dm = \int y\,dm = \int z\,dm = \int xy\,dm = \int xz\,dm = \int yz\,dm = 0 \tag{4.13}$$

$$\int (y^2 + z^2)\,dm = A, \qquad \int (x^2 + z^2)\,dm = B, \qquad \int (x^2 + y^2)\,dm = C \tag{4.14}$$

$$\int (y^2 - z^2)\,dm = C - B, \qquad \int (z^2 - x^2)\,dm = A - C$$

$$\int (x^2 - y^2)\,dm = B - A \tag{4.15}$$

where A, B, C are the moments of inertia of the Earth with respect to the space-fixed axes (O, x), (O, y), (O, z). By taking into account all these equations, we finally get the following simplified expressions for the three components:

$$L_S = 3n^2 \frac{Y_S Z_S}{a^2}(C - B), \qquad M_S = -3n^2 \frac{Z_S X_S}{a^2}(C - A)$$

$$N_S = 3n^2 \frac{X_S Y_S}{a^2}(B - A) \tag{4.16}$$

Here we still make an approximation consisting in assuming that $A = B$, which means implicitly that the Earth is rigorously axisymmetric with respect to the $(0, z)$ axis. Thus the equations above become

$$L_S = 3n^2 \frac{Y_S Z_S}{a^2}(C - A), \qquad M_S = -3n^2 \frac{Z_S X_S}{a^2}(C - A), \qquad N_S = 0 \quad (4.17)$$

The Sun is moving along the ecliptic plane (O, x_0, y_0). Then it is possible to replace the celestial rectangular coordinates X_S, Y_S, Z_S of the Sun by their expression as function of its distance $a = OS$ from the Earth, of its ecliptic longitude $\lambda_\odot$ counted from γ (Fig. 4.6) and the obliquity ε, taking into account the rotation with angle ε from the ecliptic to the equatorial frames:

$$X_S = a \cos \lambda_\odot, \qquad Y_S = a \cos \varepsilon \sin \lambda_\odot, \qquad Z_S = a \sin \varepsilon \sin \lambda_\odot \qquad (4.18)$$

Inserting these expressions in (4.17) we get

$$L_S = 3n^2(C - A) \cos \varepsilon \sin \varepsilon \sin^2 \lambda_\odot = \frac{3n^2}{4}(C - A) \sin 2\varepsilon (1 - \cos 2\lambda_\odot) \quad (4.19)$$

$$M_S = -3n^2(C - A) \sin \varepsilon \sin \lambda_\odot \cos \lambda_\odot = -\frac{3n^2}{2}(C - A) \sin \varepsilon \sin 2\lambda_\odot \qquad (4.20)$$

$$N_S = 0 \qquad (4.21)$$

4.3.4 Expression for the Lunar Torque

By analogy with the calculations carried out above for the Sun, we can start from Eqs. (4.16) to calculate the components of the moment of the torque exerted by the Moon on the Earth by

- replacing X_S, Y_S and Z_S by the rectangular coordinates X_M, Y_M, Z_M of the Moon
- replacing the scaling factor n^2 by the corresponding one n'^2 related to the orbital motion of the Moon around the Earth

Indeed, calling $M_\circ$ the mass of the Moon and a' its semi-major axis, we have

$$GM_\odot = n^2 a^3, \qquad GM_\circ = n'^2 a'^3 \qquad (4.22)$$

which gives

$$n'^2 = n^2 \left(\frac{a}{a'}\right)^3 \left(\frac{M_\circ}{M_\odot}\right) \qquad (4.23)$$

Still here, we consider at first approximation that the orbital motion of the perturbing body (the Moon) is circular, with radius a', and $a/a' = 388.93$. Moreover:

$$\frac{M_\circ}{M_\odot} = \frac{M_\circ}{M_\oplus} \frac{M_\oplus}{M_\odot} = \left(\frac{1}{81.3}\right) \times \left(\frac{1}{332\,946}\right) \approx \frac{1}{27\,068\,500} \qquad (4.24)$$

where $M_\oplus$ stands for the mass of the Earth. Finally this gives $n'^2 = kn^2$ with $k = 2.174$. The next step consists in expressing the rectangular coordinates X_M, Y_M, Z_M of the Moon as a function of its distance a' from the Earth's barycenter, the ecliptic longitude of the Moon λ_M, the longitude Ω_M of the Moon's ascending node of its orbit with respect to the ecliptic, the inclination of its orbit i_M with respect to the ecliptic, and the obliquity ε. These expressions are more complex for the Moon than for the Sun, because by contrast to the Sun, which is moving along the ecliptic, our satellite is moving along an orbit inclined with $i_M \approx 5°$ with respect to this plane. Through classical geometrical transformations we get easily:

$$X_M = a' \cos \lambda_M \tag{4.25.1}$$

$$Y_M = a' \sin \lambda_M \cos \varepsilon - a' \sin \varepsilon \sin i_M \sin(\lambda_M - \Omega_M) \tag{4.25.2}$$

$$Z_M = a' \sin \lambda_M \sin \varepsilon + a' \cos \varepsilon \sin i_M \sin(\lambda_M - \Omega_M) \tag{4.25.3}$$

Finally, by substituting these coordinates of the Moon to that of the Sun in Eq. (4.17) and by neglecting the components in i_M^2, we find:

$$L_M = \frac{3}{4} n'^2 (C - A)$$
$$\times \left[\sin 2\varepsilon (1 - \cos 2\lambda_M) + 2 \sin i_M \cos 2\varepsilon \left(\cos \Omega_M - \cos(2\lambda_M - \Omega_M) \right) \right] \tag{4.26.1}$$

$$M_M = -\frac{3}{2} n'^2 (C - A)$$
$$\times \left[\sin \varepsilon \sin 2\lambda_M - \sin i_M \cos \varepsilon \left(\sin \Omega_M - \sin(2\lambda_M - \Omega_M) \right) \right] \tag{4.26.2}$$

$$N_M = 0 \tag{4.26.3}$$

Then the total lunisolar torque with components L, M, N is obtained by combination of the components: $\mathbf{L} = (L_S + L_M, M_S + M_M, N_S + N_M)$. We note that the component along (O, x) contains a constant term, with amplitude

$$L_{const.} = \frac{3}{4} (C - A)(n^2 + n'^2) \sin 2\varepsilon = \frac{3}{4}(C - A)(1 + k)n^2 \sin 2\varepsilon \tag{4.27}$$

4.3.5 Equations for the Rotational Motion of the Earth

Now we have an explicit formulation of the lunisolar torque exerted on the Earth, we apply the fundamental equation related to the angular momentum $\boldsymbol{\sigma}$:

$$\left(\frac{d\boldsymbol{\sigma}}{dt} \right)_{\mathfrak{R}_0} = (\mathbf{L})_{\mathfrak{R}_0} \tag{4.28}$$

In the moving body-fixed frame $\mathfrak{R}'$, this equation becomes

$$\left(\frac{d\boldsymbol{\sigma}}{dt} \right)_{\mathfrak{R}'} + (\boldsymbol{\omega} \wedge \boldsymbol{\sigma})_{\mathfrak{R}'} = (\mathbf{L})_{\mathfrak{R}'} \tag{4.29}$$

Let us define the Earth's rotation vector $\boldsymbol{\omega}$ with coordinates $\boldsymbol{\omega} = (\omega_1, \omega_2, \omega_3)$ in $\mathfrak{R}'$. In matrix notation, we have

$$\sigma = \begin{pmatrix} A & 0 & 0 \\ 0 & B & 0 \\ 0 & 0 & C \end{pmatrix} \begin{pmatrix} \omega_1 \\ \omega_2 \\ \omega_3 \end{pmatrix} = \begin{pmatrix} A\omega_1 \\ B\omega_2 \\ C\omega_3 \end{pmatrix} \tag{4.30}$$

Thus the combination of (4.29) and (4.30) leads to

$$\begin{pmatrix} A\frac{d\omega_1}{dt} \\ B\frac{d\omega_2}{dt} \\ C\frac{d\omega_3}{dt} \end{pmatrix} + \begin{pmatrix} \omega_1 \\ \omega_2 \\ \omega_3 \end{pmatrix} \wedge \begin{pmatrix} A\omega_1 \\ B\omega_2 \\ C\omega_3 \end{pmatrix} = \begin{pmatrix} L' \\ M' \\ N' \end{pmatrix} \tag{4.31}$$

where L', M' and N' are the components of $\mathbf{L}$ in $\mathfrak{R}'$.

Projected along the axes of the body-fixed frame $\mathfrak{R}'$ corresponding to the principal axes of inertia, these equations give, still considering that $A = B$:

$$A\frac{d\omega_1}{dt} + (C - A)\omega_2\omega_3 = L' \tag{4.32}$$

$$A\frac{d\omega_2}{dt} - (C - A)\omega_1\omega_3 = M' \tag{4.33}$$

$$C\frac{d\omega_3}{dt} = N' \tag{4.34}$$

Moreover the components L', M' and N' of the lunisolar torque in $\mathfrak{R}'$ are related to the corresponding ones L, M and N in $\mathfrak{R}$ through a rotation with angle φ around $(O, z) = (O, z')$ (cf. Fig. 4.6):

$$L = L' \cos\varphi - M' \sin\varphi \tag{4.35}$$

$$M = L' \sin\varphi + M' \cos\varphi \tag{4.36}$$

A combination of Eqs. (4.32), (4.33), (4.35) and (4.36) gives immediately the following set of differential equations:

$$A\left(\frac{d\omega_1}{dt}\cos\varphi - \frac{d\omega_2}{dt}\sin\varphi\right) + (C - A)\omega_3(\omega_1\sin\varphi + \omega_2\cos\varphi) = L \tag{4.37}$$

$$A\left(\frac{d\omega_1}{dt}\sin\varphi + \frac{d\omega_2}{dt}\cos\varphi\right) - (C - A)\omega_3(\omega_1\cos\varphi - \omega_2\sin\varphi) = M \tag{4.38}$$

$$C\frac{d\omega_3}{dt} = N \tag{4.39}$$

Now we consider the components ω_x, ω_y and ω_z of the rotation vector with respect to the equatorial non rotating frame $\mathfrak{R}$. They can be expressed by the following elementary rotations:

$$\omega_x = -\frac{d\varepsilon}{dt}, \qquad \omega_y = \frac{d\psi}{dt}\sin\varepsilon, \qquad \omega_z = \frac{d\varphi}{dt} - \frac{d\psi}{dt}\cos\varepsilon \tag{4.40}$$

At the same time, these components are also deduced from ω_1, ω_2 and ω_3 in $\mathfrak{R}'$ by the rotation with angle φ:

$$\omega_x = \omega_1 \cos\varphi - \omega_2 \sin\varphi, \qquad \omega_y = \omega_1 \sin\varphi + \omega_2 \cos\varphi \qquad (4.41)$$

After derivation of the equations above we get easily:

$$\frac{d\omega_1}{dt}\cos\varphi - \frac{d\omega_2}{dt}\sin\varphi = \frac{d\omega_x}{dt} + \omega_y\frac{d\varphi}{dt} \qquad (4.42)$$

$$\frac{d\omega_1}{dt}\sin\varphi + \frac{d\omega_2}{dt}\cos\varphi = \frac{d\omega_y}{dt} - \omega_x\frac{d\varphi}{dt} \qquad (4.43)$$

Substituting the transformations of Eqs. (4.42) and (4.43) in Eqs. (4.37) and (4.38) leads to the following equations:

$$A\left[\frac{d\omega_x}{dt} + \omega_y\frac{d\varphi}{dt}\right] + (C - A)\omega_z\omega_y = L \qquad (4.44)$$

$$A\left[\frac{d\omega_y}{dt} - \omega_x\frac{d\varphi}{dt}\right] - (C - A)\omega_z\omega_x = M \qquad (4.45)$$

$$C\frac{d\omega_z}{dt} = C\frac{d\omega_3}{dt} = N = 0 \qquad (4.46)$$

Then replacing ω_x, ω_y and ω_z by their values in function of ε, φ and ψ thanks to (4.40), and after combination of terms,

$$-A\frac{d^2\varepsilon}{dt^2} - (C - A)\sin\varepsilon\cos\varepsilon\left(\frac{d\psi}{dt}\right)^2 + C\sin\varepsilon\frac{d\varphi}{dt}\frac{d\psi}{dt} = L \qquad (4.47)$$

$$A\sin\varepsilon\frac{d^2\psi}{dt^2} - (C - 2A)\cos\varepsilon\left(\frac{d\psi}{dt}\right)\left(\frac{d\varepsilon}{dt}\right) + C\frac{d\varphi}{dt}\frac{d\varepsilon}{dt} = M \qquad (4.48)$$

$$\frac{d^2\varphi}{dt^2} - \frac{d^2\psi}{dt^2}\cos\varepsilon + \frac{d\psi}{dt}\frac{d\varepsilon}{dt}\sin\varepsilon = 0 \qquad (4.49)$$

The third equation is equivalent to $d\omega_z/dt = 0$, which means that the component of the vector rotation along the figure axis (O, z) is constant. Moreover we know that the rates $d\psi/dt$ and $d\varepsilon/dt$ are very small in comparison to $d\varphi/dt$, and also that the expressions $d^2\psi/dt^2$ $d^2\psi/dt^2$ and $d\psi/dt \times d\varepsilon/dt$ are negligible at first approximation. Therefore the system of Eqs. (4.47) to (4.49) can be simplified to

$$C\sin\varepsilon\frac{d\psi}{dt}\frac{d\varphi}{dt} = L \qquad (4.50)$$

$$C\frac{d\varepsilon}{dt}\frac{d\varphi}{dt} = M \qquad (4.51)$$

$$\frac{d^2\varphi}{dt^2} = 0 \qquad (4.52)$$

where L (respectively M) is the sum of a solar contribution L_S (respectively M_S) and a lunar one L_M (respectively M_M), given by Eqs. (4.19) and (4.26.1) (respectively (4.20) and (4.26.2)). The integration of both Eqs. (4.50) and (4.51) will naturally give the theoretical expression of the precession-nutation in longitude ψ and in obliquity ε. The lunisolar precession in longitude, which is the linear variation $\psi_1 t$ of ψ comes from the constant part L^0 of L:

$$L^0 = L_S^0 + L_M^0 = \frac{3}{4}n^2(C - A)\sin 2\varepsilon + \frac{3}{4}n'^2(C - A)\sin 2\varepsilon$$

$$= \frac{3}{4}[n^2(1 + k)](C - A)\sin 2\varepsilon \tag{4.53}$$

Inserting this value in Eq. (4.50) we get

$$\psi_1 = \frac{3}{2}\frac{n^2(1 + k)}{\dot{\varphi}}\frac{(C - A)}{C}\cos \varepsilon \tag{4.54}$$

The physical parameter $(C - A)/C$ characterizes the relative difference between the moments of inertia C along the figure axis, and A perpendicular to it: it is called the *dynamical ellipticity*, in the case $A = B$ considered here. A priori all the quantities at the right hand side of this equation are known with very good accuracy, excepted this last parameter. n is the mean motion of the Earth: $n = 2\pi rd./y$. Moreover $n/\dot{\varphi} = T_{s.d.}/T_y = 1/366.24$ where $T_{s.d.}$ is the period of a sidereal day and T_y is the period of the sidereal revolution of the Earth. The obliquity ε can be considered as constant when not taking into account the small variations $\Delta\varepsilon$ calculated in the following and due to the nutation $\varepsilon \approx 23°26'21''$, which gives $\cos \varepsilon \approx 0.9174$. On the other side, the lunisolar precession ψ_1 is well known and determined with high precision from observational data. Its value is set at $\psi_1 = 50''.37/y$. Therefore a remarkable fact is that from (4.54) we can deduce theoretically the value of the dynamical flattening $(C - A)/C$ from the value of ψ_1 determined observationally.

$$\frac{C - A}{C} = \frac{2}{3}\frac{\psi_1}{n}\frac{\dot{\varphi}}{n}\frac{1}{1 + k}\frac{1}{\cos \varepsilon}$$

$$= \frac{2}{3} \times \frac{50.37}{360 \times 3600} \times \frac{366.24}{1 + 2.74} \times \frac{1}{0.9174} = \frac{1}{306.8} \tag{4.55}$$

Noting $\Delta\psi$ and $\Delta\varepsilon$ the nutations respectively in longitude and in obliquity, that is to say the periodic components of ψ and ε, we have, from (4.50) and (4.51),

$$C\sin\varepsilon\dot{\varphi}\frac{d\Delta\psi}{dt} = L_S^{per.} + L_M^{per.} \tag{4.56}$$

$$C\dot{\varphi}\frac{d\Delta\varepsilon}{dt} = M_S^{per.} + M_M^{per.} \tag{4.57}$$

where the symbol *per.* stands fore the periodic part of each corresponding component of the torque.

By substituting to $L_S^{per.}$, $L_M^{per.}$, $M_S^{per.}$, $M_M^{per.}$ their expressions in Eqs. (4.19), (4.26.1) (4.20) and (4.26.2) we find

$$\frac{d\Delta\psi}{dt} = -\frac{3}{2}\frac{n^2}{\dot{\varphi}}\left(\frac{C - A}{C}\right)\cos\varepsilon\cos 2\lambda_\odot - \frac{3}{2}\frac{kn^2}{\dot{\varphi}}\left(\frac{C - A}{C}\right)\cos\varepsilon\cos 2\lambda_M$$

$$+ \frac{3}{2}\frac{kn^2}{\dot{\varphi}}\left(\frac{C - A}{C}\right)\frac{\cos 2\varepsilon}{\cos\varepsilon}\sin i_M[\cos\Omega_M - \cos(2\lambda_M - \Omega_M)] \tag{4.58}$$

$$\frac{d\Delta\varepsilon}{dt} = -\frac{3}{2}\frac{n^2}{\dot{\varphi}}\left(\frac{C-A}{C}\right)\sin\varepsilon\sin 2\lambda_\odot - \frac{3}{2}\frac{kn^2}{\dot{\varphi}}\left(\frac{C-A}{C}\right)$$
$$\times\left[\sin\varepsilon\sin 2\lambda_M - \sin i_M\cos\varepsilon\left(\sin\Omega_M - \sin(2\lambda_M - \Omega_M)\right)\right] \quad (4.59)$$

Finally we can separate the nutations in longitude coming from the solar and lunar parts:

$$\Delta\psi = \Delta\psi^{Sun} + \Delta\psi^{Moon}, \qquad \Delta\varepsilon = \Delta\varepsilon^{Sun} + \Delta\varepsilon^{Moon} \quad (4.60)$$

with

$$\Delta\psi^{Sun} = -\frac{3}{4}\frac{n^2}{\dot{\varphi}}\left(\frac{C-A}{C}\right)\cos\varepsilon\frac{\sin 2\lambda_\odot}{n} \quad (4.61)$$

$$\Delta\psi^{Moon} = \frac{3}{2}\frac{kn^2}{\dot{\varphi}}\frac{(C-A)}{C}\left[\frac{\sin i_M\cos 2\varepsilon}{\sin\varepsilon}\frac{\sin\Omega_M}{\dot{\Omega}_M} - \frac{\cos\varepsilon\sin 2\lambda_M}{2n'}\right.$$
$$\left. - \frac{\sin i_M\cos 2\varepsilon}{\sin\varepsilon}\frac{\sin(2\lambda_M - \Omega_M)}{2n' - \dot{\Omega}_M}\right] \quad (4.62)$$

And for the nutation in obliquity

$$\Delta\varepsilon = \Delta\varepsilon^{Sun} + \Delta\varepsilon^{Moon} \quad (4.63)$$

with

$$\Delta\varepsilon^{Sun} = -\frac{3}{4}\frac{n^2}{\dot{\varphi}}\left(\frac{C-A}{C}\right)\sin\varepsilon\frac{\cos 2\lambda_\odot}{n} \quad (4.64)$$

$$\Delta\varepsilon^{Moon} = -\frac{3}{2}\frac{kn^2}{\dot{\varphi}}\frac{(C-A)}{C}\left[\sin i_M\cos\varepsilon\frac{\cos\Omega_M}{\dot{\Omega}_M} - \frac{\sin\varepsilon\cos 2\lambda_M}{2n'}\right.$$
$$\left. - \sin i_M\cos\varepsilon\frac{\cos(2\lambda_M - \Omega_M)}{2n' - \dot{\Omega}_M}\right] \quad (4.65)$$

Numerically this gives

$$\Delta\psi = -17''.16\sin\Omega_M - 1''.263\sin 2\lambda_\odot - 0''.205\sin 2\lambda_M$$
$$- 0''.034\sin(2\lambda_M - \Omega_M) \quad (4.66)$$
$$\Delta\varepsilon = 9''.17\cos\Omega_M + 0''.548\cos 2\lambda_\odot + 0''.089\cos 2\lambda_M$$
$$+ 0''.018\cos(2\lambda_M - \Omega_M) \quad (4.67)$$

In this section we have explained in a simplified manner, following Danjon [16] the various steps which lead to the theoretical expressions of the combined precession-nutation of the Earth in space, that is to say the motion of its figure axis in space, when undergoing the lunisolar torque. For that purpose we have made several approximations:

– We neglected the variations of distance of the perturbing bodies (Moon and Sun) considering that their relative motion is circular.
– We assumed that the Earth is axisymmetric ($A = B$), whereas in reality the Earth is triaxial, although the relative difference of moments of inertia $(A - B)/C$ is much smaller than the difference $(C - A)/C$.

- We considered here the Earth as a rigid body whereas in reality we must take into account several effects of non-rigidity: those due to the elastic mantle with a fluid outer core and a solid inner core as well as the influence of the oceans and the atmosphere, which can no more be neglected in comparison with the accuracy of modern observations.
- We neglected 2nd-order terms in Eqs. (4.47), (4.48) and (4.49).
- We did not take into account the gravitational effects of the planets which, although being considerably much smaller than the lunisolar ones, are not negligible when compared with up-to-date observational accuracy.
- The problem has been considered in the Newtonian framework while we must take into account the *geodetic precession* and *geodetic nutation* which are a time-dependent rotation of the geocentric celestial reference system (GCRS) with respect to the barycentric celestial reference system (BCRS) due to General relativity.
- All our calculations were done with a small precision, to 3 significant digits.

It is clear that all these simplifications, although allowing a straightforward and clear demonstration, are not satisfactory as soon as a good accuracy is requested. In the following we describe how the theory of the precession-nutation was pushed to a remarkable precision thanks to recent developments, and in particular by taking into account all the corrections mentioned above.

4.4 Alternative Theories of Precession-Nutation for a Rigid Earth Model

Best modeling the precession-nutation of the real Earth supposes at first step a very accurate determination of this motion when considering the simplified case of a rigid Earth. This will serve as a basis for a more complete and accurate theory including geophysical, atmospheric and oceanic contributions. After pioneering works done by Woolard [79] and Kinoshita [37, 38] to elaborate a very complete theory for rigid Earth precession-nutation, the drastic improvement of observational techniques such as VLBI, reaching the sub-milliarcsecond accuracy during the 1980's required new investigations to develop theories available up to the same level of precision. Competitive works appeared in the 1990's to accomplish this challenge. They consisted essentially in an improvement of the already well established theories mentioned above: Kinoshita's theory based on Hamiltonian formalism with the help of canonical variables [39, 65–67]. Woolard's theory based on the equivalent principles of the theorem of angular momentum and Lagrangian equations [4, 5, 54]. In the following we describe the theoretical basis of these two theoretical ways of calculation.

The latter approach was used also by Capitaine et al. [12] with a new parameterization replacing the traditional Euler angles (see Sect. 4.4.1.1); this refers to the CIP (Celestial intermediate pole) and the CIO (Celestial intermediate origin) as introduced by the IAU 2000 Resolutions (see Sect. 4.4.1.6).

Fig. 4.8 Parametrization of
the rotation of a rigid body
with Eulerian variables.
$\mathfrak{R}_0 = (O, X_0, Y_0, Z_0)$ is a
fixed inertial reference frame
and $\mathfrak{R} = (O, X, Y, Z)$ is the
body fixed moving reference
frame. The obliquity θ is the
angle between the axes
(O, Z_0) and (O, Z). The
precession ψ enables one to
determine the position of the
nodal line (O, γ) between the
body fixed equatorial plane
(O, X, Y) and the fixed
reference plane (O, X_0, Y_0).
The angle of proper rotation
ϕ enables one to determine
the position of the prime
meridian (O, X, Z) with
respect to the nodal axis
(O, γ) (from [79])

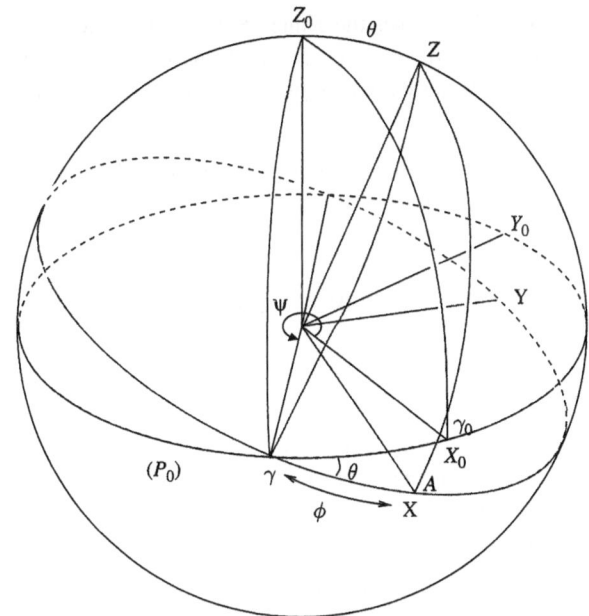

4.4.1 Dynamical Equations of the Rotation of the Rigid Earth with Lagrangian Formalism

All the calculations in this chapter are taken from Woolard [79]. The most classical
way to represent the rotation of the rigid Earth with respect to a fixed reference frame
is done through the Eulerian angles (Fig. 4.8). The two reference frames necessary
for the calculations are an inertial one $\mathfrak{R}_0 = (O, X_0, Y_0, Z_0)$ where O stands for
the center of mass of the Earth, and a body-fixed one $\mathfrak{R} = (O, X, Y, Z)$ in such a
way that (O, Z) is directed towards the axis of maximum moment of inertia C of
the Earth, also defined as the figure axis. (O, X) is oriented towards a point on the
equator of figure and (O, Z) completes the triad. The three axes (O, X), (O, Y), and
(O, Z) coincide respectively with the principal axes of inertia of the Earth, namely
A, B and C, with $A < B < C$.

4.4.1.1 Eulerian Parametrization

The Eulerian angles can be defined as follows (Fig. 4.8):

- θ, the *obliquity angle* (or simply the obliquity) represents the inclination of the
  equator of figure with respect to the fixed plane $(P_0) = (O, X_0, Y_0)$. It is generally
  reckoned positively.
- ψ, the *precession angle*, is defined in the fixed plane (O, X_0, Y_0) between a reference point γ_0 on (P_0) and the line (O, γ) along which the equator of figure

(O, X, Y) crosses the plane (P_0). For the sake of simplicity we can make γ_0 coinciding with (O, X_0). Therefore we can write $\psi = \gamma_0\gamma$.

- ϕ, sometimes called the *proper rotation*, is the angle between the axes (O, γ) and (O, X).

Thus the position of the axis of figure of the Earth in space is given by the set of the three angles (ψ, θ, ϕ).

4.4.1.2 Euler Kinematical Equations

The rotational motion of the Earth, considered as a rigid body, about its center of mass with respect to $\mathfrak{R}_0$ is the combination of three independent rotations:

- a rotation at rate $\dot{\psi}$ around Z_0.
- a rotation at rate $\dot{\theta}$ around the moving line of the node $(0, \gamma)$ of the equator of figure on the fixed plane (P_0).
- a rotation at rate $\dot{\phi}$ around the axis of figure.

These three individual rotations compound into a resultant rotation vector $\boldsymbol{\omega}$ around the *instantaneous axis of rotation* passing through the center of mass O. Its amplitude ω is the angular velocity around this axis. Therefore the position of the axis of rotation with respect to the Earth-fixed coordinate system $\mathfrak{R}$ is given at any instant by the coordinates $(\omega_1, \omega_2, \omega_3)$ of the rotation vector $\boldsymbol{\omega}$, which are linked to the derivatives of the Eulerian angles through a system of equations called the *Euler's kinematical equations*. These are

$$\omega_1 = -\dot{\theta}\cos\phi - \dot{\psi}\sin\theta\sin\phi \tag{4.68}$$

$$\omega_2 = \dot{\theta}\sin\phi - \dot{\psi}\sin\theta\cos\phi \tag{4.69}$$

$$\omega_3 = \dot{\psi}\cos\theta + \dot{\phi} \tag{4.70}$$

And reciprocally, this can be written:

$$\dot{\psi}\sin\theta = -\omega_1\sin\phi - \omega_2\cos\phi \tag{4.71}$$

$$\dot{\theta} = -\omega_1\cos\phi + \omega_2\sin\phi \tag{4.72}$$

$$\dot{\phi} = \omega_3 + \cot\theta(\omega_1\sin\phi + \omega_2\cos\phi) \tag{4.73}$$

4.4.1.3 Lagrange Formalism

A straightforward way to determine the equations of the rotational motion consists in applying Lagrange's equations. The kinetic energy of the Earth is expressed in the following classical form:

$$T = \frac{1}{2}\left(A\omega_1^2 + B\omega_2^2 + C\omega_3^2\right) \tag{4.74}$$

Now we choose the Eulerian angles ψ, θ and ϕ as the generalized coordinates q_i ($i = 1, 2, 3$). Then the Lagrangian function is

$$L = T + U = \frac{1}{2}\left(A\omega_1^2 + B\omega_2^2 + C\omega_3^2\right) + U \tag{4.75}$$

where U is the force function (potential) representing the lunisolar perturbing potential, which will be explicated in Sect. 4.4.3. The system can be considered as conservative, so that Lagrange equations can be applied:

$$\frac{d}{dt}\left(\frac{\partial L}{\partial \dot{q}_i}\right) - \frac{\partial L}{\partial q_i} = 0 \tag{4.76}$$

which gives, after expansion [79]:

$$A\frac{d\omega_1}{dt} + (C - B)\omega_2\omega_3 = \frac{\sin\phi}{\sin\theta}\left(\cos\theta\frac{\partial U}{\partial \phi} - \frac{\partial U}{\partial \psi}\right) - \cos\phi\frac{\partial U}{\partial \theta} \tag{4.77}$$

$$B\frac{d\omega_2}{dt} - (C - A)\omega_1\omega_3 = \frac{\cos\phi}{\sin\theta}\left(\cos\theta\frac{\partial U}{\partial \phi} - \frac{\partial U}{\partial \psi}\right) + \sin\phi\frac{\partial U}{\partial \theta} \tag{4.78}$$

$$C\frac{d\omega_3}{dt} + (B - A)\omega_1\omega_2 = \frac{\partial U}{\partial \phi} \tag{4.79}$$

These equations are generally called the *Euler's dynamical equations*

4.4.1.4 Method of Variation of Parameters

The external forces that act to affect the rotational motion of the Earth are so comparatively small that the equations of motion may be integrated efficiently by the method of variation of parameters: in a first step we determine a simplified solution that would occur were the external forces to vanish ($U = 0$). Then the solution is approximatively modified (through the parameters involved) to get the motion in actual conditions. As we have already mentioned we consider that

$$\frac{A - B}{C} \ll \frac{C - A}{C} \tag{4.80}$$

so that, at first approximation, the set of Eqs. (4.77) to (4.79) becomes, taking into account that $\frac{\partial U}{\partial \phi} = 0$, due to the symmetry:

$$\frac{d\omega_1}{dt} + \left(\frac{C - A}{A}\right)\omega_2\omega_3 = -\frac{\sin\phi}{A\sin\theta}\frac{\partial U}{\partial \psi} - \frac{\cos\phi}{A}\frac{\partial U}{\partial \theta} \tag{4.81.1}$$

$$\frac{d\omega_2}{dt} - \left(\frac{C - A}{A}\right)\omega_1\omega_3 = -\frac{\cos\phi}{A\sin\theta}\frac{\partial U}{\partial \psi} + \frac{\sin\phi}{A}\frac{\partial U}{\partial \theta} \tag{4.81.2}$$

$$\omega_3 = cte. \tag{4.81.3}$$

According to the method of variation of parameters, we first consider that $U = 0$. Therefore the right-hand side of Eqs. (4.81.1) and (4.81.2) reduces to zero and putting

$$\mu = \frac{C - A}{A}\omega_3 \tag{4.82}$$

leads to the trivial equations:

$$\frac{d\omega_1}{dt} + \mu\omega_2\omega_3 = 0 \tag{4.83.1}$$

$$\frac{d\omega_2}{dt} - \mu\omega_1\omega_2 = 0 \tag{4.83.2}$$

$$\omega_3 = cte. \tag{4.83.3}$$

with the obvious solutions

$$\omega_1 = f_0 \cos\mu t + g_0 \sin\mu t \tag{4.84.1}$$

$$\omega_2 = f_0 \sin\mu t - g_0 \cos\mu t \tag{4.84.2}$$

where f_0 and g_0 are constants of integration. Thus, we show from these equations that when external forces vanish ($U = 0$), and by accepting the condition of ax-isymmetry ($A = B$), the axis of rotation of the Earth describes with respect to the Earth-fixed reference frame $\Re$ a motion circular and uniform around the axis of fig-ure, represented by the axis (O, Z). This motion, called the *free polar motion*, is described with a frequency $\mu = (C - A/C)\omega_3$.

The second step in the method of variation of parameters consists in adopting the same kind of formalism as in (4.84.1) and (4.84.2) for ω_1 and ω_2 but by replacing the constants f_0 and g_0 by functions f and g depending on time:

$$\omega_1 = f \cos\mu t + g \sin\mu t \tag{4.85.1}$$

$$\omega_2 = f \sin\mu t - g \cos\mu t \tag{4.85.2}$$

Inserting these expressions in Eqs. (4.81.1) and (4.81.2) enables to determine f and g by quadrature:

$$f = f_0 - \int \left(\frac{\sin(\phi + \mu t)}{A \sin\theta} \frac{\partial U}{\partial \psi} + \frac{\cos(\phi + \mu t)}{A} \frac{\partial U}{\partial \theta} \right) dt \tag{4.86.1}$$

$$g = g_0 + \int \left(\frac{\cos(\phi + \mu t)}{A \sin\theta} \frac{\partial U}{\partial \psi} - \frac{\sin(\phi + \mu t)}{A} \frac{\partial U}{\partial \theta} \right) dt \tag{4.86.2}$$

Then the combination of Eqs. (4.71), (4.72) and (4.73) with (4.86.1) and (4.86.2) gives the variations of the Eulerian angles which determine the position of the axis of figure in space:

$$\sin\theta \frac{d\psi}{dt} = -f \sin(\phi + \mu t) + g \cos(\phi + \mu t) \tag{4.87.1}$$

$$\frac{d\theta}{dt} = -f \cos(\phi + \mu t) - g \sin(\phi + \mu t) \tag{4.87.2}$$

$$\frac{d\phi}{dt} = \omega_3 - \cos\theta \left(\frac{d\psi}{dt} \right) \tag{4.87.3}$$

The last equation can be rewritten, using Eq. (4.82), as

$$\frac{d(\phi + \mu t)}{dt} = \frac{C}{A}\omega_3 - \cos\theta \left(\frac{d\psi}{dt} \right) \tag{4.88}$$

4.4.1.5 The Motion of the Axis of Figure in Space

Equations (4.87.1) and (4.87.2) can be transformed into a more advantageous form, by substituting the expressions of f and g given by Eqs. (4.86.1) and (4.86.2) in the right-hand side, and after differentiating both members with respect to t. We find

$$\frac{d}{dt}\left(\sin\theta\frac{d\psi}{dt}\right) = (\dot{\phi}+\mu)\frac{d\theta}{dt} + \frac{1}{A\sin\theta}\frac{\partial U}{\partial\psi} \tag{4.89}$$

$$\frac{d}{dt}\left(\frac{d\theta}{dt}\right) = -(\dot{\phi}+\mu)\sin\theta\frac{d\psi}{dt} + \frac{1}{A}\frac{\partial U}{\partial\theta} \tag{4.90}$$

Using Eq. (4.88) and after re-combination, we finally obtain the derivative of the two precession-nutation angles:

$$\frac{d\theta}{dt} = -\frac{1}{C\omega_3\sin\theta}\frac{\partial U}{\partial\psi} + \frac{A}{C\omega_3}\frac{d}{dt}\left(\sin\theta\frac{d\psi}{dt}\right) + \frac{A}{C\omega_3}\cos\theta\frac{d\psi}{dt}\frac{d\theta}{dt} \tag{4.91}$$

$$\frac{d\psi}{dt} = \frac{1}{C\omega_3\sin\theta}\frac{\partial U}{\partial\theta} - \frac{A}{C\omega_3}\frac{d}{dt}\left(\frac{d\theta}{dt}\right) + \frac{A}{C\omega_3}\cos\theta\left(\frac{d\psi}{dt}\right)^2 \tag{4.92}$$

4.4.1.6 Modern Parametrization Based on IAU 2000 Resolutions

The Euler dynamical equations and the method of variation of parameters described in the previous sections have been used by Capitaine et al. [12] for a modern semi-analytical resolution (analytical representation with numerical coefficients) of the precession-nutation equations based on the CIO based parameters.

The celestial and terrestrial intermediate origins (CIO and TIO respectively), have been defined as origins on the equator of the CIP, based on the concept of the "non-rotating origin" [25] when the CIP moves in space and in the Earth, respectively. Their kinematical property provides a very straightforward definition of the Earth's diurnal rotation based on the *Earth Rotation Angle* (ERA) along the equator of the CIP (Celestial Intermediate Pole) between those two origins, which is linearly related to UT1. The ERA replaces the third Euler angle ϕ from the nodal axis (O, γ).

The CIP is defined as being the intermediate pole, in the transformation between the celestial and terrestrial systems, separating nutation from polar motion by a specific convention in the frequency domain. That convention is such that (i) the GCRS (Geocentric Celestial Reference System) CIP motion includes all the terms with periods greater than 2 days in the GCRS (i.e. frequencies between -0.5 cycles per sidereal day (cpsd) and $+0.5$ cpsd); (ii) the ITRS (International Terrestrial Reference System) CIP motion, includes all the terms outside the retrograde diurnal band in the ITRS (i.e. frequencies less than -1.5 cpsd or greater than -0.5 cpsd).

The CIO based precession-nutation parameters consist in the GCRS coordinates of the CIP unit vector, either in their polar form, E and d, or their rectangular form, $X = \sin d \cos E$, $Y = \sin d \sin E$ (see Fig. 4.9); they contain precession and nutation of the CIP, frame bias between the equator and equinox frame at J2000.0 and

Fig. 4.9 Parametrization of
the precession-nutation of the
equator using the CIO (σ)
based parameters: the
coordinates of the CIP unit
vector (either E and d, or
$X = \sin d \cos E$,
$Y = \sin d \sin E$)

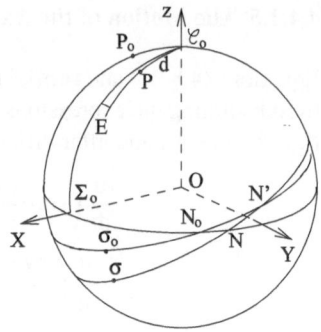

the GCRS, plus the cross terms between precession and nutation [9, 10]. Y and X replace respectively the first and second Euler angles, θ and ψ.

The CIO (Celestial Intermediate Origin) based precession-nutation equations for a rigid axially symmetric Earth are as follows [12]:

$$-\ddot{Y} + (C/A)\Omega\dot{X} = L_\Sigma/A + F'' \tag{4.93.1}$$

$$\ddot{X} + (C/A)\Omega\dot{Y} = M_\Sigma/A + G'' \tag{4.93.2}$$

Ω being the mean Earth's angular velocity, L_Σ and M_Σ the equatorial components of the torque referred to Σ (such that $\Sigma N = \Sigma_0 N$), and F'', G'' functions of X, Y and of their first and second time derivatives.

4.4.2 Dynamical Equations of the Rotation of the Rigid Earth with Hamiltonian Formalism

An alternative way to construct a theory of the rotation for a rigid Earth model consists in starting from Andoyer variables [1] instead of Eulerian angles. One of the advantages of such a choice comes from the fact that Andoyer variables are canonical. Thus it is rather easy to apply a perturbation theory based on canonical transformations to the rotational motion, and to separate precessional from nutational motion, or, in other words, secular perturbations from periodic ones [38]. Moreover it is easy to treat separately the motion of the figure, rotation and angular momentum axes.

4.4.2.1 Andoyer Angles Parametrization

We start from the two reference frames, the inertial one $\Re_0 = (O, X_0, Y_0, Z_0)$ and the Earth-fixed one $\Re = (O, X, Y, Z)$ as defined previously. We call **L** the angular momentum vector of the Earth. The plane (P_0) has the same meaning as in the previous section, whereas (P_L) stands for the plane perpendicular to **L**. The Andoyer variables consist in 3 action variables (L, G, H) and 3 angle variables (l, g, h) defined as follows (Fig. 4.10):

Action variables
They are defined with respect to the angular momentum vector.

- G is the amplitude of the angular momentum vector $\mathbf{L}$
- L is the component of $\mathbf{L}$ along the axis (O, Z)
- H is the component of $\mathbf{L}$ along the axis (O, Z_0)

From these definitions we can already introduce two angles which play a fundamental role in the theory: I is the inclination of $\mathbf{L}$ with respect to (O, Z_0) and J its inclination with respect to (O, Z) in such a way that

$$L = G \cos J, \qquad H = G \cos I \qquad (4.94)$$

Notice that I represents the obliquity of the axis of angular momentum with respect to the inertial axis (O, Z) and must not be confused with the classical obliquity, i.e. the angle between the axis of figure and the basic plane (P_0) (generally the ecliptic).

Angle variables
As in the case of the Eulerian angles they enable one to give the orientation of $\mathfrak{R}$ with respect to $\mathfrak{R}_0$, but with the intermediary of the plane (P_L) which is not involved in this first case.

- h is the angle measured along the reference plane (P_0) between the fixed point γ_0 and the node Q of (P_L) with respect to (P_0). Notice that it represents the angle of precession but for the equator of angular momentum (P_L) instead of the precession ψ of the equator of figure.
- g is the angle along the equator of angular momentum (P_L) between Q defined previously and the ascending node P of the equator of figure with respect to (P_L).
- l is the angle along the equator of figure between the ascending node P and the Earth-fixed origin axis (O, X).

4.4.2.2 Relationships Between Eulerian and Andoyer Variables

In Fig. 4.10 we represent the reference planes and axes defined together with the Eulerian angles and Andoyer action and angles. The relationships between these two set of variables can be derived from the spherical triangle (P, Q, γ) defined previously. These are

$$\cos \theta = \cos I \cos J - \sin I \sin J \cos g \qquad (4.95.1)$$

$$\frac{\sin(\psi - h)}{\sin J} = \frac{\sin(\phi - l)}{\sin I} = \frac{\sin g}{\sin \theta} \qquad (4.95.2)$$

In the case of the Earth, the angle J is very small. Its amplitude does not exceed $1''$. This means that the angular momentum axis is nearly coinciding with the

Fig. 4.10 Parametrization of
the rotation of a rigid body
with Andoyer variables with
respect to the fixed inertial
reference frame
$\mathfrak{R}_0 = (O, X_0, Y_0, Z_0)$. The
three angle variables l, g, h
enable one to determine the
body fixed reference frame
$\mathfrak{R} = (O, X, Y, Z)$ with
respect to $\mathfrak{R}_0$. The three
action variables are L and H,
respectively projections of the
vector angular momentum $\mathbf{L}$
on the body fixed axis (O, Z)
and the inertial axis (O, Z_0),
and G, the norm of $\mathbf{L}$
(from [38])

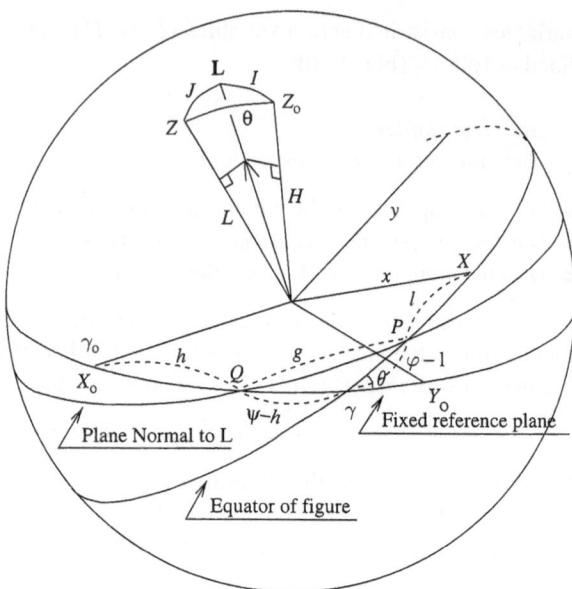

figure axis. Therefore at the first order in J we can write from the two equations
above

$$\psi = h + \frac{J}{\sin I} \sin g + O\left(J^2\right) \tag{4.96.1}$$

$$\theta = I + J \cos g + O\left(J^2\right) \tag{4.96.2}$$

$$\phi = l + g - J \cot I \sin g + O\left(J^2\right) \tag{4.96.3}$$

4.4.2.3 Andoyer Variables Referred to the Fixed and Moving Ecliptics

In practical case, the inertial reference frame $\mathfrak{R}_0$ is defined in such a way that the
(O, X_0) axis is directed towards the fixed mean equinox of a reference epoch T_0 (for
instance J2000.0), the plane $(P_0) = (O, X_0, Y_0)$ coinciding with the mean ecliptic
of T_0. For the study of the rotation of the Earth, it is convenient to adopt instead
of $\mathfrak{R}_0$ a moving reference frame $\bar{\mathfrak{R}} = (O, \bar{X}, \bar{Y}, \bar{Z})$ in such a way that the plane
$(\bar{P}) = (O, \bar{X}, \bar{Y})$ coincides to the moving ecliptic of the date, $(O, \bar{Z})$ being directed
towards the axis of this ecliptic, and $(O, \bar{X})$ towards the *departure point* as defined
by Kinoshita [38]. Now we can define the motions of the moving ecliptic of date
with respect to the fixed ecliptic of epoch through the set of two variables π_e and
Π_e: they stand respectively for the inclination and the longitude of the node of $(\bar{P})$
with respect to (P_0).

Kinoshita [38] showed that it is possible to transform the variables h, I, and
g referred to the fixed plane (P_0) into a corresponding set of variables h', I' and
g' referred to a slightly moving reference plane $(\bar{P})$, in such a way that the old

set of canonical variables (l, g, h, L, G, H) is transformed in a new canonical one (l', g', h', L', G', H'). If $\mathcal{H}$ designates the Hamiltonian of the system of the rotational motion of the rigid Earth, the new Hamiltonian, called $\mathcal{K}$, satisfies

$$G\, dg + H\, dh - \mathcal{H}\, dt = G'\, dg' + H'\, dh' - \mathcal{K}\, dt \tag{4.97}$$

with

$$G' = G, \qquad H' = G \cos I' \tag{4.98}$$

We can remark that the variables l and L do not depend on the reference frame: $l' = l$, and $L' = L$ The new Hamiltonian becomes

$$\mathcal{K} = \mathcal{H} + E \tag{4.99}$$

with

$$
\begin{aligned}
E = {}& H'(1 - \cos \pi_e) \frac{d\Pi_e}{dt} \\
& + G \sin I' \left[\frac{d\Pi_e}{dt} \sin \pi_e \cos(h' - \Pi_e) - \frac{d\pi_e}{dt} \sin(h' - \Pi_e) \right]
\end{aligned} \tag{4.100}
$$

4.4.2.4 Hamiltonian of the System and Canonical Equations of the Rotational Motion

According to Kinoshita [37, 38], the Hamiltonian $\mathcal{K}$ for the rotational motion of the rigid Earth referred to the slightly moving reference frame $\bar{\mathfrak{R}}$ defined in the previous section is

$$\mathcal{K} = F_0 + E + U \tag{4.101}$$

- F_0 represents the kinetic energy of the rotational motion. It is written

$$F_0 = \frac{1}{2} \left(\frac{\sin^2 l}{A} + \frac{\cos^2 l}{B} \right) (G^2 - L^2) + \frac{1}{2C} L^2 \tag{4.102}$$

- E is the complementary Hamiltonian component due to the change of reference frame as explained in the previous section.

$$
\begin{aligned}
E = {}& H'(1 - \cos \pi_e) \frac{d\Pi_e}{dt} \\
& + G \sin I' \left[\frac{d\Pi_e}{dt} \sin \pi_e \cos(h' - \Pi_e) - \frac{d\pi_e}{dt} \sin(h' - \Pi_e) \right]
\end{aligned} \tag{4.103}
$$

- U is the disturbing potential for the rotational motion due to the external bodies (Moon, Sun and planets). An exhaustive study of the determination of U will be given in the next section.

4.4.2.5 Equations of Motion

The equations of motion of the rotation of the rigid Earth are directly derived from the property of canonicity of the adopted Andoyer variables (l, g', h', L, G, H'). For the sake of simplicity, the prime symbols are removed from the variables, and in the moving reference frame $\bar{\Re}$, the canonical equations can be written

$$\frac{d}{dt}(L, G, H) = -\frac{\partial \mathcal{K}}{\partial (l, g, h)} \tag{4.104.1}$$

$$\frac{d}{dt}(l, g, h) = \frac{\partial \mathcal{K}}{\partial (L, G, H)} \tag{4.104.2}$$

4.4.2.6 Precession-Nutation of the Axis of Angular Momentum

In the scope of this chapter dealing with the precession-nutation motion, we essentially focus on the two variables h and I which represent respectively the precession angle and the obliquity of the axis of the plane perpendicular to the angular momentum vector. We have already explained that I is defined starting from the canonical variables through the relationship $H = G \cos I$. In addition we call Δh and ΔI the periodic variations h and I. By writing W the determining function defined by

$$W = \int \left(U_1^{per} + U_2^{per} \right) dt \tag{4.105}$$

where U_1^{per} and U_2^{per} stand for the periodic parts of respectively U_1 and U_2, we have

$$\Delta h = -\frac{\partial W}{H} = \frac{\partial W}{\partial I} \frac{\partial I}{\partial H} = -\frac{1}{G \sin I} \frac{\partial W}{\partial I} \tag{4.106}$$

$$\Delta I = \frac{1}{G} \left(-\frac{1}{\sin I} \Delta H + \cot I \Delta G \right) = \frac{1}{G} \left(\frac{1}{\sin I} \frac{\partial W}{\partial h} - \cot I \frac{\partial W}{\partial g} \right) \tag{4.107}$$

4.4.2.7 Precession-Nutation of the Figure Axis

The nutation of the figure axis $(\Delta \psi_f, \Delta \varepsilon_f)$ are directly deduced from the geometrical relationships (4.96.1) and (4.96.2) linking the axis of angular momentum and the axis of figure[4]:

$$\Delta \psi_f = \Delta h + \Delta \left[\frac{J \sin g}{\sin I} \right] + O\left(J^2 \right) \tag{4.108}$$

$$\Delta \varepsilon_f = \Delta \theta = \Delta I + \Delta [J \cos g] + O\left(J^2 \right) \tag{4.109}$$

---

[4]The reader must pay attention to the sign conventions: here ψ and h as well as ε and I have the same signs. This follows the classical geometrical rules of a positive sign in the trigonometric (counter clockwise) sense, whereas for conventional astronomical rules ψ and ε are counted positively in the clockwise sense.

with

$$\Delta\left(\frac{J\sin g}{\sin I}\right) = \frac{1}{\sin I}(\sin g\Delta J + J\cos g\Delta g) - \frac{J\sin g}{\sin^2 I}\Delta I$$

$$= \frac{1}{G\sin I}\left[\frac{\partial W}{\partial g}\sin g - W\cos g\right] \qquad (4.110)$$

and

$$\Delta(J\cos g) = \frac{1}{G}\left[\frac{\partial W}{\partial g}\cos g + W\sin g\right] \qquad (4.111)$$

These two last expressions are called the *Oppolzer terms*.

4.4.3 The Determination of the Disturbing Potential U

As seen in Sects. 4.4.1 and 4.4.2, whatever be the theory used to determine the rotational motion, and more precisely the precession-nutation of the Earth, it necessitates the precise calculation of the disturbing potential U exerted by the external body (Moon, Sun, planet). This disturbing potential can be represented by expansion in spherical harmonics of first order (U_1) and second one (U_2) [69]

$$U = U_1 + U_2 \qquad (4.112)$$

with

$$U_1 = \frac{GM}{r^3}\left[\frac{2C - A - B}{2}P_2(\sin\delta) + \frac{A - B}{4}P_2^2(\sin\delta)\cos 2\alpha_E)\right] \qquad (4.113)$$

$$U_2 = \sum_{n=3}^{\infty}\frac{GMM_E a_E^n}{r^{n+1}}$$

$$\times\left[J_n P_n(\sin\delta) - \sum_{m=1}^{n}P_n^m(\sin\delta)(C_{n,m}\cos m\alpha_E + S_{nm}\sin m\alpha_E)\right] \qquad (4.114)$$

where α_E stands for the geocentric longitude of the perturbing body as measured from a prime meridian on the Earth[5] and δ its declination. M is the mass of the perturbing body, r its distance from the center of the Earth. J_n, C_{nm} and S_{nm} are the coefficients of the geopotential, which characterize the repartition of mass inside the Earth. P_i^j are the Legendre polynomials of degree i and order j.

4.4.3.1 Use of Ecliptic Coordinates

The reference plane to measure the precession-nutation motion being the ecliptic of the date, it is convenient to express the Legendre polynomials $P_2(\sin\delta)$ and

---

[5] α_E must not be confused with the classical right ascension α measured from an equinox.

$P_2^2(\sin \delta) \cos 2\alpha$ as a function of the ecliptic longitude λ and latitude β of the perturbing body. Kinoshita et al. [40] showed how this step is possible trough the intermediary of the modified Jacobi polynomials. After expansion, we find [38]:

$$
\begin{aligned}
P_2(\sin \delta) = \frac{1}{2}(3\cos^2 J - 1) \Big[& \frac{1}{2}(3\cos^2 I - 1)P_2(\sin \beta) \\
& - \frac{1}{2}\sin 2I \, P_2^1(\sin \beta)\sin(\lambda - h) - \frac{1}{4}\sin^2 I P_2^2(\sin \beta)\cos 2(\lambda - h) \Big] \\
& + \sin 2J \Big[-\frac{3}{4}\sin 2I \, P_2(\sin \beta)\cos g \\
& - \frac{1}{4}\sum_{\varepsilon=\pm 1}(1 + \varepsilon\cos I)(-1 + 2\varepsilon\cos I)P_2^1(\sin \beta)\sin(\lambda - h - \varepsilon g) \\
& - \sum_{\varepsilon=\pm 1}\frac{1}{8}\varepsilon\sin I(1 + \varepsilon\cos I)P_2^2(\sin \beta)\cos(2\lambda - 2h - \varepsilon g) \Big] \\
& + \sin^2 J \Big[\frac{3}{4}\sin^2 I P_2(\sin \beta)\cos 2g \\
& + \frac{1}{4}\sum_{\varepsilon=\pm 1}\varepsilon\sin I(1 + \varepsilon\cos I)P_2^1(\sin \beta)\sin(\lambda - h - 2\varepsilon g) \\
& - \frac{1}{16}\sum_{\varepsilon=\pm 1}(1 + \varepsilon\cos I)^2 P_2^2(\sin \beta)\cos 2(\lambda - h - \varepsilon g) \Big]
\end{aligned}
\tag{4.115}
$$

The same kind of expansion as a function of I, J, the coordinates λ, β and the Andoyer variables l, g, h is done for the expression $P_2^2(\sin \delta) \cos 2\alpha_E$ which takes place in the part of the potential, depending on the triaxiality in Eq. (4.113).

Analytically, Woolard [79] and Kinoshita [38] showed that it is possible for the Moon as well as for the Sun to express the functions $P_2(\sin \beta)$, $P_2^1(\sin \beta)\sin(\lambda - h)$ and $P_2^2(\sin \beta)\cos 2(\lambda - h)$ etc. as Fourier series with arguments Θ_ν themselves combination of the five Delaunay arguments, which are

- l the mean anomaly of the Moon
- l' the mean anomaly of the Sun
- Ω the longitude of the node
- $F = \lambda_M - \Omega$, where λ_M is the mean longitude of the Moon
- $D = \lambda_M - \lambda_S$, where λ_S is the mean longitude of the Sun.

Kinoshita and Souchay [39] as well as Souchay et al. [67] generalized this kind of formulation by introducing the mean longitudes of the planets (excepted Neptune whose influence is negligible) when including the direct and indirect planetary perturbations, as well as a the general precession on longitude p_A.

In these works, Θ_ν are written

$$
\begin{aligned}
\Theta_\nu = i_1 l + i_2 l' + i_3 F + i_4 D + i_5 \Omega + i_6 \lambda_{Me} + i_7 \lambda_{Ve} + i_8 \lambda_{Ea} \\
+ i_9 \lambda_{Ma} + i_{10}\lambda_{Ju} + i_{11}\lambda_{Sa} + i_{12}\lambda_{Ur} + i_{13}p_A
\end{aligned}
\tag{4.116}
$$

Thus we can adopt a generic formula for the expansions used:

$$\frac{1}{2}\left(\frac{a}{r}\right)^3 (1 - 3\sin^2\beta) = \sum_\nu A_\nu^0 \cos\Theta_\nu \tag{4.117.1}$$

$$\left(\frac{a}{r}\right)^3 \sin\beta\cos\beta\sin\lambda = \sum_\nu A_\nu^1 \cos\Theta_\nu \tag{4.117.2}$$

$$\left(\frac{a}{r}\right)^3 \sin\beta\cos\beta\cos\lambda = -\sum_\nu A_\nu^1 \sin\Theta_\nu \tag{4.117.3}$$

$$\left(\frac{a}{r}\right)^3 \cos^2\beta\cos 2\lambda = \sum_\nu A_\nu^2 \cos\Theta_\nu \tag{4.117.4}$$

$$\left(\frac{a}{r}\right)^3 \cos^2\beta\sin 2\lambda = -\sum_\nu A_\nu^2 \sin\Theta_\nu \tag{4.117.5}$$

4.4.3.2 Generic Formula for the Expressions of the Potential U

According to (4.113) the determination of U presupposes the knowledge of $(\frac{a}{r})^3 P_2(\sin\delta)$ and $(\frac{a}{r})^3 P_2^2(\sin\delta)\cos 2\alpha_E$. Kinoshita [38] has shown that these expressions can be conveniently expanded in the following form:

$$\begin{aligned}\left(\frac{a}{r}\right)^3 P_2(\sin\delta) &= \frac{3}{2}(3\cos^2 J - 1)\sum_\nu B_\nu \cos\Theta_\nu \\ &\quad - \frac{3}{2}\sin 2J \sum_{\varepsilon=\pm 1}\sum_\nu C_\nu(\varepsilon)\cos(g - \varepsilon\Theta_\nu) \\ &\quad + \frac{3}{4}\sin^2 J \sum_{\varepsilon=\pm 1}\sum_\nu D_\nu(\varepsilon)\cos(2g - \varepsilon\Theta_\nu)\end{aligned} \tag{4.118}$$

and

$$\begin{aligned}\left(\frac{a}{r}\right)^3 & P_2^2(\sin\delta)\cos 2\alpha_E \\ &= -\frac{9}{2}\sin^2 J \sum_\nu B_\nu \cos(2l - \varepsilon\Theta_\nu) \\ &\quad - 3\sum_{\rho=\pm 1}\sin J(1 + \rho\cos J)\sum_{\varepsilon=\pm 1}\sum_\nu C_\nu(\varepsilon)\cos(g + 2\rho l - \varepsilon\Theta_\nu) \\ &\quad - \frac{3}{4}\sum_{\varepsilon=\pm 1}\sum_{\rho=\pm 1}(1 + \rho\cos J)^2 \sum_\nu D_\nu(\varepsilon)\cos(2g + 2\rho l - \varepsilon\Theta_\nu)\end{aligned} \tag{4.119}$$

in which

$$B_\nu = -\frac{1}{6}(3\cos^2 I - 1)A_\nu^0 - \frac{1}{2}\sin 2I A_\nu^1 - \frac{1}{4}\sin^2 I A_\nu^2 \tag{4.120}$$

$$C_\nu(\varepsilon) = -\frac{1}{4}\sin 2I A_\nu^0 + \frac{1}{2}(1 + \varepsilon \cos I)(-1 + 2\varepsilon \cos I)A_\nu^1$$
$$+ \frac{1}{4}\varepsilon \sin I (1 + \varepsilon \cos I)A_\nu^2 \qquad (4.121)$$

$$D_\nu(\varepsilon) = -\frac{1}{2}\sin^2 I A_\nu^0 + \varepsilon \sin I (1 + \varepsilon \cos I)A_\nu^1$$
$$- \frac{1}{4}(1 + \varepsilon \cos I)^2 A_\nu^2 \qquad (4.122)$$

4.4.4 Generic Formula for the Expressions of the Nutations $\Delta\psi$ and $\Delta\varepsilon$

Once the potential U has been expressed as a Fourier series the nutations are determined in a straightforward manner by simple integration and partial derivatives with respect to I and h, following Eqs. (4.105), (4.106) and (4.107) [38]

• Nutation of the angular momentum axis.

For the angular momentum the nutations in longitude $\Delta\psi_{AM}$ and in obliquity $\Delta\varepsilon_{AM}$ are given by

$$\Delta\psi_{AM} = \Delta h = -\frac{1}{G \sin I}\left(\frac{\partial W}{\partial I}\right) + O(J) = k \sum \frac{E_\nu}{N_\nu}\sin\Theta_\nu \qquad (4.123)$$

with

$$E_\nu = \left[A_\nu^0 - \frac{1}{2}A_\nu^2\right]\cos I - \frac{\cos 2I}{\sin I}A_\nu^1 \qquad (4.124)$$

and

$$\Delta\varepsilon_{AM} = \Delta I = \frac{1}{G \sin I}\left(\frac{\partial W}{\partial h}\right) + O(J) = \frac{k}{\sin I}\sum i_5 \frac{B_\nu}{N_\nu}\cos\Theta_\nu \qquad (4.125)$$

where i_5 is the coefficient of Ω in the argument Θ_ν (see Eq. (4.116)) and $N_\nu = \dot{\Theta}_\nu$. k is a scaling factor given by

$$k = 3\frac{GM}{a^3 \omega_E}\frac{2C - A - B}{2C} = 3\frac{GM}{a^3 \omega_E}H_d \qquad (4.126)$$

where M is the mass of the perturbing body, a the semi-major axis of its orbit (in the case of the solar potential, M is the mass of the Sun and a the semi-major axis of the Earth), and ω_E the sidereal angular velocity of the Earth. $H_d = (2C - A - B)/2C$ is called the *dynamical ellipticity* of the Earth.[6] We will discuss below how it is determined from the precession deduced from observations.

• Nutations of the figure axis.

The nutations in longitude $\Delta\psi_f$ and in obliquity $\Delta\varepsilon_f$ of the figure axis of the Earth are deduced from the Oppolzer terms whose expressions have already been

---

[6] In the case of axisymmetry, $H_d = (C - A)/C$.

given in Eqs. (4.110) and (4.111). We have, at the first order in J [38]:

$$\Delta\psi_f = \Delta\psi_{AM} + \frac{1}{G\sin I}\left(\frac{\partial W}{\partial g}\sin g - W\cos g\right) \tag{4.127}$$

$$= \Delta\psi_{AM} + \frac{k}{\sin I}\sum_v\sum_{\varepsilon=\pm 1}\frac{\varepsilon C_v(\varepsilon)}{n_g - \varepsilon N_v}\sin\Theta_v \tag{4.128}$$

$$\Delta\varepsilon_f = \Delta\varepsilon_{AM} - \frac{1}{G}\left(\frac{\partial W}{\partial g}\cos g + W\sin g\right) \tag{4.129}$$

$$= \Delta\varepsilon_{AM} + k\sum_v\sum_{\varepsilon=\pm 1}\frac{C_v(\varepsilon)}{n_g - \varepsilon N_v}\cos\Theta_v \tag{4.130}$$

4.5 Modern Precession-Nutation Theories for a Rigid Earth Model

Face to the tremendous improvement by an order 2 or 3 of the precision in the determination of the coefficients of nutation thanks to the VLBI technique, it became necessary, at the end of the 1980's, to construct a new rigid Earth nutation model with an extreme accuracy, at the level of the sub-milliarcsecond (mas). During the 1990's several groups undertook this work, which was reckoned as fundamental. These efforts lead to the a definitive publication of the three different models called SMART97 [5], RDAN97 [54] and REN2000 [67]. These three models based on different theoretical foundations give very close results for the nutation coefficients when each of them is compared to the others [64]. The level of truncature for each coefficient of the related series of nutation in the three works was set at least to 0.1 μas instead of 0.1 mas, that is a factor 1000, with respect to the previous series constructed by Kinoshita [38] about two decades earlier. This new truncature level required to take into account more than one thousand components of nutation instead of the 106 ones in this last paper. It forced also the authors above to include new kinds of contribution, which although being very small, cannot be ignored. After giving a brief review of the way of construction of the three kinds of series of nutation above, we present in detail each of these second-order contributions

The three nutation series SMART97, RDAN97, and REN2000 differ by the methodology used for their construction. Nevertheless they give all very close results for precession-nutation of the three axes concerned: the axis of angular momentum, the axis of rotation and the axis of figure. All the related theories use the analytical solution VSOP87 [3] for the motion of the Sun and the planets, and the analytical solution ELP2000 [15] for the orbital motion of the Moon.

- For the construction of SMART97, Bretagnon et al. [5] used an iterative analytical method based on Eulerian dynamical equations, already described in Sect. 4.4.1.3. They also used a numerical integration to test the validity of the analytical developments, finding a remarkable agreement, of 16 μas for ψ and 8 μas for ε.

- For RDAN97, Roosbeek and Dehant [54] used the torque approach. The Lagrangian equations expressing the rigid Earth response to the torque induced by the external bodies, as seen in Sect. 4.4.1, are solved analytically. In order to validate and further test their analytical model, they have also computed a benchmark series called RDNN97 built from the DE403/LE403 ephemerids [68] and from a numerical integration. Their comparison between RDAN97 and RDNN97 shows that, in the time domain, the maximum difference is 62 μas for $\Delta\psi$ and 29 μas for $\Delta\varepsilon$, whereas in the frequency domain they are respectively 6 μas and 4 μas.
- For REN2000, Souchay et al. [67] up-dated the theory set up by Kinoshita [38] based on Hamiltonian equations and described in details in Sect. 4.4.3. Souchay [64] compared also the analytical nutation given by this series with numerical integration. The r.m.s. of the residuals do not exceed 5 μas both for $\Delta\psi \sin\varepsilon$ and $\Delta\varepsilon$.

The three independent models of nutation for a rigid Earth model mentioned above show a remarkable agreement both between themselves (at the level of 1 μas for the amplitude of each individual coefficient) as with numerical integration of the equations of motion. We can conclude that any of these models is well suited to serve as a basis for a more sophisticated theory of nutation involving a real Earth with non rigid aspects.

Note that an iterative semi-analytical method based on Eulerian dynamical equations similar to that of Bretagnon (1997) was proposed [12] for integrating the equations directly as functions of the coordinates of the CIP in the GCRS.

4.5.1 The Construction of a Highly Accurate Rigid Earth Precession-Nutation Model

The construction of a highly accurate rigid Earth precession-nutation model requires a very accurate determination of the dynamical ellipticity of the Earth as well as the investigation of second-order contributions which cannot be neglected anymore, as given the level of truncature (0.1 μas) of the Fourier series of nutation. They can be enumerated as:

- the direct planetary effects
- the indirect planetary effects
- the effects of the triaxiality of the Earth
- the contributions due to second-order geopotential (J_3, J_4)
- the crossed-nutation effects
- the J_2 and planetary tilt effects
- the geodetic precession

4.5.1.1 The Observed Precession and the Determination of the Dynamical Ellipticity of the Earth

The fit between the observed value of the lunisolar precession in longitude and its theoretical formula allows the determination of the dynamical ellipticity of the Earth $H_d = (2C - A - B)/2C$ which is the fundamental parameter for the calculation of the potential U in Eq. (4.113), and as a consequence for the calculation of the amplitude of all the nutation coefficients. The general precession in longitude p_A up-dated by Williams [78] is $5028''.7700/\text{cy}$, which represents a $-0''.3266/\text{cy}$ correction with respect to a previous value adopted by the IAU 1976 General Assembly [41]. As it was explained by Kinoshita and Souchay [65], this very accurate value coming from modern observations, in particular from VLBI data, and established with respect to the moving ecliptic of the date, includes not only the lunisolar precession, but also a combination of second-order effects. These are a spin-orbit coupling effect in the Earth-Moon system ($0''.380/\text{cy}$), the effects due to the J_4 geopotential ($-0''.0026/\text{cy}$), to the direct planetary gravitational influence ($-0''.0321/\text{cy}$), to the *geodetic precession* as given by Barker and O'Connell [2]. This geodetic precession, also called the *De Sitter precession* is the relativistic rotation of the geocentric fixed celestial system with respect to the barycentric one. Its amplitude is $1''.9194/\text{cy}$ [78].

Moreover, the difference due to the adoption of a fixed ecliptic or a moving ecliptic, called *planetary precession* amounts to $11''.8745/\text{cy}$. All these contributions must be removed from the observed value of the general precession in longitude p_A, to isolate the sole lunisolar contribution ψ_A with respect to a fixed ecliptic. We find $\psi_A = 5040.6445''/\text{cy}$. Now ψ_A is also given by the following formula [65]:

$$\psi_A = \psi_A^{Moon} + \psi_A^{Sun}$$

$$= 3H_d \left(\left(\frac{M_\circ}{M_\circ + M_\oplus} \right) \left(\frac{n_M^2}{\Omega} M_0 + \left(\frac{M_\odot}{M_\odot + M_\circ + m_\oplus} \right) \left(\frac{n_E^2}{\Omega} \right) S_0 \right) \right) \cos \varepsilon_A$$

$$\tag{4.131}$$

where $M_\odot$, $M_\oplus$ and $M_\circ$ are respectively the masses of the Sun, the Earth and the Moon, n_E and n_M the mean motions of the Earth-Moon barycenter and of the Moon, and Ω the sidereal angular rate of rotation of the Earth. M_0 and S_0 are quantities coming directly from computations of the lunisolar potential. They are respectively the constant terms in the expressions $\frac{1}{2}(\frac{a}{r})^3 (1 - 3 \sin^2 \beta)$ for the Moon and for the Sun, where λ, β and r are the ecliptic coordinates of the perturbing body (Sun or Moon) with respect to the moving equinox and ecliptic of the date (for the Sun $\beta \approx 0$).

The correspondence between ψ_A as deduced from the observational value of p_A, at the left hand side of Eq. (4.131), and its theoretical expression at the right hand side enables to determine H_d given its status of sole unknown parameter. In fact each of the nutation theories above is associated by its own estimation of H_d. Bretagnon et al. [4] in SMART97 find $H_d = 0.0032737668$ while Souchay and Kinoshita [66] in REN2000 have $H_d = 0.0032737548$ and Roosbeek and Dehant in RDAN97 have $H_d = 0.0032737674$.

4.5.1.2 The Main Lunar Terms

The major contribution to the nutation, as considering the importance of the effect and the number of coefficients, comes from the Main Problem of the Moon, that is to say from the three-body problem involving the Moon orbiting around the Earth in a quasi-Keplerian motion greatly perturbed by the Sun [15]. As a consequence the influence of the planetary perturbations on the Moon's orbit are treated independently and will be considered later. Thus the only arguments entering in the expansions of the angles Θ_ν are the Delaunay's arguments l, l', F, D and Ω. The leading nutation components are those with arguments Ω and 2Ω and respective periods 18.6 y and 9.3 y [67]: $\Delta\psi = -17''.2805921 \sin\Omega + 0''.2090296 \sin 2\Omega$ and $\Delta\varepsilon = 9''.22289220 \cos\Omega - 0''.0903611 \sin 2\Omega$. For the nutation figure axis of the rigid Earth, Souchay et al. [67] find 583 coefficients in longitude and 486 in obliquity, when adopting a truncature level of 0.1 µas.

4.5.1.3 The Main Solar Terms

The main solar terms are those due to the quasi-Keplerian motion of the Earth. In other words, it comes from the expansions of the geocentric ecliptic coordinates of the Sun $\lambda_\odot$, $\beta_\odot \approx 0$ and $r_\odot$ involved in the expression of the solar potential given by Eqs. (4.113) and (4.115) following classical expansions of $a/r_\odot$ and $\lambda_\odot$ as a function of the eccentricity of the Earth. These terms, with arguments Θ_ν linear combinations only of the five Delaunay's arguments in Eq. (4.116), must be separated from those with arguments Θ_ν including also the mean longitude of the planets, which are coming from the perturbations of the planets and are called indirect planetary effects (see Sect. 4.5.1.5). The larger term in the category of the main solar terms is the semi-annual one, with period 182.621 days (see Eqs. (4.66) and (4.67)); its amplitude is $\Delta\psi_{s.a.} = -1''.317090 \sin(2F - 2D + 2\Omega)$ in longitude and $\Delta\varepsilon_{s.a.} = 0''.573034 \cos(2F - 2D + 2\Omega)$ in obliquity.

4.5.1.4 The Direct Planetary Effects

Like the Moon and the Sun, the planets induce also nutations of the Earth's axes. Vondrak [70] calculated for the first time these direct influences of the planets on the nutation, showing that they could reach the 0.1 mas level for individual components. Independently of the three theories considered here, Williams [78] calculated all the coefficients related to the direct influence of the planets, up to 0.5 µas, both for $\Delta\psi \cos\varepsilon$ and for $\Delta\varepsilon$. At this level of truncature he found 1, 103, 26, 22, 5 and 1 terms respectively for Mercury, Venus, Mars, Jupiter, Saturn and Uranus, the influence of Neptune being negligible. Exhaustive tables of the direct influences of the planets included in REN2000 can be found in Souchay and Kinoshita [66] which show a perfect agreement with Williams [78]. The argument of each component is a linear combination of the longitude of the Earth λ_{Ea}, of the longitude of the perturbing planet considered and of the general precession in longitude p_A. The leading terms in longitude and obliquity are by far due to Venus and Jupiter. They are, in µas:

$$\Delta\psi = 215.0\sin(3\lambda_{Ve} - 5\lambda_{Ea} - 2p_A) + 84.6\sin(\lambda_{Ve} - \lambda_{Ea})$$
$$- 50.4\sin(4\lambda_{Ve} - 6\lambda_{Ea} - 2p_A) + 34.9\sin(2\lambda_{Ve} - 4\lambda_{Ea} - 2p_A)$$
$$+ 35.0\sin(2\lambda_{Ve} - 2\lambda_{Ea}) - 106.2\sin(2\lambda_{Ju} + 2p_A) + 33.4\sin\lambda_{Ju}$$

$$(4.132)$$

$$\Delta\varepsilon = 93.2\cos(3\lambda_{Ve} - 5\lambda_{Ea} - 2p_A) - 21.9\sin(4\lambda_{Ve} - 6\lambda_{Ea} - 2p_A)$$
$$+ 15.1\cos(2\lambda_{Ve} - 4\lambda_{Ea} - 2p_A) + 46.0\cos(2\lambda_{Ju} + 2p_A) \qquad (4.133)$$

4.5.1.5 The Indirect Planetary Effects

The indirect planetary effects, first pointed out and estimated by Vondrak [71, 72] originate from the small perturbations of the planets on the orbital motion of the Moon around the Earth and of the Earth around the Sun. These perturbations affect the relative ecliptic coordinates λ and β of the body causing the nutation (the Moon or the Sun). In their turn, these little changes cause a change in the perturbing potential exerted by the body. In REN2000 [67] the corresponding terms of nutation can be recognized easily by the nature of their arguments, as a linear combination of the Delaunay variables l, l', F, D and Ω, of the general precession in longitude p_A and of the mean longitudes of the planets λ_{Me}, λ_{Ve} etc. At high frequency the indirect planetary effects due to Moon are dominated by two components with arguments $-l_M + 2F + 2\Omega + 18\lambda_{Ve} - 16\lambda_{Ea}$ and $l_M + 2F + 2\Omega - 18\lambda_{Ve} + 16\lambda_{Ea}$ and the same amplitude of 14.1 µas in $\Delta\psi$. Moreover for a 100 y time interval, the peak-to-peak amplitude of these planetary effects are of the order of 1 mas both for the Moon's and the Sun's parts.

4.5.1.6 The Crossed Nutation Effects

When computing the coefficients of nutation at first order, through the intermediary of $P_2(\sin\delta)$ as expressed in Eq. (4.115), the obliquity angle I as well as the longitude λ of the perturbing body are determined without taking into account the nutations. In fact, they must be replaced respectively by $I + \Delta I$ and $\lambda - \Delta h$. In other words, the nutation itself causes a slight modification of the position of the equator which in its turn provokes a slight modification of the potential U exerted by the perturbing body and in the determining function W given by Eq. (4.105), which results in *crossed nutation effects* when applying Eqs. (4.106) and (4.107). They concern 68 components for $\Delta\psi$ and 40 components for $\Delta\varepsilon$, up to 0.1 µas [67]. The leading contribution concerns the component with argument 2Ω resulting from the crossed nutations of the leading term with argument Ω. It amounts to 1.220 mas for $\Delta\psi$ and -0.238 mas for $\Delta\varepsilon$.

4.5.1.7 The J_2 Tilt Effects

The J_2 tilt effect comes from the particular perturbation on the motion of the Moon around the Earth due to the shape of the Earth, i.e. its equatorial bulge. These perturbations involving J_2 modify in their turn the potential exerted by the Moon on the Earth. This change in the potential generates some additional second-order contributions which affect in a significant manner the leading nutation coefficients of lunar origin, with arguments Ω and 2Ω.

4.5.1.8 The Planetary Tilt Effect

This effect was pointed out by Williams [78]: the orbit planes of the planets have small inclinations with respect to the ecliptic plane. As a consequence of the planetary attractions, the ecliptic planes moves. The Moon's mean plane of orbital precession follows the moving ecliptic closely, but not perfectly. This motion causes a $1''.4$ tilt of the plane of orbital precession to the ecliptic. This result in an additional torque on the oblate Earth.

4.5.1.9 Effects due to the Triaxiality of the Earth

The triaxiality of the Earth is characterized by the relative difference $(B - A)/C$ between the moments of inertia along the principal axes perpendicular to the figure axis. Were the Earth perfectly axisymmetric, the triaxiality is zero. For the real Earth we have $(B - A)/(2C - A - B) = 0.0033536$. The triaxiality takes part in the perturbing potential U_2 through the expression $(A - B/4) \times P_2^2(\sin \delta) \cos 2\alpha_E$ in Eq. (4.113). The presence of the component $\cos 2\alpha_E$ with semi-diurnal period combined with long periodic components in $P_2^2(\sin \delta)$ results after integration in quasi-semi-diurnal terms of nutation. They are listed in Souchay and Kinoshita [66] up to 0.1 µas. Two reasons lead to the relatively small values of the nutations due to the triaxiality: first the smallness of the ratio above; second the fact that when carrying out the integration according to Eq. (4.105) a large frequency value appears at the denominator, due to the very high semi-diurnal frequency. This contribution is dominated by 3 coefficients at periods 0.518 d, 0.500 d, 0.499 d with respective arguments $2\Phi - 2F - 2\Omega$, $2\Phi - 2F + 2D - 2\Omega$ and 2Φ, where Φ is the angle of sidereal rotation of the Earth. The respective amplitudes are 27.1 µas, 12.5 µas, -37.8 µas for $\Delta\psi$ and 11.0 µas, 4.7 µas and 15.0 µas for $\Delta\varepsilon$. Note that the largest coefficient originates both from the influence of the Moon and of the Sun. At last the combination of these sinusoidal terms with very close frequencies leads to a beating.

4.5.1.10 Effects due to Second-Order Potential J_3

The second order geopotential coefficient J_3 acts on the second order potential exerted by the Moon with the intermediary of the component $J_3 P_3(\sin \delta)$ in

Eq. (4.114). Because of the scaling factor $(J_3/J_2) \times (a_E/a_M)$ which characterizes the amplitudes of the corresponding nutations with respect to the first order ones depending on J_2, these amplitudes are comparatively much smaller. 17 coefficients both for $\Delta\psi$ and $\Delta\varepsilon$ are found larger than 0.1 µas [27, 66]. This contribution is characterized by a very large set of frequencies, the smallest period being 6.8 d, and the largest one 20935 y. The leading coefficient, in mas, is $-0.105 \sin(-l_M + F + \Omega)$ for $\Delta\psi$ and $-0.1089 \cos(-l_M + F + \Omega)$ for $\Delta\varepsilon$.

4.6 Modern Nutation Theory for a Non-rigid Earth Model

The nutation is almost entirely due to the torques resulting from the gravitational action of celestial bodies on the equatorial bulge of the Earth. At first approximation, one can use as a proxy, the so called rigid Earth nutation series representing the action of the torques on a hypothetical rigid Earth, having the same moments of inertia and high order moments as the real Earth. This rigid Earth nutation has been largely discussed in the previous sections. Nevertheless, with the appearance of modern observational techniques, and particularly the VLBI (Very Long Baseline Interferometry) in the early 1980's the precision of estimates obtainable for the Earth orientation parameters, among which the nutation $(\Delta\psi, \Delta\varepsilon)$, has increased greatly. This fact, combined with the increasing volume and longer time span of the data sets available, has made it possible to estimate the amplitudes of a significantly larger number of nutation components and a much accurate value of the precession rate. Face with these developments, the rigid-Earth nutation theory, starting from the early 1980's, could no more match the accuracy of observations. In other words, the various effects due to the non rigidity of the Earth, such as changes in matter distribution, atmospheric pressure variations, oceanic motions, frictions between the core and the mantle, etc., although remaining all relatively small, could no more be neglected in view of the quality of observational data.

4.6.1 Definition of Prograde and Retrograde Circular Nutations

In order to deal with non-rigid Earth nutations, experts in this topic, as geophysicians, introduce the concept of prograde and retrograde circular components of nutation. In the following, we present their definition. Generally astronomers are concerned with the lunisolar nutations in longitude $\Delta\psi$ and in obliquity $\Delta\varepsilon$ as a Fourier series in the form

$$\Delta\psi = \sum_{\nu} \Delta\psi_{\nu I} \sin \nu \Omega_0 t + \Delta\psi_{\nu O} \cos \nu \Omega_0 t \qquad (4.134)$$

$$\Delta\varepsilon = \sum_{\nu} \Delta\varepsilon_{\nu I} \cos \nu \Omega_0 t + \Delta\varepsilon_{\nu O} \sin \nu \Omega_0 t \qquad (4.135)$$

where $\Delta\psi_{\nu I}$ and $\Delta\psi_{\nu O}$ are, respectively, the in-phase and out-of-phase coefficients of the nutation in longitude, and $\Delta\varepsilon_{\nu I}$ and $\Delta\varepsilon_{\nu O}$ in obliquity. Here $\Omega_0 = 7.292115 \times 10^{-5}$ rad/s stands for the mean sidereal angular rotation rate. Notice that the out-of-phase components are generally very small with respect to the in-phase ones, for they characterize dissipative processes. Moreover we have shown in Eq. (4.116) that the frequency $\Theta_\nu = \nu\Omega_0$ is a combination of fundamental astronomical arguments.

Once the formula above have been established, the in-phase parts of the components of $\Delta\psi$ and $\Delta\varepsilon$ for any particular frequency $\Theta_\nu = \nu\Omega_0$ constitutes an elliptical nutation, whereas the out-of-phase parts constitutes another one. In their turn, these paired elliptical nutations can be resolved into two circular components, one representing a uniform rotation of the figure axis around an inertial (space-fixed) axis in the prograde sense, and the other, in the retrograde sense. The combination of the two prograde and retrograde circular nutations results respectively in the complex components $\eta^{(pro)}$ and $\eta^{(ret)}$. This is materialized by the following relationships [45]:

$$\eta^{(pro)} = -\frac{1}{2}\left(\Delta\varepsilon_{\nu I} - \frac{\nu}{|\nu|}\Delta\psi_{\nu I}\sin\varepsilon_A\right) + \frac{i}{2}\frac{\nu}{|\nu|}\left(\Delta\varepsilon_{\nu O} + \frac{\nu}{|\nu|}\Delta\psi_{\nu O}\sin\varepsilon_A\right)$$
(4.136)

$$\eta^{(ret)} = -\frac{1}{2}\left(\Delta\varepsilon_{\nu I} + \frac{\nu}{|\nu|}\Delta\psi_{\nu I}\sin\varepsilon_A\right) - \frac{i}{2}\frac{\nu}{|\nu|}\left(\Delta\varepsilon_{\nu O} - \frac{\nu}{|\nu|}\Delta\psi_{\nu O}\sin\varepsilon_A\right)$$
(4.137)

where ε_A is the mean obliquity.

4.6.2 Early Non-rigid Earth Nutation Theories

The earliest nutation theories started from the Earth as a rigid ellipsoid [38, 79]. We have seen previously that in this approximation only the Earth's principal moments of inertia and the amplitude and frequency of the tidal force are important. In pioneer works dealing with non-rigidity, Jeffreys and Vicente [35, 36] and Molodensky [50] greatly extended these results by including the effects of a fluid core (already introduced by Poincaré [53]) and of the elasticity within the mantle. They found differences from rigid Earth results of as much as 0.02″ for both the principal nutation with period 18.6 y and the leading nutation of solar origin, with semi-annual period. In comparison with the precision of the observations in the 1970's these effects could no more be neglected. These analytical theories take into account a simplified model of core and mantle deformation computed from a spherical non-rotating shell. Shen and Manshina [59], in a numerical way, as well as Sasao et al. [55], analytically, started from these previous works to include more complete dynamical and structural models of fluid core.

4.6.3 The Nutation Series of Wahr

The nutation series of Wahr [74] has been the standard of reference for roughly 20 years. It was adopted by the International Astronomical Union (IAU) as the basic nutation series named the IAU 1980 nutation [58]. It was computed by solving the equations for the field of displacements produced by the action of the tide generating potential (TGP) throughout the Earth, as applied to an oceanless elastic, ellipsoidal Earth model derived on the assumption of a hydrostatic equilibrium. Wahr theory can be considered as a further extension of the previous investigations mentioned above, with accounting more completely for the Earth's ellipticity and rotation. For that purpose he used techniques developed by Smith [62] and Wahr [73]. The first author described the linearization of infinitesimal motion for a rotating, slightly elliptical, self-gravitating, elastic, hydrostatically prestressed and oceanless Earth. Wahr [73] demonstrated that the forced motion of a rotating Earth could be expanded as a decoupled sum of normal modes of the Earth. On the opposite of what was done previously, elliptical and rotational effects were considered by Wahr [74, 75] to compute the rotational motion.

   Wahr adopted a model of Earth interior called the model 1066A of Gilbert and Dziewonski [24], based on the assumption of a hydrostatic equilibrium. In order to compute semi-analytically the nutation coefficients, Wahr [73] established a formula expressing the transfer function between a given coefficient for the rigid Earth model, with frequency ω, and the corresponding non rigid Earth coefficient. The ratio between these two coefficients is given by $\eta(a, \omega)/\eta_r(\omega)$ so that

$$
\begin{aligned}
\frac{\eta(a, \omega)}{\eta_r(a, \omega)} &- 1 \\
= &\left[B_0 + (\omega - 0.927\Omega) \left[\frac{B_1}{\omega_1 - \omega} + \frac{B_2}{\omega_2 - \omega} + \frac{1.06}{\omega + 3.28 \times 10^{-3}\Omega} \right] \right] \\
&\times \left[\frac{\Omega - \omega}{\Omega} \right] \left[\frac{\omega}{\Omega} + 3.28 \times 10^{-3} \right]
\end{aligned}
\tag{4.138}
$$

where B_0, B_1 and B_2 are frequency-independent constants. ω_1 and ω_2 are the eigenfrequencies of the Chandler Wobble (CW) and the Free Core Nutation (FCN) respectively. Ω is the frequency of the sidereal rotation of the Earth.

   Soon after its establishment, the predictions of the Wahr theory have been found to differ from VLBI observational data by much more than the uncertainties in the data itself. Face to this unsatisfactory result, an empirical series, called IERS96 series, constructed on the basis of some corrections to leading nutation coefficients from O–C discrepancies, was established, giving close agreement to the data [28, 49]. This series was still improved further by Shirai and Fukushima [60, 61], noticeably by introducing an estimated exponentially decaying free core nutation amplitude. As in Wahr [74] these two empirical series express the nutation amplitudes in terms of a resonant formula for the transfer function.

4.6.4 Further Improvements

An exhaustive review of all the improvements done in the fields of non rigid Earth nutation was done by Mathews et al. [48]. An important step towards a better geophysical accounting of nutation was taken successively by Gwinn et al. [26] and Herring et al. [29, 30], finding that a value higher than approximatively 5 % than that scheduled by the hydrostatic equilibrium state is needed for the dynamical ellipticity of the fluid core to close a gap of approximatively 2 mas (milliarcseconds) found between the observed nutation and the IAU 1980 values. This concerns the in phase part of the amplitude of the annual retrograde nutation. In parallel some studies were devoted to the computation of the effects of the ocean tides [57, 76] as well as those coming from the mantle anelasticity [77].

Alternative theoretical investigations start from the torque equations for the ellipsoidally stratified deformable Earth and its core. First developed by Molodensky [50] they were improved by Sasao et al. [55] and generalized by Mathews et al. [46, 47]. They are well suited for taking into account the dynamics of the inner core. In the last work, the torque equations and an accompanying kinematical equation reduce to a set of simultaneous linear algebraic equations. Such formula are very efficient to take into account the nonhydrostatic ellipticity and the use of an electromagnetic coupling at the core mantle boundary explaining the residuals of approximatively 0.4 mas remaining in the out-of-phase part of the retrograde annual component after taking into account anelasticity and ocean tides effects.

An independent approach was developed abundantly and exhaustively by Getino and Ferrandiz [21–23], starting from the same canonical equations as Kinoshita [38]. These authors introduced modified canonical variables to apply the theory to a non rigid Earth model taking into account an elastic mantle, a FOC (fluid outer core), a SIC (solid inner core) and a delay in the elastic response of the Earth with oceanic corrections. Although a final model with observations based on this work would have offered a better fit with the observations, as the IERS96 nutation series did previously, Mathews et al. [48] underlined the lack of explicit information concerning the fit of several parameters and their physical interpretation.

4.6.5 The Normal Modes of the Rotation of the Earth

To determine an accurate non rigid Earth nutation theory, it looks fundamental to know the normal modes of free rotational motions of the Earth as well as the eigenfrequencies σ_α which are associated with these normal modes. We will see later in Sect. 4.7.1 that those eigenfrequencies play a leading role in the transfer function from rigid Earth to non rigid Earth nutations. The principal normal modes acting as resonance modes in the transfer function are enumerated below. The list is not exhaustive: a quasi-infinite list of other normal modes exist, as the elastic vibration modes, which should not bring wobble components.

4.6.5.1 The Chandler Wobble (CW)

The only normal mode concerning the rigid Earth (in the approximation of the axisymmetric case) has already been described in Sect. 4.4.1.4. It is called the *free polar motion* or *Eulerian free wobble*. If the Earth were perfectly rigid, the frequency of this free wobble should be $(C - A/C)\Omega$ (where Ω already defined in Sect. 4.6.3 is the mean sidereal rotation rate), and its period $C/C - A = 305$ d. Chandler identified for the first time this free wobble from observational data in 1891, fixing its period to 14 months. The difference was soon interpreted as due to the deformability of the Earth [42], as well as the existence of a fluid core [34, 53] Much more recently, Smith and Dahlen [63] showed that the pole tide produced by the oceans should bring also a significant contribution. The amplitude of the Chandler wobble is variable, never exceeding $1''$.

4.6.5.2 The Retrograde Free Core Nutation (RFCN)

The free core nutation (FCN) is a normal mode of the Earth, associated with the existence of a rotating ellipsoidal fluid core inside a rotating elastic mantle. It occurs due to the excitement of a mis-alignment of the instantaneous rotation axes of the core and the mantle. More precisely the non spherical shape of the core-mantle boundary (CMB) has the consequence that any rotational motion of the fluid core relative to the mantle, with the core's rotation axis inclined to the symmetry axis of the CMB, causes imbalance of fluid pressure on the boundary, and a resultant torque which tends to bring the two axes into alignment. For the PREM model of the Earth [19], the theoretical FCN period, computed for an Earth in hydrostatic equilibrium, is 458 sidereal days in the retrograde direction. By analyzing the nutation amplitudes determined from VLBI observations, it has been shown that the FCN period is around 432 sidereal days [17, 26]. Moreover, analyses of gravimetric recordings in the diurnal frequency band also give a FCN period around 430 sidereal days [51]. The significant difference between the theoretical and observational values above can be explained by the increase of about 5 % of the core flattening with respect to the hydrostatic equilibrium value [26] One characteristic of the FCN is its variable amplitude and phase.

4.6.5.3 The Prograde Free Core Nutation (PFCN)

In the early 90's, several authors have shown that the presence of a solid inner core (SIC) gives rise to another diurnal wobble mode which corresponds to a prograde nutation [46, 47], De Vries and Wahr [18]. These last authors refer to this new mode as the *free inner core nutation* (FICN), whereas it is also sometimes called the *prograde free core nutation* (PFCN) Mathews and Shapiro [45]. Mathews et al. [47] found that the relation between the wobble motion of the fluid outer core and the mantle, in this mode, are very close to that in the already well known retrograde FCN.

4.6.5.4 The Inner Core Wobble (ICW)

The last free motion taking part in the transfer function is the inner core wobble
(ICW). First mentioned by Mathews et al. [46, 47], it is predominantly a rotation of
the figure axis of the inner core relative to the mantle. Moreover it would reduce to
the free wobble of the solid inner core (SIC) if the forces between the SIC and the
rest of the Earth could vanish.

4.7 The IAU 2006/2000 Precession Nutation

The IAU 2006/2000 Precession-nutation [10, 13] is composed of the IAU 2000
nutation and the IAU 2006 precession that replaced the precession component of
the IAU 2000 precession-nutation. That component consisted only in corrections,
$\delta\psi_A = -0.29965''$/century and $\delta\omega_A = -0.02524''$/century, to the precession rates
(in longitude and obliquity referred to the J2000.0 ecliptic), of the IAU 1976 precession and hence did not correspond to a dynamical theory [11].

4.7.1 The IAU 2000 (MHB2000) Nutation

The present conventional model of nutation adopted by the International Astronomical Union in 2000, called MHB2000, has been developed by Mathews et al. [48].
It is based on the REN2000 rigid Earth nutation series [67] of the axis of figure.
The rigid Earth nutation was transformed to the non rigid Earth nutation by applying the MHB2000 transfer function to the full REN2000 series of the corresponding
prograde and retrograde nutations and then converting back into elliptical nutation
components. This transfer function is based on the solution of the linearized dynamical equations of the wobble-nutation problem and makes use of estimated values of
seven of the parameters appearing in the theory called the BEP (Basic Earth Parameters).

The BEP were preliminary defined by Mathews et al. [46]. They consist of el-
lipticities e, e_F and e_S, and the mean equatorial moments of inertia A, A_F, and A_S
of the Earth, the fluid outer core (FOC), and the solid inner core (SIC). Other BEP
are compliance parameters κ, γ, ζ, β ... which represent the deformabilities of the
Earth and of its core regions under different kinds of forcing. Another BEP is the
density ρ_F of the FOC at the inner core boundary (ICB). Others BEP characterize the gravitational coupling between the SIC and the rest of the Earth. They are
obtained from a least-squares fit of the theory to an up-to-date precession-nutation
VLBI data set Herring et al. [32]. The MHB2000 model improves the IAU 1980
theory of nutation by taking into account the effects of mantle unelasticity, ocean
tides, electromagnetic couplings produced between the FOC and the mantle as well
as between the SIC and the FOC [8]. Moreover it takes in consideration non linear

terms which have hitherto been ignored in previous formulations. The axis of reference, often called *axis of figure* is the axis of maximum moment of inertia of the Earth in steady rotation (ignoring time dependent deformations).

4.7.1.1 Analytical Formulation in MHB2000

At the basis of the formulation, consider a forced or free nutation having an angular frequency of τ cycles per sidereal day (cpsd) in space, 1 cpsd corresponding to the mean sidereal rotation period with angular velocity Ω. This nutation itself corresponds to a wobble of the Earth's mantle, which is a circular motion of its rotation axis around its geometric axis, with frequency σ cpsd with respect to an Earth-fixed frame. Thus we have $\sigma = \tau - 1$. The amplitude $\bar{m}(\sigma)$ of this wobble, the amplitudes $\bar{m}_F(\sigma)$ and $\bar{m}_S(\sigma)$ of accompanying wobbles relative to the mantle of the FOC and the SIC, as well as the amplitudes $\bar{n}_S(\sigma)$ of the effect of the polar axis of the SIC from that of the mantle are the dynamical variables of the wobble-nutation problem on the frequency domain. According to Mathews et al. [46, 47] the amplitude $\bar{\eta}(\sigma)$ of the nutation associated with the wobble of frequency σ cpsd is related to $\bar{m}(\sigma)$ by

$$\bar{\eta}(\sigma) = -\frac{\bar{m}(\sigma)}{1 + \sigma} \qquad (4.139)$$

It follows that the transfer function $T(\sigma, e)$ from the amplitude for the rigid Earth to that for the non rigid Earth is the same for the wobble and the corresponding nutation. It is presented as a resonance expansion of the form

$$T(\sigma, e) = R + R'(1 + \sigma) + \sum_\alpha \frac{R_\alpha}{\sigma - \sigma_\alpha} \qquad (4.140)$$

where the resonance frequencies σ_α are associated with four normal modes described in Sect. 4.6.5: the Chandler wobble (CW) the retrograde free core nutation (RFCN), the prograde free core nutation (PFCN) due to the presence of an elliptical solid inner core, and a free wobble of the inner core (ICW). Mathews et al. [48] modified in some extent this formula and adopted a generalized transfer function expressed in the form

$$T(\sigma, e/e_R) = \frac{e_R - \sigma}{e_R - 1} \cdot N_0 \big(1 + (1 + \sigma)\big) \sum_{k=1}^{4} \frac{N_\alpha}{\sigma - \sigma_\alpha} \qquad (4.141)$$

with

$$N_0 = \frac{H_d}{H_{dR}} = \frac{e/1 + e}{e_R/1 + e_R} \qquad (4.142)$$

where H_d is the dynamical ellipticity of the Earth, already defined earlier, and H_{dR} its value in the rigid case.

4.7.1.2 Mantle Anelasticity Effects

Mantle anelasticity causes a small frequency-dependent phase lag in the Earth's response to periodic forcing besides altering the magnitude of the response. This anelasticity is characterized by the presence of a complex and frequency dependent shear and bulk moduli for each point of the mantle. The compliances appearing in nutation theory are computed initially for an elastic Earth model, such as the Preliminary Reference Earth Model (PREM) of Dziewonski and Anderson [19] by integrating the equations of tidal deformation, together with small contributions as the compliances coming from the Earth's ellipticity, the Coriolis force due to the Earth rotation, the differential rotations of the FOC and the SIC with respect to the mantle [7]. The anelasticity contributions to the compliances at a given excitation frequency are then computed from the same deformation equations, by evaluation of the changes in deformations resulting from the variations of the shear modulus $\mu(r)$ as studied by Wahr and Bergen [77].

4.7.1.3 Electromagnetic Coupling

The presence of an internal magnetic field influences the Earth's nutation through the effects of electromagnetic torques at the boundaries of the fluid core. Electromagnetic coupling is a consequence of the Lorentz force, which represents the force experienced by current-carrying matter in the presence of a magnetic field. The interaction between a magnetic field that crosses the outer core boundaries and the motion of conducting matter on either side of these boundaries induces an electric current, which locally perturbs the magnetic field. An increase of the Lorentz force opposes relative motion across the outer core boundaries, thereby coupling the motion of the inner core, outer core and mantle. Buffett et al. [8] calculated these effects on nutation by combining a solution for full hydrodynamic response of the fluid core. The coupling of the fluid outer core (FOC) to the mantle and the solid inner core (SIC) is described by two complex constants K^{CMB} and K^{ICB} that characterize the electromagnetic torques at the core-mantle boundary (CMB) and the inner core boundary (ICB). Predictions for K^{CMB} and K^{ICB} are compared with estimates inferred from observations of the Earth's nutation. The estimate of K^{CMB} can be explained by the presence of a thin conducting layer at the base of the mantle whose conductance has been estimated. The value of K^{ICB} can be explained with a mixture of dipole and non dipole components.

4.7.1.4 Ocean Tides Effects

Ocean tides affect nutations through changes in the inertia tensors of the Earth as well as its core regions due to the loading of the crust. Another cause is the contribution to the global angular momentum of the Earth. Ocean tidal motions in the diurnal band of frequencies are influenced by the FCN resonance. Wahr and Sasao

Table 4.1 Principal terms of nutation $\Delta\psi$ and $\Delta\varepsilon$ in the theory MHB2000 [48]

Argument	Period day	$\Delta\psi\sin('')$	$\Delta\psi\cos('')$	$\Delta\varepsilon\sin('')$	$\Delta\varepsilon\cos('')$
Ω	6798.384	−17.206416	0.0033338	0.001537	9.205233
2Ω	3399.192	0.207455	−0.001369	−0.000029	−0.089749
l'	365.260	0.147587	0.001181	−0.000192	0.007387
$2F - 2D + 2\Omega$	182.621	−1.317090	−0.001369	−0.000458	0.573034
$l' + 2F - 2D + 2\Omega$	121.749	−0.051682	−0.000052	−0.000017	0.022438
$2F + 2\Omega$	13.661	−0.227641	0.000279	0.000137	0.097846
$2F + \Omega$	13.633	−0.038730	0.000038	0.000032	0.020073
$l + 2F + 2\Omega$	9.133	−0.030146	0.000082	0.000037	0.012902

[76] have first made theoretical estimates of the contributions from ocean tides to nutation amplitudes, using as inputs the retrograde FCN eigenfrequency from Wahr [75] and tide heights from ad hoc models [52, 56]. To evaluate the role of the angular momentum $\bar{h}$ carried by the ocean tidal current it is enough to introduce it in the dynamical equations of the angular momentum. In MHB2000, values of $\bar{h}$ are taken from Chao et al. [14]. In fact accurate computation of the ocean angular momentum from ocean tide maps is difficult because of large contributions coming from small areas where the ocean is very deep.

4.7.2 The MHB2000 Nutation Series

The MHB2000 series of nutation includes 678 lunisolar terms and 687 planetary terms which are expressed as 'in-phase' and 'out-of-phase' components, together with their time-variations. That model is expected to guarantee an accuracy of about 10 µas for most of his terms. In Table 4.1 we show the principal terms of nutation $\Delta\psi$ and $\Delta\varepsilon$, with their argument, their period, their in-phase and out-of-phase amplitudes. As already calculated roughly in Eqs. (4.66) and (4.67) of Sect. 4.3.5, the largest terms have a 18.6 y period for the lunar contribution and a semi-annual period for the solar contribution, with respective arguments the longitude of the node of the Moon Ω and $2\lambda_{Earth} = 2F - 2D + 2\Omega$. The respective in-phase amplitudes are $17.206''$ and $1.317''$ for $\Delta\psi$, $9.205''$ and $0.573''$ for $\Delta\varepsilon$.

In Table 4.2 we represent the differences between the amplitudes of the principal terms nutation for a non rigid Earth model, taken from MHB2000 [48], and for a rigid Earth model, taken from REN2000 [67]. These differences reach 74 mas in $\Delta\psi$ and 23 mas in $\Delta\varepsilon$, for the leading term with argument Ω. They are also very important (40 mas and 20 mas) for the semi-annual component with argument $2F - 2D + 2\Omega$.

Table 4.2 Differences between the amplitudes of the principal terms in $\Delta\psi$ and $\Delta\varepsilon$ obtained from the non-rigid Earth theory MHB2000 [48] and the rigid Earth theory REN2000 [67]

Argument	Period day	$\Delta\psi\sin$(mas)	$\Delta\psi\cos$(mas)	$\Delta\varepsilon\sin$(mas)	$\Delta\varepsilon\cos$(mas)
Ω	6798.384	74.1760	2.9560	1.5408	−22.6769
2Ω	3399.192	−1.5742	−0.0757	−0.0323	0.5877
l'	365.260	22.0841	1.1817	−0.1924	7.5220
$2F-2D+2\Omega$	182.621	−39.6154	−1.3696	−0.4587	19.6988
$l'+2F-2D+2\Omega$	121.749	−1.7443	−0.0524	−0.0174	0.8189
$2F+2\Omega$	13.661	−6.1301	0.2796	0.1374	2.9250
$2F+\Omega$	13.633	−0.8772	0.0380	0.0318	0.6603
$l+2F+2\Omega$	9.133	−0.5633	0.0816	0.0367	0.2928

4.7.3 The IAU 2006 (P03) Precession

The IAU 2006 precession [10, 33] provides improved polynomial expressions up to the 5th degree in time t, both for the precession of the ecliptic (previously named "planetary precession") and the precession of the equator (previously named "luni-solar precession").

The precession of the equator was derived from the dynamical equations expressing the motion of the mean pole about the ecliptic pole. The convention for separating precession from nutation, as well as the integration constants used in solving the equations, has been chosen in order to be consistent with the IAU 2000A nutation. This includes corrections for the perturbing effects in the observed quantities.

In particular, the IAU 2006 value for the precession rate in longitude is such that the corresponding Earth's dynamical flattening is consistent with the MHB value for that parameter. This required applying a multiplying factor to the IAU 2000 precession rate of $\sin\varepsilon_{IAU2000}/\sin\varepsilon_{IAU2006}=1.000000470$ in order to compensate for the change (by 42 mas) of the J2000 mean obliquity of the IAU 2006 model with respect to the IAU 2000 value (i.e. the IAU 1976 value). Moreover, the IAU 2006 precession includes the Earth's J_2 rate effect (*i.e.* $\dot{J}_2=-3\times10^{-9}$/century), mostly due to the post-glacial rebound, which was not taken into account in the IAU precession models previously.

The contributions to the IAU 2006 precession rates for the 2nd order effects, the J_3 and J_4 effects of the luni-solar torque, the J_2 and planetary tilt effects, as well as the tidal effects are from Williams [78], and the non-linear terms are from MHB2000.

The geodetic precession is from Brumberg et al. [6], i.e. $p_g=1.919883''$/cy. It is important to note that including the geodetic precession and geodetic nutation in the precession-nutation model ensure that the GCRS (Geocentric celestial reference system) is without any time-dependent rotation with respect to the BCRS (Barycentric celestial reference system).

4.7.4 The Agreement of the IAU 2006/2000 Precession-Nutation with Highly Accurate VLBI Observations

The accuracy with which Earth orientation in general and precession-nutation in particular can be determined as a function of time has increased tremendously over the three past decades, due to the advances in the VLBI technology, in techniques of data analysis, and also to the expanding volume of data over a lengthening time span. VLBI (Very Long Baseline Interferometry) can be considered as the most powerful technique to measure the Earth Orientation Parameters (EOP) [31]. These parameters are related to the changes of the position of the Earth's rotation axis or more precisely the axis of the *celestial intermediate pole* (CIP) with respect to its crust, so-called polar motion, and with respect to inertial space, i.e. the precession-nutation motion. One additional EOP is related to changes in the rotation rate of the Earth, and is usually expressed as the difference between UT1 and the time standard UTC (Universal Time Coordinate).

The VLBI technique measures the differential arrival times of radio-signals from extragalactic radio-sources, which provide in particular the most stable definition of inertial system currently available, as it is materialized by the successive updates of the ICRS [43, 44]. A classical VLBI session uses a set of four to eight radio telescopes, with separations of several thousands of kilometers, which make a large amount of measurements of time delays and delay rates from usually 20 to 40 extragalactic radio-sources. Of the various factors which limit the accuracy of the determination, one of the most important is the atmospheric contribution to group delays, especially the part due to water vapor which is the most difficult to estimate reliably [20].

Nevertheless with the basic and well reckoned assumption that rigid Earth nutation is modeled with an optimal accuracy (at the level of 1 µas), the VLBI observations allow a very accurate determination of the non rigid effects of the Earth on the largest nutation coefficients. Herring et al. [32] showed that the analysis of over 20 years of VLBI data yields estimates of the nutation amplitudes with standard deviations of ≈5 µas for the nutations with periods smaller than 400 days. They show that at this level of uncertainty, the estimated amplitudes are consistent with the IAU 2006/2000 precession-nutation model which has been described previously.

Figure 4.11 shows the differences O–C (observed-calculated) between the overall nutations components dX and dY determined from combined VLBI sessions and the same nutations calculated from the series MHB2000. The remarkable agreement at the level of a few 0.1 mas (a few 100 µs) is clearly shown after 1995, whereas the residuals are much larger before that date. This is clearly due to a drastic improvements of the quality VLBI observations around this date. Notice a very dominant systematic oscillation in the residuals: it is interpreted as the retrograde Free Core Nutation (RFCN) whose origin has been explained in Sect. 4.6.5. In Fig. 4.12 we show the residuals after eliminating empirically this systematic oscillation, taking into account its changes in amplitude and phase. The very flat residuals enable to conclude that the general agreement between the theoretical and observational data is remarkable.

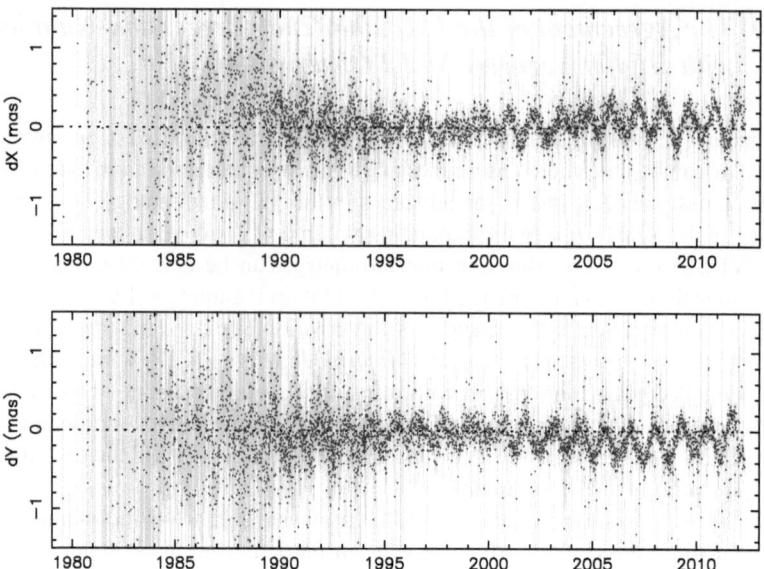

Fig. 4.11 O–C difference between the celestial motion of the pole dX and dY observed from VLBI sessions and the theoretical motion calculated from the IAU 2006/2000 precession-nutation model (credit: IVS OPA Analysis Center, Observatoire de Paris)

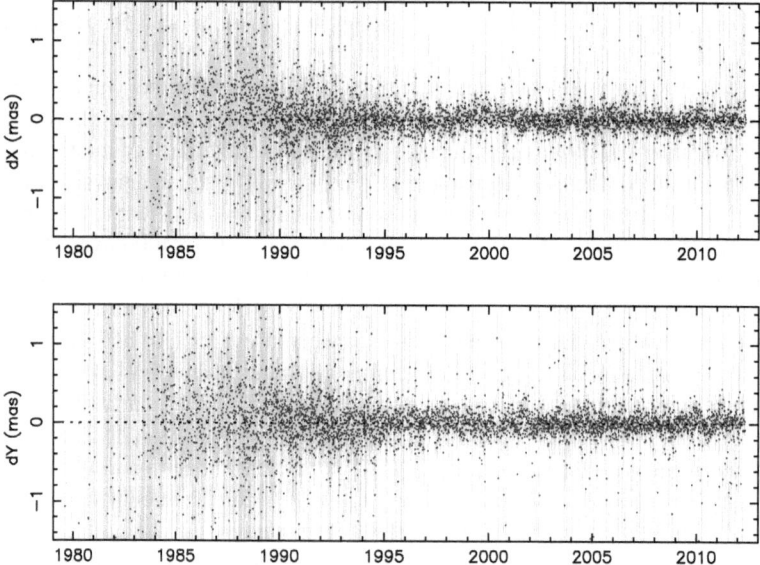

Fig. 4.12 O–C difference between the celestial motion of the pole dX and dY observed from VLBI sessions and the theoretical motion calculated from the IAU 2006/2000 precession-nutation model. The curves correspond to those in Fig. 4.10 after the FCN signal has been empirically subtracted (credit: IVS OPA Analysis Center, Observatoire de Paris)

References

1. Andoyer, H.: Bull. Astron. Paris **28**, 66 (1923)
2. Barker, B.M., O'Connell, R.F.: Phys. Rev. D **12**, 329 (1975)
3. Bretagnon, P., Francou, G.: Astron. Astrophys. **202**, 309 (1988)
4. Bretagnon, P., Rocher, P., Simon, J.L.: Astron. Astrophys. **319**, 305 (1997)
5. Bretagnon, P., Francou, G., Rocher, P., Simon, J.L.: Astron. Astrophys. **329**, 329 (1998)
6. Brumberg, V.A., Bretagnon, P., Francou, G.: In: Capitaine, N. (ed.) Proceedings Journées 1991 Systèmes de Référence Spatio-temporels, pp. 141–148. Obs. de Paris, Paris (1992)
7. Buffett, B.A., Mathews, T.A., Herring, T.A., Shapiro, I.I.: J. Geophys. Res. **98**, 21659–21676 (1993)
8. Buffett, B.A., Mathews, P.M., Herring, T.A.: J. Geophys. Res. (2002). doi:10.1029/2000JB000056
9. Capitaine, N.: Celest. Mech. Dyn. Astron. **44**, 127 (1990)
10. Capitaine, N., Wallace, P.T., Chapront, J.: Astron. Astrophys. **412**, 567 (2003)
11. Capitaine, N., Wallace, P.T.: Astron. Astrophys. **450**, 855 (2006)
12. Capitaine, N., Folgueira, M., Souchay, J.: Astron. Astrophys. **445**, 347 (2006)
13. Capitaine, N., Mathews, P.M., Dehant, V., Wallace, P.T., Lambert, S.B.: Celest. Mech. Dyn. Astron. **103**, 179 (2009)
14. Chao, B.F., Ray, J.M., Gipson, M., Egbert, G.D., Ma, C.: J. Geophys. Res. **101**, 20151–20163 (1996)
15. Chapront-Touzé, M., Chapront, J.: Astron. Astrophys. **190**, 342 (1988)
16. Danjon, A.: In: Blanchard, A. (ed.) Astronomie Générale (1959), 2nd edn. rev. and corr. (1980)
17. Defraigne, P., Dehant, V., Paquet, P.: Celest. Mech. Dyn. Astron. **62**, 363–372 (1995)
18. de Vries, D., Wahr, J.M.: J. Geophys. Res. **96**, 8275–8293 (1991)
19. Dziewonski, A.M., Anderson, D.L.: Phys. Earth Planet. Inter. **25**, 297–356 (1981)
20. Elgered, G., Davis, J.L., Herring, T.A., Shapiro, I.I.: J. Geophys. Res. **96**, 6541–6555 (1991)
21. Getino, J., Ferrandiz, J.M.: Mon. Not. R. Astron. Soc. **306**, L45–L49 (1999)
22. Getino, J., Ferrandiz, J.M.: Geophys. J. Int. **142**, 703–715 (2000)
23. Getino, J., Ferrandiz, J.M.: In: Johnston, K.J., et al. (eds.) Proceedings of IAU Colloquium 180, pp. 236–241. USNO, Washington (2000)
24. Gilbert, F., Dziewonski, A.M.: Philos. Trans. R. Soc. Lond. Ser. A **278**, 187–269 (1975)
25. Guinot, B.: In: McCarthy, D.D., Pilkington, J.D. (eds.) Time and the Earth's Rotation, p. 7. Reidel, Dordrecht (1979)
26. Gwinn, C.R., Herring, T.A., Shapiro, I.I.: J. Geophys. Res. **91**, 4755–4786 (1986)
27. Hartmann, T., Williams, J.G., Soffel, M.: Astron. J. **111**, 1400 (1996)
28. Herring, T.A.: Highlights Astron. **10**, 222–227 (1995)
29. Herring, T.A., Gwinn, C.A., Shapiro, I.I.: J. Geophys. Res. **91**, 4755–4765 (1986)
30. Herring, T.A., Buffett, B.A., Mathews, P.M., Shapiro, I.I.: J. Geophys. Res. **96**, 8259–8273 (1986)
31. Herring, T.A., Buffett, B.A., Mathews, P.M., Shapiro, I.I.: J. Geophys. Res **96**, 8258–8265 (1991)
32. Herring, T.A., Mathews, P.M., Buffett, B.A.: J. Geophys. Res. **107**(B4), 101029 (2002)
33. Hilton, J., Capitaine, N., Chapront, J., et al.: Celest. Mech. Dyn. Astron. **94**, 3351 (2006)
34. Hough, S.S.: Philos. Trans. R. Soc. Lond. Ser. A **186**, 469–506 (1895)
35. Jeffreys, H., Vicente, R.O.: Mon. Not. R. Astron. Soc. **117**, 142–161 (1957)
36. Jeffreys, H., Vicente, R.O.: Mon. Not. R. Astron. Soc. **117**, 162–173 (1957)
37. Kinoshita, H.: Publ. Astron. Soc. Jpn. **24**, 423 (1972)
38. Kinoshita, H.: Celest. Mech. Dyn. Astron. **13**, 277–326 (1977)
39. Kinoshita, H., Souchay, J.: Celest. Mech. Dyn. Astron. **48**, 187–266 (1990)
40. Kinoshita, H., Hori, G., Nakai, H.: Ann. Tokyo Astron. Obs. **14**, 14 (1974)
41. Lieske, J.H., Lederle, T., Fricke, W., Morando, B.: Astron. Astrophys. **58**, 1–16 (1977)
42. Love, A.E.H.: Proc. R. Soc. Lond. Ser. A **82**, 73–88 (1909)

43. Ma, C., Arais, E.F., Fey, T.M., et al.: Astron. J. **136**, 735L (1998)
44. Ma, C., Arias, E.F., Bianco, G., et al.: IERS Technical Note 35, 1M (2009)
45. Mathews, P.M., Shapiro, I.I.: Annu. Rev. Earth Planet. Sci. **20**, 469–500 (1992)
46. Mathews, P.M., Buffett, B.A., Herring, T.A., Shapiro, I.I.: J. Geophys. Res. **96**, 8219–8242 (1991)
47. Mathews, P.M., Buffett, B.A., Herring, T.A., Shapiro, I.I.: J. Geophys. Res. **96**, 8243–8257 (1991)
48. Mathews, P.M., Herring, T.A., Buffett, B.A.: J. Geophys. Res. **107**(B4), 3-1–3-26 (2002) doi:10.129/2001JB0000390
49. McCarthy, D.D.: IERS Conventions, IERS Tech. Note 21, Int. Earth Rot. Service (1996)
50. Molodensky, M.S.: Commun. Obs. R. Belg. **188**, 23–56 (1961)
51. Neuberg, J., Hinderer, J., Zurn, W.: Geophys. J. R. Astron. Soc. **91**, 853–868 (1987)
52. Parke, M.E., Henderschott, M.C.: Mar. Geod. **3**, 379–408 (1979)
53. Poincaré, H.: Bull. Astron. **27**, 321–356 (1910)
54. Roosbeek, F., Dehant, V.: Celest. Mech. Dyn. Astron. **70**, 215–255 (1998)
55. Sasao, T., Okubo, S., Saito, M.: In: Fedorov, E.P., Smith, M.L., Bender, P.L. (eds.) Proceedings IAU Symp. No. 78, pp. 165–183 (1980)
56. Schwiderski, E.W.: Rev. Geophys. **18**, 243–268 (1980)
57. Sasao, T., Wahr, J.M.: Geophys. J. R. Astron. Soc. **64**, 729–746 (1981)
58. Seidelmann, P.K.: Celest. Mech. Dyn. Astron. **27**, 79–106 (1982)
59. Shen, P.Y., Manshina, L.: Geophys. J. R. Astron. Soc. **46**, 467–496 (1976)
60. Shirai, T., Fukushima, T.: Astron. J. **119**, 2475–2480 (2000)
61. Shirai, T., Fukushima, T.: In: Johnston, K.J. et al. (eds.) Proceedings IAU Colloquium, Astron. J. **180**, 223–229 (2000)
62. Smith, M.L.: Geophys. J. R. Astron. Soc. **37**, 491–526 (1974)
63. Smith, M.L., Dahlen, F.A.: Geophys. J. R. Astron. Soc. **64**, 223–282 (1981)
64. Souchay, J.: Astron. J. **116**, 503–515 (1998)
65. Souchay, J., Kinoshita, H.: Astron. Astrophys. **312**, 1017–1030 (1996)
66. Souchay, J., Kinoshita, H.: Astron. Astrophys. **318**, 639–652 (1997)
67. Souchay, J., Loysel, B., Kinoshita, H., Folgueira, M.: Astron. Astrophys. Suppl. Ser. **135**, 111–131 (1999)
68. Standish, E.M., Newhall, X.X., Williams, J.G., Folkner, W.F.: JPL Planetary and Lunar Ephemerides DE403/LE403, JPL IOM314, pp. 10–127 (1995)
69. Tisserand, F.: Mécanique Céleste, Gauthier-Villars, Paris (1892). Chaps. 16–19
70. Vondrak J.: Bull. Astron. Inst. Czechoslov. **33**, 26 (1983)
71. Vondrak J.: Bull. Astron. Inst. Czechoslov. **34**, 184 (1983)
72. Vondrak J.: Bull. Astron. Inst. Czechoslov. **34**, 311 (1983)
73. Wahr, J.M.: Geophys. J. R. Astron. Soc. **64**, 651–675 (1981)
74. Wahr, J.M.: Geophys. J. R. Astron. Soc. **64**, 677–703 (1981)
75. Wahr, J.M.: Geophys. J. R. Astron. Soc. **64**, 705–725 (1981)
76. Wahr, J.M., Sasao, T.: Geophys. J. R. Astron. Soc. **64**, 747–765 (1981)
77. Wahr, J.M., Bergen, Z.: Geophys. J. R. Astron. Soc. **87**, 633–668 (1986)
78. Williams, J.G.: Astron. J. **108**, 711–724 (1994)
79. Woolard, E.W.: Astr. Pap. Am. Ephem. **15**(1), 11–165 (1953)

Chapter 5
Tides on Satellites of Giant Planets

Nicolas Rambaux and Julie Castillo-Rogez

Abstract The discovery of the satellites of the giant planets started in 1610 when Galileo Galilei pointed his telescope toward Jupiter. Since then observations from Earth- and space-based telescopes and outstanding in-situ observations by several space missions have revealed worlds of great richness and extreme diversity. One major source of energy driving the evolution of these satellites is the gravitational pull exerted by their planets. This force shapes and deforms the satellites and the resulting dissipation of mechanical energy can heat their interiors and drive spectacular activity, such as volcanic eruptions, as for Io or Enceladus. In addition, tides drive orbital evolution by circularizing the satellites' orbits and synchronizing their rotational motions.

5.1 Introduction

The giant planets of the solar system, Jupiter, Saturn, Uranus, and Neptune, have many satellites. So far, astronomers have identified 168 giant-planet satellites[1]: 66 of Jupiter, 62 of Saturn, 27 of Uranus, and 13 of Neptune. All these satellites display a large variety of dynamical configurations and geophysical properties that have been studied by continual ground-based telescopic observations and dedicated space missions *Voyager*, *Pioneer*, *Galileo*, and *Cassini-Huygens* sent by NASA and ESA, with international participation. *Pioneer* and *Voyager* achieved in the 70s and 80s a formidable trip across the outer solar system. They sent the first images of the satellites surfaces, revealing an extraordinary geological richness. Then, *Galileo*

---

[1] See the regularly update of satellite's number at IMCCE web service http://www.imcce.fr/hosted_sites/saimirror/Nomenclaf.html.

N. Rambaux (✉)
IMCCE, Observatoire de Paris, CNRS UMR 8028, Université Pierre et Marie Curie,
UPMC - Paris 06, 77 avenue Denfert-Rochereau, 75014 Paris, France
e-mail: Nicolas.Rambaux@imcce.fr

J. Castillo-Rogez
Jet Propulsion Laboratory, Caltech, Pasadena, CA 91109, USA
e-mail: julie.c.castillo@jpl.nasa.gov

J. Souchay et al. (eds.), *Tides in Astronomy and Astrophysics*,
Lecture Notes in Physics 861, DOI 10.1007/978-3-642-32961-6_5,
© Springer-Verlag Berlin Heidelberg 2013

Table 5.1 Parameters characterizing outer planet satellites; a denotes the semi-major axis, e the eccentricity, P_{orb} the orbital period and P_{rot} the rotational period, GM the gravity mass, R the radius, and H the equilibrium tide of the satellites. Source: JPL (Jet Propulsion Laboratory) Solar System Dynamics website http://ssd.jpl.nasa.gov. The period of rotation is indicated with a "C" when the rotation is chaotic [120]. The rotational period of Nereid is not accurately determined but certainly in the range 0.8–3 days [96]

Body	a (km)	e	P_{orb} (days)	P_{rot} (days)	GM (km^3 s^{-2})	R (km)	H (m)
Io	421800	0.0041	1.769	1.769	5959.916	1821.6	3118.8
Europa	671100	0.0094	3.551	3.551	3202.739	1560.8	776.7
Ganymede	1070400	0.0013	7.155	7.155	9887.834	2631.2	500.7
Callisto	1882700	0.0074	16.69	16.69	7179.289	2410.3	89.2
Mimas	185539	0.0196	0.942	0.942	2.5026	198.20	3662.0
Enceladus	238037	0.0047	1.370	1.370	7.2027	252.10	1577.1
Tethys	294672	0.0001	1.888	1.888	41.2067	533.00	2903.5
Dione	377415	0.0022	2.737	2.737	73.1146	561.70	960.6
Rhea	527068	0.0010	4.518	4.518	153.9426	764.30	574.2
Titan	1221865	0.0288	15.95	15.95	8978.1382	2575.50	102.0
Hyperion	1500934	0.0232	21.28	C	0.3727	135.00	10.0
Iapetus	3560851	0.0293	79.33	79.33	120.5038	735.60	2.0
Phoebe	12947913	0.1634	550.30	0.45	0.5532	106.60	0.004
Miranda	129900	0.0013	1.413	1.413	4.4	235.8	1857.3
Ariel	190900	0.0012	2.520	2.520	86.4	578.9	1082.6
Umbriel	266000	0.0039	4.144	4.144	81.5	584.7	441.5
Titania	436300	0.0011	8.706	8.706	228.2	788.9	118.4
Oberon	583500	0.0014	13.46	13.46	192.4	761.4	50.9
Triton	354759	0.0000	5.877	5.877	1427.6	353.4	1.7
Nereid	5513818	0.7507	360.13	< 3	2.06	170	0.02

(1993–2003) and *Cassini* (2004-today) were dedicated to the Jupiter and Saturnian systems, respectively, performing extensive observations and permitting a greater understanding of the relationships between planets, rings, and satellites. Here, we focus on large regular satellites with radii larger than 100 kilometers. The satellites' main physical and dynamical properties are gathered in Table 5.1. Figure 5.1 represents the satellites as a function of their relative sizes, densities, and distance to the parent-planet expressed in planetary radius. The sizes of the satellites range from 2631.2 km for Ganymede (larger than Mercury) to 106.6 km for Phoebe. Satellite densities reflect their internal composition ranging from 3.6 g cm^{-3} for Io, dominated by silicates and a large metallic core, to 0.97 g cm^{-3} for water-dominated Tethys. Intermediate densities reflect variations in the relative fractions of ice, silicates, and porosity. The rock mass fraction determines in part the amount of tidal

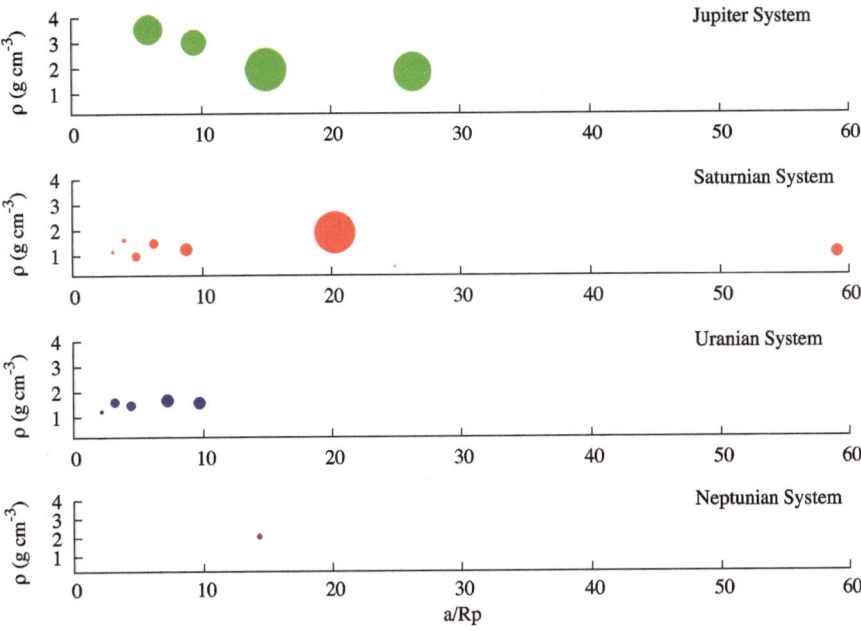

Fig. 5.1 Satellite density as a function of the distance to the parent-planet (expressed in planetary radius). The relative size of the satellites is respected but is not to scale with the distance. From *left* to *right* and *top* to *bottom*: Io, Europa, Ganymede, Callisto; Mimas, Enceladus, Tethys, Dione, Rhea, Titan, Hyperion, Iapetus; Miranda, Ariel, Umbriel, Titania, Oberon; Triton. The satellites Phoebe and Nereid are not represented because they are far from their planets (215 Saturn's radii and 223 Neptune's radii, respectively)

dissipation expected in these objects, as ice is generally more dissipative than silicates.

From Table 5.1 and Fig. 5.1, we can see that most satellites are close to their parent-planets and thus may experience significant tidal stressing. The tidal force results from the amplitude of the gradient of the external gravitational field between the sub-planet and the anti-planet hemispheres. This tidal force distorts the satellite, if it is not rigid, so that the amplitude of the *equilibrium tide* is expressed through [79]

$$H = R_s \frac{M_p}{M_s} \left(\frac{R_s}{d} \right)^3 \tag{5.1}$$

where R_s and M_s are the radius and mass of the satellite, M_p is the mass of the planet, and d is the distance between the satellite and its parent-planet. Here, the amplitude is expressed for a particle where the tide-generating body is at the zenith. The *equilibrium tide* represents the ratio between the external gravitational potential and the gravity of the body. The displacement at the surface is obtained by multiplying H by the secular Love number that represents the ability of the body to deform when in hydrostatic equilibrium (see Sects. 5.4 and 5.5). In the case of a homogeneous body this factor, labeled h_2, is equal to $5/2$. It decreases with increasing concentration of the density toward the center of the object. The tidal bulge H increases with the size of the object and decreases with the distance to the primary. The equilibrium tide is 3.6 km for Mimas, then around 3 km for Io and Tethys, and then decreases to a few meters for those small satellites located beyond 10 planetary radii, as shown in Table 5.1.

The planet is also subject to an equilibrium tide exerted by its satellites. The combined planet and satellite tides drive the evolution of the satellite's orbit, making it contract or expand depending on the dissipation within each body. For most satellites, the tides lead to orbit circularization. If the satellite is close to its parent-planet, then the orbit evolves toward the planet's equatorial plane, whereas for distant satellites subject to little dissipation the equilibrium plane is intermediate between the planet's equatorial and orbital planes [87]. In addition, the tides raised on the satellites lead to despinning. Most large satellites (apart from Phoebe and Nereid) are in synchronous spin-orbit resonance, i.e. the orbital and rotational periods are equal on average. As a consequence, the satellites show on average the same face toward the planet, like the Moon toward the Earth.

For a satellite in spin-orbit synchronous resonance, the secular part of the tidal potential elongates the body along the planet-satellite axis. In cooperation with the centrifugal potential this flattens the poles of the satellite; the resulting equilibrium figure is then a triaxial ellipsoid. Under the assumption of hydrostaticity, the equilibrium figure brings information on the density structure of the body. Departure from hydrostaticity may inform on the geophysical and dynamical evolution of the object (e.g. fossil shape, mass anomalies, etc.).

The periodic part of the tidal potential deforms the body continuously and leads to solid friction within the material. The amount of friction is a function of the orbital eccentricity. The consequences of that process can be spectacular, such as volcanic activity as observed on Enceladus or Io. Other outstanding signatures of tides can be found on the surface of Europa, related to faults and cycloid cracks. The heating resulting from tidal friction is also believed to play a role in the origin and/or preservation of subsurface oceans in Europa, Ganymede, Titan, Triton ([49] and references therein). Callisto is far from Jupiter so the tidal dissipation in that object is small. However, the presence of an internal ocean has been suggested based on *Galileo*'s magnetometer data. Its long-term preservation is explained by slow heat loss [69]. Oceans inside Rhea, Titania, Oberon have been suggested but this is still debated, in absence of observational constraints [49]. These geological and geophysical consequences are described in many very good reviews on satellites (e.g. [16, 49, 87, 105]; and the book on icy satellites by Grasset et al. [38]).

This chapter is divided in four sections following this introductory Sect. 5.1. In Sect. 5.2, we outline a simple version of the tidal theory that is a toy model useful for conveying the main concepts and illustrating the consequences of tidal friction. Section 5.3 describes the influence of the tides on the dynamical evolution of satellites. The equilibrium figure of a satellite resulting from tidal distortion is described in Sect. 5.4, and in Sect. 5.5 we describe and discuss the consequences of tidal dissipation in icy satellites.

5.2 Tidal Potential

The historical developments leading to the modern formulation of tidal modeling can be found in Chap. 2 of the present volume. The modern treatment of the solid body tides began with a seminal series of papers written by Darwin [19, 20]. Since this pioneering work, tidal modeling has been extensively explored in the literature (e.g. [35, 37, 51, 55, 65, 72–74]). Recent laboratory measurements of the response of planetary materials to cyclic forcing (e.g. [67] for a review) has lead to reviewing the tidal theory (e.g. [28, 29, 31]). Traditionally, the tidal theory is developed from the tide-generated disturbing potential into Fourier series, and a dissipative component is related to each term. Here, for the sake of simplicity, we follow the approach of MacDonald [65], in which dissipation is modeled in the form of a constant phase lag. However one has to keep in mind that this approach implicitly assumes a certain rheology for the material (response to stress) that can lead to unphysical situations (see the review in [31] and [29]).

Now, we outline the main aspects of the tidal theory used in this chapter. The giant planets and their natural satellites are not point-mass bodies, as generally assumed in ideal mechanical systems, and they deform under the gravitational acceleration of external bodies. For a satellite S of radius R_s, the mean gravitational acceleration due to the planet P is the vector $GM_p\mathbf{SP}/SP^3$ where G is the gravitational constant and M_p the mass of the perturbing body, i.e. the parent-planet in the present case. For each element of the satellite M the relative distance between the element and the planet is the vector $\mathbf{MP}$. Consequently the net tidal acceleration $\mathbf{g}^T$ experienced by the element is

$$\mathbf{g}^T = GM_p\left(\frac{\mathbf{MP}}{MP^3} - \frac{\mathbf{SP}}{SP^3}\right). \tag{5.2}$$

It is a differential acceleration. By setting $\mathbf{SP} = \mathbf{d}$ and $\mathbf{SM} = \mathbf{x}$ the vector position of an element in the satellite, $\mathbf{MP}$ may be decomposed as $\mathbf{MP} = \mathbf{SP} - \mathbf{SM} = \mathbf{d} - \mathbf{x}$, leading to the approximation for small values of $|\mathbf{x}|$

$$MP^{-3} \approx d^{-3}\left(1 + 3\frac{\mathbf{d}}{d}\cdot\frac{\mathbf{x}}{d}\right). \tag{5.3}$$

Injecting this expression into Eq. (5.2), we obtain

$$\mathbf{g}^T = \frac{GM_p}{d^3}(3(\mathbf{x}.\mathbf{e})\mathbf{e} - \mathbf{x}), \tag{5.4}$$

where $\mathbf{e} = \mathbf{d}/d$ is the unit cosine vector. The Cartesian expression of the tidal force in the rotating reference frame of the satellite is then

$$\mathbf{g}^T = \frac{GM_p}{d^3}(2x_p, -y_p, -z_p). \tag{5.5}$$

The gravitational force is then stronger on the x_p direction that points toward the perturbing body and negative in the y_p and z_p directions. This tidal acceleration

Fig. 5.2 Geometry of the tidal problem in the body reference frame $(\mathbf{e}_x, \mathbf{e}_y, \mathbf{e}_z)$. **d** is the vector pointing towards the disturbing body and **r** is the vector pointing towards the perturbed body. $\mathbf{R}_s$ targets at the surface of the body

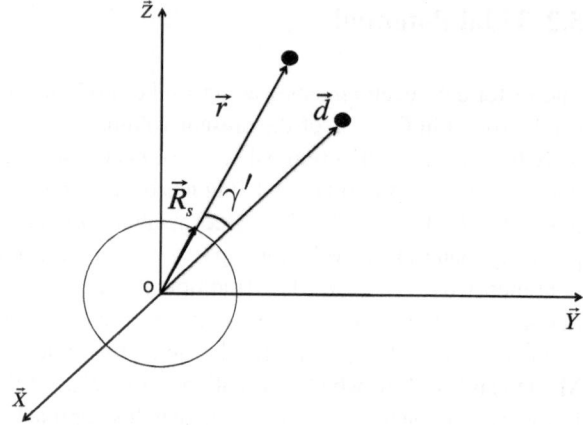

implies that the satellite is elongated in the direction towards the planet and flattened in the perpendicular direction.

The tidal acceleration can be expressed through a tidal potential defined by

$$\mathbf{g}^T = \nabla U_T \tag{5.6}$$

where the tidal potential U_T is

$$U_T = \frac{GM_p}{2d^3}\left(2x_p^2 - y_p^2 - z_p^2\right) \tag{5.7}$$

in Cartesian coordinates, or

$$U_T = \frac{GM_p}{d}\left(\frac{R_s}{d}\right)^2 \frac{3\cos\gamma'^2 - 1}{2} \tag{5.8}$$

in spherical coordinates. The parameter γ' is the angle between the position vectors **d** and **x** as shown in Fig. 5.2. The last factor in the previous expression corresponds to the Legendre polynomial of degree 2, and the tidal potential is then written in synthetic form as

$$U_T = \frac{GM_p}{d}\left(\frac{R_s}{d}\right)^2 P_2(\cos\gamma'). \tag{5.9}$$

The Legendre polynomial of degree 2 results from the development performed in Eq. (5.3). The development at higher order in $|\mathbf{x}|$ leads to the introduction of higher degrees in the Legendre polynomial, and the generalized potential is then expressed as

$$U_T = \frac{GM_p}{d}\sum_{l=2}^{\infty}\left(\frac{R_s}{d}\right)^l P_l(\cos\gamma'). \tag{5.10}$$

The non-rigid satellite is distorted by the tidal potential. According to the degree-2 development in the potential, the satellite is elongated in two opposite directions. The resulting bulge follows the tidal acceleration and, in the case of an elastic body, the bulge lies along the relative direction of the distorted satellite to the perturbing planet, as shown by Eq. (5.5). However, if the distorted material is not purely elastic, then the tidal bulge is offset with respect to the satellite–planet axis, i.e., the response of the material to stress is delayed as a consequence of internal friction. This phase lag with respect to the position of the tide-generating body induces dissipation inside the system, and the tidal bulge modifies the gravitational potential of the satellite. For small deformations a linear theory may be assumed, for which the tidal response of the distorted body is proportional to the external tidal potential evaluated at the surface. The coefficient of proportionality is called the dynamic Love number and it depends on the density and rheological structure of the body, and of the frequency of the excitation. Therefore, the dynamic Love number is different at each degree l, and the additional potential of the distorted satellite at a point $\mathbf{r}$ in space is then equal to

$$U = \frac{GM_p}{d} \sum_{l=2}^{+\infty} k_l \left(\frac{R_s}{r}\right)^{l+1} \left(\frac{R_s}{d}\right)^l P_l(\cos \gamma') \tag{5.11}$$

(e.g. [29, 57]). The ratio $(R_s/r)^{(l+1)}$ comes from the Dirichlet theorem for external potential $r > R_s$. As a consequence, the external potential decreases quickly as a function of distance. For example, in the case of Miranda, the medium-sized satellite closest to Uranus, the ratio (R_s/r) is equal to 0.0018 and the error in the potential truncated at degree 2 is around 1/50. We then limit the description of the potential to the second degree; the simplified potential U is then

$$U = k_2 \frac{GM_p}{R_s} \left(\frac{R_s}{r}\right)^3 \left(\frac{R_s}{d}\right)^3 P_2(\cos \gamma'). \tag{5.12}$$

In this potential, the quantities $(R_s/d)^3$, γ' are related to the tide-raising potential whereas the $(R_s/r)^3$ quantity represents the response of the satellite's potential at degree 2 (Eq. (5.11)).

In the case of a rigid homogeneous body, the Love number k_2 may be expressed as

$$k_2 = \frac{3/2}{1 + \frac{19\mu}{2\rho g R_s}} \tag{5.13}$$

where μ is the rigidity, ρ the density, g is the surface gravity acceleration, and R_s is the radius of the body. It is customary to introduce the dimensionless rigidity $\tilde{\mu}$

$$\tilde{\mu} = \frac{19\mu}{2\rho g R_s} \tag{5.14}$$

that represents the ratio of the elasticity to the cohesive force of the body's self-gravity. For small icy satellites the dimensionless rigidity $\tilde{\mu}$ is of the order of 10^2 and the rigidity dominates. The Love number can be thus simplified as

$$k_2 \sim \frac{3}{19} \frac{\rho g R_s}{\mu} \tag{5.15}$$

Fig. 5.3 Love number k_2 for different interior models assumed for Ganymede (from Moore and Schubert [76]). The *solid line* represents models without an internal ocean, whereas the thickness of the ocean is equal to 200 km (*dotted line*), and 20 km (*dashed line*). Each model is shown for two different assumptions on the value of the mean rigidity of the ice: $\mu = 10^9$ Pa and $\mu = 10^{10}$ Pa

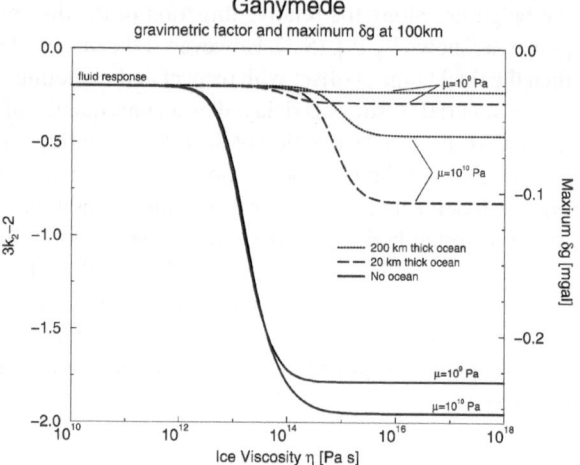

whereas for a fluid body $\tilde{\mu}$ is equal to zero and the Love number is simply

$$k_2 = \frac{3}{2}. \tag{5.16}$$

However, most satellites are not homogeneous, with radial variations in composition and temperature (and also possibly in porosity in the smaller representatives). In this case, the Love numbers are smaller than the values predicted by Eqs. (5.15)–(5.16) and numerical integration is required to estimate these parameters (e.g. [13, 76, 112] and references therein). In addition, some large satellites like Europa, Ganymede, Callisto, and Titan might hold an internal ocean beneath their surface. In this case the tidal Love numbers increase toward the fluid limit as illustrated in Fig. 5.3 coming from Moore and Schubert [76].

When the tide-raising body located at **d** and the perturbed body disturbed by the potential **r** are the same, then **r** coincides with **d** in the elastic case. In the inelastic case, the vector **r** is out of phase with respect to **d** due to friction created by the motion of defects in the material. There are two main approaches for introducing into the equations the delay due to friction. The first approach proposed by Darwin [19, 20] and implemented by Kaula [57], Efroimsky and Williams [29], Ferraz-Mello et al. [31] is in four steps. (i) The tidal potential is developed in the form of Fourier series by expressing explicitly **r** and **d**, (ii) then, for each term of the series, a phase lag is introduced, and (iii) the gradient of the potential is computed with respect to the position of **r**, and finally (iv) **d** is replaced by **r**. This scheme is described and discussed in details in Efroimsky and Williams [29]. The phase lag ϵ is determined by the inelasticity of the body material and is a function of the forcing frequencies. The second approach is presented below in more details. This approach has the advantage to be simple, because the tidal lag is represented by a time delay Δt. This approach is however limited by the fact that the dissipation factor, labeled Q, and the time delay Δt are linked through the following relationship that assumes that the object behaves like a Maxwell body, i.e.

$$\Delta t = (\omega Q)^{-1} \qquad (5.17)$$

with ω is the synodic (or tidal) forcing frequency. However, according to laboratory measurements the frequency-dependence of the dissipation factor depends on the forcing frequency to the power α with α between 0.1 and 0.5 (see [67]). Keeping this limitation in mind, the lag is introduced by a Taylor development of $\mathbf{d} = \mathbf{r}(t - \Delta t)$ for each frequency, that leads to

$$\mathbf{d} \sim \mathbf{r}(t) - \Delta t \frac{d\mathbf{r}(t)}{dt}. \qquad (5.18)$$

Therefore, $\mathbf{d}$ can be seen as the position of $\mathbf{r}(t)$ with a time delay Δt in the past relatively to the coordinate system linked to the body, and the tidal bulge is dragged by the rotation of the body.

The phase shift between the action and the response of the body leads to energy dissipation due to friction that can be expressed through the dissipation factor Q introduced in Eq. (5.17) [65]. The dissipation factor is defined as the maximum energy E stored during one cycle over the energy dissipated ΔE during that cycle [37]

$$Q = 2\pi \frac{E}{\Delta E}. \qquad (5.19)$$

This definition is related to the damped harmonic oscillator model and the limitation of this analogy has been discussed in Greenberg [39] and Efroimsky and Williams [29].

The potential, Love number, and dissipation factor have been defined for a satellite deformed by a tide-raising planet. These expressions are still valid in the case of a planet deformed by a satellite by substituting all satellite parameters by planet parameters and the planet parameters by those corresponding to the tide-raising body (satellite or the Sun). However, in the case of giant planets, the power law equation (5.17) is not applicable because the full dynamics of the atmospheric response to the tide raising potential must be accounted for (e.g. [52]).

5.3 Tidal Dynamics

5.3.1 Introduction

Tidal interaction implies an evolution in the rotational motion and orbital parameters of satellites mainly due to the transfer of angular momentum between the satellite's orbit and planet's rotation as well as energy dissipation inside the satellite. The equilibrium configuration for an isolated two-body problem is a satellite in synchronous spin-orbit resonance and a circular orbit [87]. For a moon around the giant planet, the mutual gravitational interactions with the other satellites lead to equilibrium states close to this equilibrium configuration, as for example in the case of Io, for which the Laplace resonance sustains a non-zero eccentricity and then a high dissipation regime as discussed below. Here we describe the tidal interaction between isolated body and we do not introduce the effect of the orbital resonances.

5.3.2 Transfer of Angular Momentum

The tidal potential generates a tidal bulge on the perturbed body. In the purely elastic case the tidal bulge is always aligned toward the perturbing body and, by symmetry, the resulting torque is null. Consequently, there is no transfer of angular momentum between the two bodies. In the inelastic case, the bulge is offset with respect to the direction between the satellite and the perturbing body and the resulting tidal torque drives an exchange of angular momentum.

A simple description of the angular momentum transfer can be investigated by considering a system composed of two rotating bodies, P and S, orbiting around a center of mass G, in circular orbit, and isolated in space. Here, we assume that P is a planet of mass M_p larger than the mass of the second body M_s, the satellite. The total angular momentum $\mathbf{H}$ of this system is the sum of the planet's angular momentum $\mathbf{H}_p$ and the satellite's angular momentum $\mathbf{H}_s$,

$$\mathbf{H} = \mathbf{H}_p + \mathbf{H}_s. \tag{5.20}$$

The planetary angular momentum is expressed in the barycentric reference frame of this system as

$$\mathbf{H}_p = M_p \mathbf{GP} \wedge \mathbf{v}_p + \mathbf{I}_p \boldsymbol{\Omega}_p \tag{5.21}$$

where $\mathbf{GP}$ is the direction vector between G and P, $\mathbf{v}_p$ is the orbital velocity of the planet around the center of mass, $\mathbf{I}_p$ the tensor of inertia of the planet, and $\boldsymbol{\Omega}_p$ its rotational velocity. Similarly, the angular momentum of the satellite is expressed as

$$\mathbf{H}_s = M_s \mathbf{GS} \wedge \mathbf{v}_s + \mathbf{I}_s \boldsymbol{\Omega}_s \tag{5.22}$$

where the indices s refer to the satellite. By using the barycenter definition of G, we simplify the expression of the total angular momentum as

$$\mathbf{H} = M_s \mathbf{PS} \wedge \mathbf{v}_s + \mathbf{I}_p \boldsymbol{\Omega}_p + \mathbf{I}_s \boldsymbol{\Omega}_s. \tag{5.23}$$

Consequently, the angular momentum is composed of the rotational angular momentum of each body with the satellite's orbital angular momentum around the planet but with the barycentric velocity. By assuming that the spins are normal to the orbital planes the vectorial equation is expressed as a scalar equation, where C_p and C_s are the polar moments of inertia of each body. In addition, for circular orbits, the velocity v_s can be simply expressed as $v_s = an$ with a the orbital radius and n the mean motion. Here we assume that the orbit of the satellite and the spin of the planet are rotating in the same sense. Triton is in a retrograde orbit, so v_s is then equal to $-an$. Table 5.2 presents physical parameters of the planets and satellites considered in this chapter. It appears that the rotational angular momenta of the planets and the

Table 5.2 Parameters of giant planets

Giant planet	Radius (km)	Mass (10^{24} kg)	P_{rot}
Jupiter	71398	1898.6	9h55m27.3s
Saturn	60330	568.46	10h39m22.4s
Uranus	26200	86.832	17.24 ± 0.01 h
Neptune	25225	102.43	16.11 ± 0.01 h

orbital angular momenta are much larger than the rotational angular momenta of the satellites by four to five orders of magnitude. Equation (5.23) is then simplified as

$$H = M_s a^2 n + C_p \Omega_p \qquad (5.24)$$

that is a constant function of the motion for the isolated two body problem. This equation shows the relationship between the variation of the semi-major axis of the satellite (contained also in n through Kepler's third law) and the rotation of the planet.

To obtain the relationship between the semi-major axis and rotational rates, we derive Eq. (5.24),

$$M_s \frac{d(a^2 n)}{dt} + C_p \frac{d\Omega_p}{dt} = 0. \qquad (5.25)$$

However, the derivation of the first term on the right-hand side is not straightforward, because a and n are related through Kepler's third law $n^2 a^3 = G(M_p + M_s) \simeq GM_p$. After introducing Kepler's third law and expressing n as a function of a, the variation of the semi-major axis a is directly related to the variation of the rotational velocity of the planet through

$$\frac{1}{a}\frac{da}{dt} = -2\frac{C_p}{M_p R_p^2}\left(\frac{M_p}{M_s}\right)\left(\frac{R_p}{a}\right)^2 \frac{1}{n}\frac{d\Omega_p}{dt} \qquad (5.26)$$

or

$$\frac{da}{dt} \propto -a^{1/2}\frac{d\Omega_p}{dt}. \qquad (5.27)$$

The satellite's semi-major axis and the planet's angular velocity evolve in opposite directions, that is the planet's angular velocity decreases if the satellite's orbit expands, and vice-versa due to the transfer of angular momentum between the orbit and the rotation. In the case of Triton, which is in retrograde orbit, the sign of the right-hand side is positive meaning that acceleration in the spin corresponds to an expansion of the orbit, and deceleration in the spin implies a contraction of the orbit. The evolution of the Triton dynamics is described in Correia [18].

The variation in the planet's angular velocity can be computed from the gravitational torque exerted by the satellite on the offset planetary bulge. The rotational $\mathbf{H}_p^R$ angular momentum of the planet is equal to the torque applied to the planet

$$\frac{d\mathbf{H}_p^R}{dt} = \mathbf{T}. \qquad (5.28)$$

As $H_p^R = C_p \Omega_p$, under the assumption that the spin vector is aligned along the polar axis $\mathbf{e}_z$ and C_p is assumed to be constant, we have

$$\frac{d\Omega_p}{dt} = \frac{1}{C_p} T_z. \tag{5.29}$$

The vectorial torque is the cross-product of the radial vector between the centers of mass of the planet and of the satellite with the tidal force resulting from the tidal potential ∇U expressed in Eq. (5.12):

$$\frac{d\Omega_p}{dt} = \frac{M_s}{C_p} (\mathbf{r} \wedge \nabla U)_z. \tag{5.30}$$

By using the potential equation (5.12) and Q defined in Eq. (5.17), we obtain the relation

$$\frac{d\Omega_p}{dt} = -\frac{3}{2} \left(\frac{k_2}{Q}\right)_p \frac{G M_s^2 R_p^5}{C_p a^6} \text{sign}(\Omega_p - n) \tag{5.31}$$

where the sign represents the effect of the torque braking the spin of the planet and leading to spin synchronization (because here the orbit is assumed to be circular). Combining Eqs. (5.26) and (5.31) with $\Omega_p > n$, we obtain the final expression of the evolution of the orbital motion of the satellite due to the tides raised in the planet (e.g. [37]):

$$\frac{da}{dt} = +3 \left(\frac{k_2}{Q}\right)_p \frac{M_s}{M_p} \left(\frac{R_p}{a}\right)^5 na, \tag{5.32}$$

or

$$\frac{da}{dt} \propto \left(\frac{k_2}{Q}\right)_p a^{-11/2}. \tag{5.33}$$

So far these equations have been developed for a circular orbit. If the eccentricity is non-zero, the tides raised by the satellites on the planets increase the orbital eccentricity of the satellite as shown by Jeffreys [55], Goldreich [35]. By expanding the tidal potential U, Eq. (5.12), in Fourier series and then working out the equations of orbital element variations, the evolution of the eccentricity is derived as [37]

$$\frac{de}{dt} = +\frac{57}{8} \left(\frac{k_2}{Q}\right)_p \frac{m_s}{m_p} \left(\frac{R_p}{a}\right)^5 ne. \tag{5.34}$$

The effect is to increase the eccentricity because for almost satellites we are in the case where $\Omega_p > n$, i.e., outside the synchronous orbit (where $\Omega_p = n$). In this case, the impulse due to the planetary tidal bulge on the satellite is larger at the periapsis, increasing the apoapses distance and then the eccentricity.

5.3.3 Tides Raised on the Satellites

5.3.3.1 Introduction

The consequences of the tides acting on the planet have been described in the previous section and now we tackle the question of the tides acting on the satellites.

These drive despinning and generally lead to synchronization of the satellite's rotation with its orbital motion in a time shorter than the age of the solar system. Besides, the orientation of the satellite's spin converges toward an equilibrium value called the *Cassini state* that is generally close to the normal to the orbital plane ([10, 17]; see review in [18, 87]).

Exceptions are Phoebe and Nereid that are too far from their parent-planets, more than 200 planetary radii (Table 5.1). Also, Hyperion stands out as the only satellite presenting a chaotic rotation due to its strong non-spherical shape [120].

Other consequences of the tides raised on the satellites include circularization of the orbits and contraction or expansion of the semi-major axes, depending on the relative amounts of dissipation inside the planet and in the satellite and if orbital resonances are present [87].

5.3.3.2 Despinning

The initial spin rate of the main satellites is estimated to be a few hours (e.g. [87]). However, at present time most satellites are rotating synchronously (see Table 5.1). Spin rate evolution is the consequence of the tides raised in the satellite by the parent-planet gravitational potential. This torque may be computed from Eq. (5.31), in which the orbit of the satellite is assumed to be circular, in the equatorial plane of parent-planet, and the obliquity is equal to zero (e.g. [31, 35, 65])

$$\Gamma = \frac{3}{2}\left(\frac{k_2}{Q}\right)_s \frac{GM_p^2}{R_s}\left(\frac{R_s}{a}\right)^6 \mathrm{sign}(\Omega_s - n). \tag{5.35}$$

This torque acts to despin the satellites on a relatively short timescale, of a few million years. The evolution of the angular momentum yields

$$\frac{dC_s\Omega_s}{dt} = \Gamma \tag{5.36}$$

and

$$\frac{d\Omega_s}{dt} = -\frac{3}{2}\left(\frac{k_2}{Q}\right)_s \frac{GM_p^2 R_s^5}{C_s a^6}\mathrm{sign}(\Omega_s - n) \tag{5.37}$$

and by introducing the Love number defined in Eq. (5.15)

$$\frac{d\Omega_s}{dt} = -\frac{45}{76}\left(\frac{\rho_s R_s^2}{\mu_s Q_s}\right)n^4\mathrm{sign}(\Omega_s - n). \tag{5.38}$$

The damping timescale is then estimated to be equal to [87]

$$\tau \approx \frac{76}{45}\left(\frac{\Omega_s\mu_s Q_s}{\rho_s R_s^2}\right)\frac{1}{n^4}. \tag{5.39}$$

Here, Q_s is the dissipation factor, C_s the polar moment of inertia, ρ_s the mean density, and μ_s the mean bulk modulus of the satellite. For typical values of $Q_s = 100$ and $\mu_s = 10^{10}$ Pa, we infer from Eq. (5.39) that the despinning timescale decreases when the satellite's size increases. After the tidal modeling of MacDonald [65] the resulting satellite despinning time is around a few million years, except for the distant Iapetus whose despinning timescale is of the order of the age of the solar system due to the large distance from Saturn [1]. However, more realistic models of dissipation gives for Iapetus a despinning time around 0.9 Gyr [13]. This leads to the general observation that the consequences of tidally-induced stressing may significantly vary from one modeling framework to another, and this is particularly the case for the MacDonald approach and the more elaborate modeling developed by Efroimsky and Williams [29] or Ferraz-Mello et al. [31] based after [19, 20].

In the case of a circular orbit, the final spin state is at the exact synchronous resonance. However, if the orbit is eccentric the final rotational period is slightly larger than the orbital period. For the tidal model based on [65] the spin frequency is expressed as [63, 119]

$$\Omega_s^e = n\left(1 + \frac{19}{2}e^2\right). \tag{5.40}$$

The spin frequency is larger than the orbital frequency, because the tidal torque is larger at the periapsis and leads to a positive torque on the satellite that accelerates its spin rate.

5.3.3.3 Spin-Orbit Resonance

Most large satellites are in synchronous resonance with a non-zero eccentricity. The mechanism leading to the capture of the satellites in spin-orbit resonance is the gravitational torque exerted by the parent-planet on the asymmetrical shapes of the satellites. Therefore Eq. (5.36) becomes

$$\frac{dC_s\Omega_s}{dt} + \frac{3}{2}\left(\frac{a}{d}\right)^3 n^2(B_s - A_s)\sin 2\gamma = \Gamma \tag{5.41}$$

where the second term on the left-hand side represents the restoring torque. The angle γ is the orientation of the satellite's long axis relative to the direction of the satellite to the planet (e.g. [36]). The stability criterion for capture into resonance is that the average tidal torque must be smaller than the maximum possible restoring torque due to the parent-planet [36]. This is expressed as

$$|\langle\Gamma\rangle| < \frac{3}{2}(B_s - A_s)n^2\left(\frac{a}{d}\right)^3. \tag{5.42}$$

By using the expression of the tidal torque equation (5.35) and for a small eccentricity, we obtain the constraint on the triaxiality of the satellite $(B - A)/C$ for synchronization to occur:

$$\frac{B_s - A_s}{C_s} > \frac{1}{C_s}\left(\frac{k_2}{Q}\right)_s \frac{R_s^5}{G}. \tag{5.43}$$

Therefore if the permanent asymmetrical bulge $(B_s - A_s)$ is large enough the satellite can be captured into synchronous resonance. It appears that for most satellites the hydrostatic value of $(B_s - A_s)$ (Sect. 5.4) is always larger than the critical value, implying systematic capture in resonance, as observed in nature. However, the direction of the long axis of the satellite is slightly shifted from zero at the pericenter because of the balance between the non-zero tidal torque and the permanent torque [123]. That offset in the direction of the satellite is relatively small. For example, in the case of Enceladus the shift is estimated at a maximum 0.57 degrees [92]. Measuring such a small displacement is challenging, even with an orbiter, but the required measurement accuracy may be achieved with an in situ tracking device, such as a transponder. In addition, it has been suggested that a transient regime may occur. Indeed, since the figure axis is shifted, the shape of the satellite may relax by creep in order to adjust to the external potential resulting in the satellite's rotating slightly faster than the synchronous rotation to maintain an equilibrium orientation, and implying a non-synchronous motion of the surface [41, 121]. While there is no direct evidence for such a motion from *Voyager* and *Galileo* images, the interpretation of tidally-induced tectonic patterns at Europa seems to support this scenario ([42], [44, Sect. 5.5.5]).

5.3.3.4 Satellite Orbital Evolution

The rotational equilibrium configuration of most satellites is the synchronous spin-orbit rotation regime, since the despinning time is generally shorter than the age of the solar system. In this case, on average, the tidal bulge of the satellite is aligned with the gravitational force of the planet. As a consequence, there is no transfer of angular momentum between the satellite's orbit and rotation. However, if the satellite has a non-zero orbital eccentricity, it is deformed more strongly at the periapsis than at the apoapsis leading to a time-varying tidal potential called *radial tides*. In addition, the long-axis of the satellite oscillates around its mean value because the velocity of the orbital motion varies along the elliptical orbit [79]: it accelerates at the periapsis and decelerates at the apoapsis according to Kepler's second law. This oscillation is called the *optical libration*. Its amplitude is equal to twice the eccentricity, and it leads to a second time-varying potential term called *librational tides*. Due to the radial and librational tides, the orbital energy is still transferred between the satellite's orbit and rotation, which affects the satellite's orbit over long timescales [35].

Here, the energy source is the orbital energy $-GM_sM_p/2a$ whose dampening by the tidal dissipation decreases the semi-major axis. As the orbit angular momentum $M_s na^2\sqrt{1-e^2}$ is conserved, the decrease in the semi-major axis a is associated with a decrease in the eccentricity e leading to the circularization of the orbit. The temporal evolution of the semi-major axis and eccentricity have been expressed by Goldreich and Soter [37] for a satellite not involved in any orbital resonance:

$$\frac{da}{dt} = -21\left(\frac{k_2}{Q}\right)_s \frac{m_p}{m_s}\left(\frac{R_s}{a}\right)^5 nae^2, \tag{5.44}$$

$$\frac{de}{dt} = -\frac{21}{2}\left(\frac{k_2}{Q}\right)_s \frac{m_p}{m_s}\left(\frac{R_s}{a}\right)^5 ne. \tag{5.45}$$

We note that the semi-major axis and eccentricity evolution rates are negative lead-ing to a decrease of both quantities as long as the eccentricity is non-zero. As the evolution of the eccentricity is faster than the evolution of the semi-major axis (which depends on e^2), the orbit is circularized well before the semi-major axis a is significantly modified.

The combined evolution of the semi-major and eccentricity result in the combi-nation of the tides raised on the planet, Eqs. (5.32) and (5.34), and the tides raised on the satellite, Eqs. (5.44) and (5.45) for satellites outside orbital resonances. The tides raised on the satellites lead to a decrease in both the semi-major axis and eccen-tricity, whereas the tides raised on the planet lead to an increase of both quantities, since all major satellites of the giant planets evolve beyond the synchronous orbit (defined by the distance to the primary at which the orbital period equals the ro-tation period of the planet, $\Omega_p = n$). The orbital evolution of the satellites around the same parent-planet depends on their semi-major axes, sizes, material properties, and thermal evolution. Consequently, the satellites cross many orbital resonances, as for example Dione and Enceladus that are currently in 2:1 resonance or the Galilean satellites, Io, Europa, Ganymede, that are in 4:2:1 resonance (the *Laplace resonance* discussed in more details below). Resonance crossing leads to an additional transfer of angular momentum (e.g. [71]) that strongly influences the orbital evolution of the satellites. In addition, when the satellites are in orbital resonances, eccentricity pumping prevails over the circularization of the orbit, which further sustains tidal heating inside the satellites as discussed in Sect. 5.5.

Tidal dissipation also affects the inclinations of the satellites orbits. The equilib-rium configuration depends on the distance of the moon to the parent-planet [87]. If a moon is close to its planet, the orbital precession is mainly driven by the oblateness of the planet, and the equilibrium orbital plane coincides with the equatorial plane of the planet. This is the case for the majority of the satellites and thus explains why these objects have a small inclination (usually around 1 degree or less). On the other hand, if the moon is far from its planet, the orbital precession may be dominated by the action of the Sun, in which case the equilibrium orbital plane coincides with the mean orbital plane of the planet (i.e. the orbital plane of the Sun seen from the planet). In the intermediate position, the equilibrium plane is between the equatorial and orbital planetary planes, as seen for example in the case of Iapetus [117].

5.3.3.5 Measurements of Tidal Accelerations

Measurement of the dissipation in giant planet systems through the tracking of nat-ural satellite orbital motion started about one century ago with de Sitter [24]. The method consists in fitting astrometric measurements of the satellites positions observed over long periods with orbital model including the tides. However, assessing the influence of the tides on the orbital motions of the satellites is complex because the satellites orbits are perturbed by mutual gravitational attraction with other satellites as well as by the non-spherical shape of the parent-planet. Consequently, the determination of the tides requires accurate numerical models that account for the many dynamical perturbations expected over long duration observations, in order to decorrelate the various dynamical effects (long periods with secular effects due to the tides). Such studies have been performed by Lainey et al. [61, 62] on the Galilean and Saturnian satellites.

For the Galilean system, the accelerations due to the tides have induced a cumulative shift in the satellite orbital positions of 55 km, −125 km and −365 km for Io, Europa, and Ganymede, respectively over the past 116 years analyzed by Lainey et al. [61]. This means that Io's orbit is contracting, while the orbits of Europa and Ganymede are expanding.

The orbital motions of Io, Europa, and Ganymede are driven by the Laplace resonance, i.e. their mean motions are related through (e.g. [40]):

$$n_1 - 3n_2 + 2n_3 = 0 \qquad (5.46)$$

where 1, 2, 3 correspond to Io, Europa, and Ganymede, respectively and n is the mean motion. This resonance results in the excitation of the satellites eccentricities and is thus instrumental in maintaining significant tidal dissipation in Io and Europa. However, since these satellites are evolving in opposite directions, it is expected that they would eventually escape from the Laplace resonance [61, 97].

In addition, from the determination of the semi-major axis evolution of these satellites Lainey et al. [61] could infer the ratios k_2/Q characteristic of Jupiter and Io. The relationship between the semi-major axis and the eccentricity evolution rates with k_2/Q are shown in Eqs. (5.32), (5.34), (5.44), and (5.45). The tides within Europa and Ganymede are not measurable because of the orbital correlations caused by the Laplace resonance. The tides related to Callisto are negligible because the satellite is too far from Jupiter ($H = 89.2$ meters for Callisto whereas it is 3119 meters for Io, see Table 5.1). The k_2/Q ratio for Io is then equal to 0.015 ± 0.003, which is consistent with the value inferred from heat flow mapping. Such a result implies that Io's interior is close to thermal equilibrium and that the heat flow radiated at the surface is mainly due to tidal heating [61].

Jupiter's k_2/Q is equal to $(1.102 \pm 0.203) \times 10^{-5}$ [61]. On top of this, geophysical model of Jupiter's interior that yields the value of k_2 leads to the determination of Jupiter's dissipation factor. Gavrilov and Zharkov [34] predicted a value of 0.379 implying a Q factor of $(3.56 \pm 0.66) \times 10^4$. This value is close to the lower bound on Q determined from the tidally-induced orbital migration of the satellites over the age of the solar system that is in the range $6 \times 10^4 < Q < 2 \times 10^6$ [123]. The dissipation determined by Lainey et al. [61] is consistent with the dissipation models of Jupiter. This result shows that dissipation within giant planets is much stronger than anticipated for the past four decades.

A similar study performed by Lainey et al. [62] for the Saturnian system lead to the determination of Saturn's (k_2/Q) equal to $(2.3 \pm 0.7) \times 10^{-4}$. This value is one order smaller (so larger dissipation) than the usual value estimated from theoretical arguments [83]. In addition, Lainey et al. [62] found that the orbit of Mimas moves toward Saturn at a rate of $da/dt = -(15.7 \pm 4.4) \times 10^{-15}$ au/days. It is not possible to derive directly Mimas' k_2/Q from that rate as was done for Io because the satellite is in resonance with Tethys and also interacts with the Saturnian rings.

The measurement of the tides expressed in the Saturnian system brings new information on the understanding of this system. For example, Enceladus presents plumes and heat emerging from the south pole. The associated energy is estimated to be about 15.8 ± 3.1 GW [45]. Meyer and Wisdom [71] using an obsolete ancient determination of Saturn's k_2/Q pointed out that this power can not be produced from tidal dissipation because it is then inconsistent with the long-term preservation of Enceladus' eccentricity [71]. However, by taking into account the additional transfer of angular momentum resulting from the 2:1 resonance with Dione [71] and the new k_2/Q value, Lainey et al. [62] explained both the observed heat and the preservation of Enceladus' eccentricity. In addition, in that framework, Saturn's dissipation factor is inconsistent with the scenario assuming that the moons formed outside the synchronous orbit and then migrated to their current positions. A recent model of accretion of the moons inside and at the outer edge of Saturn's rings appears more consistent with the observed dissipation as well as geological observations and satellite surface composition [14, 62].

In summary, astrometric measurements have led to a quantification of the dissipation inside Jupiter and Saturn by combining accurate modern numerical models and a historical astrometric record spanning more than one century. The dissipation in the Uranian and Neptunian systems has not been estimated at this time. The strong correlation due to the Laplace resonance makes it more difficult to infer the tidal dissipation in Europa and thus complementary methods are required to determine that parameter, for example, through accurate characterization of the satellite's rotation, as shown by Rambaux et al. [92] for Enceladus, or by direct measurement of the gravity field and surface displacement [116]. Both of these techniques require in situ observations, with a dedicated orbiter or surface tracking instruments (e.g. beacons, very broad-band seismometer).

5.4 Static Tides and the Shape of the Moons

5.4.1 Introduction

The secular shapes of major satellites in synchronous rotation can be well approximated by a triaxial ellipsoid under the assumption of hydrostatic equilibrium [6, 21, 125]. This shape results from the deformation of the body in response to the centrifugal and tidal forces. The tidal force acts to elongate the moon along the parent-planet-satellite direction. This is due to the synchronous resonance as the satellites keep the same face towards the parent-planet. The centrifugal force acts to flatten the satellite's shape along its rotation axis. The combination of these two forces leads to a triaxial shape because the parent-planet generating the tidal force is usually in the equatorial plane of the satellite. The amplitude of the resulting distortion depends on distribution of mass inside the body. Hence shape data can be used to obtain information on the interior.

In practice, the forces shaping the satellites present a secular and a periodic component. The consequences of the periodic component are discussed in the next section. Here, we focus on the time-independent (secular) contribution of the forces and the resulting equilibrium shape of the satellite, i.e., when the satellite's shape had time to relax to an equilibrium ellipsoid with the long axis pointing toward the parent-planet and the short axis aligned with the rotating axis.

The steady rotational potential determining the equilibrium shape is

$$U_c = \frac{\Omega_s^2 r^2}{3} \left(P_{20}(\cos\theta) - 1 \right) \tag{5.47}$$

where the potential acts at a point located at (r, θ, λ) with r the radial component, θ the colatitude, and λ the longitude. The parameter P_{20} is the Legendre polynomial at degree 2 and order 0. Due to the axial symmetry, the potential is independent from the longitude λ. Ω_s is the mean angular rotation of the satellite and is equal to the mean motion due to the synchronous rotation. The tidal potential is expressed in spherical coordinates as in Eq. (5.8) that we recall here

$$U_T = \frac{GM_p}{d} \left(\frac{R_s}{d} \right)^2 \frac{3\cos^2\gamma' - 1}{2}. \tag{5.48}$$

Here, the tidal potential contains both a secular and a periodic component in the development of the radial distance d and in the orientation angle γ'. The secular part in d is obtained by assuming that the orbit of the satellite is circular; the secular part of the orientation is evaluated for an equatorial orbit and assuming that the moon is in exact spin-orbit synchronous rotation. Thus the angle γ' is expressed by

$$\cos\gamma' = \sin\theta \cos\lambda. \tag{5.49}$$

Combining the centrifugal and tidal potentials results in

$$\phi_2 = \Omega_s^2 R_s^2 \left(-\frac{5}{6} P_{20}(\cos\theta) + \frac{1}{4} P_{22}(\cos\theta) \cos 2\lambda \right). \tag{5.50}$$

5.4.2 Moments of Inertia

The secular potential equation (5.50) entails a permanent deformation of the satellite. The induced potential at the surface R_s of the satellite is assumed to be linear in ϕ_2 with a coefficient of proportionality, the secular Love number labeled here as k_f:

$$\delta\phi_2 = k_f \phi_2. \tag{5.51}$$

The secular Love number is equal to $3/2$ for a homogeneous body in hydrostatic equilibrium. Its value decreases as density increases with depth (see Sect. 5.2).

The response of the satellite to the secular potential induces a potential that can be developed in spherical harmonics to degree 2 as:

$$\delta\phi_2 = \frac{GM}{a} \sum_{n=0}^{\infty} \left(\frac{a}{r}\right)^{n+1} \sum_{m=0}^{n} (C_{n,m} \cos m\lambda + S_{n,m} \sin m\lambda) P_{n,m}(\sin\varphi) \quad (5.52)$$

and the identification between the potential equations (5.51) and (5.52) of each degree-2 term leads to the relation

$$C_{20} = -\frac{5}{6} k_f q \quad (5.53)$$

$$C_{22} = \frac{1}{4} k_f q, \quad (5.54)$$

where we have introduced the dimensionless parameter $q = \frac{\Omega_s^2 R_s^3}{GM_s}$ corresponding to the ratio of the centrifugal to the gravitational potential at the equator. The C_{22} coefficient is a purely tidal term, whereas C_{20} can be decomposed into a component induced by the centrifugal potential ($1/3$) and another one from the tidal potential ($1/2$). The relations (5.53) and (5.54) can be simply combined as

$$C_{20} = -\frac{10}{3} C_{22}. \quad (5.55)$$

As noticed by Moore et al. [77], this relation indicates that the body responds to the sum of the time-averaged centrifugal and tidal potentials and there is no additional deviation from spherical symmetry. This relation is often shortly assimilated as a consequence of the hydrostatic equilibrium. However, in order to assess the equilibrium state of the object it is necessary to compare its shape and gravity data, as discussed in more details below.

If the body is in hydrostatic equilibrium, another step toward understanding its interior comes from the Radau-Darwin approximation. The gravitational coefficients C_{20} and C_{22} are related to the satellite's principal moments of inertia A, B, C (with $C > B > A$) through [122]

$$C_{20} = -\frac{2C - (B + A)}{2MR^2}, \quad (5.56)$$

$$C_{22} = \frac{B - A}{4MR^2} \quad (5.57)$$

and the axial moment of inertia C/MR^2 can be deduced from the Radau-Darwin approximation for hydrostatic bodies [79]:

$$\frac{C}{MR^2} = \frac{2}{3} \left[1 - \frac{2}{5} \left(\frac{4 - k_f}{1 + k_f} \right)^{1/2} \right]. \quad (5.58)$$

In theory, the inferred moment of inertia is a simple function of the internal mass distribution inside an object.

In practice, satellites shapes and interiors depart from hydrostaticity due to mass concentrations, large variations in topography at various scales, or even sometimes a fossil bulge relic of an earlier stage in the evolution of the object. That bulge may be acquired before a moon became locked in spin-orbit resonance or during the tidal migration of the satellite. A famous example is Saturn's satellite Iapetus that presents a large equatorial bulge frozen when the object had a rotation period of 16 hours that strongly differs from its current 80 day rotation period [12]. In the case of endogenic sources of non-hydrostaticity (e.g. mass anomalies, stress associated with internal activity, such as convection) the impact on the gravity and shape is expressed at degrees higher than two [33].

5.4.3 Satellite Shapes

Under the assumption of hydrostatic equilibrium the shapes of synchronous satellites may be approximated as triaxial ellipsoids with principal axes denoted (a, b, c), where a is the long axis pointing toward the parent-planet, c the short axis along the polar axis, and b the intermediate (equatorial) axis. The proportional factor between the excitation and the radial response of the satellite is determined by the fluid Love number, h_f, defined as [78, 116]

$$u = h_f \frac{\phi_2}{g} \qquad (5.59)$$

where u is the vertical tidal surface displacement and g is the gravitational acceleration at the satellite's surface. If the body is in hydrostatic equilibrium, then h_f is related to k_f by the following relationship (e.g. [125])

$$h_f = k_f + 1. \qquad (5.60)$$

Therefore, a strengthless and homogeneous body is characterized by k_f equal to $3/2$ and h_f to $5/2$. As for k_f, the Love number h_f depends on the density profile. Departure from the equality (5.60) implies that the object is not in hydrostatic equilibrium, a crucial piece of information on the evolution of the object. Indeed, the ability of an object's shape to relax or to preserve non-hydrostatic anomalies over the long term is a function of the maximum temperature reached within the object and the mechanism driving heat transfer.

The Love number h_f may be deduced from the principal axes of the ellipsoid through [22, 125]

$$a = R_s \left(1 + \frac{7}{6} q h_f \right), \qquad (5.61)$$

$$b = R_s \left(1 - \frac{1}{3} q h_f \right), \qquad (5.62)$$

$$c = R_s \left(1 - \frac{5}{6} q h_f \right) \qquad (5.63)$$

neglecting terms of order 2 in q. Thus, by determining k_f from gravity data, it is possible to estimate the principal axis of a given satellite in hydrostatic equilibrium. Comparison with actual shape data, if available, leads to constraints on the departure of hydrostaticity of the object and then on its internal and surface evolution. In addition, these relations imply that

$$(a - c) = 4(b - c) \tag{5.64}$$

assuming that higher orders in the parameter q can be neglected.

5.4.4 Gravity and Shape Observations

5.4.4.1 Observational Methods

The gravity field of a satellite is measured by accurately tracking the trajectory of a spacecraft approaching the object and accounting for orbital perturbations acting on the spacecraft. Accurate measurement of the spacecraft's position is inferred from the shift in the radio signal Doppler frequency tracked from Earth's ground stations. The requirement in the accuracy of the Doppler shift is a few hundred meters per second.

The determination of the gravity field, and especially the C_{20} and C_{22} coefficients, requires enough flybys distributed in equatorial and polar orbits in order to determine each coefficient independently and verify whether the hydrostaticity assumption (5.55) applies to the object. When the gravity data are too sparse, it is still possible to determine the gravity field by assuming the relation (5.55). Such approach imposes a strong constraint on the geophysics of these bodies that has to be kept in mind during the interpretation of the data. The gravity data reduction technique is described in [98] review on the Galilean satellites, and in [60] for the Saturnian satellites.

The global shape of a satellite is determined by combining the various limb profiles of wide-angle images and then searching for an ellipsoid that can match these observations [22, 109, 110], or by using an altimeter such as *Cassini*'s RADAR altimeter for Titan [124]. The gravity and topography fields can be combined to infer constraints on the interior, such as non-hydrostatic anomalies. However, shape and gravity observations have been obtained only for a few bodies: Io, Europa, Ganymede, Callisto by the *Galileo* spacecraft and Enceladus, Rhea, and Titan by the *Cassini* spacecraft (see Table 5.3).

5.4.4.2 Galilean Satellites

The density of Io is relatively high (3.530 g cm^{-3}) indicating that this satellite is primarily rocky. Europa also presents a high density 3.013 g cm^{-3} but its surface is totally covered with ice, suggesting the presence at depth of a large rocky core. Ganymede and Callisto have lower densities consistent with an ice mass fraction around 30 %. Magnetometer data suggest that the three icy satellites shelter deep oceans beneath their icy surfaces [98]. This hypothesis is also supported by the

Table 5.3 Gravity data provided by the *Galileo* (a) and *Cassini* (b) missions. (a) Schubert et al. [98], (b) MacKenzie et al. [66]. Rhea is not in hydrostatic equilibrium and the Radau-Darwin approximation can not be applied, (c) Iess et al. [53]. The data for the gravity field of Enceladus are not available at this time

Satellites	$C_{20}(10^{-6})$	$C_{22}(10^{-6})$	C/MR^2	k_f
Io[a]	-1859.5 ± 2.7	558 ± 0.8	0.37824 ± 0.00022	1.3043 ± 0.0019
Europa[a]	-435.5 ± 8.2	131.5 ± 2.5	0.346 ± 0.005	1.048 ± 0.0020
Ganymede[a]	-12753 ± 2.9	38.26 ± 0.87	0.3115 ± 0.0028	0.804 ± 0.018
Callisto[a]	-32.7 ± 0.8	10.2 ± 0.3	0.3549 ± 0.0042	1.103 ± 0.035
Rhea[b]	-931 ± 12	237.0 ± 4.5	$-$	$-$
Titan[c]	-31.808 ± 0.404	9.983 ± 0.039	0.3414 ± 0.0005	1.0097 ± 0.0039

geological record and can be explained by thermal evolution models ([9, 49, 59, 86, 98] and references therein).

The gravity fields of the Galilean satellites have been determined during the *Galileo* mission that dedicated 4–5 flybys to each satellite (see a review in [98]). As *Galileo* performed 4 equatorial flybys and 1 polar flyby of Io, Anderson et al. [4] managed to decorrelate the C_{20} from the C_{22} coefficients. In the case of Ganymede even with the equatorial and orbital flybys [2], it is not possible to decorrelate the two coefficients and the relationship (5.55) has to be assumed [98]. Indeed, the gravity field of Ganymede includes components of degree and order 4 due to mass concentration that could be detected by disk-cap mass anomaly modeling [85]. The situation is even worse in the case of Europa and Callisto because the gravity passes at these objects where all in near equatorial orbit, so that only C_{22} could be determined [3, 5]. However, the gravity field of Callisto presents a non-zero S_{22} coefficient suggesting that an anomaly (interior, surface) may affect its potential [5].

At first order, the shape data available for the Galilean satellites are mostly consistent with ellipsoids in hydrostatic equilibrium [5, 98]. As a consequence, constraints on the density profile may be obtained from inferring the secular Love number k_f from C_{22} through Eq. (5.54) and the Radau-Darwin equation (5.58). Results displayed in Table 5.3 indicate that the satellites are not homogeneous because their C/MR^2 values are smaller than 0.4, i.e. the upper limit corresponding to a homogeneous spherical body. Models of these satellites matching both their axial moments of inertia C/MR^2 and mean densities indicate that Io, Europa, and Ganymede present a core enriched in rock, while Callisto is partially differentiated ([98] and references therein).

5.4.4.3 Saturnian Satellites

The gravity and topography of the Saturnian satellites have been inferred from observations obtained by the *Cassini-Huygens* mission that arrived in the system on July 1st, 2004. The *Cassini* orbiter performed several flybys of all major satellites but only a few of these flybys have been dedicated to radio science tracking that enables gravity field measurement. So far, only the gravity fields of Titan, Enceladus, and Rhea have been obtained to degree two. For the other medium-sized satellites, only the mass has been determined from radio tracking of *Cassini*, so far (see a review in [60]).

Titan has a particular place in the family of the Saturnian satellites. It is the largest with a radius of 2575 km, i.e. 3.4 times the radius of Rhea (764 km) and it is the only moon with a thick atmosphere, which is composed mainly of Nitrogen and Methane. This thick atmosphere precludes the direct observation of Titan's surface, and only Cassini RADAR, VIMS (Visual and Infrared Mapping Spectrometer), and the in-situ *Huygens* probe that revealed a fascinating world revolving around rich geological features such as dunes, channels, lakes, impact craters, and putative cryo-volcanos. The gravity and topography measurements bring constraints on Titan's interior, which will help assess the relative contribution of endogenic activity and atmospheric processes to the evolution of the surface.

The gravity field of Titan has been determined by Iess et al. [53] based on four dedicated gravity science flybys by *Cassini*. In an earlier study based on 3 flybys, Rappaport et al. [94] inferred the gravity field to degree two and found the ratio of C_{20}/C_{22} to be different from the $-10/3$ value expected for an object in hydrostatic equilibrium. However, by using one more flyby and introducing the degree 3 coefficients, Iess et al. [53] inferred C_{20}/C_{22} around $-10/3$, hence demonstrating the importance of including higher degree terms in the inversion of gravity data. So Titan's quadrupole field is consistent with that expected for a hydrostatically relaxed body shaped by tidal and rotational potentials. By applying the Radau-Darwin approximation, Iess et al. inferred Titan's polar moment of inertia C/MR^2 equal to 0.3414 ± 0.0005. This information, combined with the mean density, is an important constraint on interior models. The relatively large value of Titan's C/MR^2 (as a reference, Ganymede's mean moment of inertia is equal to ~ 0.3115 [98]) suggests that it is only partially differentiated, and that its core may contain a large fraction of water, either in the form of ice mixed with rock [7] or as water of hydration, i.e. water trapped in the silicate structure [11, 32].

In addition, the topography of Titan has been measured by radar altimetry data [124]. These authors determined a ratio of $(a-c)/(b-c) \approx 2.2$ that differs from the hydrostatic equilibrium inferred by [53] from their gravity data under the assumption of hydrostatic equilibrium. Thus Titan appears more flattened than predicted for a hydrostatically relaxed body. Nimmo and Bills [81] suggested that the discrepancy could be related to large lateral variations in the icy shell thickness. Choukroun and Sotin [15] showed that this difference might be imputed to meteorological and chemical processes acting on the icy surface as part of a methane-ethane substitution cycle. Therefore, this result highlights the importance of clearly separating the hydrostatic contribution of the shape resulting from the secular tidal potential in order to quantify the non-hydrostatic contributions due, in this case, to atmospheric processes.

Another Saturnian satellite of major interest is Enceladus, due to its active south polar region that may be associated with a liquid water reservoir [91, 106]. Only three gravity passes have been dedicated to Enceladus. While the results are not available at this time, the preliminary data indicate that the gravity field of Enceladus contains a non-negligible degree-three component [26]. These authors reported that the degree 2 coefficients dominate, as expected for a synchronous satellite, but the C_{20} and C_{22} present a small departure from the values predicted under the assumption of hydrostatic equilibrium. In addition, the C_{30} is negative, corresponding to a negative gravity anomaly at the south pole. The interpretation of these results is currently under investigation and will reveal crucial information on Enceladus' icy shell structure.

The third Saturnian satellite for which gravity measurements have been performed is Rhea, although the limited dataset (only one pass) makes it difficult to infer robust constraints on the interior of the object. Still, it appears that, like in the case of Enceladus, Rhea's gravity field contains a degree-three component [66, 82]. These authors suggested that the source of that component is the impact basin Tirawa. The mass anomaly associated with the large crater induced a reorientation of the moon's principal axes in order to minimize the rotational energy so that the smallest principal axis moment of inertia is oriented toward the parent-planet, while the largest principal axis moment of inertia presents a small angle from the normal to the orbit, following the Cassini states [87]. More gravity passes of Rhea are required in order to better understand the relationship between its gravity field and topography.

Thomas [108] published the triaxial shapes measured for 20 Saturnian satellites from limb profiles. The global shape of Rhea matches a hydrostatic figure; while for Mimas, Enceladus, and Tethys the degree 2 shapes are not consistent with hydrostatic equilibrium [82, 108]. Consequently, lateral variations in topography or internal structure (e.g. mass concentrations) need to be accounted for in the interpretation of the gravity measurements as these features can bear a non-negligible signature at high-degree spherical harmonics [82].

5.5 Internal Stress

5.5.1 Introduction

In the previous section, we focused on the constant part of the tidal potential that determines a satellite's triaxial shape. We now focus on the time-varying potential induced by eccentricity, obliquity, or physical librations (i.e. oscillations superimposed on the uniform rotation component, e.g. [46, 93]). This source of stress has profound impact on the interior and surface of the satellite. The deformation of the satellite results in friction within the material resulting in the satellite's response being out of phase with respect to the tidal forcing. This friction generates heating. When the tidal heat production exceeds the amount of heat that can be transferred to the satellite's surface, partial melting of the material ensues, and volcanism becomes the more efficient vector of heat [77]. The most spectacular expression of tidal heating is certainly the active volcanism on Io (Sect. 5.5.3). In the case of icy bodies, cryovolcanism acts at lower temperature, as observed on Enceladus, and suggested for Titan, Triton, Miranda, and Ganymede (Sect. 5.5.4). Also, the stress incurred by the periodic tidal distortion of a satellite's surface can lead to fracturing and drive tectonic activity (Sect. 5.5.5).

5.5.2 Tidal Heating from Mechanical Energy Dissipation

There are two approaches for the computation of tidal heating in satellites. The first one consists in computing the heat produced at each point of the body by using the strain-stress tensor resulting from the tidal distortion. This method has been employed in a series of paper (e.g. [89, 99, 112]) and it enables the quantification of the heat production in each part of the body. However modeling realistic, radially and laterally heterogeneous bodies with this method requires sophisticated numerical codes. In this chapter we focus on another, simpler approach that applies at the global scale of the object, but is equivalent to the former. It is based on the computation of the work performed by the tides (e.g. [118]).

The dissipated energy is equal to the work rate of the tidal force. That work rate is equal to the scalar product of the tidal force $\rho \nabla U$ (where ρ is the material density and U the tidal perturbed potential defined in Sect. 5.2) and the velocity $\mathbf{v}$ of an element of the body integrated over the volume:

$$\frac{dE}{dt} = -\int_{body} \rho \mathbf{v} . \nabla U \, dV. \tag{5.65}$$

The volume integral can be transformed into a surface integral by assuming that the interior is incompressible and homogeneous. Then we obtain

$$\frac{dE}{dt} = -\rho \int_{body} U \, \mathbf{v} . \mathbf{n} \, dS \tag{5.66}$$

by Gauss' theorem. Here $\mathbf{n}$ is the normal to the surface and the quantity $\mathbf{v} . \mathbf{n}$ is the rate at which the surface is elevated. The phase lag between the potential U and the elevation of the surface is then introduced as

$$\zeta = h_2 \frac{U'}{g} \tag{5.67}$$

where h_2 is the dynamic Love number quantifying the deformation of the object (integrated over its radius) and U' is the tidal potential lagged because of friction inside the body. Then, after computing the tidal potential and averaging over the orbital period (short period), tidal dissipation is found as (e.g. [31, 50, 65, 89, 99, 119])

$$\frac{dE}{dt} = -\frac{21}{2} \frac{k_2}{Q} \frac{R_s^5 n^5}{G} e^2. \tag{5.68}$$

Fig. 5.4 Global view of Jupiter's moon Io taken by Galileo in September 1997. This composite image has color enhanced in order to highlight different regions such as *big red ring* of Pele volcano at the *bottom left*. The *dark spot* close to the *center* of the image was a new distinctive structure illustrating the ongoing activity on Io (http://photojournal.jpl.nasa.gov/catalog/PIA01667), courtesy of NASA

This expression applies to a satellite in synchronous spin-orbit resonance with negligible obliquity, orbital inclination, as well as physical libration. The general expression taking into account these additional perturbations can be found in Wisdom [119] and Levrard [63]. The power of 5 applying to the moon's radius and mean motion implies increased heating in large satellites and/or satellites close to their parent-planets. In addition, the amount of dissipated energy depends on the orbital eccentricity (e^2). This dependency is related to the source of the tidal work being the radial and diurnal tides presented in Sect. 5.3.3.4. The contribution of the librational tides is 4/3 larger than the contribution of the radial tides and the sum of the two contributions leads to the factor 21/2. Finally, the energy rate depends on the ratio k_2/Q that is function of the capacity of the satellite interior to deform and dissipate mechanical energy, which is itself a function of the forcing frequency (Sect. 5.3).

5.5.3 A Hot Satellite: Io

The spacecraft *Voyager 1* revealed in 1979 a unique volcanic world, Io, with active lava flows, volcanic plumes several hundred kilometer high, and a young surface devoid of impact craters [68, 101]. The *Galileo* image displayed in Fig. 5.4, shows Io in false color in order to enhance the different geological structures, where the white and gray features represent sulfur dioxide frost, whereas the bright red and black features are related to recent volcanic activity. In addition, *Galileo* detected eruptions and identified a recent ring of reddish material deposits around an area called Tvashtar Catena [58, 68].

Io's thermal emission has been inferred from *Galileo*'s data between 1.2 W m^{-2} and 3 W m^{-2} [95, 115] and the global power has been found between 50 and 125 TW. For comparison, Earth's global dissipation budget is around 3.3–4 TW [30], i.e., 25 times smaller. In comparison, other internal heat sources (accretional, radiogenic heating and specific gravitational energy) are several orders of magnitude less significant and insufficient to drive partial melting leading to volcanic activity. This scenario was predicted by Peale et al. [90] a few weeks before the arrival of *Voyager* and developed in more details by, e.g. Moore [77]. The source of energy is related to tides raised by Jupiter on Io that are big due to Io proximity (421 000 km to compare with the Earth-Moon distance of 384 000 km), large Jupiter's mass (300 times the Earth's mass), and a relatively high Io's eccentricity (0.0041). For illustration, the elevation of Io's surface due to the diurnal tides is $3eh_2H$ [16] where e is the eccentricity, h_2 the surface Love number, and H the equilibrium amplitude defined in Eq. (5.1) leading to a surface elevation of about 300 meters height.

It could be surprising that Io kept a non-zero eccentricity because the effect of tides is to circularize the orbit. The eccentricity may be separated in two terms: a free and a forced eccentricity (e.g. [40]). The free eccentricity depends on the initial condition and it is damped to a very small value around 10^{-5} as expected from tidal theory [40]. The forced eccentricity is related to the Laplace resonance Eq. (5.46), and the eccentricity of Io is then pumped by the resonance with Europa and Ganymede that allows the preservation of a non-zero eccentricity until today [40, 88, 121].

This huge amount of energy poses the problem of the transfer of energy from the interior of Io to the surface and the moon's thermal equilibrium. Moore [75] investigated the question of energy transfer by studying convection in a partially molten core and he deduced that the heat that can be transferred is one order lower than the observed flux. This suggests that Io is either out of thermal equilibrium or another heat transport mechanism is taking place. Indeed, the recent determination of Io's dissipation factor determined by Lainey et al. [61] leads to a heat flux equal to $2.24 \pm 0.45 \text{ W m}^{-2}$ that is within the range of the observed surface heat flux. In addition, the (k_2/Q) values for Io and Jupiter suggest that Io is close to thermal equilibrium with the energy produced by tidal dissipation being radiated at the surface.

The tidal heating on Io is very unique due to its spectacular consequences that can be seen from Earth (e.g. [115]). However, tidal heating also plays a role on other satellites like Europa where it contributes to the preservation of a deep ocean beneath the icy shell [47, 84, 104, 111]. Ganymede and Miranda may have encountered past resonances that have enhanced their eccentricities and increased their tidal heat budget leading to a possible phase of resurfacing of these satellites (e.g. [23, 100]).

Fig. 5.5 Global view of
Enceladus taken by Cassini
space mission. The *south pole*
presents the famous plumes
of water ice burst into the
Saturnian system
(http://photojournal.jpl.nasa.
gov/catalog/PIA12733),
courtesy of NASA

5.5.4 Cryovolcanism

Tidal heating can lead to an exotic form of volcanism called *cryovolcanism*, where
the volcano erupts liquid or vapor phases of volatile elements such as water ([54]
and references therein). The *Voyager*'s observations of Geyser on Neptune's moon
Triton [102] or active plumes at the south pole of Enceladus observed by Cassini
spacecraft (Fig. 5.5) [91, 106] may suggest that this process, that has no equivalent
on Earth, could be widespread in the icy satellites of the outer solar system. However, the existence of cryovolcanism on icy satellites is still debated, as for example
in the case of Titan where some cryovolcanic-looking features may actually have
a tectonic origin [16, 54, 64, 80, 103]. Nevertheless, signatures of past cryovolcanism have been identified on Europa and Ganymede by the Galileo mission and on
Miranda from Voyager imaging (e.g. a review [16]).

Like for Io, Enceladus's cryovolcanism is certainly driven by tidal dissipation
[107]. The amount of energy dissipated within Enceladus has been measured by
Cassini's Composite Infrared Spectrometer at 5.8 ± 1.9 GW [106] and updated at
15.8 ± 3.1 GW by Howett et al. [45]. Based on the latest measurements of Saturn's
dissipation factor, Enceladus is certainly in thermal equilibrium [62]. For Triton,
Geyser-like plumes of 8 km height have been observed by *Voyager 2* [102] and the
mechanism driving the plumes appears to be recent cryovolcanism that could also
explain some enigmatic features and the young age of the Triton's surface [16].

Fig. 5.6 Cycloidal cracks at the surface of Europa (from Hoppa et al. [44])

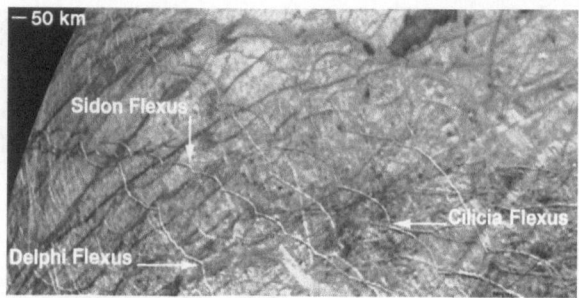

5.5.5 Tidally Driven Tectonics

Tidal stressing leads to surface fractures at various scales, some of which may run along the entire surfaces of icy satellites. By studying these structures geologists can deduce properties of the surface layer such as its thickness and thermal structure. In addition, the patterns displayed by the geological fractures can reveal key information on the tidal history of the satellites, such as their orbital evolution, reorientation of the outer shell as consequence of decoupling, or non-synchronous rotation of the icy shell (see review in [16], and [56]). Helfenstein and Parmentier [43] first pointed out a correlation between global-scale lineaments patterns and tidal stress in *Voyager* images for Europa. Since then, three types of tectonic features induced by tidal stress have been identified: lineaments, cycloidal cracks, and strike-slip faults (see a review in [8] and references therein). Similar geological structures have recently been identified on Enceladus and Triton, whereas Ganymede and Miranda show past evidence of active tectonics [16].

The icy crust is generally modeled as an elastic layer overlying the tidally deformed body and the tidal stress σ is computed following the formalism developed by Vening-Meinisz [114], Melosh [70]. Compressive stresses are defined when σ is positive and tensional stresses when σ is negative. Fracturation occurs when the tensile stress exceeds the tensile strength of the crust. Then the cracks propagate following the ever-changing stress field and stop where the tensile stress becomes insufficient. Each source of the time-varying potential presents a different stress field whose pattern can be compared against tectonic patterns. Matching the computed field stress with surface cracks can help identify the origin of the tidal stress source and constrain the elastic properties of the crust. For example Fig. 5.6 shows cycloidal cracks on Europa's surface, observed by *Voyager* [101]. These cycloidal structures have been interpreted as the geological consequence of diurnal variations in the tidal stress field [44].

5.6 Conclusion

Tides rhythm the evolution of giant planets satellites like the tides on Earth rhythm the flux and reflux on coasts. Their actions lead to despinning and endogenic activity, such as spectacular volcanism on Io and Enceladus. Tides also bear a signature in satellites shapes that may be measured if these objects are in hydrostatic equilibrium. However, each satellite also appears to be unique, making the investigation of these systems so fascinating. Many questions are still open and some of them are still under development thanks to the Cassini space mission that has been providing astonishing results. ESA's JUpiter ICy moons Explorer, JUICE, under development [25] will visit the Jupiter system in the next decades (2028) and especially Ganymede and Europa. Even the tidal theory is under an active reevaluation motivated by new laboratory experiments [67] allowing to introduce a more complex rheological response of satellites [13, 27]. In addition, the tidal theory has to introduce the presence of a subsurface ocean, expected in some satellites, that could lead to increased dissipation energy [113].

Acknowledgements   Part of this work has been conducted at the Jet Propulsion Laboratory, California Institute of Technology, under contract to the National Aeronautics and Space Administration.

References

1. Aleshkina, E.Y.: Sol. Syst. Res. **43**, 71–78 (2009). doi:10.1134/S0038094609010079
2. Anderson, J.D., Lau, E.L., Sjogren, W.L., Schubert, G., Moore, W.B.: Nature **384**, 541 (1996)
3. Anderson, J.D., Schubert, G., Jacobson, R.A., et al.: Science **281**, 2019 (1998)
4. Anderson, J.D., Jacobson, R.A., Lau, E.L., Moore, W.B., Schubert, G.: J. Geophys. Res. **106**, 32963 (2001)
5. Anderson, J.D., Jacobson, R.A., McElrath, T.P., et al.: Icarus **153**, 157 (2001)
6. Balmino, G.: Artif. Satell. **42**, 141 (2007)
7. Barr, A.C., Citron, R.I., Canup, R.M.: Icarus **209**, 858 (2010)
8. Bills, B.G., Nimmo, F., Karatekin, Ö., Van Hoolst, T., Rambaux, N., Levrard, B., Laskar, J.: Europa. In: Pappalardo, R.T., McKinnon, W.B., Khurana, K.K. (with the assistance of René Dotson with 85 collaborating authors) (eds.) The University of Arizona Space Science Series, vol. 119, p. 119. University of Arizona Press, Tucson (2009). ISBN 9780816528448
9. Carr, M.H., Belton, M.J.S., Chapman, C.R., et al.: Nature **391**, 363 (1998)
10. Cassini, G.D.: Traite de l'Origine et de Progres de l'Astronomie. Paris (1693)
11. Castillo-Rogez, J.C., Lunine, J.I.: Geophys. Res. Lett. **37**, L20205 (2010)
12. Castillo-Rogez, J.C., Matson, D.L., Sotin, C., et al.: Icarus **190**, 179 (2007)
13. Castillo-Rogez, J.C., Efroimsky, M., Lainey, V.: J. Geophys. Res. (Planets) **116**, 9008 (2011)
14. Charnoz, S., Crida, A., Castillo-Rogez, J.C., et al.: Icarus **216**, 535 (2011)
15. Choukroun, M., Sotin, C.: Geophys. Res. Lett. **39**, 4201 (2012)
16. Collins, G., McKinnon, W., Moore, J., Nimmo, F., Pappalardo, R., Prockter, L., Schenk, P.: Tectonics of the outer planet satellites. In: Planetary Tectonics pp. 264–350. Cambridge University Press, Cambridge (2010)
17. Colombo, G.: Astron. J. **71**, 891 (1966)
18. Correia, A.C.M.: Astrophys. J. Lett. **704**, L1 (2009)
19. Darwin, G.H.: Proc. Soc. Lond. Ser. I **30**, 1 (1879)
20. Darwin, G.H.: Philos. Trans. R. Soc. Lond. Ser. I **171**, 713 (1880)
21. Dermott, S.F.: Icarus **37**, 575 (1979)
22. Dermott, S.F., Thomas, P.C.: Icarus **73**, 25 (1988)
23. Dermott, S.F., Malhotra, R., Murray, C.D.: Icarus **76**, 295 (1988)
24. de Sitter, W.: Mon. Not. R. Astron. Soc. **91**, 706 (1931)

25. Dougherty, M., et al.: JUICE: exploring the emergence of habitable worlds around gas giants. ESA/SRE, p. 18 (2011)
26. Ducci, M., Iess, L., Armstrong, J.W., et al.: In: Lunar and Planetary Institute Science Conference Abstracts, vol. 43, p. 2200 (2012)
27. Efroimsky, M.: Celest. Mech. Dyn. Astron. 112, 283 (2012)
28. Efroimsky, M., Lainey, V.: J. Geophys. Res. 112, E12003 (2007). doi:10.1029/2007JE002908
29. Efroimsky, M., Williams, J.G.: Celest. Mech. Dyn. Astron. 104, 257–289 (2009). doi:10.1007/s10569-009-9204-7
30. Egbert, G.D., Ray, R.D.: Nature 405, 775 (2000)
31. Ferraz-Mello, S., Rodríguez, A., Hussmann, H.: Celest. Mech. Dyn. Astron. 101, 171 (2008)
32. Fortes, A.D.: Planet. Space Sci. 60, 10 (2012)
33. Gao, P., Stevenson, D.J.: In: 43rd Lunar and Planetary Institute Science Conference Abstracts, p. 1701 (2012)
34. Gavrilov, S.V., Zharkov, V.N.: Icarus 32, 443 (1977)
35. Goldreich, P.: Mon. Not. R. Astron. Soc. 126, 257 (1963)
36. Goldreich, P., Peale, S.: Astron. J. 71, 425 (1966)
37. Goldreich, P., Soter, S.: Icarus 5, 375 (1966)
38. Grasset, O., Coustenis, A., Durham, W.B., et al.: Space Sci. Rev. 153, 5 (2010)
39. Greenberg, R.: Astrophys. J. Lett. 698, L42 (2009)
40. Greenberg, R.: Rep. Prog. Phys. 73, 036801 (2010)
41. Greenberg, R., Weidenschilling, S.J.: Icarus 58, 186 (1984)
42. Greenberg, R., Hoppa, G.V., Bart, G., Hurford, T.: Celest. Mech. Dyn. Astron. 87, 171 (2003)
43. Helfenstein, P., Parmentier, E.M.: Icarus 53, 415 (1983)
44. Hoppa, G.V., Tufts, B.R., Greenberg, R., Geissler, P.E.: Science 285, 1899 (1999)
45. Howett, C.J.A., Spencer, J.R., Pearl, J., Segura, M.: J. Geophys. Res. (Planets) 116, 3003 (2011)
46. Hurford, T.A., Bills, B.G., Helfenstein, P., et al.: Icarus 203, 541 (2009)
47. Hussmann, H., Spohn, T., Wieczerkowski, K.: Icarus 156, 143–151 (2002)
48. Hussmann, H., Sohl, F., Spohn, T.: Icarus 185, 258 (2006)
49. Hussmann, H., Sotin, C., Lunine, J.I.: In: Schubert, G., Spohn, T. (eds.) Treatise on Geophysics—Planets and Moons, pp. 509–539. Elsevier, Amsterdam (2007)
50. Hussmann, H., Choblet, G., Lainey, V., et al.: Space Sci. Rev. 153, 317 (2010)
51. Hut, P.: Astron. Astrophys. 99, 126 (1981)
52. Ioannou, P.J., Lindzen, R.S.: Astrophys. J. 406, 266–278 (1993)
53. Iess, L., Rappaport, N.J., Jacobson, R.A., et al.: Science 327, 1367 (2010)
54. Jaumann, R., Clark, R.N., Nimmo, F., et al.: Saturn from Cassini-Huygens, p. 637 (2009)
55. Jeffreys, H.: Mon. Not. R. Astron. Soc. 122, 339 (1961)
56. Kattenhorn, S.A., Hurford, T.: Europa. In: Pappalardo, R.T., McKinnon, W.B., Khurana, K.K. (with the assistance of René Dotson with 85 collaborating authors) (eds.) The University of Arizona Space Science Series, vol. 199, p. 199. University of Arizona Press, Tucson (2009). ISBN 9780816528448
57. Kaula, W.M.: Rev. Geophys. Space Phys. 2, 661 (1964)
58. Keszthelyi, L., et al.: J. Geophys. Res. (Planets) 106(E12), 33025–33052 (2001)
59. Khurana, K.K., Kivelson, M.G., Stevenson, D.J., et al.: Nature 395, 777 (1998)
60. Krupp, N., Khurana, K.K., Iess, L., et al.: Space Sci. Rev. 153(1–4), 11–59 (2010)
61. Lainey, V., Arlot, J.-E., Karatekin, Ö., Van Hoolst, T.: Nature 459, 957 (2009)
62. Lainey, V., Karatekin, Ö., Desmars, J., et al.: Astrophys. J. 752, 14 (2012)
63. Levrard, B.: Icarus 193, 641 (2008)
64. Lopes, R.M.C., Mitchell, K.L., Stofan, E.R., et al.: Icarus 186, 395 (2007)
65. MacDonald, G.J.F.: Rev. Geophys. Space Phys. 2, 467 (1964)
66. Mackenzie, R.A., Iess, L., Tortora, P., Rappaport, N.J.: Geophys. Res. Lett. 35, 5204 (2008)

67. McCarthy, C.M., Castillo-Rogez, J.C.: Planetary ices attenuation properties. In: Gudipati, M.S., Castillo-Rogez, J.C. (eds.) Science of Solar System Ices, pp. 183–226. Springer, New York (2011)

68. McEwen, A.S.: Active volcanism on Io. Science **297**, 2220–2221 (2002)

69. McKinnon, W.B.: Icarus **183**, 435 (2006)

70. Melosh, H.J.: Icarus **43**, 334 (1980)

71. Meyer, J., Wisdom, J.: Icarus **188**, 535 (2007)

72. Mignard, F.: Moon Planets **20**, 301 (1979)

73. Mignard, F.: Moon Planets **23**, 185 (1980)

74. Mignard, F.: Moon Planets **24**, 189 (1981)

75. Moore, W.B.: J. Geophys. Res. **108**, E8 (2003). doi:10.1029/2002JE001943

76. Moore, W.B., Schubert, G.: Icarus **166**, 223 (2003)

77. Moore, W.B., Schubert, G., Anderson, J.D., Spencer, J.R.: In: Io After Galileo: A New View of Jupiter's Volcanic Moon, p. 89 (2007)

78. Munk, W.H., MacDonald, G.J.F.: Cambridge [Eng.] University Press, Cambridge (1960)

79. Murray, C.D., Dermott, S.F.: In: Murray, C.D., Dermott, S.F. (eds.) Solar System Dynamics. Cambridge University Press, Cambridge (2000). ISBN 0521575974

80. Nelson, R.M., Kamp, L.W., Lopes, R.M.C., et al.: Geophys. Res. Lett. **36**, 4202 (2009)

81. Nimmo, F., Bills, B.G.: Icarus **208**, 896 (2010)

82. Nimmo, F., Bills, B.G., Thomas, P.C., Asmar, S.W.: J. Geophys. Res. (Planets) **115**, 10008 (2010)

83. Ogilvie, G.I., Lin, D.N.C.: Astrophys. J. **610**, 477 (2004)

84. Ojakangas, G.W., Stevenson, D.J.: Thermal state of an ice shell on Europa. Icarus **81**, 220–241 (1989)

85. Palguta, J., Schubert, G., Zhang, K., Anderson, J.D.: Icarus **201**, 615 (2009)

86. Pappalardo, R.T., Head, J.W., Greeley, R., et al.: Nature **391**, 365 (1998)

87. Peale, S.J.: Annu. Rev. Astron. Astrophys. **37**, 533 (1999)

88. Peale, S.J.: Celest. Mech. Dyn. Astron. **87**, 129 (2003)

89. Peale, S.J., Cassen, P.M.: Icarus **36**, 245–269 (1978)

90. Peale, S.J., Cassen, P., Reynolds, R.T.: Science **203**, 892–894 (1979)

91. Porco, C.C., Helfenstein, P., Thomas, P.C., et al.: Science **311**, 1393 (2006)

92. Rambaux, N., Castillo-Rogez, J.C., Williams, J.G., Karatekin, Ö.: Geophys. Res. Lett. **37**, 4202 (2010)

93. Rambaux, N., Van Hoolst, T., Karatekin, Ö.: Astron. Astrophys. **527**, A118 (2011)

94. Rappaport, N.J., Iess, L., Wahr, J., et al.: Icarus **194**, 711 (2008)

95. Rathbun, J.A., Spencer, J.R., Tamppari, L.K., Martin, T.Z., Barnard, L., Travis, L.D.: Icarus **169**, 127–139 (2004). 12.021. doi:10.1016/j.icarus.2003

96. Schaefer, B.E., Tourtellotte, S.W., Rabinowitz, D.L., Schaefer, M.W.: Icarus **196**, 225 (2008)

97. Schubert, G.: Nature **459**, 920 (2009)

98. Schubert, G., Anderson, J.D., Spohn, T., McKinnon, W.B.: Jupiter: The Planet, Satellites and Magnetosphere, p. 281 (2004)

99. Segatz, M., Spohn, T., Ross, M.N., Schubert, G.: Icarus **75**, 187–206 (1988)

100. Showman, A.P., Malhotra, R.: Icarus **127**, 93 (1997)

101. Smith, B.A., Soderblom, L.A., Beebe, R., et al.: Science **206**, 927 (1979)

102. Soderblom, L.A., Becker, T.L., Kieffer, S.W., Brown, R.H., Hansen, C.J., Johnson, T.V.: Science **250**, 410–415 (1990)

103. Sotin, C., Jaumann, R., Buratti, B.J., et al.: Nature **435**, 786–789 (2005)

104. Sotin, C., Tobie, G., Wahr, J., McKinnon, W.B.: Europa, In: Pappalardo, R.T., McKinnon, W.B., Khurana, K.K. (with the Assistance of René Dotson with 85 collaborating authors) (eds.) The University of Arizona Space Science Series, vol. 85, p. 85. University of Arizona Press, Tucson (2009). ISBN 9780816528448

105. Sotin, C., Mitri, G., Rappaport, N., Schubert, G., Stevenson, D.: Titan from Cassini-Huygens, p. 61 (2010)

106. Spencer, J.R., Pearl, J.C., Segura, M., et al.: Science **311**, 1401 (2006)

107. Spencer, J.R., Barr, A.C., Esposito, L.W., et al.: Saturn from Cassini-Huygens, p. 683 (2009)
108. Thomas, P.C.: Icarus **208**, 395 (2010)
109. Thomas, P.C., Davies, M.E., Colvin, T.R., et al.: Icarus **135**, 175 (1998)
110. Thomas, P.C., Burns, J.A., Helfenstein, P., et al.: Icarus **190**, 573 (2007)
111. Tobie, G., Choblet, G., Sotin, C.: J. Geophys. Res. **108** (2003). doi:10.1029/2003JE002099
112. Tobie, G., Mocquet, A., Sotin, C.: Icarus **177**, 534–549 (2005)
113. Tyler, R.H.: Nature **456**, 770 (2008)
114. Vening-Meinisz, F.A.: Trans. AGU **28**, 1–61 (1947)
115. Veeder, G.J., Davies, A.G., Matson, D.L., et al.: Icarus **219**, 701 (2012)
116. Wahr, J.M., Zuber, M.T., Smith, D.E., Lunine, J.I.: J. Geophys. Res. (Planets) **111**, 12005 (2006)
117. Ward, W.R.: Icarus **46**, 97 (1981)
118. Wisdom, J.: Astron. J. **128**, 484–491 (2004)
119. Wisdom, J.: Icarus **193**, 637 (2008)
120. Wisdom, J., Peale, S.J., Mignard, F.: Icarus **58**, 137 (1984)
121. Yoder, C.F.: Nature **279**, 767 (1979)
122. Yoder, C.F.: In: Global Earth Physics. A Handbook of Physical Constants (1995)
123. Yoder, C.F., Peale, S.J.: Icarus **47**, 1 (1981)
124. Zebker, H.A., Stiles, B., Hensley, S., et al.: Science **324**, 921 (2009)
125. Zharkov, V.N., Leontjev, V.V., Kozenko, V.A.: Icarus **61**, 92 (1985)

Chapter 6
Recent Developments in Planet Migration Theory

Clément Baruteau and Frédéric Masset

Abstract Planetary migration is the process by which a forming planet undergoes a drift of its semi-major axis caused by the tidal interaction with its parent protoplanetary disc. One of the key quantities to assess the migration of embedded planets is the tidal torque between the disc and the planet, which has two components: the Lindblad torque and the corotation torque. We review the latest results on both components for planets in circular orbits, with special emphasis on the various processes that give rise to a large corotation torque and those contributing to its saturation. The additional corotation torque could help address the shortcomings that have recently been exposed in models of planet population synthesis. We also review recent results concerning the migration of giant planets that carve gaps in the disc (type II migration) and the migration of sub-giant planets that open partial gaps in massive discs (type III migration).

6.1 Introduction

The extraordinary diversity of extrasolar planetary systems has challenged our understanding of how planets form and how their orbits evolve as they form. Among the many processes contemplated thus far to account for the observed properties of extrasolar planets, the gravitational interaction between planets and their parent protoplanetary disc plays a prominent role. Considered for a long time as the key ingredient in shaping planetary systems, planet–disc interactions, which drive the well-known planetary migration (drift of a planet's semi-major axis during the lifetime of the gaseous disc), have recently been judged by many as being over-emphasised. On one hand, observational data show evidence for vigorous migration

C. Baruteau (✉)
DAMTP, University of Cambridge, Wilberforce Road, Cambridge CB3 0WA, UK
e-mail: C.Baruteau@damtp.cam.ac.uk

F. Masset
Instituto de Ciencias Físicas, Universidad Nacional Autónoma de México (UNAM), Apdo. Postal 48-3, 62251 Cuernavaca, MO, Mexico
e-mail: masset@fis.unam.mx

J. Souchay et al. (eds.), *Tides in Astronomy and Astrophysics*,
Lecture Notes in Physics 861, DOI 10.1007/978-3-642-32961-6_6,
© Springer-Verlag Berlin Heidelberg 2013

in many planetary systems, as attested by the existence of hot Jupiters, Neptunes, and Super-Earths (the recently discovered Kepler-20 planetary system, with coplanar rocky and icy planets alternating at periods less than 80 days [31], provides a good example), or by the existence of many mean-motion resonances. On the other hand, there is also compelling evidence that other processes are capable of altering the orbits as dramatically as planet–disc interactions (the existence of highly-eccentric or retrograde planets is an example). Also, although many systems seem to have undergone orbital migration, many others display planets at distances from their star that are of same order of magnitude as the distances of the planets in our Solar System to the Sun. One may be tempted to conclude from this that theories of planet–disc interactions are overrated, to the point that they could be just ignored in scenarios of formation of planetary systems. Yet, as will be detailed in the following sections, the facts below are difficult to circumvent:

- Each component of the torque exerted by the disc on a planet is so large that it can halve or double the planet's semi-major axis in a time that is usually 2 or more orders of magnitude shorter than the lifetime of the protoplanetary disc.
- These torque components do not cancel out. The residual torque amounts to a fair fraction of each torque component, so that one should in general expect that planet–disc interactions have a strong effect on planet orbits over the lifetime of the disc.

The central difficulty in planetary migration theories lies precisely in predicting the value of the residual torque. In addition to being a fair difference between several large-amplitude torques, it is very sensitive to the disc properties near the planet's orbit (e.g. density, temperature profiles). This by no means implies that the total torque is negligible, but it helps understand why migration theories are slowly maturing.

One of the main purposes of this review is to provide the reader with an up-to-date presentation of the state of planet–disc interactions, with emphasis on the torque formulae that govern the migration of low-mass planets. The reasons for this special emphasis are three-fold:

- Low-mass planets are the most critical in planetary population synthesis, as they potentially undergo the fastest migration.
- The sensitivity of detection methods has increased to a point where we can find a plethora of Neptune-sized planets or below, which exclusively underwent the migration processes typical of low-mass planets during their formation.
- The subject has been the focus of significant efforts in the recent past.

The torque acting on a low-mass planet in circular orbit can be decomposed into two components: (i) the differential Lindblad torque, arising from material passing by the planet at supersonic velocities, which is deflected and therefore exchanges angular momentum and energy with the planet, and (ii) the corotation torque, arising from material slowly drifting with respect to the planet, in the vicinity of its orbit. The differential Lindblad torque has been extensively studied from the early times of planetary migration theories, and is known in much greater detail than the corotation torque. In fact the corotation region, which has been under intense scrutiny over the last five years, has proved to have a much more complex dynamics than previously thought. In particular, the value of the corotation torque depends sensitively on the radiative properties of the gas disc, and may exhibit large values when the gas is radiatively inefficient (as is generally expected in regions of planet formation). In addition to this complexity, new problems emerge as the computational resources render tractable the task of simulating a planet embedded in realistic discs, namely 3-dimensional discs invaded by turbulence.

This review is organised as follows. After a brief description of the physical model and notation in Sect. 6.2, we present in Sect. 6.3 the migration of low-mass planets (type I migration). We detail some recent results on the differential Lindblad torque in Sect. 6.3.1, and we put special emphasis on the recent developments on the corotation torque in Sect. 6.3.2. The migration of gap-opening planets is then examined in Sect. 6.4, with type III migration in Sect. 6.4.2, followed by type II migration in Sect. 6.4.3. Finally, in Sect. 6.5, we discuss some recent themes related to planet–disc interactions, such as the discovery of massive planets at large orbital separations, and recent models of planetary population synthesis. Most sections end with a summary of their content.

6.2 Physical Model and Notation

In most of the following we shall consider 2-dimensional discs, considering vertically averaged or vertically integrated quantities where appropriate. At the present time, most of the recent investigation on the migration of low-mass planets has been undertaken in 2 dimensions (with a list of exceptions that includes, but is not restricted to [3, 12, 23, 24, 47, 86, 87, 104, 113]), and much insight can be gained into the mechanisms of the different components of the torque exerted by the disc on the planet through a 2-dimensional analysis. It should be remembered, however, that 2-dimensional results are plagued by the unavoidable use of a softening length for the planet's gravitational potential, which fits a two-fold purpose:

- It mimics the effects of the finite thickness of a true disc, by lowering the magnitude of the planet's potential well.
- In numerical simulations, it avoids the potential divergence at the scale of the mesh zone.

For this reason, the reader should bear in mind that the ultimate torque expressions should be sought by means of 3-dimensional calculations, and that 2-dimensional calculations are only used in a first step to elucidate the mechanisms that contribute to the torque. At the time of writing this manuscript, most of the physics of the torque in 2 dimensions is fairly well understood, which is why we put special emphasis on the 2-dimensional analysis.

We consider a planet of mass M_p orbiting a star of mass $M_\star$ with orbital frequency Ω_p. We denote by q the planet-to-primary mass ratio. The planet is assumed to be on a prograde circular orbit of semi-major axis a, coplanar with the disc, so that we do not consider in this work eccentric, inclined, or retrograde planets. The protoplanetary disc in which the planet is embedded is modelled as a 2-dimensional viscous disc in radial equilibrium, with the centrifugal acceleration and the radial acceleration related to the pressure gradient balancing the gravitational acceleration due to the central star. We use P to denote the vertically integrated pressure, and s to denote (a measure of) the gas entropy, which we express as

$$s = \frac{P}{\Sigma^\gamma},$$

(6.1)

where Σ is the surface density of the gas and γ the ratio of specific heats C_p/C_v. We denote with T the vertically averaged temperature. We assume in most of what follows that the surface density and temperature profiles are power laws of radius, with indices α and β, respectively:

$$\Sigma \propto r^{-\alpha}$$

(6.2)

and

$$T \propto r^{-\beta}.$$

(6.3)

The disc pressure scale length is $H = c_s/\Omega$, with Ω the gas orbital frequency and c_s the sound speed. We define the disc aspect ratio by $h = H/r$. When T is a power law of radius, so is h, with an index f dubbed the flaring index:

$$h \propto r^f,$$

(6.4)

which satisfies $\beta = 1 - 2f$. In almost all studies of planet–disc interactions, the disc is modelled with a stationary kinematic viscosity ν, aimed at modelling the disc's turbulent properties. We will consider that ν can be written as $\nu = \alpha_v H^2 \Omega$ [102], with α_v denoting the alpha viscous parameter associated with the turbulent stresses in the disc.

We will also make use of Oort's constants. The first Oort's constant scales with the shear of the flow,

$$A = \frac{1}{2} r \frac{d\Omega}{dr},$$

(6.5)

while the second Oort's constant scales with the vertical component of the vorticity of the flow ω_z:

$$B = \frac{1}{2r} \frac{d(r^2 \Omega)}{dr} = \frac{\omega_z}{2}.$$

(6.6)

Whenever used, A and B are implicitly meant to be evaluated at the planet's orbital radius.

The governing equations of the flow are the equation of continuity, the Navier-Stokes equations and the energy equation (except when dealing with isothermal discs, as specified below), together with the closure relationship provided by the equation of state, which is that of an ideal gas. We do not reproduce these governing equations here, but refer the interested reader to, for example, [14, 18, 22, 24, 69, 91].

Fig. 6.1 Disc's surface
density perturbed by a
low-mass planet. Streamlines
in the frame corotating with
the planet are over plotted by
solid curves. The set of
streamlines that librate with
respect to the planet delimits
the planet's horseshoe region

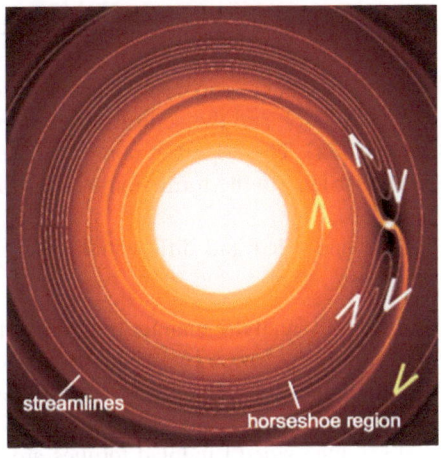

6.3 Migration of Low-Mass Embedded Planets: Type I Migration

Up until recently, type I migration referred to the regime of migration of low-mass planets that could be tackled through a linear analysis [48, 104]. Recently, however, [89] have shown that one of the torque components, namely the corotation torque, can become non-linear at all planetary masses, provided the disc viscosity is sufficiently small. We shall nevertheless still qualify type I as the migration of low-mass planets, up to a threshold mass that we shall specify in Sect. 6.4.1. We entertain below the two components of the tidal torque: the differential Lindblad torque and the corotation torque. In the following, we address the properties of type I migration by a direct inspection of the tidal torque Γ, where the planet migration rate $\dot{a}$ is given by

$$2Ba\dot{a}M_p = \Gamma. \tag{6.7}$$

6.3.1 Differential Lindblad Torque

The differential Lindblad torque accounts for the exchange of angular momentum between the planet and the trailing density waves (spiral wakes) that it generates in the disc (see illustration in Fig. 6.1). The density waves propagating inside the planet's orbit carry away negative angular momentum, and thus exert a positive torque on the planet, named the inner Lindblad torque. Similarly, the spiral density waves propagating outside the planet's orbit carry away positive angular momentum, which corresponds to a negative torque on the planet (the outer Lindblad torque). The angular momentum of a planet on a circular orbit scales with the square root of its semi-major axis. The inner Lindblad torque thus tends to make the planet move outwards, while the outer Lindblad torque tends to make it move inwards. The residual torque, called differential Lindblad torque, results from a balance between the inner and outer torques. In the absence of viscosity, the angular momentum taken away by the wakes is conserved until wave breaking occurs, resulting in the formation of a shock and the deposition of the wave's energy and angular momentum to the disc [34]. The impact of non-linearities induced by the planet's wakes, which in particular lead to the formation of a gap about the planet's orbit, will be described in Sect. 6.4.1.

The one-sided and differential Lindblad torques can be evaluated in different manners:

1. In a fully analytic manner upon linearization of the flow equations [32]. In this framework, waves propagate away from Lindblad resonances with the planet [108], and they constructively interfere into a one-armed spiral pattern [85], which begins where the Keplerian flow is supersonic with respect to the planet [34]. One-sided Lindblad torques are then evaluated as the sum of the torques arising at each Lindblad resonance. The locations of Lindblad resonances are shifted with respect to their nominal location (given by the condition of mean-motion resonance with a test particle) by pressure effects. In particular, resonances with high azimuthal wavenumber have accumulation points at $\pm 2H/3$ from the planetary orbit, instead of accumulation at the orbit. This provides a torque cutoff [2, 33], which can only be evaluated approximately [1]. This renders fully analytic methods of Lindblad torque calculations only approximate.
2. The torque can be evaluated by solving numerically the linearised equations of the flow. This approach was initially undertaken by [48], and recently revisited by [88] and [91].
3. An intermediate approach may be used, in which one solves numerically linear equations obtained by an expansion of the flow equations in H/r, where H is the disc thickness [104].

Figure 6.2 illustrates a number of properties of the Lindblad torque that provide some insight into its scaling with the disc and planet parameters. In this figure, the torque value has been obtained through the use of analytical formulae similar to that of [110] (Eqs. (3) to (7) therein), except for the introduction of a softening length for the planet's potential, and for the minor correction consisting of the introduction of a factor Ω/κ in the forcing terms (see [76], after their Eq. (13); κ denotes the horizontal epicyclic frequency). This figure shows that the torque undergoes a sharp cutoff past a peak value, which is found to be of order $m_{max} \sim (2/3)(r/H)$. Also, the dashed line shows that, up to the cutoff, the one-sided Lindblad torque approximately scales with m^2—as expected from the WKB analysis of [33]—from which we infer the one-sided Lindblad torque, summed over m, to scale approximately as m_{max}^3, i.e. as $(r/H)^3$. Besides, the torque naturally scales with the disc's surface density and with the square of the planet mass, and dimensional arguments further imply that it ought to scale as:

$$\Gamma_0 = \Sigma q^2 \Omega^2 a^4 / h^3,$$ (6.8)

which is indeed the scaling of the one-sided Lindblad torque [110].

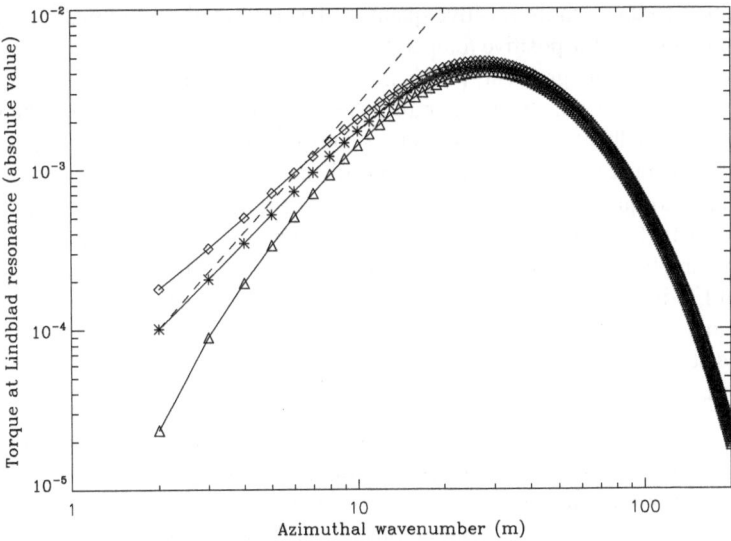

Fig. 6.2 Torques at individual outer Lindblad resonances (*diamonds*) and inner Lindblad resonances (*triangles*), in absolute value. Results are obtained for a disc with aspect ratio $H/r = 0.02$, and with a softening length $\varepsilon = 0.5H$ for the planet potential. Torques are normalised to the torque value given by Eq. (6.8). *Stars* show the average value of the inner and outer Lindblad torques at a given azimuthal wavenumber m, and the *dashed line* illustrates its m^2 dependence at small m

Figure 6.2 also shows that there exists an asymmetry between the outer and inner torques, the former being systematically larger than the latter. The reasons for this asymmetry are examined in depth in [110]. The relative asymmetry is found to scale with the disc thickness: in thinner discs, the relative asymmetry of the inner and outer torque is smaller (which can be understood as due to the accumulation points of the inner and outer resonances lying closer to the orbit). As a consequence, the differential Lindblad torque scales with:

$$\Gamma_{\text{ref}} = \Sigma q^2 \Omega^2 a^4 / h^2. \tag{6.9}$$

The asymmetry between the inner and outer torques also depends upon:

- The temperature gradient, since it affects the location of Lindblad resonances, and therefore the magnitude of the forcing potential at a given resonance [108]. For instance, a steeper (decreasing) temperature profile decreases the disc's angular frequency (by increasing the magnitude of the radial pressure gradient), which shifts all Lindblad resonances inwards. Outer resonances get closer to the planet orbit, which strengthens the outer torque, whereas inner Lindblad resonances are shifted away from the planet orbit, which decreases the inner torque. The net effect of a steeper temperature gradient is therefore to make the differential Lindblad torque a more negative quantity. In the same vein, a shallower temperature gradient would shift all Lindblad resonances outwards, making the differential

Lindblad torque a more positive quantity. The differential Lindblad torque may become positive for positive temperature gradients [110].

- The surface density gradient, which affects the location of Lindblad resonances in the exact same way as the temperature gradient, but now also because the torque at a given Lindblad resonance directly scales with the underlying surface density [108]. A steeper (decreasing) density profile naturally increases the magnitude of the inner torque compared to that of the outer torque, but this effect is mostly compensated for by an inward shift of all Lindblad resonances (just like when steepening the temperature profile, as described above). It implies that the differential Lindblad torque is quite insensitive to the density gradient near the planet location.
- The disc's self-gravity, which also impacts the location of Lindblad resonances, essentially by changing the wave's dispersion relation [9, 98].

An accurate determination of the asymmetry between the inner and outer torques yields a dimensionless factor to be put in front of Γ_{ref} to give the expression for the differential Lindblad torque. This issue has triggered a lot of theoretical efforts in the last three decades, and it is not completely solved yet. As of the writing of this review, the two main results are:

- An expression obtained by solving numerically the linearised equations of the flow in 2 dimensions with a softened planet potential, in discs with arbitrary gradients of surface density and temperature [88, 91]. It takes the form:

$$\frac{\Gamma_L}{\Gamma_{\text{ref}}} = -(2.5 + 1.7\beta - 0.1\alpha)\left(\frac{0.4}{\varepsilon/H}\right)^{0.71}, \tag{6.10}$$

with α and β defined in Eqs. (6.2) and (6.3). The expression in Eq. (6.10) is most accurate for softening lengths $\varepsilon \sim 0.4H$.

- An expression obtained by a semi-analytic method for globally isothermal, 3-dimensional discs with arbitrary gradients of surface density [104]:

$$\frac{\Gamma_L}{\Gamma_{\text{ref}}} = -(2.34 - 0.10\alpha). \tag{6.11}$$

We note that Eqs. (6.10) and (6.11) exhibit a similar behaviour (for the case $\beta = 0$, exclusively contemplated by [104]), that is to say a constant term of similar magnitude, and a weak dependence on the surface density gradient. Yet, the latter depends strongly on the softening length, as can be noticed by comparing to the 2-dimensional, unsmoothed, globally isothermal expression also provided by [104]:

$$\frac{\Gamma_L}{\Gamma_{\text{ref}}} = -(3.20 + 1.47\alpha). \tag{6.12}$$

This raises the question of whether the dependence on the temperature gradient, in a 3-dimensional disc, would be as steep as that of Eq. (6.10). Thus far this is an unanswered question, even if recent numerical simulations seem to indicate that the dependence of the differential Lindblad torque on the temperature gradient in a 3-dimensional disc is comparable to that of a 2-dimensional disc with a smoothing length $\varepsilon \simeq 0.4H$ [15]. We also comment that self-gravity slightly enhances the amplitude of the differential Lindblad torque by a factor approximately equal to $(1 + Q_p^{-1})$, with Q_p the Toomre Q-parameter at the planet's orbital radius [9].

Unless the disc has a temperature profile that strongly increases outward, the differential Lindblad torque is a negative quantity which, by itself, would drive type I migration on timescales shorter than a few $\times 10^5$ yrs for an Earth-mass object in a disc with a mass comparable to that of the Minimum Mass Solar Nebula [110]. Note however that Eqs. (6.10) and (6.11) have been derived under the assumption that the profiles of surface density and temperature are power laws of the radius. Local variations in the disc's temperature and/or density profiles, due for example to opacity transitions [76] or to dust heating [36] may change the sign and magnitude of the differential Lindblad torque. An approximate generalisation of the Lindblad torque expression, valid in non-power law discs, has been derived by [68] for 2-dimensional discs with a softening length $\varepsilon = 0.6H$. This expression reads:

$$\frac{\Gamma_L}{\Gamma_{\text{ref}}} = -\left(2.00 - 0.16\alpha + 1.11\beta - 0.80[\beta_2^+ - \beta_2^-]\right), \qquad (6.13)$$

where

$$\beta_2 = h\frac{d^2 \log T}{d(\log r)^2}, \qquad (6.14)$$

and where a quantity with a $\pm$ subscript is to be evaluated in $r = a \pm H/5$. Obviously, if the temperature profile is a power-law of radius, one has $\beta_2^+ = \beta_2^- = 0$, and Eq. (6.13) reduces to a standard linear combination of α and β.

The differential Lindblad torque has also been investigated in strongly magnetised, non-turbulent discs. The case of a 2-dimensional disc with a toroidal magnetic field has been studied by [105] through a linear analysis. She found that the differential Lindblad torque is reduced with respect to non-magnetised discs (as waves propagate outside the Lindblad resonances at the magneto-sonic speed $(c_s^2 + v_A^2)^{1/2}$, with v_A the Alfvén speed, rather than the sound speed c_s). Additional angular momentum is taken away from the planet by the propagation of slow MHD waves in a narrow annulus near magnetic resonances. These results were essentially confirmed by [29] with non-linear 2D MHD simulations in the regime of strong toroidal field (the plasma β-parameter, $\beta = c_s^2/v_A^2$, was taken equal to 2 in their study). More recently, 2D and 3D disc models with a poloidal magnetic field were investigated by [81] with a linear analysis in the shearing sheet approximation. While the differential Lindblad torque is reduced similarly as with a toroidal magnetic field, extraction of angular momentum by slow MHD and Alfvén waves is found to occur in 3 dimensions only.

We sum up the results presented in this section.

- The differential Lindblad torque corresponds to the net rate of angular momentum carried away by density waves (wakes) the planet generates in the disc at Lindblad resonances.
- The sign and magnitude of the differential Lindblad torque arise from a slight asymmetry in the perturbed density distribution associated to each wake.

- The differential Lindblad torque is a stationary quantity, largely independent of
  the disc's turbulent viscosity. Alone, it would drive the migration of Earth mass
  embedded planets in as short a time as a few $\times 10^5$ yrs in typical protoplanetary
  discs.

6.3.2 Corotation Torque

The other component of the tidal torque, the corotation torque, has long been neglected in studies of planetary migration. Firstly, it was shown to have a lower
absolute value than the differential Lindblad torque for typical (decreasing) radial
profiles of the disc's surface density [104]. Secondly, this torque component, for
reasons that will be presented at length in Sect. 6.3.3, should tend to zero after a finite time in an inviscid disc (the corotation torque is said to "saturate"). However, as
indicated by [109], some amount of turbulence should prevent the corotation torque
from saturating, and in the last decade the asymptotic value of the torque at large
time in the presence of dissipative processes has been tackled either analytically
[5, 65], by means of numerical simulations [66], or both [70, 92]. Besides, it was
discovered by [86] that in radiative discs, the corotation torque could, under certain
circumstances, be so large and positive that it could largely counteract the differential Lindblad torque, thereby leading to outward planetary migration. This was
subsequently interpreted as a new component of the corotation torque arising from
the disc's entropy gradient [8, 69, 88, 91].

The corotation torque on low-mass planets is usually linked to the so-called
horseshoe drag, which corresponds to the exchange of angular momentum between
the planet and its horseshoe region. The planet's horseshoe region encompasses the
disc region where fluid elements are on horseshoe streamlines with respect to the
planet orbit (see Fig. 6.1). We therefore start by explaining the concept of horseshoe
dynamics and horseshoe drag.

6.3.2.1 Horseshoe Dynamics

In the restricted three-body problem (RTBP), it is useful to write the Hamiltonian
of the test particle in the frame that corotates with the secondary. In this frame, the
potential does not depend explicitly on time and the Hamiltonian H is therefore
conserved. It reads:

$$H = E - \Omega_p L, \tag{6.15}$$

where E is the total energy of the test particle as seen in an inertial frame centred
on the primary, and L its angular momentum. The Hamiltonian of Eq. (6.15) is usually called the Jacobi constant. A sketch of the lines of constant Jacobi value when
the kinetic energy is zero (E therefore exclusively amounts to the potential energy),
named zero velocity curves (ZVC), reveals the existence of a horseshoe-like region

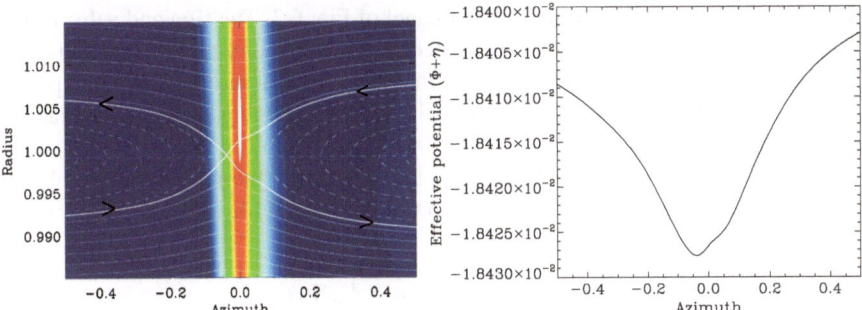

Fig. 6.3 *Left*: streamlines in the vicinity of an Earth-mass planet embedded in a disc with $H/r = 0.05$. The softening length of the potential is $\varepsilon = 0.6H$. The *thick lines* show the separatrices of the horseshoe region (the frontiers between the *dashed streamlines* that exhibit horseshoe-like motion and the *solid ones* corresponding to circulating material with respect to the planet orbit). A few *arrows* show the flow's direction with respect to the planet. The *horizontal dashed line* shows the corotation radius, where the angular velocities of the disc and the planet are equal. *Right*: effective potential at the corotation radius, as a function of azimuth

encompassing the secondary's orbit [80]. Furthermore, in the limit of a small secondary's mass, the trajectory of the guiding centre of the test particle in this area is shown to have a radial displacement from the orbit that is twice that of the associated ZVC [80], so that the guiding centre of a test particle can also exhibit a horseshoe-like motion in the vicinity of the orbit. In a similar fashion, a gas parcel in a disc with pressure can exhibit a horseshoe-like motion, very similar to that of the RTBP. There are, however, important differences between these two cases. Firstly, in the RTBP, the test particle can exhibit epicyclic motion on top of its global horseshoe-like trajectory. In the horseshoe region of a low mass planet, where no shocks are present, the gas parcels cannot cross each other's orbits, and therefore essentially follow nearly circular streamlines far from the planet. Secondly, the width of the horseshoe region is quite different in the two cases. It is much more narrow, for the same planetary mass, in the gaseous case than in the RTBP. Also, it has a different scaling with the planetary mass in the two cases: in the RTBP, the width of the horseshoe region scales with the cube root of the planetary mass, whereas in the gaseous case, provided the planet mass is not too large (we will specify how large below), it scales with the square root of the planet mass. To understand the reasons for this difference, we depict in the left-hand panel of Fig. 6.3 the streamlines in the vicinity of the coorbital region of a low-mass planet. The planet is located at radius $r = 1$, and azimuth $\varphi = 0$. The separatrices of the planet's horseshoe region are depicted by thick curves. We see that, quite to the contrary of the RTBP, there is no equivalent to the Roche lobe region around the planet (no circulating fluid material bound to it). Another difference is that the fixed point (or stagnation point) at which the separatrices intersect lies on the orbit (whereas in the RTBP, they intersect at the Lagrange points, away from the orbit, on a line joining the central star and the planet [80]). The azimuth of the stagnation point corresponds to the azimuth where, at corotation, the effective potential (the sum of the gravitational potential and fluid enthalpy) is minimum. This is illustrated in the right panel of Fig. 6.3. The sign and value of the stagnation point's azimuth is closely related to the asymmetry of the inner and outer wakes generated by the planet [90]. Note that there may be several stagnation points near the planet's corotation radius, depending on the softening length of the planet's potential [14].

Assuming that the fluid motion is in a steady state in the vicinity of the planet, we may use a Bernoulli invariant in the corotating frame [14, 71, 73]. This invariant can be cast as:

$$B_J = \frac{v^2}{2} + \eta + \Phi - \frac{1}{2}r^2\Omega_p^2, \tag{6.16}$$

where v is the fluid velocity in the corotating frame, η the fluid's specific enthalpy, Φ the sum of the star's and the planet's gravitational potentials, and Ω_p the planet's angular frequency. Equating the value of the Bernoulli invariant at the stagnation point (where by definition $v = 0$), and far from the planet on a separatrix, one finds the following expression for the half-width x_s of the planet's horseshoe region, away from the planet:

$$x_s \propto \left| \Phi_p + \eta' \right|_s^{1/2}, \tag{6.17}$$

where Φ_p is the planetary potential and η' the perturbation of the gas specific enthalpy introduced by the planet. The s subscript on the right hand side of Eq. (6.17) means that these variables are to be evaluated at the stagnation point. When the planet mass is sufficiently small, the streamlines are found to be in good agreement with those inferred from the linear expansion of the perturbed velocity field. Note that this does not imply that the corotation torque is in general a linear process. As was shown indeed by [89], the torque exerted by the coorbital material on the planet eventually becomes non-linear, no matter how small the planet mass is, provided dissipative effects are sufficiently small. However, the fact that important non-linear processes take place in this region hardly affects the streamlines themselves. In this low-mass regime, the stagnation point has therefore a location independent of the planet mass, necessarily at the corotation radius. In that case, the effective perturbed potential $\Phi_p + \eta'$ scales with the planet mass (M_p), and Eq. (6.17) implies that $x_s \propto M_p^{1/2}$. This is no longer true when the location of the stagnation point depends on the planet mass. In particular, in the high-mass regime, the planet gravity dominates over the perturbed enthalpy, and the situation resembles that of the RTBP. The Bernoulli invariant at the inner and outer stagnation points (which are L_1-like and L_2-like, respectively) is dominated by the planetary potential term and so scales as M_p/R_H, where $R_H = a(M_p/M_\star)^{1/3}$ is the planet's Hill radius. In the high-mass region, the width of the horseshoe region therefore scales with the cubic root of the planet mass, as in the RTBP [73].

Lastly, another important difference between the RTBP and the case of a low-mass embedded planet is that of the U-turn timescale. Firstly, it should be noted that the expression *U-turn timescale* is ambiguous, for it depends on the horseshoe streamline under consideration. The closer to corotation, the longer it takes to perform a horseshoe U-turn, and the U-turn time reaches its minimum value close to the separatrix. It is usually this minimum value that is meant by the ambiguous expression *U-turn timescale*. Note that the corotation torque nearly reaches a constant value (notwithstanding saturation considerations, that we shall contemplate later) after this timescale, as it corresponds to the fluid elements that most contribute to the torque, because their angular momentum jump is the largest, and because their mass flow-rate is large.

In the RTBP, the U-turn timescale is of the order of the dynamical timescale τ_{dyn}, i.e. a planet orbital period. If one regards, in a crude approximation, the case of a low-mass embedded planet as an expurgated version of the RTBP, where most of the initial horseshoe streamlines are made circulating, and where only those lying close to corotation keep their horseshoe character, one expects the U-turn timescale in this case to be significantly longer than the dynamical timescale. This is indeed the case: the U-turn timescale is approximately $h_p \tau_{\text{lib}}$ [8], with h_p the disc's aspect ratio at the planet's orbital radius, and where τ_{lib} is the libration timescale, i.e. the time it takes to complete a closed horseshoe trajectory. The libration timescale reads

$$\tau_{\text{lib}} = \frac{8\pi a}{3\Omega_p x_s}. \tag{6.18}$$

An alternate, equivalent expression for the U-turn timescale is $\tau_{\text{U-turn}} \sim \tau_{\text{dyn}} H/x_s$, which is corroborated by numerical simulations in which one monitors the advection of a passive scalar. This expression shows that the U-turn timescale can indeed be longer than the dynamical timescale by a significant factor, as x_s can be much smaller than H for deeply embedded, low-mass objects.

We sum up the results presented in this section:

- The coorbital region of a low-mass embedded planet in a gaseous disc exhibits a horseshoe-like region.
- This region is much more narrow than in the restricted three-body problem, and its radial width scales with the square root of the planetary mass.
- The stagnation points are located at the corotation radius. There is no equivalent to the Roche lobe region for low-mass objects.
- The horseshoe U-turn timescale is significantly longer than the dynamical timescale.

6.3.2.2 Horseshoe Drag: An Overview

Far from the horseshoe U-turns in the vicinity of the planet, a fluid element or test particle essentially follows a nearly circular orbit, and therefore has a nearly constant angular momentum. When it reaches a U-turn, a fluid element is either sent inward or outward, thereby crossing the planet orbit. It does so by exchanging angular momentum and energy with the planet. The torque resulting from the interaction of the planet with all the fluid elements in the course of performing their horseshoe U-turn is called horseshoe drag [109].

Upon insertion of the planet in the disc, it takes some time to establish the horseshoe drag, namely a time of the order of the horseshoe U-turn timescale [89]. This

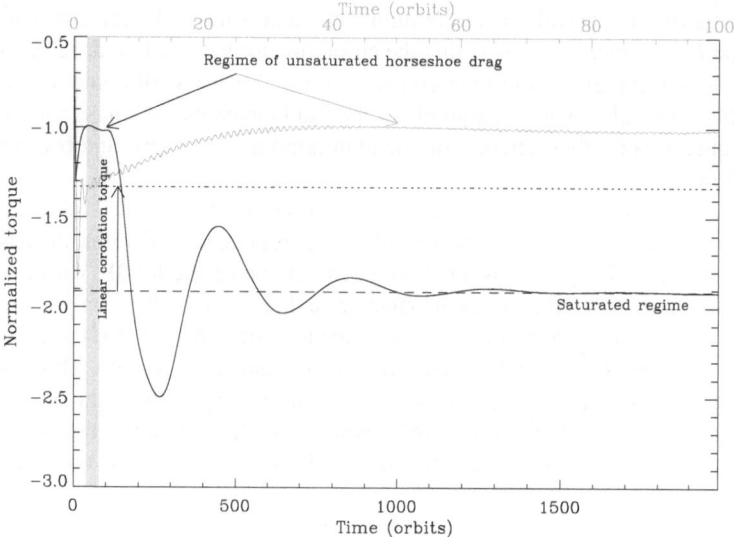

Fig. 6.4 Time evolution of the total torque (sum of the differential Lindblad torque and corotation torque) on a $M_p = 10^{-6} M_\star$ planet mass embedded in an inviscid isothermal disc. The *black curve* (*bottom x-axis*) shows the torque evolution over 2000 planet orbits, while the *grey curve* (*top x-axis*) focuses on the evolution over the first 100 planet orbits. The *dashed line* shows the value of the differential Lindblad torque, and the *dotted curve* highlights the corotation torque predicted with a linear analysis. Taken from [68]

is illustrated in Fig. 6.4, which displays the time evolution of the total torque on a $M_p = 10^{-6} M_\star$ mass planet embedded in a thin ($h = 0.05$) isothermal disc with uniform density profile. The disc is inviscid in this example. Once established (after ~ 30 planet orbits in our example), the horseshoe drag remains approximately constant over a longer timescale, which corresponds to the time it takes for a fluid element to drift from one end of the horseshoe to the other (that is, about half a libration time, given by Eq. (6.18)). The value of the horseshoe drag that exists between the horseshoe U-turn time, and half the horseshoe libration time, is called the *unsaturated horseshoe drag*. Beyond this stage, subsequent U-turns may cause further time evolution of the horseshoe drag depending on the disc viscosity, which will be described in Sect. 6.3.3. In the particular case depicted here, where the disc is inviscid, the horseshoe drag eventually saturates (it cancels out) after a few libration timescales. Until Sect. 6.3.3, we focus on the properties of the fully unsaturated horseshoe drag.

6.3.2.3 Horseshoe Drag in Barotropic Discs

A hint of the torque exerted by the coorbital material on the planet can be obtained by the examination of the perturbed surface density. Nonetheless, this examination

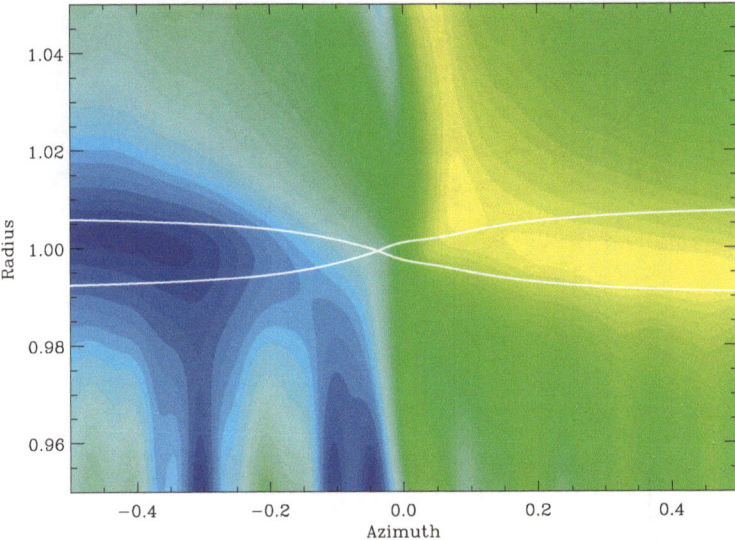

Fig. 6.5 Perturbed surface density in the coorbital region of an Earth-mass planet in a disc with $h = 0.05$ and $\alpha = 0$ (uniform background surface density). The planet is located at $r = 1$, $\varphi = 0$. The *solid curves* show the separatrices of the planet's horseshoe region. In order to remove the planet's wakes and to render this map more legible, we have subtracted the density perturbation obtained in a situation where no corotation torque is expected (namely $\alpha = 3/2$, as will be shown below). This cancellation is imperfect, however, as the wakes of the two cases are not strictly identical

is rather difficult, because the density perturbation in the planet's coorbital region is very small, typically one or two orders of magnitude smaller than the density perturbation associated to the wakes. Yet, as can be seen in Fig. 6.5, an approximate subtraction of the wakes density perturbation reveals two regions of opposite signs: a region of positive perturbed density ahead of the planet ($\phi > 0$) and a region of negative perturbed density behind the planet ($\phi < 0$), which both yield a positive torque on the planet. The sign of the perturbed density in the coorbital region depends on the background density profile, here it is uniform ($\alpha = 0$). The largest perturbations can be seen to originate near the downstream separatrix in either case (the outer separatrix at negative azimuth, and the inner separatrix at positive azimuth), but the perturbation is spread radially and extends much beyond the horseshoe region. This is to be expected on general grounds: in a barotropic disc, where the gas pressure depends only on its mass density, any disturbance near corotation excites evanescent pressure waves, which extend typically over the disc pressure length scale (here $H = 0.05a$).

Even if some insight into the corotation torque can be gained by the examination of the perturbed density maps, a much more useful quantity is the vortensity (the ratio of the vertical component of the vorticity to the surface density, also known as potential vorticity), which is materially conserved away from shocks in inviscid, barotropic, 2-dimensional discs. This is illustrated in Fig. 6.6. Ward [109] has evalu-

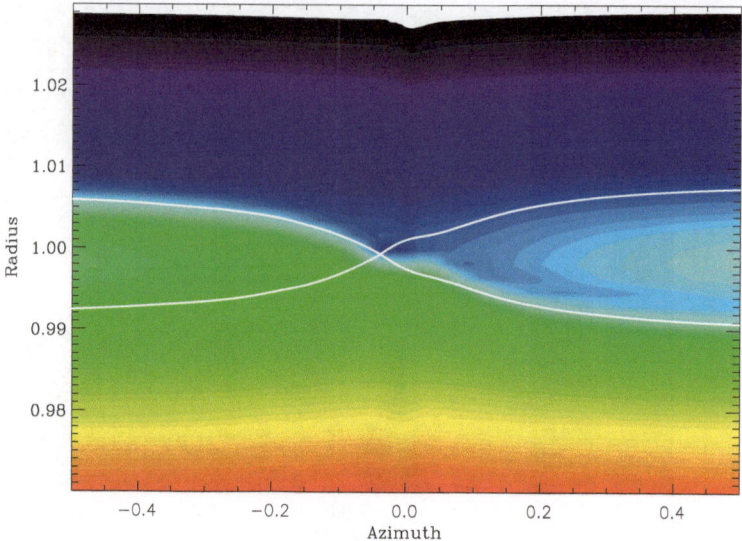

Fig. 6.6 Advection of vortensity in the horseshoe region of an embedded, low-mass protoplanet. As in previous panels, the planet is at $r = 1$, $\varphi = 0$, and the separatrices of the planet's horseshoe region are depicted by *white curves*. The unperturbed disc's vortensity decreases with radius. The (high) vortensity in the inner side of the horseshoe region (at $\varphi < 0$, $r < 1$) is brought to the outer side (at $\varphi < 0$, $r > 1$) by the outward U-turns behind the planet. Similarly, the (low) vortensity in the outer side of the horseshoe region (at $\varphi > 0$, $r > 1$) is brought to the inner side (at $\varphi > 0$, $r < 1$) by the inward U-turns ahead of the planet

ated the torque exerted on the planet by test particles embarked on horseshoe motion, by making use of the Jacobi invariant of these particles (see Sect. 6.3.2.1). A similar calculation can be performed for fluid motion, provided one uses a Bernoulli invariant by adding the enthalpy to the Jacobi constant [14, 71]. In both cases one finds that the horseshoe drag has the following expression:

$$\Gamma_{HS} = 8|A|B^2 a \left[\int_{-x_s}^{x_s} \left(\frac{\Sigma}{\omega_z} \bigg|_F - \frac{\Sigma}{\omega_z} \bigg|_R \right) x^2 \, dx \right], \tag{6.19}$$

where x denotes the radial distance to the planet orbit. In Eq. (6.19), the subscript F indicates that the (inverse of) the vortensity Σ/ω_z has to be evaluated away from the planet, in Front of the latter ($\phi > 0$), and the subscript R indicates that it has to be evaluated at the Rear of the planet, away from it ($\phi < 0$). The integral in Eq. (6.19) is usually called the horseshoe drag integral, and in a barotropic disc it can be simplified to yield the following expression [14, 70, 91]:

$$\Gamma_{HS} = \frac{3}{4} \Sigma \mathcal{V} \Omega_p^2 x_s^4, \tag{6.20}$$

where the quantity $\mathcal{V}$, called the (inverse) vortensity gradient for short, is defined by

$$\mathcal{V} = \frac{d \log(\Sigma/B)}{d \log r},$$ (6.21)

and can be recast as $3/2 - \alpha$ for density profiles that can be approximated as power-law functions of radius over the radial width of the planet's horseshoe region. In Eq. (6.20), all terms are to be evaluated at the planet's orbital radius. This equation shows that in 2-dimensional barotropic discs, the horseshoe drag cancels out when the surface density profile decreases locally as $r^{-3/2}$, while it is positive for density profiles shallower than $r^{-3/2}$. For density profiles strongly increasing outward, the horseshoe drag can be sufficiently positive to counteract the (negative) differential Lindblad torque, and therefore stall the migration of low-mass planets [74]. Such density jumps may be encountered near the star's magnetospheric cavity, or near the inner edge of a dead zone, across which the disc's effective turbulence decreases outward (the dead zone refers to the region near the midplane of protoplanetary discs that is sandwiched together by partially ionised surface layers).

The horseshoe drag expression in Eq. (6.20) exclusively holds in the case of barotropic discs. Those, naturally, are an idealised concept, and true discs have a more complex physics, which yields a more complex expression for the corotation torque. However, in any case, as we shall see, a common component of the corotation torque is given by Eq. (6.20), so that baroclinic effects yield additional terms to this expression.

We sum up the results presented in this section. In 2-dimensional barotropic discs, where the gas pressure only depends on the surface density:

- The horseshoe drag is powered by the advection of the fluid's vortensity along horseshoe streamlines inside the planet's horseshoe region.
- It is proportional to the inverse vortensity gradient across the horseshoe region (that is, the quantity $3/2 + d \log \Sigma/d \log r$ for power law discs). It can therefore be negative, zero, or positive depending on the surface density gradient across the horseshoe region. For typical discs density and temperature profiles, its magnitude is smaller than that of the (negative) differential Lindblad torque.

6.3.2.4 Horseshoe Drag in Locally Isothermal Discs

A long considered framework, both in analytical and numerical studies is that of locally isothermal discs, in which the temperature is a fixed function of radius. No energy equation is considered in this case, but the flow is no longer barotropic: the pressure becomes a function of the density *and* position (through the temperature). The vortensity is no longer materially conserved. Its Lagrangian derivative features a source term arising from misaligned density and pressure gradients, or misaligned temperature and density gradients [60]:

$$\frac{D}{Dt}\left(\frac{\vec{\omega}_z}{\Sigma}\right) = \frac{\nabla\Sigma \times \nabla P}{\Sigma^3} = \frac{\nabla\Sigma \times \nabla T}{\Sigma^2}.$$ (6.22)

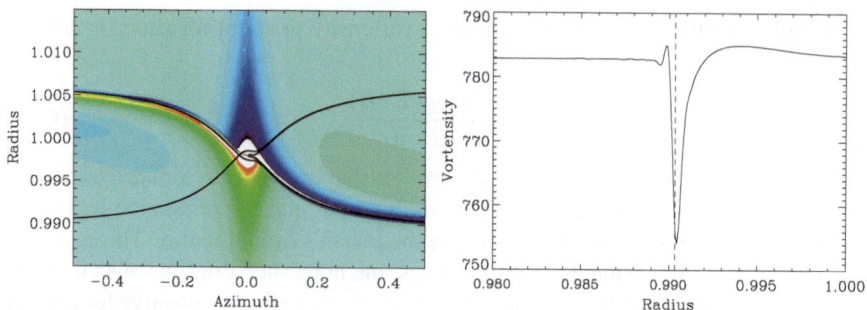

Fig. 6.7 Vortensity field in the coorbital region of a low-mass planet (*left*) and radial profile of vortensity at $\phi = +0.5$ rad (*right*), 30 orbital periods after the insertion of the planet. The disc has no background vortensity gradient, and a flat aspect ratio $H/r = 0.05$. *Thin stripes* of vortensity of opposite signs are clearly visible at the downstream separatrices. One can also see a mild production of vortensity in the wake, which fades away as one recedes from corotation, because of the winding of the wake and because fluid elements move faster away from corotation. Contrary to the adiabatic case that we shall present in Sect. 6.3.2.5, the vortensity cut is not singular at the separatrix, as can be seen in the *right panel*. The radial resolution in the run presented is $9.3 \times 10^{-5} a$

As the temperature gradient has a radial direction and sensibly the same magnitude everywhere in the coorbital region, the strength of the source term depends on the density gradient: wherever the azimuthal density gradient is large, the source term is large. This occurs at the tip of the horseshoe U-turns where we have a strong azimuthal density gradient owing to the density enhancement in the planet's immediate vicinity. The time derivative in Eq. (6.22) can be expressed as a derivative with respect to the curvilinear abscissa s along the streamline:

$$\frac{D}{Ds}\left(\frac{\vec{\omega}_z}{\Sigma}\right) = \frac{\nabla\Sigma \times \nabla T}{v\Sigma^2}, \qquad (6.23)$$

where v is the norm of the fluid velocity in the corotating frame, where we have used $ds = v\,dt$. As the fluid stagnates in the vicinity of the stagnation point (i.e. v can be arbitrarily small, provided one chooses a streamline sufficiently close to the stagnation point), the source term of Eq. (6.23) formally diverges in the vicinity of the stagnation point. The total amount of vortensity created, integrated over the horseshoe streamlines, however, remains finite. Figure 6.7 shows a vortensity map in the vicinity of a low-mass planet for a disc with $\alpha = 3/2$ (no background vortensity gradient) and $\beta = 1$ (uniform aspect ratio).

Thus far there is no rigorous mathematical proof that the horseshoe drag expression of Eq. (6.19) still holds in locally isothermal discs, as all demonstrations of that expression rely on the existence of a Bernoulli invariant, which does not exist in the locally isothermal case. Yet, data from numerical simulations suggest that this expression is still valid in that regime. Assuming its validity from now on, we infer that the horseshoe drag must exhibit a dependence on the temperature gradient. The rational for this being that the outgoing vortensity, accounted for by the horseshoe drag integral, includes the vortensity produced in the vicinity of the planet, which depends on the temperature gradient. 2-dimensional numerical simulations have confirmed the existence of an additional component of the corotation torque that depends on the temperature gradient [6, 14, 91]. The sign and value of this temperature-related corotation torque have a complex, and rather steep dependence with the softening length of the planet's potential [14]. Indeed, the topology of the horseshoe region depends heavily on the softening length: at large softening lengths, only one X-stagnation point is observed at corotation, in the planet vicinity, whereas at low softening lengths, two X-stagnation points are usually observed at corotation, on either side of the planet [14, 90]. The vortensity produced along a streamline depends on the path followed by the streamline. This situation is therefore much more complex than in the barotropic case or the adiabatic case that we present below, in which the existence of invariants under certain circumstances allows to get rid of the dependence on the actual path followed by fluid elements during their U-turns. Quite ironically, the locally isothermal case, which has served as a standard framework for more than two decades, is very difficult to tackle analytically.

The steep dependence of the temperature related corotation torque on the softening length appeals for a 3-dimensional study of this torque, which is not plagued by softening issues. Such study has been undertaken by [15], who find a linear dependence of the 3-dimensional horseshoe drag on the temperature gradient $\beta = -d \log T / d \log r$, as steep as the dependence on the vortensity gradient $\mathcal{V}$ given by Eq. (6.20).

6.3.2.5 Horseshoe Drag in Adiabatic Discs

In the previous sections, the set of governing equations of the fluid did not include an energy equation, and the disc temperature, which was set as a prescribed function of radius, did not evolve in time. The first calculations undertaken with an energy equation were those of [79] and [86]. The former were devised in the shearing sheet framework, so that no net torque could be experienced by the planet, owing to the symmetry properties of the shearing sheet. Still, these authors found that radiative cooling could significantly affect the perturbed surface density pattern associated with the wakes, thus changing the magnitude of the one-sided Lindblad torque. Paardekooper and Mellema [86] considered a planet in a 3-dimensional disc, with an energy equation and thermal diffusion, and nested grids around the planet to achieve a very high resolution. They found that the migration of a low-mass planet could be reversed in sufficiently opaque discs, under the action of the corotation torque. The same result was in particular obtained in the adiabatic limit, which we are now going to focus on, as thermal diffusion yields an additional complexity, not needed at this stage. We will take thermal diffusion and other dissipative processes into account in Sect. 6.3.3.

It was soon realised that the results of [86] were due to a new component of the corotation torque, linked to the entropy gradient [8, 88]. This is illustrated in Fig. 6.8, in which we compare the torque results for 69 different random disc profiles (the density slope α being a random variable uniformly distributed over the interval

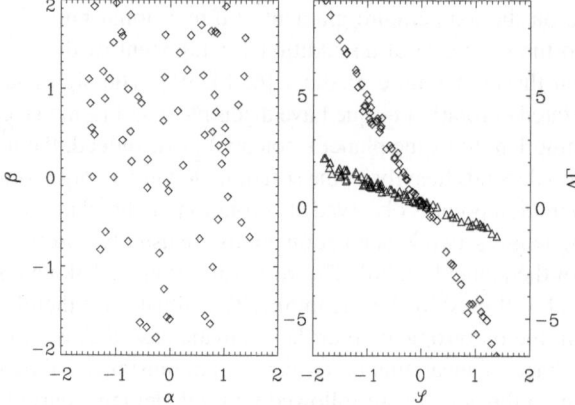

Fig. 6.8 Torque difference between the adiabatic and locally isothermal cases, given by Eq. (6.24), as a function of the entropy gradient (*right panel*), at early times (linear stage, *triangles*) and during the horseshoe drag stage (*diamonds*). The quantity $\mathscr{S}$ in x-axis is defined in Eq. (6.30). Results have been obtained with calculations with random values of the surface density slope ($-\alpha$) and temperature slope ($-\beta$), shown in the *left plot*. The total number of runs is 138

$[-3/2, +3/2]$, and the temperature slope β being an independent random variable uniformly distributed over the interval $[-2, +2]$). Each calculation has a smoothing length of the planet's potential $\varepsilon = 0.3H$, and an aspect ratio at the planet location $h_p = 0.05$.

For each pair of α and β, we ran two calculations: a locally isothermal one, and an adiabatic one with a ratio of specific heats $\gamma = 1.4$. The torque difference, dubbed *adiabatic torque excess*, is then obtained by:

$$\Delta\Gamma_{\text{HS}}^{\text{entr}} = \Gamma_{\text{ad}} - \frac{\Gamma_{\text{iso}}}{\gamma}. \tag{6.24}$$

The correction of the isothermal torque Γ_{iso} by a factor γ^{-1} is necessary as both the differential Lindblad torque and the barotropic part of the horseshoe drag (the vorticity-related corotation torque) scale with the inverse square of the sound speed, which turns out in the adiabatic case to be $c_s^{\text{adi}} = c_s^{\text{iso}}\gamma^{1/2}$. The right part of Fig. 6.8 shows a clear one-to-one relationship between the adiabatic torque excess and the entropy gradient, irrespective of the individual values of α and β, which justifies the name *entropy-related corotation torque* given to the adiabatic torque excess. This torque is shown in two different regimes: the linear regime, soon after the insertion of the planet in the disc, and the horseshoe drag regime, reached after a longer timescale, as discussed in previous sections. As pointed out by [88], the non-linear corotation torque can be much larger (in this example, by a factor of about 5) than its linear counterpart, depending on the planet's softening length.

An interpretation of the entropy-related corotation torque was given soon after its discovery by [8] and [88]. Although this early interpretation was shown not be quite correct when the horseshoe drag expression in the adiabatic case was worked

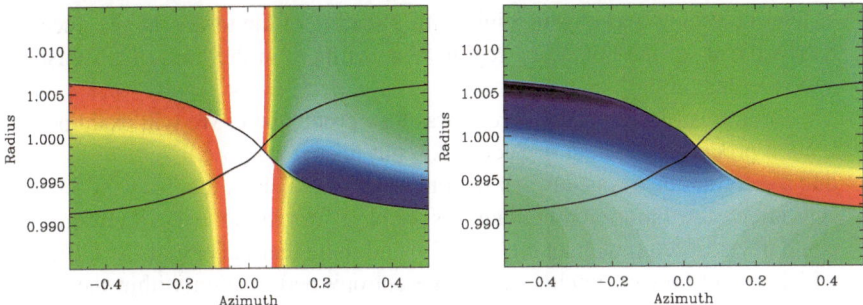

Fig. 6.9 Surface density perturbation in the planet's coorbital region of an adiabatic disc (*left*) and perturbation of entropy (*right*). The sign of the density lobes is opposite that of the entropy lobes. In addition to the lobes, the *left plot* also shows the wake, saturated in this representation. The background entropy profile increases with radius in this example

out subsequently, we describe it here briefly as the mechanism on which it is based allows for a clear understanding of the correct origin of the entropy torque.

In an adiabatic disc, the entropy is materially conserved along the path of fluid elements, as long as they do not cross shocks, in a strict analogy with vortensity for barotropic flows. If we assume, for instance, a disc that has a positive radial entropy gradient prior to the planet insertion, outward horseshoe U-turns (behind the planet) bring to the outer side of the horseshoe region the low entropy of the inside. Similarly, inward horseshoe U-turns (ahead of the planet) bring to the inner side of the horseshoe region the high entropy of the outside. As the disc maintains a pressure equilibrium, the relative variations of the pressure across the coorbital region can be neglected, so that the surface density features relative variations opposite to that of the entropy, by virtue of the first order expansion of Eq. (6.1):

$$\gamma \frac{\delta \Sigma}{\Sigma} + \frac{\delta s}{s} = \frac{\delta P}{P} \approx 0. \qquad (6.25)$$

As a consequence, two lobes of perturbed surface density appear in the horseshoe region [8, 45, 47, 88], that both yield a torque of same sign. These two lobes are shown in Fig. 6.9.

The early interpretations identified the entropy related torque as the torque arising from the above density lobes. This explanation was appealing at first, because it gives the correct sign for the entropy related corotation torque, and the correct order of magnitude: [8] performed an approximate, horseshoe-drag like integration to evaluate the impact of these lobes on the torque, whereas [88] performed an approximate direct summation. Both results were in rough agreement with the magnitude of the adiabatic torque excess, at least for the value of the potential softening length used in these studies. Yet, this explanation of the entropy torque quickly turned out not to be fully satisfactory, for the following reasons:

- Numerical explorations performed at different smoothing values showed that the entropy torque was approximately scaling as ε^{-1}, down to very low values of the smoothing length ($\varepsilon \sim 0.05H$). This apparent divergence of the entropy torque at low smoothing was incompatible with a scaling of the torque in x_s^4, since the half-width x_s of the horseshoe region remains finite in the limit of a vanishing softening length [90].

- The saturation of the entropy torque was also problematic. We will examine saturation processes in detail in Sect. 6.3.3, but for our purpose it suffices to know that the corotation torque always saturates in inviscid discs, that is to say tends to zero after a few libration timescales (as we have seen in Fig. 6.4). Therefore, one could devise a setup with an *inviscid* disc and finite thermal or entropy diffusion, which would forever maintain the same entropy perturbation within the horseshoe region (finite thermal diffusion is required in order to avoid phase mixing of entropy, as we shall see in Sect. 6.3.3.2). As expected, the entropy related corotation torque is found to saturate as the disc is inviscid, while the (approximate) same density lobe structure is maintained within the horseshoe region [70]. This implies the density lobe structure would exert a torque at early times, but not at late times, which is contradictory.
- Finally, as we have seen in Sect. 6.3.2.3, the density perturbation responsible for the corotation torque is not bound to the horseshoe region, but can extend further radially by the excitation of evanescent waves. In the barotropic case, the vortensity-related horseshoe drag is actually fully accounted for by an evanescent density distribution within the coorbital region. In the adiabatic case, attributing the entropy torque (i.e. the whole difference between an adiabatic and an isothermal calculation) to the density lobes was therefore tantamount to assuming that the evanescent wave structure in the coorbital region was the same in the adiabatic and isothermal cases, which is not obvious.

The identification of a convenient invariant of the flow for adiabatic discs with uniform temperature profile allowed [69, 70] to demonstrate that the horseshoe drag expression was exactly the same as that of the barotropic case, given by Eq. (6.19). In this case, the evaluation of the horseshoe drag amounts again to a budget of the vortensity entering or leaving the vicinity of the planet on horseshoe streamlines. An important consequence of this is that the torque due to the density lobes must not be incorporated manually, separately, into the corotation torque expression, and the whole problem of determining the horseshoe drag amounts to an evaluation of the vortensity distribution within the horseshoe region. Before we clarify this point, we stress that the vortensity distribution within the horseshoe region has the following features:

- Since the flow is baroclinic, vortensity is not materially conserved along streamlines. However, contrary to the locally isothermal case, the existence of a flow invariant in adiabatic discs with flat temperature profile allows an estimate of the vortensity acquired by a streamline during a U-turn, independently of its actual path [69].
- The vortensity created over the interior of the horseshoe region is very small, and has no impact on the torque, because it has same sign on both sides of the planet [69, 91].

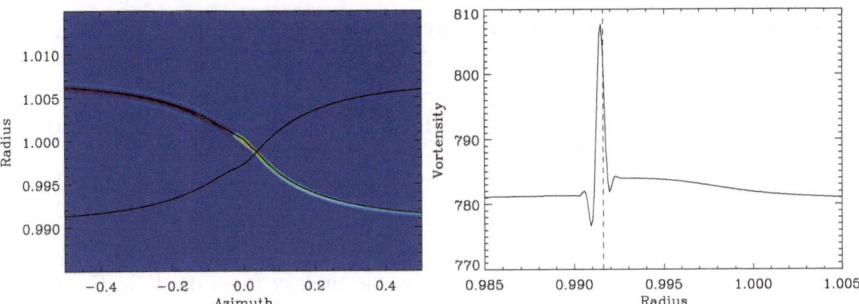

Fig. 6.10 Vortensity map in the coorbital region of an Earth-mass planet embedded in an adiabatic disc with $H = 0.05$ at the orbital radius of the planet (*left*), and radial profile of vortensity at $\phi = 0.5$ rad. (*right*), 60 orbital periods after the insertion of the planet in the disc. The grid resolution and disc gradients are the same as in Fig. 6.7 ($\alpha = 3/2$, $\beta = 0$). The vortensity peak has a much more compact profile than that of Fig. 6.7. This is to be expected as we have a singular vortensity sheet in the adiabatic case, and a continuous one, peaked at the separatrix, in the locally isothermal case

- The main difference arises from a (formally) singular production of vortensity (or vorticity) on downstream separatrices, due to the entropy discontinuity at this location (which results from the entropy advection within the horseshoe region). This (formally) singular production of vortensity is readily apparent in the source term of Eq. (6.22), and is illustrated in Fig. 6.10. It can be evaluated analytically either using the flow invariant introduced in [69], or directly using Eq. (6.22) as in [91]. The first approach is self-contained and yields the amount of singular vortensity as a function of the flow properties at the stagnation point. The second is not restricted to flat temperature profiles, but it requires knowledge of the fluid velocity along horseshoe streamlines, which depends on the exact geometry of the horseshoe region, much like in locally isothermal discs discussed in Sect. 6.3.2.4.

We now clarify the contribution of the density lobes to the corotation torque. The corotation torque is directly related to the density perturbation within the corotation region, which can be written as

$$\frac{\delta \Sigma}{\Sigma} = \frac{1}{\gamma} \left(\frac{\delta P}{P} - \frac{\delta s}{s} \right), \tag{6.26}$$

where, counter to Eq. (6.25), we shall not assume $\delta P = 0$. The pressure perturbation δP can be shown to satisfy a second-order partial differential equation [69], with solution of the form:

$$\frac{\delta P}{P} = \gamma K * \frac{\delta u}{u}, \tag{6.27}$$

where $u = s^{1/\gamma} \times \Sigma / \omega_z$, $K \propto \exp(-|x|/H)$ is a Green kernel normalised to unity ($\int K(x)\, dx = 1$, with x the radial distance to the planet orbit), and H is the local pressure scale height. Further denoting the inverse vortensity by $l = \Sigma/\omega_z$, Eqs. (6.26) and (6.27) yield

$$\frac{\delta\Sigma}{\Sigma} = K * \left(\frac{\delta l}{l} + \frac{1}{\gamma}\frac{\delta s}{s}\right) - \frac{1}{\gamma}\frac{\delta s}{s}. \tag{6.28}$$

The first term on the right-hand side of Eq. (6.28) corresponds to the perturbed surface density associated to evanescent pressure waves (like in the barotropic case, where it reduces to $K * \delta l/l$), and the second term to the density lobes resulting from entropy advection. Since the convolution by the unitary function K in Eq. (6.28) does not change the linear mass of the perturbation ($\int \delta\Sigma(x)\,dx$), the corotation torque is the same as if, in the expression for the density perturbation in Eq. (6.28), the convolution product were actually discarded [69]. In the barotropic case for instance, this leads to the torque expression given by Eq. (6.19). In the adiabatic case, it shows that, counter-intuitively, the density lobes exert no *net* corotation torque. This further explains why, akin to the barotropic case, the calculation of the corotation torque comes to evaluating the vortensity distribution within the horse shoe region. Since the main difference in the vortensity field between the adiabatic and barotropic cases is the appearance of a singular sheet of vorticity at the downstream separatrices, and given that the magnitude of this sheet scales with the entropy gradient, this singular vorticity sheet can be unambiguously identified as the origin of the entropy-related torque.

Upon evaluation of the magnitude of the vorticity sheet at the separatrices, [69] inferred the following expression for the entropy-related corotation torque:

$$\Delta\Gamma_{\mathrm{HS}}^{\mathrm{entr}} = -\frac{1.3\mathscr{S}}{\varepsilon/H}\Sigma\Omega_p^2 q^2 a^4 h^{-2}, \tag{6.29}$$

where the above expression has been derived in the framework of a flat temperature profile, and assuming a ratio of specific heats $\gamma = 1.4$. In Eq. (6.29), all disc quantities are to be evaluated at the planet's orbital radius, and the quantity $\mathscr{S}$ is defined by

$$\mathscr{S} = \frac{1}{\gamma}\frac{d\log s}{d\log r}, \tag{6.30}$$

and can be recast as $[\beta + (\gamma - 1)\alpha]/\gamma$ for surface density and temperature profiles that can be approximated as power-law functions of radius over the planet's horseshoe region.

Considering discs with arbitrary temperature profiles, [91] also evaluated the production of vortensity at downstream separatrices, which required estimating the velocity along streamlines through a fit of numerical simulations. Unlike [69], they manually added the torque contribution from the density lobe structure. In the end, the latter remains small compared to the torque contribution from the singular sheet of vorticity. This explains why, overall, the derivations of the entropy-related corotation torque by [69] and [91] are in broad numerical agreement, within 30 %.

The generalisation to an arbitrary temperature profile of Eq. (6.29) cannot be tackled fully analytically, much as in the locally isothermal case. Yet, Eq. (6.24) shows that the adiabatic corotation torque is the sum of the entropy related term, given by Eq. (6.29), and the locally isothermal corotation torque (corrected by a factor γ). The latter is itself made up of two terms, as we have seen in Sects. 6.3.2.3 and 6.3.2.4. The corotation torque is therefore, in a general situation, the sum of three terms:

- The vortensity related torque, proportional to the vortensity gradient, and given by Eq. (6.20).
- The temperature related torque, proportional to the temperature gradient, discussed in Sect. 6.3.2.4.
- The entropy related torque, proportional to the entropy gradient, given by Eq. (6.29).

It can be observed that there are only two degrees of freedom for the disc profiles (the density and temperature gradients α and β, or the vortensity and entropy gradients $\mathcal{V}$ and $\mathcal{S}$, etc.), so that for a specific setup one may simplify the torque expression as a linear combination of the two independent parameters. This simplification is not desirable, however, because it blurs the distinct physical origin and characteristic of each of the three terms. Besides, one can disentangle these three terms by varying parameters such as the smoothing length ε or the ratio of specific heats γ. Any simplification of the torque expression is thus highly setup dependent.

Finally, we summarise the main message to take away about the corotation torque.

- In all cases it features a term that scales with the gradient of vortensity across the horseshoe region, given by Eq. (6.20). It has one or two additional terms, depending on whether an energy equation is taken into consideration. The first of those scales with the temperature gradient, and if an energy equation is included, there is a second one that scales with the entropy gradient.
- In all cases the corotation torque comes from the *vortensity* distribution in the horseshoe region. The additional contributions arise from the vortensity *created* by the temperature gradient and/or the entropy gradient.

6.3.3 Saturation Properties of the Horseshoe Drag

We have described in Sect. 6.3.2 the physical origin and properties of the corotation torque in inviscid discs, with a special emphasis on the fully unsaturated horseshoe drag, which is the maximum value the corotation torque may take. This value is obtained about one horseshoe U-turn timescale after the planet insertion in the disc, and is maintained over about half a libration timescale. Its sign and magnitude are determined by the gradients of vortensity, temperature and entropy across the horseshoe region.

In the absence of diffusion processes, after about half a libration timescale, the vortensity and entropy advected along the downstream separatrices of the horseshoe

region reach the planet again, undergo another U-turn, and phase mixing starts to occur. Vortensity and entropy are progressively stirred up within the horseshoe region,
and the horseshoe drag oscillates with time with a decreasing amplitude, as shown
in Fig. 6.4. The horseshoe drag ultimately cancels out as both vortensity and entropy
get uniformly distributed after several libration times [70, 92]. This is known as the
horseshoe drag saturation.

Diffusion processes (viscosity, thermal diffusion) may maintain respectively the
vortensity and entropy gradients across the horseshoe region, and thus sustain the
horseshoe drag to a non-vanishing value. We review below the saturation properties of the horseshoe drag in barotropic discs (Sect. 6.3.3.1) and in radiative discs
(Sect. 6.3.3.2).

6.3.3.1 Saturation Properties of the Vortensity-Related Horseshoe Drag in Barotropic Viscous Discs

In barotropic discs, the horseshoe drag saturates as vortensity is strictly advected
along horseshoe streamlines. Viscosity acting as a diffusion source term in the
vortensity equation can sustain a non-zero vortensity gradient across the horseshoe region. The vortensity-related horseshoe drag then attains a steady-state value,
which arises from a net exchange of angular momentum between the horseshoe
region and the rest of the disc [65, 66]. This steady-state value depends on how
the viscous diffusion timescale across the horseshoe region (τ_{visc}) compares with
the horseshoe libration timescale (τ_{lib}) and the horseshoe U-turn timescale ($\tau_{\text{U-turn}}$).
Denoting by ν_p the kinematic viscosity at the planet location, $\tau_{\text{visc}} \sim x_s^2/\nu_p$. The
libration timescale is given by Eq. (6.18), and the U-turn timescale is typically a
fraction H/r of the libration timescale [8].

For the corotation torque to remain close to its maximum, fully unsaturated value
in the long term, the inequality

$$\tau_{\text{U-turn}} \leq \tau_{\text{visc}} \leq \tau_{\text{lib}}/2 \tag{6.31}$$

should be verified. When the second inequality is satisfied, the vortensity at the upstream separatrices is kept stationary, which prevents phase mixing of vortensity
within the horseshoe region [5, 65]. When the first inequality is satisfied, vortensity
is approximately conserved along U-turns, which maximises the effective vortensity gradient across the horseshoe region [66, 70, 92]. Taking $x_s \sim 1.1 a\sqrt{q/h_p}$ (as
measured with a planet softening length $\approx 0.6H$), inequality (6.31) may be cast as

$$0.32 q^{3/2} h_p^{-7/2} \leq \alpha_{\text{v,p}} \leq 0.16 q^{3/2} h_p^{-9/2}, \tag{6.32}$$

where $\alpha_{\text{v,p}}$ and h_p denote the disc's alpha viscosity and aspect ratio at the planet
location, respectively. The alpha viscosity for which the corotation torque takes its
maximum value can be approximated as $0.16 q^{3/2} h_p^{-4}$ [7].

The saturation properties of the corotation torque are illustrated in Fig. 6.11 for a
2 Earth-mass planet embedded in a thin ($h_p = 0.05$) viscous disc. The background
temperature profile is uniform, and the surface density decreases as $r^{-1/2}$. The

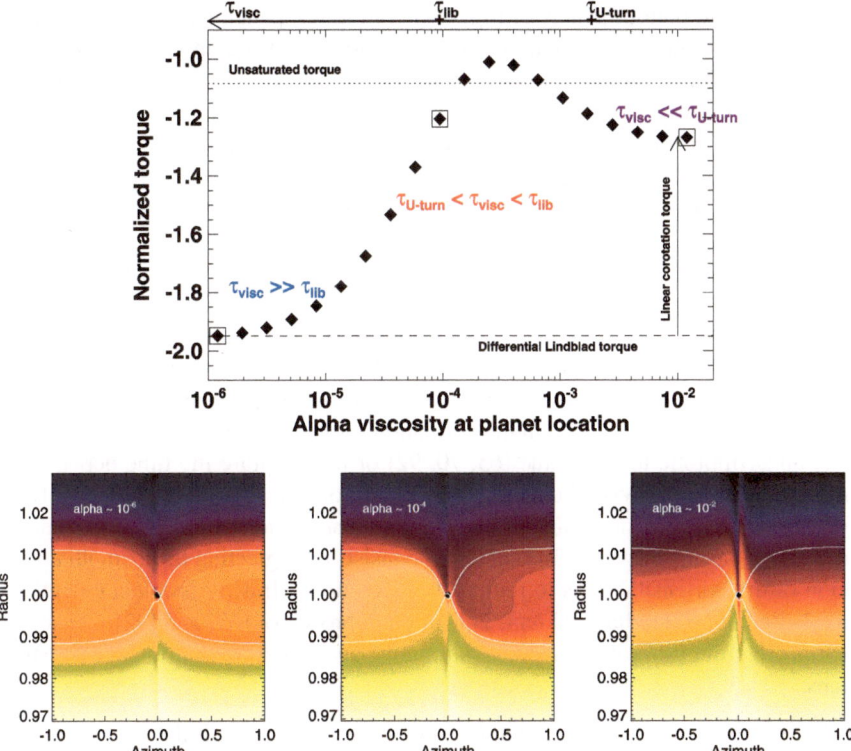

Fig. 6.11 *Top*: steady-state torque on a $M_p = 6 \times 10^{-6} M_\star$ planet mass embedded in a thin ($h = 0.05$) disc for various alpha viscous parameters at the planet location. In this series of runs, $\alpha = 1/2$ and $\beta = 0$. Different saturation regimes of the corotation torque are illustrated, depending on how the viscous timescale across the planet's horseshoe region (τ_{visc}) compares with the horseshoe U-turn timescale ($\tau_{\text{U-turn}}$) and the libration timescale (τ_{lib}). The final torque value in an inviscid case, which reduces to the differential Lindblad torque, is shown by a *dashed line*. The fully unsaturated total torque (differential Lindblad torque plus fully unsaturated horseshoe drag) in an inviscid run is depicted by a *dotted line*. *Bottom*: vorticity distribution inside the planet's horseshoe region for the three alpha viscosities shown by squares in the *top panel* (viscosity increases from *left* to *right*). The separatrices of the horseshoe region are depicted by *solid curves*, and the planet position by a *filled circle*

top panel displays the steady-state torque at different alpha viscosities (a constant kinematic viscosity ν was used in the simulations). The left-hand term in inequality (6.32) is $\approx 1.7 \times 10^{-4}$, while the right-hand term in $\approx 1.7 \times 10^{-3}$, and it is clear from Fig. 6.11 that the corotation torque is maximum between these two alpha viscosities. When the viscosity is small enough so that $\tau_{\text{visc}} \gg \tau_{\text{lib}}$, viscosity is inefficient at restoring the vortensity gradient across the horseshoe region, and the horseshoe drag takes very small values (it saturates). At very large viscosities, such that $\tau_{\text{visc}} \ll \tau_{\text{U-turn}}$, the corotation torque plateaus at its value in the linear regime [89]. The vortensity distribution inside the horseshoe region for each saturation regime is shown in the bottom panels of Fig. 6.11. In the left panel, $\alpha_{v,p} \sim 10^{-6}$, and the steady-state vortensity distribution within the horseshoe region is uniform, resulting in a vanishing corotation torque. In the middle panel, $\alpha_{v,p} \sim 10^{-4}$ maintains a maximum vortensity contrast between the rear and front parts of the planet, with the consequence that the corotation torque is close to its fully unsaturated value. In the right panel, $\alpha_{v,p} \sim 10^{-2}$ imposes the initial (unperturbed) vortensity profile along the horseshoe U-turns, and the horseshoe drag therefore reduces to the linear corotation torque.

All attempts to capture analytically the saturation of the corotation torque have been carried out using a simplified streamline model that assumes the drift of the coorbital material with the velocity of the unperturbed disc, and which does not resolve spatially nor temporally the U-turns. This model was proposed in [65] and in a more formal manner in [70], where a numerical implementation of it is also described. All analytic works on the corotation torque saturation, whether they provide an asymptotic torque value [65, 70, 92] or they capture the time dependence of the torque in an inviscid disc [111] make use of this simplified model. Solving for the torque asymptotic value in this simplified advection-diffusion model can be tacked in a variety of ways. One such way consists of neglecting the azimuthal variation of the vortensity so as to reduce the advection-diffusion problem essentially to a 1-dimensional radial problem. This is the approach of [65] and [92]. These two works are quite different in their assumptions, and suffer from orthogonal restrictions:

- Masset [65] exclusively contemplates the case of a disc with flat profiles of surface density and kinematic viscosity, so that his results must be rescaled by hand to apply to a general case. The approach used in this work considers the global angular momentum budget of the trapped horseshoe region, and relies upon the evaluation of the viscous friction of the disc on the separatrices. It also takes into account the viscous drift of material across the horseshoe region.
- Paardekooper et al. [92] consider a disc with an arbitrary surface density gradient, and directly use the horseshoe drag integral of Eq. (6.19). Their model assumes no radial drift of disc material across the horseshoe region.

Quite remarkably, these two approaches yield the exact same result, which can be cast either in terms of Airy functions [65] or in terms of Bessel functions [92].

One can also solve the advection-diffusion equation satisfied by the fluid's vortensity in 2 dimensions, the solution being exact in the limit of a small viscosity. In this limit, the problem amounts to an alternation of convolutions (viscous diffusion of vortensity between two successive horseshoe U-turns) and reflections (mapping of vortensity—or, vortensity conservation—from one tip of the horseshoe region to the other during a U-turn). This is the approach of [70], who also discard the possible radial drift of disc material across the horseshoe region. The dependence thus obtained—Eq. (119) of [70]—is broadly the same as that of [65] and [92], but reproduces more closely the results from numerical simulations. We note that the decay of the torque value found at large viscosity (see Fig. 6.11), which corresponds to a decay towards the linear corotation torque value [89], has not yet been described analytically in a self-contained manner. Masset and Casoli [70] and Paardekooper et al. [92] use an ad-hoc reduction factor, either with one free parameter [70] or two free parameters [92], the value of the free parameters being inferred from numerical simulations.

In summary, in barotropic discs, vortensity essentially obeys an advection-diffusion equation in the coorbital region. When the viscous diffusion timescale across the horseshoe region is:

- Long compared to the libration period, vortensity is progressively stirred up and the corotation torque ultimately saturates (tends to zero).
- Short compared to the libration period, but long compared to the horseshoe U-turn time, the corotation torque is close to its fully unsaturated, maximum value.
- Short compared to the horseshoe U-turn time, the corotation torque tends to its value predicted in the linear regime.

6.3.3.2 Saturation Properties of the Horseshoe Drag in Radiative Discs

Much as in barotropic discs, the estimate of the asymptotic corotation torque value in radiative discs amounts to the determination of the vortensity distribution within the horseshoe region at later times. There is an additional complexity, however, due to the fact that this is no longer an advection-diffusion problem, but an advection-diffusion-creation problem, as vortensity is created during the U-turns (see Sect. 6.3.2.5, and in particular Fig. 6.10). Furthermore, the amount of vortensity created depends on the entropy distribution, as was explained in Sect. 6.3.2.5. This analysis was undertaken by [92] in the case of a unitary thermal Prandtl number (the viscosity ν and thermal diffusion χ have same value). A corotation torque expression was proposed by these authors, as a result of a fit of numerical simulations. Under the assumption of a unitary Prandtl number, the parameter space to be explored is 1-dimensional, and for a (common) value of ν and χ, the radiative torque is found to saturate more easily than the barotropic torque. This is interpreted by the authors as due to the fact the entropy-related corotation torque is essentially due to a unique streamline, where the advection speed is maximal (that of the separatrices).

To relax the assumption of a unitary Prandtl number, one may assume that the torque dependence upon viscosity or thermal diffusion have the same shape, which allows to propose a formula with two independent parameters ν and χ, which can then be validated by checking its accuracy with numerical simulations. This is the approach adopted by [92]. Another solution consists in using a streamline model such as the one outlined in Sect. 6.3.3.1. This is the approach of [70]. As the vortensity is now determined by an advection-diffusion-creation problem, one needs to amend the barotropic model of Sect. 6.3.3.1 by adding the creation of vortensity during the U-turns, which is determined by the entropy field. Therefore, prior to the determination of the vortensity distribution, an analysis of the entropy distribution at later times is required. This preliminary determination can be done easily, because the entropy obeys an advection-diffusion problem formally equivalent to the vortensity distribution in the barotropic case, in which one replaces the vortensity with the entropy, and the viscosity ν with the thermal diffusion χ. Once the entropy distribution within the horseshoe region is known, the vortensity distribution at late times is obtained, which allows, upon the use of the horseshoe drag expression of Eq. (6.19), for an expression of the corotation torque as a function of viscosity and thermal diffusivity, and which can be checked *a posteriori* against numerical simulations.

The corotation torque expressions, as a function of viscosity and thermal diffusivity, are given by Eqs. (161)–(164) of [70], or by Eqs. (52)–(53) of [92].

6.3.4 Type I Migration in Turbulent Discs

We have examined in the previous sections the properties of planet–disc interactions assuming viscous discs, described with a stationary kinematic viscosity aimed at modelling their turbulent transport properties. Because the corotation torque may play a dramatic role in the orbital evolution of low-mass planets, and its magnitude is intimately related to diffusion processes taking place within the planet's horseshoe region, it is relevant to determine how turbulence may impact type I migration.

Turbulence in protoplanetary discs can have a variety of origins. These include hydrodynamic instabilities, such as Rossby-wave instabilities [60], the global baroclinic instability [44, 61], the sub-critical baroclinic instability [51], planetary gap instabilities [57, 58] (which we will discuss in Sect. 6.4.2), or the Kelvin-Helmholtz instability triggered by the vertical shear of the gas as dust settles into the mid plane [43]. Convective instability might also be relevant in the inner parts of massive discs, and it would be interesting to examine its impact on type I migration. Perhaps the most likely source of turbulence in protoplanetary discs is the magnetohydrodynamic (MHD) turbulence resulting from the magnetorotational instability (MRI) [4]. It relies on the coupling of the ionised gas to the weak magnetic field in the disc. Ionisation may occur in the vicinity of the central object due to the star's irradiation, or further out in the disc layers, most probably through the UV background or cosmic rays. It is currently debated which regions of planet formation near the disc mid plane are sufficiently ionised ('active') to trigger the MRI, and which ones remain neutral (which is usually referred to a 'dead zone'). In the latter case, some transport of angular momentum would still be present through the propagation of waves induced by MHD turbulence in the disc's upper layers [28]. The alpha viscous parameter associated to MHD turbulence is typically in the range $[5 \times 10^{-3} - 5 \times 10^{-2}]$ in active regions, while being about two orders of magnitude smaller in dead zones.

The properties of type I migration in weakly magnetised turbulent discs have been investigated in a couple of studies. Nelson and Papaloizou [84] performed 3D simulations of locally isothermal discs fully invaded by MHD turbulence. They found that the running time-averaged torque on a fixed protoplanet experiences rather large-amplitude oscillations over the reduced temporal range over which the simulation could be run, and that its final value differs quite substantially from the

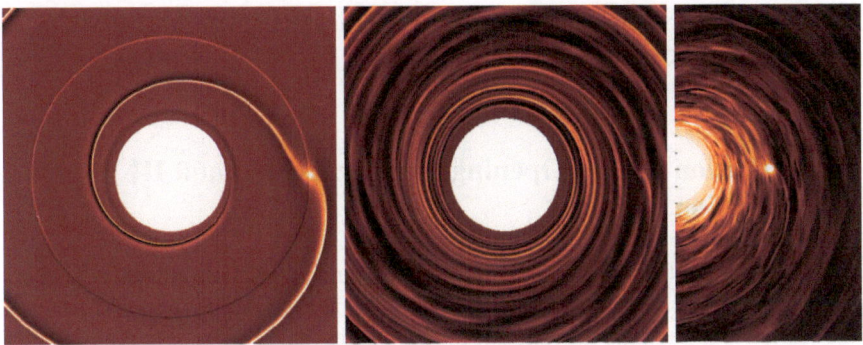

Fig. 6.12 Perturbation of the disc's density by an embedded planet. *Left*: adiabatic 2-dimensional disc. The density lobes within the coorbital region, which arise from the advection of entropy, help identify the (tiny) radial width of the planet's horseshoe region. *Middle*: case of an isothermal 2D disc with turbulence induced by stochastic stirring. *Right*: case of an isothermal 3D disc invaded by MHD turbulence due to the MRI (the density in the disc mid plane is displayed). In the *middle* and *right panels*, the turbulent density perturbation is comparable to the perturbed density associated to the planet's wakes. Images taken from [6, 7, 10], respectively

torque value expected in viscous disc models. Similar results were obtained by [82], who allowed the planet orbit to evolve. A primary reason for the observed difference between the viscous torque and the time-averaged turbulent torque is that the 3D MHD simulations were not converged in time. This was suggested by [7], who considered 2D isothermal discs subject to stochastic forcing, using the turbulence model originally developed by [50]. They showed indeed that when time-averaged over a sufficiently long time period, which may be as long as a thousand orbits, both the differential Lindblad torque and the corotation torque behave very similarly as in equivalent viscous disc models. These results were essentially confirmed by the 3D MHD simulations by [10], who adopted a locally isothermal disc model with a mean toroidal magnetic field, in which non-ideal MHD effects and vertical stratification were neglected (see illustration in Fig. 6.12). Similar agreement was obtained by [106] with vertical stratification. Nonetheless, [10] found an additional corotation torque with moderate magnitude in their 3D MHD simulations, related to the presence of a mean toroidal magnetic field. The existence and properties of this additional corotation torque have been explored by [35] in 2D weakly magnetised, non-turbulent disc models, in which the effects of turbulence are modelled by viscous and magnetic diffusivities. They find that the additional corotation torque can take large values, and even exceed the differential Lindblad torque, depending on the disc's viscous and magnetic diffusivities, and the amplitude of the background magnetic field.

The aforementioned results were for embedded planets with a horseshoe radial width that is a moderate fraction of the disc's pressure scaleheight, the latter being the typical size of turbulent eddies. The existence of horseshoe dynamics and a corotation torque is unknown for planets with a horseshoe region width that is a small fraction of the turbulent eddy size. In this case, it is possible that turbulence acts more as a source of random advection of vortensity through the horseshoe region, rather than diffusion.

6.4 Migration of Gap-Opening Planets: Type II and III Migration

The disc response to a low-mass planet has been studied in details in Sect. 6.3, where we have focused on the two components of the type I migration torque. The aim of this section is to examine the range of planet masses that is relevant to type I migration in Sect. 6.4.1, and to give a concise description of planet–disc interactions for planets that are massive enough to significantly perturb the disc's mass distribution (Sects. 6.4.2 and 6.4.3).

6.4.1 Shock Formation and Gap-Opening Criterion

The wakes generated by a planet in a disc carry angular momentum as they propagate away from the planet. This angular momentum is eventually deposited in the disc through some wave damping processes, which leads to redistributing the disc mass. An efficient wave damping mechanism relies on the non-linear wave evolution of the wakes into shocks [34]. The (negative) angular momentum deposited by the inner wake decreases the semi-major axis of the fluid elements in the disc region inside the planet's orbit (the inner disc). Similarly, the (positive) angular momentum deposited by the outer wake increases the semi-major axis of the fluid elements in the outer disc.

The distance d_s from the planet where the planet-generated wakes become shocks is given by [26, 34]:

$$d_s \approx 0.8 \left(\frac{\gamma + 1}{12/5} \frac{q}{h^3} \right)^{-2/5} H(a), \qquad (6.33)$$

where γ is the gas adiabatic index, and a denotes the planet's semi-major axis. As the magnitude of the one-sided Lindblad torque peaks at $\sim 4H(a)/3$ from the planet's orbit, a linear description of the differential Lindblad torque thus fails when $|d_s| \sim 4H(a)/3$. This condition can be recast as $q \sim 0.3h^3$ for $\gamma = 5/3$. When $|d_s| \leq 2H(a)/3$, wakes turn into shocks within their excitation region. Fluid elements just outside the planet's horseshoe region are pushed away from the planet orbit after crossing the wakes, which directly affects the planet's coorbital region by inducing asymmetric U-turns [67]. Horseshoe fluid elements therefore get progressively repelled from the planet orbit after each U-turn, and the planet slowly depletes its coorbital region. The equilibrium structure (width, depth) of the annular gap the planet forms around its orbit is determined by a balance between gravity, viscous and pressure torques [17].

Shock formation and its damping efficiency are very sensitive to the disc's viscosity, and gap-opening results from a balance between (i) a planet mass large enough to induce shocks where the wake excitation takes place, and (ii) a viscosity small enough to maximise the amount of angular momentum deposited by the shocks in the planet's immediate vicinity:

1. The first condition reads $|d_s| \leq 2H(a)/3$, which corresponds to $q \geq 1.5h^3$ for $\gamma = 5/3$. It means that the planet's Bondi radius $r_B = GM_p/c_s^2$, where the pressure distribution is most strongly perturbed by the planet, as well the planet's Hill radius become comparable to the local pressure scaleheight. This is known as the thermal criterion for gap opening [55].
2. The second condition, known as the viscous criterion, can be expressed as $q \geq 40/\mathscr{R}$, where $\mathscr{R} = a_p^2 \Omega_p / \nu$ is the Reynolds number[1] [13, 53].

The above two conditions for gap-opening have been revisited by [17], who provide a unified criterion that takes the form

$$1.1 \left(\frac{q}{h^3} \right)^{-1/3} + \frac{50 \alpha_v h^2}{q} \leq 1, \tag{6.34}$$

where we have written the disc's kinematic viscosity $\nu = \alpha_v h^2 a^2 \Omega$ [102], and where in Eq. (6.34) h and α_v are to be evaluated at the planet's semi-major axis. An illustration of the smallest planet mass opening a gap according to criterion (6.34) is shown in Fig. 6.13, where it is clear that the gap-opening mass increases with increasing disc viscosity and aspect ratio. Assuming $h \approx 0.05$, which may be typical of planet forming regions, the gap-opening mass is in the Saturn-mass range for regions with low turbulent activity (dead zones, with typically $\alpha_v \sim 10^{-4}$), and is in the Jupiter-mass range in regions where $\alpha_v \sim 10^{-2}$. Note that when disc self-gravity is included, the gap-opening criterion of Eq. (6.34) should involve the effective planet mass, that is the sum of the planet and circumplanetary disc masses, rather than the planet mass alone.

Prior to a depletion of their coorbital region due to shock formation at the wake's excitation region, planets with increasing mass experience a flow transition in their immediate vicinity. The flow passes from a low-mass planet configuration described in Sect. 6.3.2.1 and Fig. 6.3, to a high-mass configuration, where fluid elements may become trapped inside a circumplanetary disc around the planet. This flow transition is accompanied by a rapid increase in the half-width x_s of the horseshoe region, from a fluid-dominated regime (where $x_s \propto (q/h)^{1/2}$) to a gravity-dominated regime (where $x_s \sim R_H \propto q^{1/3}$) [73]. This rapid increase yields a significant increase in the corotation torque, as the latter scales as x_s^4. This effect is found to

---

[1] Although traditionally dubbed Reynolds number essentially for dimensional considerations, this ratio has little to do with the dimensionless ratio that must be considered to assess whether a flow is laminar or turbulent. If one regards the planet as an obstacle in the sheared Keplerian flow, it would be more appropriate to consider as a characteristic scale the size of its Roche lobe or $\sim x_s$, and as a characteristic velocity $2|A|x_s$.

Fig. 6.13 Minimum
planet-to-primary mass ratio
leading to gap-opening as a
function of the disc's alpha
viscous parameter (*x-axis*)
and aspect ratio (*y-axis*) at
the planet's semi-major axis.
This minimum mass is
calculated numerically using
criterion (6.34)

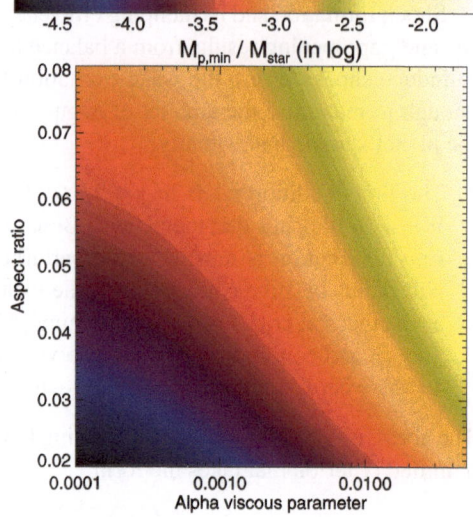

be most significant for planet-to-primary mass ratios $q \sim 0.6h^3$ [73], which corresponds to 20 Earth-mass planets in $h = 0.05$ discs. It may contribute to further slowing down, or even reversing the migration of growing planets before they carve a gap around their orbit [24, 73].

6.4.2 Partial Gap-Opening: Type III Migration in Massive Discs

So far, we have addressed the properties of planet migration through a direct analysis of the tidal torque, the latter being directly proportional to the migration rate, see Eq. (6.7). This approach is valid for low-mass planets that do not open a gap, for which migration has a negligible feedback on the tidal torque (note that a weak, negative feedback slightly decreases the magnitude of the entropy-related horseshoe drag [69]). Nevertheless, migrating planets that open a partial gap around their orbit experience an additional corotation torque due to fluid elements flowing across the horseshoe region [71]. If for instance the planet migrates inwards, fluid elements circulating near the inner separatrix of the horseshoe region enter the horseshoe region, and execute an outward U-turn when they reach the vicinity of the planet. Upon completion of the U-turn, these fluid elements leave the horseshoe region as the planet keeps migrating, and end up circulating in the outer disc. Consequently, the mass distribution within the horseshoe region may become asymmetric, as the horseshoe region adopts approximately a trapezoidal shape in the azimuth-radius plane [71]. As a consequence, in the case of an inward migrating planet, there is more mass behind the planet than ahead of it, owing to the partial depletion of the asymmetric horseshoe region. This point is illustrated in the left panel of Fig. 6.14. Similarly, if the planet migrates outwards, fluid elements circulating near the outer separatrix may embark on single inward U-turns across the horseshoe region.

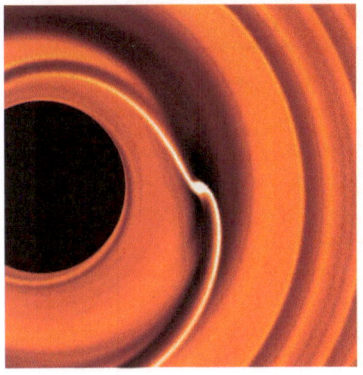
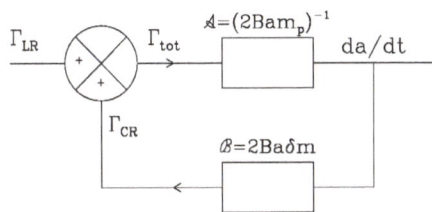

Fig. 6.14 *Left*: illustration of the flow asymmetry ahead of and behind a Saturn-mass planet undergoing rapid inward runaway migration. *Right*: type III planetary migration seen as a feedback loop. The latter remains stable if the open-loop transfer function $\mathscr{A} \times \mathscr{B} < 1$, or $\delta m < M_p$. From [67]

Assuming steady migration at a moderate rate (this point will be clarified below), the additional corotation torque experienced by the planet due to the orbit-crossing flow is, to lowest order in x_s/a,

$$\Gamma_{\text{cross}} = 2\pi a \dot{a}\, \Sigma_s \times 4Bax_s, \tag{6.35}$$

where Σ_s is the surface density at the inner (outer) horseshoe separatrix for a planet migrating inwards (outwards). The term $2\pi a \dot{a}\, \Sigma_s$ in the right-hand side of Eq. (6.35) is the mass flux across the horseshoe region. The second term ($4Bax_s$) is the amount of specific angular momentum that a fluid element near a horseshoe separatrix exchanges with the planet when performing a horseshoe U-turn. Note that the above expression for Γ_{cross} assumes that all circulating fluid elements entering the coorbital region embark on horseshoe U-turns, whereas a fraction of them may actually become trapped inside the planet's circumplanetary disc. Since Γ_{cross} is proportional to, and has same sign as $\dot{a}$, migration may become a runaway process. We now discuss under which circumstances a runaway may happen.

The planet and its coorbital material (which encompasses the horseshoe region, with mass M_{hs}, and the circumplanetary disc, with mass M_{cpd}) migrate at the same drift rate, $\dot{a}$, which we assume to be constant. The rate of angular momentum change of the planet and its coorbital region includes (i) the above contribution Γ_{cross} to the corotation torque, and (ii) the tidal torque that, for planets opening a partial gap, essentially reduces to the differential Lindblad torque Γ_{LR}:

$$2Ba\dot{a}(M_p + M_{\text{cpd}} + M_{\text{hs}}) = 2\pi a \dot{a}\, \Sigma_s \times 4Bax_s + \Gamma_{\text{LR}}. \tag{6.36}$$

Equation (6.36) can be written as

$$2Ba\dot{a}\tilde{M}_p = 2Ba\dot{a}\delta m + \Gamma_{\text{LR}}, \tag{6.37}$$

where $\tilde{M}_p = M_p + M_{\text{cpd}}$ corresponds to an *effective* planet mass, and where $\delta m = 4\pi ax_s \Sigma_s - M_{\text{hs}}$ is called the coorbital mass deficit [71]. It represents the difference

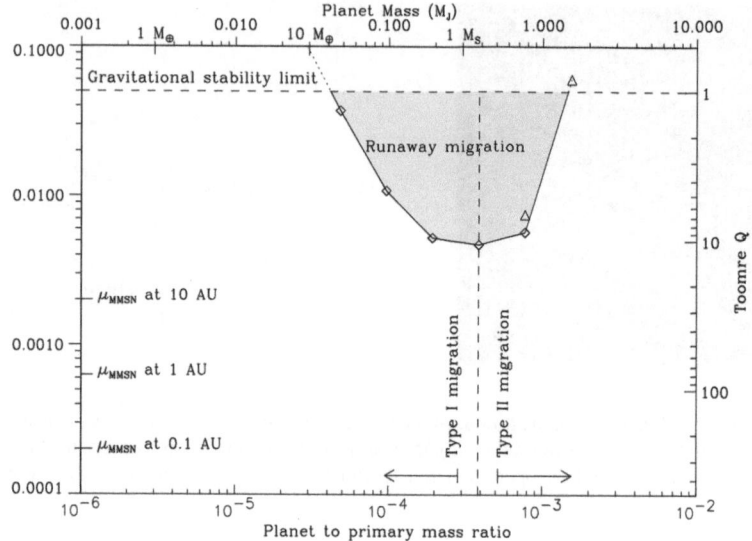

Fig. 6.15 Occurrence for type I, II and III (runaway) migrations with varying the planet-to-primary mass ratio (*bottom x-axis*) and the disc-to-primary mass ratio at the planet location (*left y-axis*). The disc's aspect ratio is $h = 0.05$ and its alpha viscosity is $\alpha_v = 4 \times 10^{-3}$. The *right y-axis* shows the Toomre Q-parameter at the planet location. The *upper part* of the plot is limited by the gravitational instability limit (*dashed line*). From [71]

between (i) the mass the horseshoe region would have if it had a uniform surface density equal to that of the separatrix-crossing flow, and (ii) the actual horseshoe region mass.

The migration rate of a partial gap-opening planet, given by Eq. (6.37), can be described as a feedback loop [67]. This is illustrated in the right panel of Fig. 6.14, where the loop input is the differential Lindblad torque, and its output is the migration rate. When $\delta m < \tilde{M}_p$, the feedback loop remains stable. The drift rate in this case is not strictly a type I nor a type II migration rate. It is rather a type I rate enhanced by coorbital effects. No special name has been assigned to this kind of migrating regime. When $\delta m > \tilde{M}_p$, the feedback loop gets unstable, and migration enters a runaway regime, which can be either inward or outward. The drift rate as a function of the disc mass undergoes a bifurcation [67], and this regime is called runaway type III migration [71, 96, 97].

Runaway migration is based on the planet's ability to build up a coorbital mass deficit by opening a gap. It does not apply to low-mass planets, for which $\delta m \ll \tilde{M}_p$. It does not apply to high-mass planets neither, which open a wide, deep gap, so that the surface density of the separatrix-crossing flow is too small to produce a significant mass deficit. It rather concerns intermediate-mass planets, marginally satisfying the gap-opening criterion in Eq. (6.34), in massive discs (the larger the disc mass, the larger the density of the orbit-crossing flow). Its occurrence is illustrated in Fig. 6.15 for a disc with aspect ratio $h = 5\%$ and alpha viscosity $\alpha_v = 4 \times 10^{-3}$, where we see that runaway migration may be particularly relevant to Saturn-mass planets in massive discs (with a Toomre-Q parameter at the planet's orbital radius typically less than about 10). Bear in mind, however, that the occurrence for runaway migration is sensitive to the values of h and α_v, since they affect the planet's ability to open a partial gap. Also, note that the criterion for runaway migration features the *effective* planet mass $\tilde{M}_p$, sum of the planet mass and the circumplanetary disc mass. The occurrence for runaway migration is therefore sensitive to the mass distribution inside the circumplanetary disc, which may be significantly affected by the assumed physical modelling, e.g. whether gas accretion on the planet is taken into account [21], the inclusion of self-gravity [114], or the treatment for the gas thermodynamics [95]. It may also be affected by grid resolution effects in numerical simulations [25, 95].

The simple model described above helps understand the condition for migration to enter a runaway regime. However, since it assumes steady migration (that is, constant $\dot{a}$), this model is no longer valid when migration actually enters the runaway regime, where the migration rate increases exponentially over a time comparable to the horseshoe libration period. A more general approach can be found in [67, 71, 96]. As long as the orbital separation by which the planet migrates over a libration period remains smaller than the radial width of the planet's horseshoe region, Γ_{cross} remains approximately proportional to $\dot{a}$ (slow runaway regime). At larger migration rates (fast runaway regime), Γ_{cross} reaches a maximum and slowly decreases with increasing $\dot{a}$ [71] (see also Fig. 16 in [67]). The precise dependence of Γ_{cross} with $\dot{a}$ in this fast runaway regime is intrinsically related to the evolution of the mass coorbital deficit, and therefore to the planet's migration history. The orbital evolution of planets subject to runaway type III migration is therefore difficult to predict. Numerical simulations find that, depending on the resolution of the gas flow surrounding the planet, the timescale for inward runaway type III migration can be as short as a few 10^2 orbits [19, 71].

The sign of Γ_{cross} is dictated by the initial drift of the planet. Runaway migration can therefore be directed inwards or outwards, depending on the sign of $\dot{a}$ before the runaway takes place. In particular, migration may be directed outwards if, despite the coorbital region being partly depleted, a (positive) horseshoe drag remains strong enough to counteract the (negative) differential Lindblad torque. Outward runaway migration could thus be an attractive mechanism to account for the recent discovery of massive planets at large orbital separations (which we will further discuss in Sect. 6.5.1). Simulations [67, 71, 97] however show that, despite the expected increase in the mass of the orbit-crossing flow as the planet moves outwards (for background surface density profiles shallower than r^{-2}), the mass coorbital deficit cannot be retained indefinitely. The increase in the mass of the circumplanetary disc, and the strong distortion of the flow within it at large migration rates, lead the planet to eventually lose its coorbital mass deficit, and the sense of migration is found to reverse.

Type III migration has been recently revisited in low-viscosity discs ($\alpha_v \leq$ a few $\times 10^{-4}$). Depending on the disc mass, the edges of the planet-induced gap may be subject to two kinds of instabilities. In low-mass discs, gap edges are unstable to vortex-forming modes [52, 58, 60]. They lead to the formation of several vortices sliding along the gap edges, which merge and form large-scale vortices. When they pass by the planet, these vortices may embark on horseshoe U-turns and exert a large corotation torque on the planet, with the consequence that the planet can be scattered inwards or outwards [56]. When the fluid's self-gravity is taken into account, only a fraction of the large-scale vortices actually embark on horseshoe U-turns, the rest of the vortices keeps on sliding along the gap edges. This provides a periodic, intermittent corotation torque on the planet. Depending on the relative strengths of the vortices embarking on inward and outward U-turns, this mechanism acts much like an intermittent type III migration regime. In massive self-gravitating discs (stable against the gravitational instability), vortex-forming modes are replaced by global edge modes, which excite spiral density waves [57]. A decreasing radial profile of the Toomre-Q parameter favours edge modes at the gap's outer edge. The periodic protrusion of edge mode-induced density waves near the gap's outer edge provides a periodic source of (positive) corotation torque on the planet, and induces an intermittent type III migration regime. Numerical simulations by [59] show that edge modes can sustain outward migration, until the planet leaves its gap.

We briefly sum up the main results of this paragraph:

- Migrating planets experience an additional corotation torque due to fluid elements flowing across the horseshoe region, and embarking on horseshoe U-turns. It is proportional to the planet's migration rate at small migration rates, which gives a positive feed back on migration. When the feedback loop diverges, the migration type is known as type III migration.
- The planet and its circumplanetary disc feel an effective corotation torque that is proportional to the coorbital mass deficit, defined through Eq. (6.37). The occurrence of a runaway feedback (i.e. of type III migration) corresponds to the coorbital mass deficit exceeding the mass of the planet and its circumplanetary disc. This applies to planets opening a partial gap around their orbit in massive protoplanetary discs.
- The orbital evolution of planets undergoing type III migration is sensitive to the time evolution of the coorbital mass deficit, which makes it difficult to predict. Numerical simulations show that runaway migration operates on very short timescales, typically in 100 to 1000 planet orbits.

6.4.3 Deep Gap-Opening: Type II Migration

Planets massive enough to clear their coorbital region and open a deep gap around their orbit (see illustration in Fig. 6.16) enter the migration regime called type II migration. Such planets satisfy the gap-opening criterion given by Eq. (6.34). Assuming, for instance, a protoplanetary disc with aspect ratio $\sim 5\%$ and alpha viscosity $\alpha_v \sim 10^{-2}$, type II migration typically applies to planets more massive than Jupiter orbiting Sun-like stars. Compared to the type I and type III migration regimes described previously, the amplitude of the corotation torque is much reduced due to

Fig. 6.16 Gap opened by a Jupiter-mass planet orbiting a Sun-like star

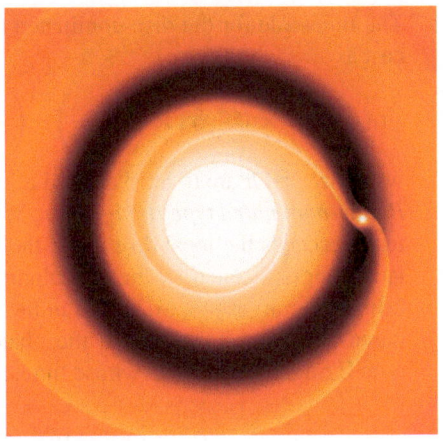

the clearing of the planet's coorbital region, and the differential Lindblad torque balances the viscous torque exerted by the disc. The net torque on the planet can be written as a fraction C_{II} of the viscous torque due to the outer disc [16]. This factor C_{II} features the time-dependent fraction of gas f_{gas} remaining in the planet's coorbital region.

The particular case with f_{gas} going to zero corresponds to what is usually referred to as standard type II migration regime. Its timescale can be approximated as

$$\tau_{\mathrm{II}} \approx \frac{2r_{\mathrm{o}}^2}{3\nu(r_{\mathrm{o}})}\left(1 + \frac{M_p}{4\pi\,\Sigma(r_{\mathrm{o}})r_{\mathrm{o}}^2}\right), \qquad (6.38)$$

where ν denotes the disc's kinematic viscosity, Σ the surface density of the disc perturbed by the planet, and r_{o} is the location in the outer disc where most of the planet's angular momentum is deposited. It can be approximated as the location of the outer separatrix of the planet's horseshoe region, which, for gap-opening planets, is $r_{\mathrm{o}} \approx a + 2.5R_{\mathrm{H}}$ [73]. The first term on the right-hand side of Eq. (6.38) corresponds to the viscous drift timescale at radius r_{o}, and the second term features the ratio of the planet mass to the local disc mass at radius r_{o}. Two migration regimes can therefore be distinguished:

1. *Disc-dominated type II migration.* When the planet mass is much smaller than the local disc mass (by which we refer to the quantity $4\pi\,\Sigma(r_{\mathrm{o}})r_{\mathrm{o}}^2$), the planet behaves much like a fluid element that the disc causes to drift viscously. The planet's migration timescale then matches the disc's viscous drift timescale $\approx 2r_{\mathrm{o}}^2/3\nu(r_{\mathrm{o}})$ [54]. In this migration regime, called disc-dominated type II migration, the planet remains confined within its gap. Should the planet migrate slightly faster than the disc near its orbit, the increased inner Lindblad torque due to the planet getting closer to the gap's inner edge would push the planet outward. Conversely, should the planet migrate at a slower pace than the (local) disc, the increased outer Lindblad torque would push the planet back inward.

The timescale for the disc-dominated type II migration regime, $\tau_{II,d}$, can be re-cast as

$$\tau_{II,d} \approx 4.7 \times 10^4 \text{ yrs} \times \left(\frac{\alpha_v}{10^{-2}}\right)^{-1} \left(\frac{h}{0.05}\right)^{-2} \left(\frac{M_\star}{M_\odot}\right)^{-1/2} \left(\frac{r_0}{5 \text{ AU}}\right)^{3/2}, \quad (6.39)$$

where α_v and h are to be evaluated at r_0.

2. *Planet-dominated type II migration.* When the planet mass becomes comparable to, or exceeds the local disc mass, the orbital evolution of a gap-opening planet is no longer dictated by the disc alone. The inertia of the planet slows down its orbital migration [41, 103], and in the limit when the planet mass is large compared to the local disc mass, the planet enters the so-called planet-dominated type II migration regime, whose timescale $\tau_{II,p}$ reads

$$\tau_{II,p} \approx \tau_{II,d} \times \left(\frac{M_p}{4\pi \Sigma(r_0)r_0^2}\right), \quad (6.40)$$

with $\tau_{II,d}$ given by Eq. (6.39), and where the planet to local disc mass ratio reads

$$\frac{M_p}{4\pi \Sigma(r_0)r_0^2} \approx 200 \left(\frac{M_p}{M_\odot}\right) \left(\frac{\Sigma(r_0)}{150 \text{ g cm}^{-2}}\right)^{-1} \left(\frac{r_0}{5 \text{ AU}}\right)^{-2}. \quad (6.41)$$

In self-gravitating discs, the distinction between the planet- and disc-dominated type II migration regimes should involve the comparison between the local disc mass and the effective planet mass $\tilde{M}_p$, that is the sum of the planet and circumplanetary disc masses. Related to this point, we comment that the planet and its circumplanetary disc migrate at the same drift rate. When its self-gravity is included, the protoplanetary disc torques both the planet and the circumplanetary disc. However, if self-gravity is discarded, as is usually the case in numerical simulations, the protoplanetary disc can only torque the planet, and the circumplanetary disc remains a passive spectator of the migration. In this case, the planet must exert an additional effort to maintain the planet and circumplanetary disc joint migration. Put another way, the circumplanetary disc artificially slows down migration when self-gravity is discarded. To avoid this artificial slowdown, [19] showed that, in simulations discarding self-gravity, the calculation of the torque on the planet must exclude the material inside the circumplanetary disc. In addition, migration rates with and without self-gravity can be in close agreement, provided that the mass of the circumplanetary disc is added to that of the planet when calculating the gravitational potential felt by the protoplanetary disc.

In the early stages of their formation and orbital evolution, most massive gap-opening planets should be subject to disc-dominated type II migration, and migrate on a timescale comparable to the disc's viscous timescale. Note from Eq. (6.38) that this corresponds to the shortest migration timescale a gap-opening planet can get. Depletion of the protoplanetary disc, or substantial migration towards the central object, should, however, slow down migration as the planet's inertial mass becomes comparable to the local disc mass. It is nonetheless interesting to note from Eq. (6.39) that in the early stages of the disc evolution, type II migration can be relatively fast in the disc's turbulent parts. This may make difficult the maintenance of massive planets at reasonably large orbital separations from their host star. Additional mechanisms, like the effect of stellar irradiation on the disc's density and temperature profiles near the planet's orbit (formation of shadow regions near the gap's inner edge, and irradiated 'puffed up' regions near the gap's outer edge) [42] could help slow down type II migration.

Gap formation and type II migration are intimately related to the disc viscosity in laminar viscous disc models, and a few studies have investigated their properties in MHD turbulent discs [83, 93, 112]. These studies have considered 3-dimensional magnetised disc models, where vertical stratification and non-ideal MHD effects were discarded for simplicity. They found that the structure of the annular gap opened by a massive planet in fully MHD turbulent discs is essentially in line with the predictions of viscous disc models with a similar alpha viscous parameter near the planet location. Gaps in turbulent discs tend, however, to be wider than in viscous discs [83, 112]. Some other differences arise between turbulent and viscous disc models, particularly in the vicinity of the planet, where magnetic field lines are compressed and ordered at the location of the wakes and the circumplanetary disc. The connection between the circumplanetary and protoplanetary discs through magnetic field lines can cause magnetic braking of the circumplanetary material [93], which may help increase gas accretion onto the planet [93, 112].

Before leaving this section, we comment that the overall properties of planet–disc interactions with gap-opening planets remain essentially unchanged when taking the disc's vertical stratification into account, and therefore the 2-dimensional approximation is valid. Nonetheless, different structures in the flow circulating around the planet in two- and 3-dimensions, and the related accretion rate onto the planet, may affect the planet's migration rate (see e.g. [25]).

We summarise the results described in this section:

- Planets massive enough to open a wide and deep annular gap around their orbit are subject to type II migration.
- When the local disc mass (roughly speaking, the mass interior to the planet orbit) remains large compared to the planet mass, the planet migration timescale corresponds to the viscous drift timescale (disc-dominated type II migration).
- When the planet mass becomes comparable to, or exceeds the local disc mass, migration is slowed down by the planet's inertia (transition to planet-dominated type II migration).

6.5 Planet Migration Theories and Observed Diversity of Exoplanets

The properties of planet–disc interactions have been examined in details in the previous sections, with a particular emphasis on the expected migration rate for planets with different masses. The migration rate is intimately related to the disc's physical properties (e.g. mass, sound speed, cooling properties, turbulent stresses) near the planet location, which underlines that the modelling of protoplanetary discs plays as much of an important role as planet–disc interaction theories in predicting the evolution of planetary systems. We continue our exposition with a brief discussion of several aspects of planet–disc interactions which could account for the observed diversity of (extrasolar) planets.

6.5.1 Massive Planets at Large Orbital Separations

Amongst the recent discoveries of exoplanets, particularly exciting is the observation by direct imaging of about 10 massive exoplanets located at separations ranging from 10 to 200 AU from their host star (e.g. [49, 63]). Most of these planets are so far observed to be the only planetary companions of their host star. Yet, it is possible that their present location results from a scattering event with another massive companion on a shorter-period orbit. A remarkable exception is the HR 8799 planetary system. It comprises four planets with masses evaluated in the range $[7 - 10]$ Jupiter masses, and estimated separations of 14, 24, 38 and 68 AU [64]. The planets are close to being in mutual mean-motion resonances, and it seems likely that planet–disc interactions could have played a major role in shaping this planetary system. We discuss below the relevance of planet–disc interactions to account for massive planets at large orbital separations.

6.5.1.1 Outward Migration of a Pair of Massive Resonant Planets

In the standard core-accretion scenario for planet formation, it is difficult to form Jupiter-like planets in isolation further than ~ 10 AU from a Sun-like star [39, 99]. As we have seen in Sect. 6.4, planets in the Jupiter-mass range orbiting Solar-type stars are expected to open an annular gap around their orbit. If a partial gap is opened, outward runaway type III migration could occur under some circumstances, but as we have discussed in Sect. 6.4.2, numerical simulations indicate that it is difficult to sustain this outward migration in the long term. If the planet opens a deep gap, inward type II migration is expected. It is therefore unlikely that a single massive planet formed through the core-accretion scenario within ~ 10 AU of its host star could migrate to several tens of AUs.

A notable exception to this generally expected trend has been recently proposed by [20], based on a migration mechanism originally studied by [72]. This mechanism relies on the joint migration of a pair of resonant massive planets embedded in a common gap. In this mechanism, the innermost planet is massive enough to open a deep gap and migrate inwards on a timescale comparable to that of type II migration. The outermost, less massive planet migrates inwards at a larger pace while carving a partial gap around its orbit. If both planets open overlapping gaps, and maintain a mean-motion resonance between their orbits, their joint migration *could* proceed outwards. The global picture is the following: as the inner planet is more massive, the torque it experiences from the inner disc (inner Lindblad torque) is larger than the (absolute value of the) torque the outer planet experiences from the outer disc (outer Lindblad torque). To maintain joint outward migration in the long term, the fluid elements outside the common gap must be funnelled to the inner disc by embarking on horseshoe trajectories. Otherwise, material would pile up at the outer edge of the common gap, much like a snow-plough, and the torque balance as well as the sense of migration would eventually reverse. An illustration of the joint outward migration mechanism is shown in Fig. 6.17.

The migration reversal described above requires an asymmetric density profile within the common gap. It is thus sensitive to the disc's aspect ratio and viscosity, which enter the gap-opening criterion. It is also sensitive to the mass ratio of the two planets. If the outer-to-inner planets mass ratio is too small, the density contrast within the common gap will be too large to affect the evolution of the innermost planet (the gas density near the outer planet's orbit remains too large to significantly decrease the outer Lindblad torque acting on the inner planet). Conversely, if the outer-to-inner planets mass ratio is too large, the Lindblad torques imbalance will favour joint inward migration. By changing the planets mass ratio during joint migration, gas accretion onto the planets could affect the possibility of sustaining outward migration in the long term. This issue requires further investigation, and accurate modelling of the gas accretion processes onto Saturn sized planets.

We also mention that the joint outward migration scenario has been recently discussed in the context of the Solar System [107]. Inward migration of Jupiter in the primordial Solar nebula down to ≈ 1.5 AU, followed by joint outward migration with Saturn to the current location of both planets (the "Grand tack") would truncate the disc of planetesimals interior to Jupiter's orbit at about 1 AU. The subsequent formation of the terrestrial planets is found to occur with the correct mass ratio between Earth and Mars, and would also account for the compositional structure of the asteroid belt [107].

6.5.1.2 Migration of Planets Formed by Gravitational Instability

An alternative to the core-accretion formation scenario involves the fragmentation of massive protoplanetary discs into clumps through the gravitational instability. Gravitational instability (GI) may typically occur at separations larger than 30 to 50 AU from a central (Sun-like) star, if the Toomre-Q parameter approaches unity and the disc's cooling timescale becomes of order the dynamical timescale (e.g. [30, 100]). While several massive planets could form by fragmentation of a massive disc at several tens of AUs from their star, they are unlikely to stay in place. The tidal interaction with the gravito-turbulent disc they are embedded in should rapidly bring planets formed by GI to the disc's inner regions [11, 77, 115], in a timescale comparable to that of type I migration [11]. The orbital evolution of a single Jupiter-mass planet embedded in a gravito-turbulent disc (where the planet is supposed to have formed by GI) is illustrated in Fig. 6.18, where we see that the planet migrates from 100 to 20 AU in typically less than 10^4 yrs.

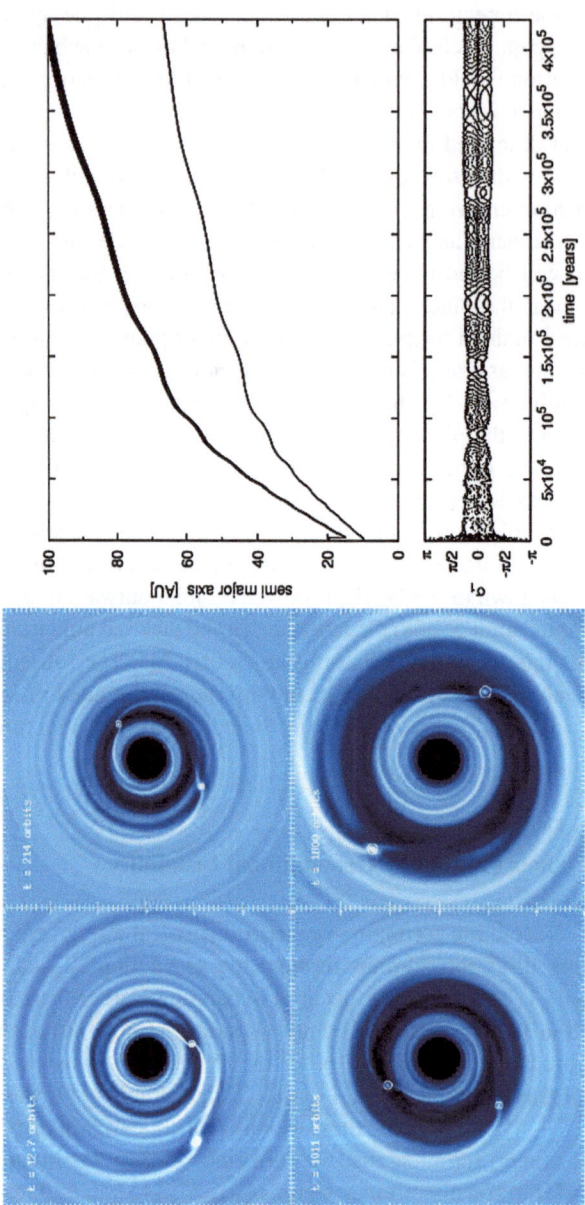

Fig. 6.17 Illustration of the joint outward migration of a pair of resonant massive planets. The *left panel* shows the evolution of the disc's surface density perturbed by a Jupiter-mass planet (inner planet) and a Saturn-mass planet (outer one). After an episode of rapid convergent migration (*top-left quadrant*) resulting in their capture into mean-motion resonance, planets open overlapping gaps (*top-right quadrant*), which leads to their joint outward migration (*lower quadrants*). The *right panels* illustrate the outcome of the same mechanism applied to an inner 3-Jupiter mass planet, and an outer 2-Jupiter-mass planet orbiting a $2M_\odot$ mass star (taken from [20]). The time evolution of the planets semi-major axis is shown in the *top-right panel*, and that of their $2:1$ critical resonance angle is in the *bottom-right panel*

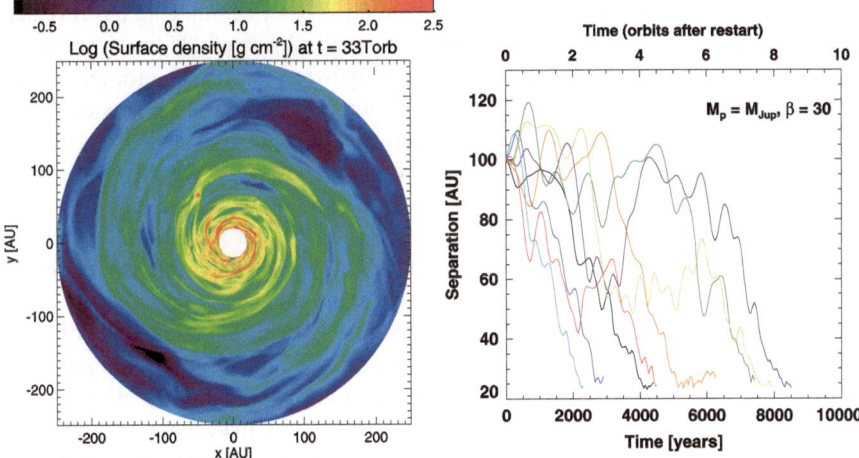

Fig. 6.18 Jupiter-mass planet embedded in a gravito-turbulent disc. After setting up a quasi steady-state gravito-turbulent disc with (gravito-turbulent) shock heating balancing disc cooling (parametrised here by a simple β-cooling function, see [11]), simulations were restarted with inserting a Jupiter-mass planet at 100 AU. The *left panel* shows the disc's surface density three orbits after restart, the planet being located at $x \sim -50$ AU, $y \sim 60$ AU. The *right panel* displays the time evolution of the planet's orbital separation in 8 different restart simulations with varying the azimuth of the planet prior to its insertion in the disc. Taken from [11]

Investigation is under way to determine the evolution of planets formed by GI when they reach the inner parts of protoplanetary discs. The latter should be too hot to be gravitationally unstable, and other sources of turbulence, such as the magnetorotational instability, could prevail, changing the background disc profiles as well as the amount of turbulence. It is thus possible that the rapid type I migration of planets formed by GI slows down in the disc inner parts and results in the formation of a gap. Gap-opening may also occur if significant gas accretion occurs during the planets fast inward migration [115]. Planet–planet interactions, which may result in scattering events, mergers or captures in resonance, should also play a prominent role in shaping planetary systems formed by GI.

6.5.2 Planet Population Syntheses

Planetary astrophysics is undergoing an epoch of explosive growth, driven by the observational discoveries of more than 750 exoplanets over the past two decades. Outstanding progress in detection techniques have uncovered planetary systems very different from ours. Since the discovery of the first Hot Jupiter [75], radial velocity surveys have made possible the detection of Earth-like planets, some in the habitable zone of their star [94]. Transit space missions CoRoT and KEPLER are digging out hundreds of close-in extrasolar planets, some in exotic environments (like Kepler-

16 b, the first circumbinary exoplanet discovered [27]). Direct imaging has revealed the existence of massive giant planets located at several tens of AU from their star.

Such diversity provides an exciting opportunity to test our theories for the formation and evolution of planetary systems. By coupling theoretical models of planet formation and migration, and of disc evolution, planet population syntheses estimate the statistical distribution of exoplanets according to their mass, semi-major axis, and eccentricity, which they compare to observed distributions [37, 39, 40, 78, 101]. At the moment, models of planet population syntheses are not able to reproduce the statistical properties of extrasolar planets. For instance, they predict a deficit of super-Earths and Neptune-like planets with orbital periods less than 50 days, while observations have revealed a significant number of exoplanets in this range of mass and period [38]. The origin for this discrepancy can be found in uncertain prescriptions for the minimum core mass for the onset of gas accretion [37], as well as in the modeling of type I migration. The difficulty raised by the excessively rapid inward type I migration, predicted by the long-time reference torque formula by [104], was circumvented by introducing a reduction factor in front of this torque formula. Population syntheses models tried to constrain this factor to reproduce the statistical properties of detected exoplanets. This reduction factor was found to range from 0.01 to 0.1. We note however that the introduction of this reduction factor to provide planetary population synthesis in agreement with the statistics of detected extrasolar planets is *ad hoc*, and that there is no reason to expect that the type I migration drift rates are systematically overestimated in theoretical studies by a factor 10 to 100. Rather, as we emphasised throughout this manuscript, type I migration is very sensitive to the disc's density and temperature profiles near the planet orbit. Large slopes of mass density and/or temperature, over a limited radial range, may reverse the tidal torque exerted on the planet. This, in turn, may create "planetary traps" at the points where the tidal torque cancels out (much like what was contemplated by [74] for the case of a positive surface density gradient), which can stop incoming protoplanets, depending on their mass. The number and location of these traps may vary as the disc evolves. This view of type I migrating objects subject to several traps on their way to the star [36, 62] sounds more compatible with the state of migration theories than an *ad hoc* reduction factor. This has motivated recent works to produce accurate, yet simple formulae for type I migration [68, 70, 92]. These formulae include a description of the corotation torque in discs with arbitrary viscosity and thermal diffusion, and corrections to the Lindblad torque for discs with non power-law profiles. Their incorporation into models of planet population synthesis will hopefully provide a better comprehension of the diversity of observed exoplanets.

6.6 Conclusions

We have reviewed the recent progress made in understanding planet–disc interactions, and the properties of planetary migration driven by such interactions. We have focused on the migration of growing protoplanets (type I migration), which has been the subject of intensive investigation over the past five years. Being for a while the second-place actor of planet migration theories, the corotation torque has been shown to play a prominent role in realistic protoplanetary discs, where it can slow down, stall, or reverse type I migration. This review is especially aimed at giving a comprehensive, detailed description of the mechanisms responsible for the corotation torque. The type II and type III migration regimes for gap-opening planets are also reviewed and discussed in the context of observed exoplanets. Being aimed at migration of planets on circular orbits, this review has set aside interesting recent developments on the tidal interactions of eccentric or inclined planets with their discs. We have also focused essentially on the mechanisms that drive the migration of a single planet in a disc, and we have therefore excluded most results about the migration of several planets. For a recent review covering these topics, the reader is referred to [46]. This list of restrictions of the present review stresses that the research on planet–disc interactions is a very active branch of planet formation, with a growing body of avenues. We finally reiterate the plea made in the introduction: planetary migration is not overrated. The tremendous value of each of the tidal torque components exerted on a given planet, associated with the great sensitivity of these torques to the underlying disc structure, appeals for a detailed knowledge of the properties of protoplanetary discs, and significant efforts toward an accurate determination of each torque component. This also reasserts tidal interactions as a prominent process in shaping planetary systems.

Acknowledgements   It is a pleasure to thank Jérôme Guilet, Sijme-Jan Paardekooper, and Stephen Thomson for their detailed reading of this manuscript, as well as John Papaloizou for stimulating discussions.

References

1. Artymowicz, P.: Disk-satellite interaction via density waves and the eccentricity evolution of bodies embedded in disks. Astrophys. J. **419**, 166 (1993). doi:10.1086/173470
2. Artymowicz, P.: On the wave excitation and a generalized torque formula for Lindblad resonances excited by external potential. Astrophys. J. **419**, 155 (1993). doi:10.1086/173469
3. Ayliffe, B.A., Bate, M.R.: Migration of protoplanets with surfaces through discs with steep temperature gradients. Mon. Not. R. Astron. Soc. **415**, 576–586 (2011). doi:10.1111/j.1365-2966.2011.18730.x
4. Balbus, S.A., Hawley, J.F.: A powerful local shear instability in weakly magnetized disks. I—Linear analysis. II—Nonlinear evolution. Astrophys. J. **376**, 214–233 (1991). doi:10.1086/170270
5. Balmforth, N.J., Korycansky, D.G.: Non-linear dynamics of the corotation torque. Mon. Not. R. Astron. Soc. **326**, 833–851 (2001). doi:10.1046/j.1365-8711.2001.04619.x
6. Baruteau, C.: Toward predictive scenarios of planetary migration. Ph.D. thesis. CEA Saclay, Service d'Astrophysique (2008)
7. Baruteau, C., Lin, D.N.C.: Protoplanetary migration in turbulent isothermal disks. Astrophys. J. **709**, 759–773 (2010). doi:10.1088/0004-637X/709/2/759
8. Baruteau, C., Masset, F.: On the corotation torque in a radiatively inefficient disk. Astrophys. J. **672**, 1054–1067 (2008). doi:10.1086/523667

9. Baruteau, C., Masset, F.: Type I planetary migration in a self-gravitating disk. Astrophys. J. **678**, 483–497 (2008). doi:10.1086/529487

10. Baruteau, C., Fromang, S., Nelson, R.P., Masset, F.: Corotation torques experienced by planets embedded in weakly magnetized turbulent discs. Astron. Astrophys. **533**, A84 (2011). doi:10.1051/0004-6361/201117227

11. Baruteau, C., Meru, F., Paardekooper, S.J.: Rapid inward migration of planets formed by gravitational instability. Mon. Not. R. Astron. Soc. **416**, 1971–1982 (2011). doi:10.1111/j.1365-2966.2011.19172.x

12. Bate, M.R., Lubow, S.H., Ogilvie, G.I., Miller, K.A.: Three-dimensional calculations of high- and low-mass planets embedded in protoplanetary discs. Mon. Not. R. Astron. Soc. **341**, 213–229 (2003). doi:10.1046/j.1365-8711.2003.06406.x

13. Bryden, G., Chen, X., Lin, D.N.C., Nelson, R.P., Papaloizou, J.C.B.: Tidally induced gap formation in protostellar disks: gap clearing and suppression of protoplanetary growth. Astrophys. J. **514**, 344–367 (1999). doi:10.1086/306917

14. Casoli, J., Masset, F.S.: On the horseshoe drag of a low-mass planet. I. Migration in isothermal disks. Astrophys. J. **703**, 845–856 (2009). doi:10.1088/0004-637X/703/1/845

15. Casoli, J., Masset, F.S.: (in prep.)

16. Crida, A., Morbidelli, A.: Cavity opening by a giant planet in a protoplanetary disc and effects on planetary migration. Mon. Not. R. Astron. Soc. **377**, 1324–1336 (2007). doi:10.1111/j.1365-2966.2007.11704.x

17. Crida, A., Morbidelli, A., Masset, F.: On the width and shape of gaps in protoplanetary disks. Icarus **181**, 587–604 (2006). doi:10.1016/j.icarus.2005.10.007

18. Crida, A., Sándor, Z., Kley, W.: Influence of an inner disc on the orbital evolution of massive planets migrating in resonance. Astron. Astrophys. **483**, 325–337 (2008). doi:10.1051/0004-6361:20079291

19. Crida, A., Baruteau, C., Kley, W., Masset, F.: The dynamical role of the circumplanetary disc in planetary migration. Astron. Astrophys. **502**, 679–693 (2009). doi:10.1051/0004-6361/200811608

20. Crida, A., Masset, F., Morbidelli, A.: Long range outward migration of giant planets, with application to Fomalhaut b. Astrophys. J. Lett. **705**, L148–L152 (2009). doi:10.1088/0004-637X/705/2/L148

21. D'Angelo, G., Lubow, S.H.: Evolution of migrating planets undergoing gas accretion. Astrophys. J. **685**, 560–583 (2008). doi:10.1086/590904

22. D'Angelo, G., Henning, T., Kley, W.: Nested-grid calculations of disk-planet interaction. Astron. Astrophys. **385**, 647–670 (2002)

23. D'Angelo, G., Henning, T., Kley, W.: Thermohydrodynamics of circumstellar disks with high-mass planets. Astrophys. J. **599**, 548–576 (2003)

24. D'Angelo, G., Kley, W., Henning, T.: Orbital migration and mass accretion of protoplanets in three-dimensional global computations with nested grids. Astrophys. J. **586**, 540–561 (2003)

25. D'Angelo, G., Bate, M.R., Lubow, S.H.: The dependence of protoplanet migration rates on co-orbital torques. Mon. Not. R. Astron. Soc. **358**, 316–332 (2005). doi:10.1111/j.1365-2966.2005.08866.x

26. Dong, R., Rafikov, R.R., Stone, J.M.: Density waves excited by low-mass planets in protoplanetary disks. II. High-resolution simulations of the nonlinear regime. Astrophys. J. **741**, 57 (2011). doi:10.1088/0004-637X/741/1/57

27. Doyle, L.R., Carter, J.A., Fabrycky, D.C., Slawson, R.W., Howell, S.B., Winn, J.N., Orosz, J.A., Prsa, A., Welsh, W.F., Quinn, S.N., Latham, D., Torres, G., Buchhave, L.A., Marcy, G.W., Fortney, J.J., Shporer, A., Ford, E.B., Lissauer, J.J., Ragozzine, D., Rucker, M., Batalha, N., Jenkins, J.M., Borucki, W.J., Koch, D., Middour, C.K., Hall, J.R., McCauliff, S., Fanelli, M.N., Quintana, E.V., Holman, M.J., Caldwell, D.A., Still, M., Stefanik, R.P., Brown, W.R., Esquerdo, G.A., Tang, S., Furesz, G., Geary, J.C., Berlind, P., Calkins, M.L., Short, D.R., Steffen, J.H., Sasselov, D., Dunham, E.W., Cochran, W.D., Boss, A., Haas, M.R., Buzasi, D., Fischer, D.: Kepler-16: a transiting circumbinary planet. Science **333**, 1602 (2011). doi:10.1126/science.1210923

28. Fleming, T., Stone, J.M.: Local magnetohydrodynamic models of layered accretion disks. Astrophys. J. **585**, 908–920 (2003). doi:10.1086/345848

29. Fromang, S., Terquem, C., Nelson, R.P.: Numerical simulations of type I planetary migration in non-turbulent magnetized discs. Mon. Not. R. Astron. Soc. **363**, 943–953 (2005). doi:10.1111/j.1365-2966.2005.09498.x

30. Gammie, C.F.: Nonlinear outcome of gravitational instability in cooling, gaseous disks. Astrophys. J. **553**, 174–183 (2001). doi:10.1086/320631

31. Gautier, T.N. III, Charbonneau, D., Rowe, J.F., Marcy, G.W., Isaacson, H., Torres, G., Fressin, F., Rogers, L.A., Désert, J.M., Buchhave, L.A., Latham, D.W., Quinn, S.N., Ciardi, D.R., Fabrycky, D.C., Ford, E.B., Gilliland, R.L., Walkowicz, L.M., Bryson, S.T., Cochran, W.D., Endl, M., Fischer, D.A., Howel, S.B., Horch, E.P., Barclay, T., Batalha, N., Borucki, W.J., Christiansen, J.L., Geary, J.C., Henze, C.E., Holman, M.J., Ibrahim, K., Jenkins, J.M., Kinemuchi, K., Koch, D.G., Lissauer, J.J., Sanderfer, D.T., Sasselov, D.D., Seager, S., Silverio, K., Smith, J.C., Still, M., Stumpe, M.C., Tenenbaum, P., Van Cleve, J.: Kepler-20: a Sun-like star with three sub-Neptune exoplanets and two Earth-size candidates. ArXiv e-prints (2011)

32. Goldreich, P., Tremaine, S.: The excitation of density waves at the Lindblad and corotation resonances by an external potential. Astrophys. J. **233**, 857–871 (1979). doi:10.1086/157448

33. Goldreich, P., Tremaine, S.: Disk-satellite interactions. Astrophys. J. **241**, 425–441 (1980). doi:10.1086/158356

34. Goodman, J., Rafikov, R.R.: Planetary torques as the viscosity of protoplanetary disks. Astrophys. J. **552**, 793–802 (2001)

35. Guilet, J., Baruteau, C., Papaloizou, J.C.B.: Type I planet migration in weakly magnetised laminar discs (submitted)

36. Hasegawa, Y., Pudritz, R.E.: Dead zones as thermal barriers to rapid planetary migration in protoplanetary disks. Astrophys. J. Lett. **710**, L167–L171 (2010). doi:10.1088/2041-8205/710/2/L167

37. Hellary, P., Nelson, R.P.: Global models of planetary system formation in radiatively-inefficient protoplanetary discs. Mon. Not. R. Astron. Soc. **419**, 2737–2757 (2012). doi:10.1111/j.1365-2966.2011.19815.x

38. Howard, A.W., Marcy, G.W., Johnson, J.A., Fischer, D.A., Wright, J.T., Isaacson, H., Valenti, J.A., Anderson, J., Lin, D.N.C., Ida, S.: The occurrence and mass distribution of close-in super-earths, Neptunes, and Jupiters. Science **330**, 653 (2010). doi:10.1126/science.1194854

39. Ida, S., Lin, D.N.C.: Toward a deterministic model of planetary formation. I. A desert in the mass and semimajor axis distributions of extrasolar planets. Astrophys. J. **604**, 388–413 (2004). doi:10.1086/381724

40. Ida, S., Lin, D.N.C.: Toward a deterministic model of planetary formation. IV. Effects of type I migration. Astrophys. J. **673**, 487–501 (2008). doi:10.1086/523754

41. Ivanov, P.B., Papaloizou, J.C.B., Polnarev, A.G.: The evolution of a supermassive binary caused by an accretion disc. Mon. Not. R. Astron. Soc. **307**, 79–90 (1999). doi:10.1046/j.1365-8711.1999.02623.x

42. Jang-Condell, H.: Planet shadows in protoplanetary disks. I. Temperature perturbations. Astrophys. J. **679**, 797–812 (2008). doi:10.1086/533583

43. Johansen, A., Henning, T., Klahr, H.: Dust sedimentation and self-sustained Kelvin-Helmholtz turbulence in protoplanetary disk midplanes. Astrophys. J. **643**, 1219–1232 (2006). doi:10.1086/502968

44. Klahr, H.H., Bodenheimer, P.: Turbulence in accretion disks: vorticity generation and angular momentum transport via the global baroclinic instability. Astrophys. J. **582**, 869–892 (2003)

45. Kley, W., Crida, A.: Migration of protoplanets in radiative discs. Astron. Astrophys. **487**, L9–L12 (2008). doi:10.1051/0004-6361:200810033

46. Kley, W., Nelson, R.P.: Planet-disk interaction and orbital evolution. ArXiv e-prints (2012)

47. Kley, W., Bitsch, B., Klahr, H.: Planet migration in three-dimensional radiative discs. Astron. Astrophys. **506**, 971–987 (2009). doi:10.1051/0004-6361/200912072

48. Korycansky, D.G., Pollack, J.B.: Numerical calculations of the linear response of a gaseous disk to a protoplanet. Icarus **102**, 150–165 (1993). doi:10.1006/icar.1993.1039
49. Lagrange, A., Bonnefoy, M., Chauvin, G., Apai, D., Ehrenreich, D., Boccaletti, A., Gratadour, D., Rouan, D., Mouillet, D., Lacour, S., Kasper, M.: A giant planet imaged in the disk of the young star β pictoris. Science **329**, 57 (2010). doi:10.1126/science.1187187
50. Laughlin, G., Steinacker, A., Adams, F.C.: Type I planetary migration with MHD turbulence. Astrophys. J. **608**, 489–496 (2004). doi:10.1086/386316
51. Lesur, G., Papaloizou, J.C.B.: The subcritical baroclinic instability in local accretion disc models. Astron. Astrophys. **513**, A60 (2010). doi:10.1051/0004-6361/200913594
52. Li, H., Finn, J.M., Lovelace, R.V.E., Colgate, S.A.: Rossby wave instability of thin accretion disks. II. Detailed linear theory. Astrophys. J. **533**, 1023–1034 (2000). doi:10.1086/308693
53. Lin, D.N.C., Papaloizou, J.: Tidal torques on accretion discs in binary systems with extreme mass ratios. Mon. Not. R. Astron. Soc. **186**, 799–812 (1979)
54. Lin, D.N.C., Papaloizou, J.: On the tidal interaction between protoplanets and the protoplanetary disk. III—Orbital migration of protoplanets. Astrophys. J. **309**, 846–857 (1986). doi:10.1086/164653
55. Lin, D.N.C., Papaloizou, J.C.B.: On the tidal interaction between protostellar disks and companions. In: Levy, E.H., Lunine, J.I. (eds.) Protostars and Planets III, pp. 749–835 (1993)
56. Lin, M.K., Papaloizou, J.C.B.: Type III migration in a low-viscosity disc. Mon. Not. R. Astron. Soc. **405**, 1473–1490 (2010). doi:10.1111/j.1365-2966.2010.16560.x
57. Lin, M.K., Papaloizou, J.C.B.: Edge modes in self-gravitating disc-planet interactions. Mon. Not. R. Astron. Soc. **415**, 1445–1468 (2011). doi:10.1111/j.1365-2966.2011.18797.x
58. Lin, M.K., Papaloizou, J.C.B.: The effect of self-gravity on vortex instabilities in disc-planet interactions. Mon. Not. R. Astron. Soc. **415**, 1426–1444 (2011). doi:10.1111/j.1365-2966.2011.18798.x
59. Lin, M.K., Papaloizou, J.C.B.: Outward migration of a giant planet with a gravitationally unstable gap edge. Mon. Not. R. Astron. Soc. **421**, 780–788 (2012). doi:10.1111/j.1365-2966.2011.20352.x
60. Lovelace, R.V.E., Li, H., Colgate, S.A., Nelson, A.F.: Rossby wave instability of Keplerian accretion disks. Astrophys. J. **513**, 805–810 (1999). doi:10.1086/306900
61. Lyra, W., Klahr, H.: The baroclinic instability in the context of layered accretion. Self-sustained vortices and their magnetic stability in local compressible unstratified models of protoplanetary disks. Astron. Astrophys. **527**, A138 (2011). doi:10.1051/0004-6361/201015568
62. Lyra, W., Paardekooper, S.J., MacLow, M.M.: Orbital migration of low-mass planets in evolutionary radiative models: avoiding catastrophic infall. Astrophys. J. Lett. **715**, L68–L73 (2010). doi:10.1088/2041-8205/715/2/L68
63. Marois, C., Macintosh, B., Barman, T., Zuckerman, B., Song, I., Patience, J., Lafrenière, D., Doyon, R.: Direct imaging of multiple planets orbiting the star HR 8799. Science **322**, 1348 (2008). doi:10.1126/science.1166585
64. Marois, C., Zuckerman, B., Konopacky, Q.M., Macintosh, B., Barman, T.: Images of a fourth planet orbiting HR 8799. Nature **468**, 1080–1083 (2010). doi:10.1038/nature09684
65. Masset, F.S.: On the co-orbital corotation torque in a viscous disk and its impact on planetary migration. Astrophys. J. **558**, 453–462 (2001)
66. Masset, F.S.: The co-orbital corotation torque in a viscous disk: numerical simulations. Astron. Astrophys. **387**, 605–623 (2002)
67. Masset, F.S.: Planet-disk interactions. EAS Publ. Ser. **29**, 165–244 (2008). doi:10.1051/eas:0829006
68. Masset, F.S.: On type-I migration near opacity transitions. A generalized Lindblad torque formula for planetary population synthesis. Celest. Mech. Dyn. Astron. **111**, 131–160 (2011). doi:10.1007/s10569-011-9364-0
69. Masset, F.S., Casoli, J.: On the horseshoe drag of a low-mass planet. II. Migration in adiabatic disks. Astrophys. J. **703**, 857–876 (2009). doi:10.1088/0004-637X/703/1/857

70. Masset, F.S., Casoli, J.: Saturated torque formula for planetary migration in viscous disks with thermal diffusion: recipe for protoplanet population synthesis. Astrophys. J. **723**, 1393–1417 (2010). doi:10.1088/0004-637X/723/2/1393

71. Masset, F.S., Papaloizou, J.C.B.: Runaway migration and the formation of hot Jupiters. Astrophys. J. **588**, 494–508 (2003)

72. Masset, F., Snellgrove, M.: Reversing type II migration: resonance trapping of a lighter giant protoplanet. Mon. Not. R. Astron. Soc. **320**, L55 (2001)

73. Masset, F.S., D'Angelo, G., Kley, W.: On the migration of protogiant solid cores. Astrophys. J. **652**, 730–745 (2006). doi:10.1086/507515

74. Masset, F.S., Morbidelli, A., Crida, A., Ferreira, J.: Disk surface density transitions as protoplanet traps. Astrophys. J. **642**, 478–487 (2006). doi:10.1086/500967

75. Mayor, M., Queloz, D.: A Jupiter-mass companion to a Solar-type star. Nature **378**, 355 (1995). doi:10.1038/378355a0

76. Menou, K., Goodman, J.: Low-mass protoplanet migration in T Tauri α-disks. Astrophys. J. **606**, 520–531 (2004). doi:10.1086/382947

77. Michael, S., Durisen, R.H., Boley, A.C.: Migration of gas giant planets in gravitationally unstable disks. Astrophys. J. Lett. **737**, L42 (2011). doi:10.1088/2041-8205/737/2/L42

78. Mordasini, C., Alibert, Y., Benz, W., Naef, D.: Extrasolar planet population synthesis. II. Statistical comparison with observations. Astron. Astrophys. **501**, 1161–1184 (2009). doi:10.1051/0004-6361/200810697

79. Morohoshi, K., Tanaka, H.: Gravitational interaction between a planet and an optically thin disc. Mon. Not. R. Astron. Soc. **346**, 915–923 (2003). doi:10.1111/j.1365-2966.2003.07140.x

80. Murray, C.D., Dermott, S.F.: Solar System Dynamics. Solar System Dynamics. Cambridge University Press, Cambridge (2000). ISBN 0521575974

81. Muto, T., Machida, M.N., Inutsuka, S.i.: The effect of poloidal magnetic field on type I planetary migration: significance of magnetic resonance. Astrophys. J. **679**, 813–826 (2008). doi:10.1086/587027

82. Nelson, R.P.: On the orbital evolution of low mass protoplanets in turbulent, magnetised disks. Astron. Astrophys. **443**, 1067–1085 (2005). doi:10.1051/0004-6361:20042605

83. Nelson, R.P., Papaloizou, J.C.B.: The interaction of a giant planet with a disc with MHD turbulence—II. The interaction of the planet with the disc. Mon. Not. R. Astron. Soc. **339**, 993–1005 (2003). doi:10.1046/j.1365-8711.2003.06247.x

84. Nelson, R.P., Papaloizou, J.C.B.: The interaction of giant planets with a disc with MHD turbulence—IV. Migration rates of embedded protoplanets. Mon. Not. R. Astron. Soc. **350**, 849–864 (2004). doi:10.1111/j.1365-2966.2004.07406.x

85. Ogilvie, G.I., Lubow, S.H.: On the wake generated by a planet in a disc. Mon. Not. R. Astron. Soc. **330**, 950–954 (2002). doi:10.1046/j.1365-8711.2002.05148.x

86. Paardekooper, S.J., Mellema, G.: Halting type I planet migration in non-isothermal disks. Astron. Astrophys. **459**, L17–L20 (2006). doi:10.1051/0004-6361:20066304

87. Paardekooper, S.J., Mellema, G.: Growing and moving low-mass planets in non-isothermal disks. Astron. Astrophys. **478**, 245–266 (2008). doi:10.1051/0004-6361:20078592

88. Paardekooper, S.J., Papaloizou, J.C.B.: On disc protoplanet interactions in a non-barotropic disc with thermal diffusion. Astron. Astrophys. **485**, 877–895 (2008). doi:10.1051/0004-6361:20078702

89. Paardekooper, S.J., Papaloizou, J.C.B.: On corotation torques, horseshoe drag and the possibility of sustained stalled or outward protoplanetary migration. Mon. Not. R. Astron. Soc. **394**, 2283–2296 (2009). doi:10.1111/j.1365-2966.2009.14511.x

90. Paardekooper, S.J., Papaloizou, J.C.B.: On the width and shape of the corotation region for low-mass planets. Mon. Not. R. Astron. Soc. **394**, 2297–2309 (2009). doi:10.1111/j.1365-2966.2009.14512.x

91. Paardekooper, S., Baruteau, C., Crida, A., Kley, W.: A torque formula for non-isothermal type I planetary migration—I. Unsaturated horseshoe drag. Mon. Not. R. Astron. Soc. **401**, 1950–1964 (2010). doi:10.1111/j.1365-2966.2009.15782.x

92. Paardekooper, S., Baruteau, C., Kley, W.: A torque formula for non-isothermal type I planetary migration—II. Effects of diffusion. Mon. Not. R. Astron. Soc. **410**, 293–303 (2011). doi:10.1111/j.1365-2966.2010.17442.x
93. Papaloizou, J.C.B., Nelson, R.P., Snellgrove, M.D.: The interaction of giant planets with a disc with MHD turbulence—III. Flow morphology and conditions for gap formation in local and global simulations. Mon. Not. R. Astron. Soc. **350**, 829–848 (2004). doi:10.1111/j.1365-2966.2004.07566.x
94. Pepe, F., Lovis, C., Ségransan, D., Benz, W., Bouchy, F., Dumusque, X., Mayor, M., Queloz, D., Santos, N.C., Udry, S.: The HARPS search for Earth-like planets in the habitable zone. I. Very low-mass planets around HD 20794, HD 85512, and HD 192310. Astron. Astrophys. **534**, A58 (2011). doi:10.1051/0004-6361/201117055
95. Pepliński, A., Artymowicz, P., Mellema, G.: Numerical simulations of type III planetary migration—I. Disc model and convergence tests. Mon. Not. R. Astron. Soc. **386**, 164–178 (2008). doi:10.1111/j.1365-2966.2008.13045.x
96. Pepliński, A., Artymowicz, P., Mellema, G.: Numerical simulations of type III planetary migration—II. Inward migration of massive planets. Mon. Not. R. Astron. Soc. **386**, 179–198 (2008). doi:10.1111/j.1365-2966.2008.13046.x
97. Pepliński, A., Artymowicz, P., Mellema, G.: Numerical simulations of type III planetary migration—III. Outward migration of massive planets. Mon. Not. R. Astron. Soc. **387**, 1063–1079 (2008). doi:10.1111/j.1365-2966.2008.13339.x
98. Pierens, A., Huré, J.M.: How does disk gravity really influence type-I migration? Astron. Astrophys. **433**, L37–L40 (2005). doi:10.1051/0004-6361:200500099
99. Pollack, J.B., Hubickyj, O., Bodenheimer, P., Lissauer, J.J., Podolak, M., Greenzweig, Y.: Formation of the giant planets by concurrent accretion of solids and gas. Icarus **124**, 62–85 (1996). doi:10.1006/icar.1996.0190
100. Rafikov, R.R.: Can giant planets form by direct gravitational instability? Astrophys. J. Lett. **621**, L69–L72 (2005). doi:10.1086/428899
101. Schlaufman, K.C., Lin, D.N.C., Ida, S.: The signature of the ice line and modest type I migration in the observed exoplanet mass-semimajor axis distribution. Astrophys. J. **691**, 1322–1327 (2009). doi:10.1088/0004-637X/691/2/1322
102. Shakura, N.I., Sunyaev, R.A.: Black holes in binary systems. Observational appearance. Astron. Astrophys. **24**, 337–355 (1973)
103. Syer, D., Clarke, C.J.: Satellites in discs: regulating the accretion luminosity. Mon. Not. R. Astron. Soc. **277**, 758–766 (1995)
104. Tanaka, H., Takeuchi, T., Ward, W.R.: Three-dimensional interaction between a planet and an isothermal gaseous disk. I. Corotation and Lindblad torques and planet migration. Astrophys. J. **565**, 1257–1274 (2002)
105. Terquem, C.E.J.M.L.J.: Stopping inward planetary migration by a toroidal magnetic field. Mon. Not. R. Astron. Soc. **341**, 1157–1173 (2003). doi:10.1046/j.1365-8711.2003.06455.x
106. Uribe, A.L., Klahr, H., Flock, M., Henning, T.: Three-dimensional magnetohydrodynamic simulations of planet migration in turbulent stratified disks. Astrophys. J. **736**, 85 (2011). doi:10.1088/0004-637X/736/2/85
107. Walsh, K.J., Morbidelli, A., Raymond, S.N., O'Brien, D.P., Mandell, A.M.: A low mass for mars from Jupiter's early gas-driven migration. Nature **475**, 206–209 (2011). doi:10.1038/nature10201
108. Ward, W.R.: Density waves in the solar nebula—differential Lindblad torque. Icarus **67**, 164–180 (1986). doi:10.1016/0019-1035(86)90182-X
109. Ward, W.R.: Horseshoe orbit drag. In: Lunar and Planetary Institute Conference Abstracts, pp. 1463 (1991)
110. Ward, W.R.: Protoplanet migration by nebula tides. Icarus **126**, 261–281 (1997)
111. Ward, W.R.: A streamline model of horseshoe torque saturation. In: Lunar and Planetary Institute Science Conference Abstracts, vol. 38, p. 2289 (2007)
112. Winters, W.F., Balbus, S.A., Hawley, J.F.: Gap formation by planets in turbulent protostellar disks. Astrophys. J. **589**, 543–555 (2003). doi:10.1086/374409

113. Zhang, H., Lai, D.: Wave excitation in three-dimensional discs by external potential. Mon. Not. R. Astron. Soc. **368**, 917–934 (2006). doi:10.1111/j.1365-2966.2006.10167.x
114. Zhang, H., Yuan, C., Lin, D.N.C., Yen, D.C.C.: On the orbital evolution of a Jovian planet embedded in a self-gravitating disk. Astrophys. J. **676**, 639–650 (2008). doi:10.1086/528707
115. Zhu, Z., Hartmann, L., Nelson, R.P., Gammie, C.F.: Challenges in forming planets by gravitational instability: disk irradiation and clump migration, accretion, and tidal destruction. Astrophys. J. **746**, 110 (2012). doi:10.1088/0004-637X/746/1/110

Chapter 7
Tides in Planetary Systems

Stéphane Mathis, Christophe Le Poncin-Lafitte, and Françoise Remus

Abstract The Solar system is the seat of many interactions between the Sun, the planets and their natural satellites. Moreover, since 1995, a large number of extra-solar planetary systems has been discovered where planets orbit around other stars, sometimes very close to them. Therefore, in such systems, tidal interactions are one of the key mechanisms that must be studied to understand the celestial bodies' dynamics and evolution. Indeed, tides generate displacements and flows in planetary (and in the host star) interiors. The associated kinetic energy is then dissipated into heat because of internal friction processes. This leads to secular evolution of orbits and of spins with characteristic time-scales that are intrinsically related to the properties of dissipative mechanisms, those latters depending both on the internal structure of the studied bodies and on the tidal frequency. This lecture is aimed to review the must advanced theories to study tidal dynamics in planetary systems and the different tidal flows or displacements that can be excited by a perturber,

S. Mathis (✉) · F. Remus
Laboratoire AIM Paris-Saclay, CEA/DSM-CNRS-Université Paris Diderot, CEA,
91191 Gif-sur-Yvette, France
e-mail: stephane.mathis@cea.fr

S. Mathis
LESIA, Observatoire de Paris-CNRS-Université Paris Diderot-Université Pierre et Marie Curie,
Observatoire de Paris, 5 place Jules Janssen, 92195 Meudon, France

C. Le Poncin-Lafitte
Syrte, Observatoire de Paris, CNRS UMR8630, Université Pierre et Marie Curie,
Observatoire de Paris, 77 avenue Denfert-Rochereau, 75014 Paris, France
e-mail: christophe.leponcin@obspm.fr

F. Remus
LUTH, Observatoire de Paris-CNRS-Université Paris Diderot, Observatoire de Paris,
5 place Jules Janssen, 92195 Meudon Cedex, France
e-mail: francoise.remus@obspm.fr

F. Remus
IMCCE, Observatoire de Paris-UMR 8028 du CNRS-Université Pierre et Marie Curie,
Observatoire de Paris, 77 avenue Denfert-Rochereau, 75014 Paris, France

J. Souchay et al. (eds.), *Tides in Astronomy and Astrophysics*,
Lecture Notes in Physics 861, DOI 10.1007/978-3-642-32961-6_7,
© Springer-Verlag Berlin Heidelberg 2013

the conversion of their kinetic energy into heat, the related exchanges of angular momentum, and the consequences for systems evolution.

7.1 Introduction

The Solar system is the seat of many interactions between the Sun, the planets and their natural satellites. Moreover, since 1995, a large number of extrasolar planets has been discovered and characterised (see [105]). In such planetary systems, bodies can orbit close to the others and thus tidal interactions are one of the key physical processes that must be studied to understand orbital and rotational evolution. This evolution is crucial for the habitability question, whether systems could host the development of life; determining factors are the presence of liquid water and of a protective magnetosphere which are closely linked to the values of the planets' orbital elements and rotation rate.

To answer to such important planetary problematics, we have thus to study in details the action of tides. Then, once a given two-body or multiple system is formed, its fate is determined by the initial conditions and the mass ratio between its components. Through tidal interaction between each of them, the system evolves either to a stable state of minimum energy (where all spins are aligned, the orbits are circular and the rotation of each body is synchronised with the orbital motion) or the companion tends to spiral into the parent body. Indeed, by converting kinetic energy into heat through internal friction, tidal interactions modify the orbital and rotational dynamics of the components of the considered system and their internal structure through internal heating [1, 7, 27, 28, 34, 53, 54, 71, 73]. This mechanism depends sensitively on the internal structure (rocky, icy or fluid) and dynamics of the perturbed body (asynchronism, orbital eccentricity and inclination, obliquity).

In this context, the main goal of this lecture is to give a detailed review on tidal interactions modeling in planetary (and in the host star) interiors. First, in Sect. 7.2, we summary the most general derivation of the tidal potential and the related evolution equations for orbital elements and spins where dissipative processes are introduced. Then, in Sect. 7.3, we show how the properties of the tidal dissipation are strongly related to bodies' internal structure. We thus review such dissipative processes first in fluid regions, then in solid or icy ones, and finally at their interfaces. Next, in Sect. 7.4, we conclude.

7.2 Advanced Tidal Dynamics

In celestial mechanics, one of the main approximation done in the modeling of tidal effects (star-innermost planet or planet-natural satellites interactions) is to consider the tidal perturber as a point mass body. However, a large number of extrasolar planets orbiting very close to their parent stars have been discovered during the past

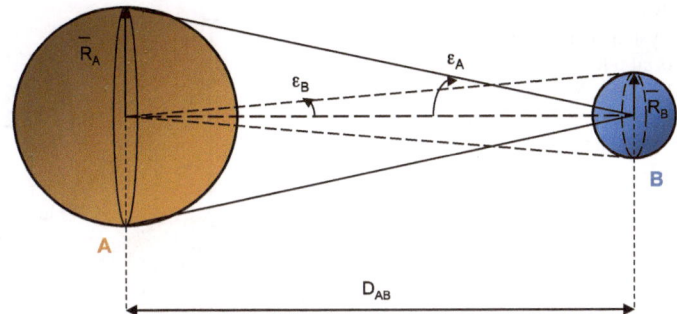

Fig. 7.1 System of two extended bodies A & B. $\overline{R}_A$ and $\overline{R}_B$ are the respective mean radius of A and B while D_{AB} is the distance between their respective center of mass. ε_A and ε_B are the conical angles with which each perturber is seen from the center of mass of the perturbed body. When $\frac{\overline{R}_i}{D_{AB}} \ll 1$ with $i = A$ or B the i body could be considered as a ponctual mass perturber. (Taken from Mathis and Le Poncin-Lafitte [77]; courtesy Astronomy & Astrophysics)

decade (see [105]). Moreover, in the Solar System, Phobos around Mars and the inner natural satellites of Jupiter, Saturn, Uranus and Neptune are very close to their parent planets. In such cases, the ratio of the perturber mean radius to the distance between the center of mass of the bodies can be not any more negligible compared to unity (cf. Fig. 7.1). In that situation, neglecting the extended character of the perturber may to be relaxed, so the tidal interaction between two extended bodies must be solved in a self-consistent way with taking into account the full gravitational potential of the extended perturber, generally expressed with some mass multipole moments, and then to consider their interaction with the tidally perturbed body. In the literature, not so many studies have been done [8–11, 44, 49, 72] and this is the reason why we here first choose to introduce one of the most general formalism to treat tidal dynamics, which is based on the results obtained by [44] and [77].

7.2.1 Gravitational Potentials

7.2.1.1 Multipole Expansion of the External Gravitational Field of an Extended Body

First, let us consider some matter distribution, corresponding to a body A, in an inertial reference frame.

The Newtonian gravitational potential of this body, $V^A(t, \mathbf{x})$ (where t is the classical time and $\mathbf{x}$ the current position vector), is obtained by solving the Poisson equation

$$\nabla^2 V^A(t, \mathbf{x}) = -4\pi G \rho_A(t, \mathbf{x}) \quad \text{with} \quad \lim_{|\mathbf{x}| \to \infty} V^A(t, \mathbf{x}) = 0, \tag{7.1}$$

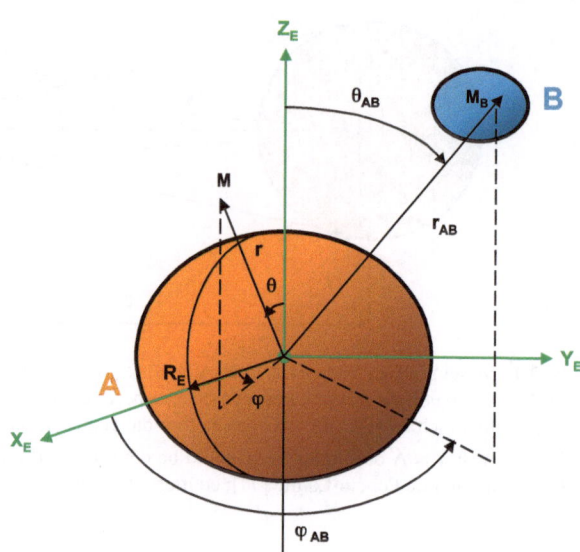

Fig. 7.2 Spherical coordinates system associated to the equatorial reference frame $\mathscr{R}_E : \{O_A, X_E, Y_E, Z_E\}$ of an extended body A; we have $\mathbf{r} \equiv (r, \theta, \varphi)$ and $\mathbf{r}_{AB} \equiv (r_{AB}, \theta_{AB}, \varphi_{AB})$ where r_{AB}, θ_{AB} and φ_{AB} are the coordinates of the center of mass of the potential extended perturber B. R_A is the equatorial radius of A. (Taken from Mathis and Le Poncin-Lafitte [77]; courtesy Astronomy & Astrophysics)

$\rho_A(t, \mathbf{x})$ being its density. This leads to the expression for $|\mathbf{x}| \geq R_A$

$$V^A(t, \mathbf{x}) = G \int_A \frac{\rho_A(t, \mathbf{x}')}{|\mathbf{x} - \mathbf{x}'|} \, d^3\mathbf{x}', \tag{7.2}$$

where R_A is the equatorial radius of A. Following [44] and [77], the final multipolar expression for V^A for $|\mathbf{r}| \geq R_A$ is then obtained:

$$V^A(t, \mathbf{r}) = G \sum_{l_A=0}^{\infty} \sum_{m_A=-l_A}^{l_A} M_{l_A, m_A} \frac{Y_{l_A, m_A}(\theta, \varphi)}{r^{l_A+1}}, \tag{7.3}$$

where we introduced the gravitational moments in the physical space

$$M_{l_A, m_A} = \frac{4\pi}{2l_A + 1} \int_{M_A} r^{l_A} Y^*_{l_A, m_A}(\theta, \varphi) \, dM_A, \tag{7.4}$$

with M_A is the mass of A and $dM_A = \rho_A r^2 \, dr \sin\theta \, d\theta \, d\varphi$, and the usual spherical harmonics

$$Y_{l,m}(\theta, \varphi) = \mathscr{N}_{l,m} P_l^{|m|}(\cos\theta) \exp[im\varphi]$$

$$\text{with } \mathscr{N}_{l,m} = (-1)^{\frac{m+|m|}{2}} \sqrt{\frac{2l+1}{4\pi} \frac{(l-|m|)!}{(l+|m|)!}} \tag{7.5}$$

with the classical Legendre Polynomials (P_l^m). The spherical coordinates (r, θ, φ) have been introduced (see Fig. 7.2) and (l, m) are the usual quantum numbers. Finally, z^* is the complex conjugate of a given complex number z.

One should note the symmetry property of M_{l_A, m_A}:

$$M_{l_A, -m_A} = (-1)^{m_A} M^*_{l_A, m_A}. \tag{7.6}$$

Moreover, M_{l_A,m_A} could be represented in its polar form

$$M_{l_A,m_A} = |M_{l_A,m_A}| \exp[i\delta M_{l_A,m_A}], \tag{7.7}$$

where the following identities are obtained from Eq. (7.6)

$$\begin{cases} |M_{l_A,-m_A}| = |M_{l_A,m_A}|, \\ \text{Arg}(M_{l_A,-m_A}) = m_A\pi - \text{Arg}(M_{l_A,m_A}) = m_A\pi - \delta M_{l_A,m_A}. \end{cases} \tag{7.8}$$

Using the classical symmetry property concerning spherical harmonics given in Eq. (7.5), $V^A(t, \mathbf{r})$ could also be expressed with the associated Legendre polynomials:

$$V^A(t, \mathbf{r}) = G \sum_{l_A=0}^{\infty} \sum_{m_A=0}^{l_A} \frac{P_{l_A}^{m_A}(\cos\theta)}{r^{l_A+1}} \left[C_{l_A,m_A} \cos(m_A\varphi) + S_{l_A,m_A} \sin(m_A\varphi) \right], \tag{7.9}$$

where the usual coefficients C_{l_A,m_A} and S_{l_A,m_A} are given by:

$$\begin{cases} C_{l_A,m_A} = \mathcal{N}_{l_A,m_A} (2 - \delta_{m_A,0}) \text{Re}(M_{l_A,m_A}), \\ S_{l_A,m_A} = -2\mathcal{N}_{l_A,m_A} (1 - \delta_{m_A,0}) \text{Im}(M_{l_A,m_A}). \end{cases} \tag{7.10}$$

The expression of M_{l_A,m_A} and $\delta M_{l_A,m_A}$ are then deduced for $m_A \geq 0$:

$$|M_{l_A,m_A}| = \frac{1}{\mathcal{N}_{l_A,m_A}} \sqrt{ \left[\frac{C_{l_A,m_A}}{(2 - \delta_{m_A,0})} \right]^2 + \left(\frac{S_{l_A,m_A}}{2} \right)^2 (1 - \delta_{m_A,0})^2 }, \tag{7.11}$$

$$\delta M_{l_A,m_A} = -\text{Arctan}\left[\frac{(1 - \delta_{m_A,0})(2 - \delta_{m_A,0})}{2} \frac{S_{l_A,m_A}}{C_{l_A,m_A}} \right]. \tag{7.12}$$

In the general case, the gravitational moments are expanded as:

$$M_{l_A,m_A} = M_{l_A,m_A}^{S_A} + M_{l_A,m_A}^{T_A}. \tag{7.13}$$

$M_{l_A,m_A}^{S_A}$ and $M_{l_A,m_A}^{T_A}$ are respectively those in the case where A is isolated (without any perturber) and those induced by the tidal perturber(s).

One can identify some special values of M_{l_A,m_A} relevant for the gravitational field of a body A. The trivial one is its mass, M_A

$$M_{0,0} = \sqrt{4\pi} M_A. \tag{7.14}$$

Furthermore, we know that the external field of an axisymmetric body A can be expressed as a function of the usual multipole moment J_{l_A}[1] (see e.g. [104])

$$V^A(t, \mathbf{r}) = \frac{G M_A}{r} \left[1 - \sum_{l_A > 0} J_{l_A} \left(\frac{R_A}{r} \right)^{l_A} P_{l_A}(\cos\theta) \right]; \tag{7.15}$$

---

[1] They are driven by two types of deformation. The first one is those induced by internal dynamical processes such that rotation (through the centrifugal acceleration) and magnetic field (through the volumetric Lorentz force). The second one is the axisymmetric permanent tidal oval shape due to a companion in close binary or multiple systems.

using Eq. (7.3), we identify in a straigthforward way:

$$V^A(t, \mathbf{r}) = G \left[\frac{M_A}{r} + \sum_{l_A=0}^{\infty} M^A_{J_{l_A};l_A,0} \frac{Y_{l_A,0}(\theta, \varphi)}{r^{l_A+1}} \right] \qquad (7.16)$$

where

$$M^A_{J_{l_A};l_A,0} = M^{S_A}_{l_A,0} + M^{T_A}_{l_A,0} = -\frac{J_{l_A} M_A R^{l_A}_A}{\mathcal{M}_{l_A,0}}. \qquad (7.17)$$

We can now focus on the second type of gravitational interaction, namely the tides between two extended bodies.

7.2.1.2 Determination of the Tidal Potential

Let us now introduce an accelerated reference frame, *i.e.* (t, X^i_A), associated with a body A which is related to a global inertial frame through the transformation

$$x^i = z^i_A(t) + X^i_A, \qquad (7.18)$$

$z^i_A(t)$ being the arbitrary motion of the local A-frame. The equations of motion with respect to the local A-frame reads (see [26]):

$$\frac{\partial \rho_A}{\partial t} + \frac{\partial (\rho_A v^i_A)}{\partial X^i_A} = 0, \qquad (7.19)$$

$$\frac{\partial (\rho_A v^i_A)}{\partial t} + \frac{\partial}{\partial X^j_A} \left(\rho_A v^i_A v^j_A + t^{ij} \right) = \rho_A \frac{\partial V^A_{\text{eff}}}{\partial X^i_A}, \qquad (7.20)$$

where $\rho_A(t, \mathbf{X}_A) \equiv \rho_A(t, \mathbf{z}_A)$ is the mass volumic density expressed in the local A-frame, v^i_A being the velocity with respect to this frame while t^{ij} denotes the stress tensor (note that we have adopt the Einstein's summation convention).

The following effective potential appears

$$V^A_{\text{eff}}(t, \mathbf{X}_A) = \sum_{B=1}^{N} V^B(t, \mathbf{z}_A + \mathbf{X}_A) - V^A_{\text{ext}}(t, \mathbf{z}_A) - \frac{d^2 \mathbf{z}_A}{dt^2} \cdot \mathbf{X}_A, \qquad (7.21)$$

where

$$V^A_{\text{ext}}(t, \mathbf{X}_A) = \sum_{B \neq A} V^B(t, \mathbf{X}_A), \qquad (7.22)$$

the considered body A being tidally interacting with $N - 1$ perturbing extended bodies B; V^B is the potential of each body B different from A. The last term of Eq. (7.21) represents the inertial effects on the accelerated local frame A. This effective potential can be split into the potential of A given in Eq. (7.3) and a tidal potential, V^A_T, as follows:

$$V^A_{\text{eff}} = V^A + V^A_T, \qquad (7.23)$$

the tidal part being given by

$$V_{\mathrm{T}}^{\mathrm{A}}(t, \mathbf{X}_A) = V_{\mathrm{ext}}^{\mathrm{A}}(t, \mathbf{z}_A + \mathbf{X}_A) - V_{\mathrm{ext}}^{\mathrm{A}}(t, \mathbf{z}_A) - \frac{d^2 \mathbf{z}_A}{dt^2} \cdot \mathbf{X}_A. \qquad (7.24)$$

Following [44] and [77], the general multipolar expression for the tidal potential for $|r| \leq R_A$ is then obtained. It is then recast in its general spectral form, using that $V_{\mathrm{T}}^{\mathrm{A}}$ is real and expanding it in the spherical harmonics for $|r| \leq R_A$:

$$V_{\mathrm{T}}^{\mathrm{A}}(t, \mathbf{r}, \mathbf{r}_{AB}) = \sum_{B \neq A} G \sum_{l_A=0}^{\infty} \sum_{m_A=-l_A}^{l_A} \left[A_{\mathrm{I};l_A,m_A}(t, \mathbf{r}_{AB}) \right.$$

$$\left. + A_{\mathrm{II};l_A,m_A}(t, \mathbf{r}_{AB}) \right] r^{l_A} Y_{l_A,m_A}(\theta, \varphi), \qquad (7.25)$$

where the coefficients $A_{\mathrm{I};l_A,m_A}$ and $A_{\mathrm{II};l_A,m_A}$ are respectively given by:

$$A_{\mathrm{I};l_A,m_A} = (-1)^{l_A} \frac{4\pi}{2l_A + 1} (1 - \delta_{l_A,0})(1 - \delta_{l_A,1})$$

$$\times \sum_{l_B=0}^{\infty} \sum_{m_B=-l_B}^{l_B} \gamma_{l_B,m_B}^{l_A,m_A} \left(M_{l_B,m_B}^{\mathrm{B}} \right)^* \frac{Y_{l_A+l_B,m_A+m_B}^*(\theta_{AB}, \varphi_{AB})}{r_{AB}^{l_A+l_B+1}}$$

$$(7.26)$$

and

$$A_{\mathrm{II};l_A,m_A} = -\frac{1}{M_A} \frac{4\pi}{3} \delta_{l_A,1} \sum_{l_A'=1}^{\infty} \sum_{m_A'=-l_A'}^{l_A'} \sum_{l_B=0}^{\infty} \sum_{m_B=-l_B}^{l_B} (-1)^{l_A'+1} (2l_A' + 2l_B + 1)$$

$$\times \gamma_{l_B,m_B}^{l_A',m_A'} \left(M_{l_A',m_A'}^{\mathrm{A}} \right)^* \left(M_{l_B,m_B}^{\mathrm{B}} \right)^* \gamma_{l_A'+l_B,m_A'+m_B}^{1,m_A}$$

$$\times \frac{Y_{l_A'+l_B+1,m_A'+m_B+m_A}^*(\theta_{AB}, \varphi_{AB})}{r_{AB}^{l_A'+l_B+2}}. \qquad (7.27)$$

We have introduced r_{AB}, θ_{AB} and φ_{AB} the coordinates of the center of mass of the extended perturber B (see Fig. 7.2) and the following coupling coefficient

$$\gamma_{j,k}^{l,m} = \gamma_{l,m}^{j,k}$$

$$= \sqrt{\frac{2l+1}{(l+m)!(l-m)!} \frac{2j+1}{(j+k)!(j-k)!}}$$

$$\times \sqrt{\frac{[(l+j)-(m+k)]![(l+j)+(m+k)]!}{4\pi[2(l+j)+1]}}. \qquad (7.28)$$

The respective physical meanings of terms I and II are identified. Term I corresponds to the gravitational interaction of B with A, while term II is the acceleration responsible for the movement of the center of mass of A. In the case of a ponctual mass perturber B, we recall that we get (see for example [80]):

$$V_{\mathrm{T}}^{\mathrm{A}}(t, \mathbf{r}, \mathbf{r}_{AB}) = V^{\mathrm{B}}(t, \mathbf{r}, \mathbf{r}_{AB}) - V_{\mathrm{orb}}^{\mathrm{A}}(t, \mathbf{r}, \mathbf{r}_{AB}) \qquad (7.29)$$

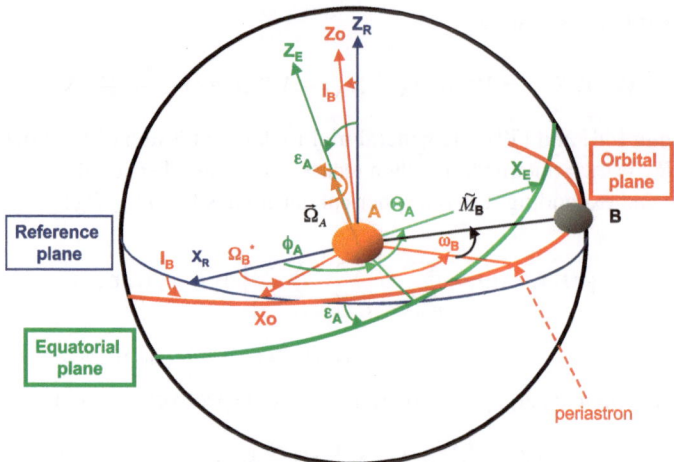

Fig. 7.3 Inertial Reference, Orbital and Equatorial rotating frames ($\mathscr{R}_R$, $\mathscr{R}_O$ and $\mathscr{R}_{E;T}$) and associated Euler's angles of orientation. (Taken from Mathis and Le Poncin-Lafitte [77]; courtesy Astronomy & Astrophysics)

where

$$V^B(t, \mathbf{r}, \mathbf{r}_{AB}) = G\frac{M_B}{|\mathbf{r} - \mathbf{r}_{AB}|} \quad \text{and} \quad V_{\text{orb}}^A(t, \mathbf{r}, \mathbf{r}_{AB}) = G\frac{M_B}{r_{AB}}\left(1 + \frac{\mathbf{r}_{AB} \cdot \mathbf{r}}{r_{AB}^2}\right),$$

(7.30)

M_B being its mass.

The more general form of the tidal potential being derived, we now express $A_{I;l_A,m_A}$ and $A_{II;l_A,m_A}$ as a function of the Keplerian orbit elements of the perturber B. Here, we take into account the relative inclinations of the spin of each body with respect to the orbital plane. It is then necessary to define three reference frames, represented in Fig. 7.3, all centered on the center of mass of the considered body A, O_A:

- an inertial frame $\mathscr{R}_R : \{O_A, \mathbf{X}_R, \mathbf{Y}_R, \mathbf{Z}_R\}$, time independent, with $\mathbf{Z}_R$ in the direction of the total angular momentum of the whole system $\mathbf{L}_{\text{Total}} = \mathbf{L}_{\text{Orbital}} + \mathbf{L}_{\text{BodyA}} + \sum_k \mathbf{L}_{\text{BodyB}_k}$ which is a first integral (we are studying here the two bodies interaction between A and each potential perturber B_k with $k \in [\![1, N]\!]$).
- an orbital frame $\mathscr{R}_O : \{O_A, \mathbf{X}_O, \mathbf{Y}_O, \mathbf{Z}_O\}$. We define here three Euler angles to link this frame to $\mathscr{R}_R : \{O_A, \mathbf{X}_R, \mathbf{Y}_R, \mathbf{Z}_R\}$:
  - I_B, the inclination of the orbital frame with respect to $(O_A, \mathbf{X}_R, \mathbf{Y}_R)$;
  - ω_B, the argument of the pericenter;
  - Ω_B^*, the longitude of the ascending node.
  Let us finally define the last three quantities associated to the elliptic elements of body B's center of mass: a_B, the semi major axis, e_B, the eccentricity and $\widetilde{M}_B$, the mean anomaly with $\widetilde{M}_B \approx n_B t$, n_B being the mean motion.

- a spin equatorial frame $\mathscr{R}_{E;T} : \{O_A, \mathbf{X}_E, \mathbf{Y}_E, \mathbf{Z}_E\}$. This frame is rotating with the angular velocity, Ω_A. This frame is linked to $\mathscr{R}_R : \{O_A, \mathbf{X}_R, \mathbf{Y}_R, \mathbf{Z}_R\}$ by three Euler angles:
  - ε_A, the obliquity, *i.e.* the inclination of the equatorial plane with respect to the reference plane $(O_A, \mathbf{X}_R, \mathbf{Y}_R)$;
  - Θ_A, the mean sideral angle where $\Theta_A = d\Omega_A/dt$. This is the angle between the minimal axis of inertia and the straight line due to the intersection of the planes $(O_A, \mathbf{X}_E, \mathbf{Y}_E)$ and $(O_A, \mathbf{X}_R, \mathbf{Y}_R)$;
  - ϕ_A, the general precession angle.

The Kaula's transform is then used to explicitly express all the generic multipole expansion in spherical harmonics in term of Keplerian elements. Using the results derived by [52], the following identity is obtained:

$$
\frac{Y_{l,m}(\theta_{AB}, \varphi_{AB})}{r_{AB}^{l+1}} = \frac{1}{a_B^{l+1}} \sum_{j=-l}^{l} \sum_{p=0}^{l} \sum_q \kappa_{l,j} d_{j,m}^l(\varepsilon_A)
$$
$$
\times F_{l,j,p}(I_B) G_{l,p,q}(e_B) \exp[i\Psi_{l,m,j,p,q}], \tag{7.31}
$$

where the $\kappa_{l,j}$ coefficients are given by:

$$
\kappa_{l,j} = \sqrt{\frac{2l+1}{4\pi} \frac{(l-|j|)!}{(l+|j|)!}}. \tag{7.32}
$$

$d_{j,m}^l(\varepsilon_A)$ is the obliquity function which is defined as follow for $j \geq m$ (see Table 7.1):

$$
d_{j,m}^l(\varepsilon_A) = (-1)^{j-m} \left[\frac{(l+j)!(l-j)!}{(l+m)!(l-m)!}\right]^{\frac{1}{2}} \left[\cos\left(\frac{\varepsilon_A}{2}\right)\right]^{j+m} \left[\sin\left(\frac{\varepsilon_A}{2}\right)\right]^{j-m}
$$
$$
\times P_{l-j}^{(j-m,j+m)}(\cos\varepsilon_A), \tag{7.33}
$$

the $P_l^{(\alpha,\beta)}(x)$ being the Jacobi polynomials. The value of the function for indices j which do not verify $j \geq m$ are deduced from

$$
d_{j,m}^l(\pi + \varepsilon_A) = (-1)^{l-j} d_{-j,m}^l(\varepsilon_A) \tag{7.34}
$$

or from their symmetry properties:

$$
d_{j,m}^l(\varepsilon_A) = (-1)^{j-m} d_{-j,-m}^l(\varepsilon_A) = d_{m,j}^l(-\varepsilon_A). \tag{7.35}
$$

On the other hand, one should note that: $d_{j,m}^l(0) = \delta_{jm}$.

The inclination function, $F_{l,j,p}(I_B)$ (see Table 7.2), is defined in a similar way:

$$
F_{l,j,p}(I_B) = (-1)^p \left[\frac{4\pi}{2l+1} \frac{(l+j)!}{(l-j)!}\right]^{\frac{1}{2}} Y_{l,l-2p}\left(\frac{\pi}{2}, 0\right) d_{l-2p,j}^l(-I_B), \tag{7.36}
$$

where

$$
Y_{l,m}\left(\frac{\pi}{2}, 0\right) = \left[\frac{2l+1}{4\pi}\right]^{\frac{1}{2}} \frac{[(l-m)!(l+m)!]^{\frac{1}{2}}}{2^l[(l-m)/2]![(l+m)/2]!} \cos\left[(l-m)\frac{\pi}{2}\right]; \tag{7.37}
$$

Table 7.1 Values of the obliquity function $d^2_{j,m}(\varepsilon)$ in the case where $j \geqslant m$ obtained from Eq. (7.33) (see [77])

j	m	$d^2_{j,m}(\varepsilon)$
2	2	$(\cos \frac{\varepsilon}{2})^4$
2	1	$-2(\cos \frac{\varepsilon}{2})^3 (\sin \frac{\varepsilon}{2})$
2	0	$\sqrt{6}(\cos \frac{\varepsilon}{2})^2 (\sin \frac{\varepsilon}{2})^2$
1	1	$(\cos \frac{\varepsilon}{2})^4 - 3(\cos \frac{\varepsilon}{2})^2 (\sin \frac{\varepsilon}{2})^2$
1	0	$-\sqrt{6}\cos \varepsilon (\cos \frac{\varepsilon}{2})(\sin \frac{\varepsilon}{2})$
0	0	$1 - 6(\cos \frac{\varepsilon}{2})^2 (\sin \frac{\varepsilon}{2})^2$

Table 7.2 Values of the inclination function $F_{2,j,p}(I)$. Values for indices $j < 0$ can be deduced from Eq. (7.38) (see [77])

j	p	$F_{2,j,p}(I)$
0	0	$\frac{3}{8} \sin^2 I$
0	1	$-\frac{3}{4} \sin^2 I + \frac{1}{2}$
0	2	$\frac{3}{8} \sin^2 I$
1	0	$\frac{3}{4} \sin I (1 + \cos I)$
1	1	$-\frac{3}{2} \sin I \cos I$
1	2	$-\frac{3}{4} \sin I (1 - \cos I)$
2	0	$\frac{3}{4}(1 + \cos I)^2$
2	1	$\frac{3}{2} \sin^2 I$
2	2	$\frac{3}{4}(1 - \cos I)^2$

Table 7.3 Values of the eccentricity function $G_{2,p,q}(e)$ (see [77])

p	q	p	q	$G_{2,p,q}(e)$
0	-2	2	2	0
0	-1	2	1	$-\frac{1}{2}e + \cdots$
0	0	2	0	$1 - \frac{5}{2}e^2 + \cdots$
0	1	2	-1	$\frac{7}{2}e + \cdots$
0	2	2	-2	$\frac{17}{2}e^2 + \cdots$
1	-2	1	2	$\frac{9}{4}e^2 + \cdots$
1	-1	1	1	$\frac{3}{2}e + \cdots$
		1	0	$(1 - e^2)^{-3/2}$

moreover, the following symmetry property is verified:

$$F_{l,-j,p}(I_B) = \left[(-1)^{l-j}\frac{(l-j)!}{(l+j)!}\right] F_{l,j,p}(I_B). \qquad (7.38)$$

The eccentricity functions $G_{l,p,q}(e_B)$, are polynomial functions having e_B^q for argument (see [52, 63] for their detailed properties). Their values for usual sets $\{l, p, q\}$ are given in Table 7.3. In the case of weakly eccentric orbits, the summation over a small number of values for q is sufficient ($q \in [\![-2, 2]\!]$).

Finally, the phase argument is given by

$$\Psi_{l,m,j,p,q} = (l - 2p + q)\tilde{M}_{\mathrm{B}} + \Phi_{l,m,j,p,q}(\omega_{\mathrm{B}}, \Omega_{\mathrm{B}}^*, \Theta_{\mathrm{A}}, \phi_{\mathrm{A}}), \qquad (7.39)$$

where

$$\Phi_{l,m,j,p,q} = (l - 2p)\omega_{\mathrm{B}} + j(\Omega_{\mathrm{B}}^* - \phi_{\mathrm{A}}) - m\Theta_{\mathrm{A}} + (l - m)\frac{\pi}{2}. \qquad (7.40)$$

This is also written as

$$\Psi_{l,m,j,p,q} = \sigma_{l,m,p,q}(n_{\mathrm{B}}, \Omega_{\mathrm{A}})t + \psi_{l,m,j,p,q}(\omega_{\mathrm{B}}, \Omega_{\mathrm{B}}^*, \phi_{\mathrm{A}}), \qquad (7.41)$$

where we have defined the tidal frequency

$$\sigma_{l,m,p,q} = (l - 2p + q)n_{\mathrm{B}} - m\Omega_{\mathrm{A}} \qquad (7.42)$$

and

$$\psi_{l,m,j,p,q} = (l - 2p)\omega_{\mathrm{B}} + j(\Omega_{\mathrm{B}}^* - \phi_{\mathrm{A}}) + (l - m)\frac{\pi}{2}. \qquad (7.43)$$

The Kaula's transform thus allows us to express each function of $\mathbf{r}_{\mathrm{AB}}$, *i.e.* of $(r_{\mathrm{AB}}, \theta_{\mathrm{AB}}, \varphi_{\mathrm{AB}})$, as a function of the Keplerian relative orbital elements of B in the A-frame. Applying Eq. (7.31) to $A_{\mathrm{I};l_{\mathrm{A}},m_{\mathrm{A}}}$ and $A_{\mathrm{II};l_{\mathrm{A}},m_{\mathrm{A}}}$ respectively given in Eqs. (7.26) and (7.27), we obtain:

$$A_{\mathrm{I};l_{\mathrm{A}},m_{\mathrm{A}}} = (-1)^{l_{\mathrm{A}}} \frac{4\pi}{2l_{\mathrm{A}} + 1}(1 - \delta_{l_{\mathrm{A}},0})(1 - \delta_{l_{\mathrm{A}},1})$$

$$\times \sum_{l_{\mathrm{B}}=0}^{\infty} \sum_{m_{\mathrm{B}}=-l_{\mathrm{B}}}^{l_{\mathrm{B}}} \gamma_{l_{\mathrm{B}},m_{\mathrm{B}}}^{l_{\mathrm{A}},m_{\mathrm{A}}} |M_{l_{\mathrm{B}},m_{\mathrm{B}}}^{\mathrm{B}}| \exp[-i\delta M_{l_{\mathrm{B}},m_{\mathrm{B}}}^{\mathrm{B}}]\frac{1}{a_{\mathrm{B}}^{l_{\mathrm{A}}+l_{\mathrm{B}}+1}}$$

$$\times \sum_{j=-(l_{\mathrm{A}}+l_{\mathrm{B}})}^{l_{\mathrm{A}}+l_{\mathrm{B}}} \sum_{p=0}^{l_{\mathrm{A}}+l_{\mathrm{B}}} \sum_{q} \kappa_{l_{\mathrm{A}}+l_{\mathrm{B}},j} d_{j,m_{\mathrm{A}}+m_{\mathrm{B}}}^{l_{\mathrm{A}}+l_{\mathrm{B}}}(\varepsilon_{\mathrm{A}}) F_{l_{\mathrm{A}}+l_{\mathrm{B}},j,p}(I_{\mathrm{B}})$$

$$\times G_{l_{\mathrm{A}}+l_{\mathrm{B}},p,q}(e_{\mathrm{B}}) \exp[-i\Psi_{l_{\mathrm{A}}+l_{\mathrm{B}},m_{\mathrm{A}}+m_{\mathrm{B}},j,p,q}] \qquad (7.44)$$

and

$$A_{\mathrm{II};l_{\mathrm{A}},m_{\mathrm{A}}} = -\frac{1}{M_{\mathrm{A}}} \frac{4\pi}{3}\delta_{l_{\mathrm{A}},1} \sum_{l_{\mathrm{A}}'=1}^{\infty} \sum_{m_{\mathrm{A}}'=-l_{\mathrm{A}}'}^{l_{\mathrm{A}}'} \sum_{l_{\mathrm{B}}=0}^{\infty} \sum_{m_{\mathrm{B}}=-l_{\mathrm{B}}}^{l_{\mathrm{B}}} (-1)^{l_{\mathrm{A}}'+1}(2l_{\mathrm{A}}' + 2l_{\mathrm{B}} + 1)$$

$$\times \gamma_{l_{\mathrm{B}},m_{\mathrm{B}}}^{l_{\mathrm{A}}',m_{\mathrm{A}}'} |M_{l_{\mathrm{A}}',m_{\mathrm{A}}'}^{\mathrm{A}}| \exp[-i\delta M_{l_{\mathrm{A}}',m_{\mathrm{A}}'}^{\mathrm{A}}]|M_{l_{\mathrm{B}},m_{\mathrm{B}}}^{\mathrm{B}}| \exp[-i\delta M_{l_{\mathrm{B}},m_{\mathrm{B}}}^{\mathrm{B}}]$$

$$\times \gamma_{l_{\mathrm{A}}'+l_{\mathrm{B}},m_{\mathrm{A}}'+m_{\mathrm{B}}}^{1,m_{\mathrm{A}}} \frac{1}{a_{\mathrm{B}}^{l_{\mathrm{A}}'+l_{\mathrm{B}}+2}} \sum_{r=-(l_{\mathrm{A}}'+l_{\mathrm{B}}+1)}^{l_{\mathrm{A}}'+l_{\mathrm{B}}+1} \sum_{s=0}^{l_{\mathrm{A}}'+l_{\mathrm{B}}+1} \sum_{u} \kappa_{l_{\mathrm{A}}'+l_{\mathrm{B}}+1,r}$$

$$\times d_{r,m_{\mathrm{A}}'+m_{\mathrm{B}}+m_{\mathrm{A}}}^{l_{\mathrm{A}}'+l_{\mathrm{B}}+1}(\varepsilon_{\mathrm{A}}) F_{l_{\mathrm{A}}'+l_{\mathrm{B}}+1,r,s}(I_{\mathrm{B}})G_{l_{\mathrm{A}}'+l_{\mathrm{B}}+1,s,u}(e_{\mathrm{B}})$$

$$\times \exp[-i \Psi_{l_{\mathrm{A}}'+l_{\mathrm{B}}+1,m_{\mathrm{A}}'+m_{\mathrm{B}}+m_{\mathrm{A}},r,s,u}], \qquad (7.45)$$

where as in Eq. (7.13) $M_{l_{\mathrm{B}},m_{\mathrm{B}}}^{\mathrm{B}} = M_{l_{\mathrm{B}},m_{\mathrm{B}}}^{\mathrm{S_B}} + M_{l_{\mathrm{B}},m_{\mathrm{B}}}^{\mathrm{T_B}}$ and $M_{l_{\mathrm{A}},m_{\mathrm{A}}}^{\mathrm{A}} = M_{l_{\mathrm{A}},m_{\mathrm{A}}}^{\mathrm{S_A}} + M_{l_{\mathrm{A}},m_{\mathrm{A}}}^{\mathrm{T_A}}$.

Like in [126, 130], the tidal potential can be splitted into two components. The first one, $V_{T;1}^A(\mathbf{r}, \mathbf{r}_{AB})$, is stationary (*i.e.* the tidal frequency vanishes: $\sigma = 0$). It corresponds to the axisymmetric permanent deformation induced by B. In the case of a ponctual mass perturber and of a system where all the spins are aligned, Zahn [126, 130] shown that $V_{T;1}^A = -\frac{GM_B}{a_B^3}\frac{1}{2}(1 - e_B^2)^{-3/2} r^2 P_2(\cos\theta)$. Then, the second component is the time dependent part of the perturbation, $V_{T;2}^A(t, \mathbf{r}, \mathbf{r}_{AB})$, for which $\sigma \neq 0$.

7.2.1.3 The Two Bodies Interaction Potential

Finally, we define the mutual gravitational interaction potential[2] of two bodies A and B is defined as:

$$V_{A-B}(t, \mathbf{r}_{AB}) = \int_{M_A} V^B(t, \mathbf{r}, \mathbf{r}_{AB}) \, dM_A. \tag{7.46}$$

Following once again [44] and [77], the following multipolar expansion is obtained:

$$V_{A-B} = G \sum_{l_A=0}^{\infty} \sum_{m_A=-l_A}^{l_A} \sum_{l_B=0}^{\infty} \sum_{m_B=-l_B}^{l_B} M_{l_A,m_A}^A M_{l_B,m_B}^B (-1)^{l_A} \gamma_{l_B,m_B}^{l_A,m_A}$$

$$\times \frac{Y_{l_A+l_B,m_A+m_B}(\theta_{AB}, \varphi_{AB})}{r_{AB}^{l_A+l_B+1}}. \tag{7.47}$$

Finally, using the Kaula transformation given in Eqs. (7.31), (7.32) and (7.39) as previously done for V_T^A, V_{A-B} is expressed as a function of the obliquity (ε_A), and of the Keplerian orbital elements of B (a_B, e_B and I_B):

$$V_{A-B} = G \sum_{l_A=0}^{\infty} \sum_{m_A=-l_A}^{l_A} \sum_{l_B=0}^{\infty} \sum_{m_B=-l_B}^{l_B} M_{l_A,m_A}^A M_{l_B,m_B}^B (-1)^{l_A} \gamma_{l_B,m_B}^{l_A,m_A}$$

$$\times \frac{1}{a_B^{l_A+l_B+1}} \sum_{v=-(l_A+l_B)}^{(l_A+l_B)} \sum_{w=0}^{l_A+l_B} \sum_{b} \kappa_{l_A+l_B,v} d_{v,m_A+m_B}^{l_A+l_B}(\varepsilon_A) F_{l_A+l_B,v,w}(I_B)$$

$$\times G_{l_A+l_B,w,b}(e_B) \exp[i \, \Psi_{l_A+l_B,m_A+m_B,v,w,b}]; \tag{7.48}$$

this interaction potential contains all multipole-multipole couplings.

Since all type of gravitational potentials have been examined, we now study the dynamics of a system of extended bodies.

---

[2]The denomination of V_{A-B} as a potential is not very pertinent since it has the dimension of the product of a mass by a potential. However, we keep it to stay coherent with [44].

7.2.2 Equations of Motion

7.2.2.1 External Gravitational Potential of a Tidally Perturbed Body

The goal of this section is to derive the external gravitational potential of a tidally perturbed extended body A by an extended body B. This potential is the sum of the structural self-gravitational potential of A, $V_S^A(t, \mathbf{r})$, and of $\widetilde{V}_T^A(t, \mathbf{r}, \mathbf{r}_{AB})$, the tidally induced gravitational potential corresponding to the response of A to the perturbing potential $V_T^A(t, \mathbf{r}, \mathbf{r}_{AB})$:

$$V_{ext}^A(t, \mathbf{r}, \mathbf{r}_{AB}) = V_S^A(t, \mathbf{r}) + \widetilde{V}_T^A(t, \mathbf{r}, \mathbf{r}_{AB}) \tag{7.49}$$

with the following definition for V_S

$$V_S^A(t, \mathbf{r}) = G \sum_{l_A=0}^{\infty} \sum_{l_A=-m_A}^{m_A} M_{l_A,m_A}^{S_A} \frac{Y_{l_A,m_A}(\theta, \varphi)}{r^{l_A+1}} \tag{7.50}$$

where $M_{l_A,m_A}^{S_A}$ are the multipole moments of A in the case where it is not tidally perturbed by any other body, in other words, in the case it is isolated. By definition the external gravitational potential is harmonic; therefore $V_{ext}^A(t, \mathbf{r}, \mathbf{r}_{AB})$ verifies the Laplace equation:

$$\nabla^2 V_{ext}^A(t, \mathbf{r}, \mathbf{r}_{AB}) = 0 \quad \text{if } |\mathbf{r}| \geq R_A, \tag{7.51}$$

that directly leads to the same equation for $\widetilde{V}_T^A(t, \mathbf{r}, \mathbf{r}_{AB})$:

$$\nabla^2 \widetilde{V}_T^A(t, \mathbf{r}, \mathbf{r}_{AB}) = 0 \quad \text{if } |\mathbf{r}| \geq R_A. \tag{7.52}$$

Following [23, 24, 62, 81], we use the classical Love numbers, $k_{l_A}^A$, which allow us to characterise the response of the body A to the tidal perturbation [70]. The boundary conditions for $\widetilde{V}_T^A(t, \mathbf{r}, \mathbf{r}_{AB})$ are:

$$\begin{cases} \widetilde{V}_T^A(t, |\mathbf{r}| \to 0, \mathbf{r}_{AB}) = 0, \\ \widetilde{V}_T^A(t, |\mathbf{r}| = R_A, \mathbf{r}_{AB}) = \sum_{l_A} k_{l_A}^A V_{l_A}(t, |\mathbf{r}| = R_A, \mathbf{r}_{AB}), \end{cases} \tag{7.53}$$

where V_{l_A} is the l_A^{th} spherical harmonic of $V_T^A(t, \mathbf{r}, \mathbf{r}_{AB})$.

We recall that $V_T^A(t, \mathbf{r}, \mathbf{r}_{AB})$ has been expanded as follow for $|r| \leq R_A$ (see Eq. (7.25))

$$V_T^A(t, \mathbf{r}, \mathbf{r}_{AB}) = G \sum_{l_A,m_A} \left[A_{I;l_A,m_A}(t, \mathbf{r}_{AB}) + A_{II;l_A,m_A}(t, \mathbf{r}_{AB}) \right] r^{l_A} Y_{l_A,m_A}(\theta, \varphi). \tag{7.54}$$

Using the well-known properties of the Laplace's equation, we search the solution for $\widetilde{V}_T^A$ where $|r| \geq R_A$ of the form

$$\widetilde{V}_T^A(t, \mathbf{r}, \mathbf{r}_{AB}) = G \sum_{l_A=0}^{\infty} \sum_{m_A=-l_A}^{l_A} M_{l_A,m_A}^{T_A}(t, \mathbf{r}_{AB}) \frac{Y_{l_A,m_A}(\theta, \varphi)}{r^{l_A+1}}. \tag{7.55}$$

Inserting Eqs. (7.54) and (7.55) into Eq. (7.53), the final solution of $\widetilde{V}_T^A$ is then derived

$$M_{l_A,m_A}^{T_A} = M_{l_A,m_A}^{T_A;I}(t,\mathbf{r}_{AB}) + M_{l_A,m_A}^{T_A;II}(t,\mathbf{r}_{AB}), \tag{7.56}$$

where $M_{l_A,m_A}^{T_A;I}$ and $M_{l_A,m_A}^{T_A;II}$ are given by

$$\begin{cases} M_{l_A,m_A}^{T_A;I} = k_{l_A}^A A_{I;l_A,m_A} R_E^{2l_A+1}, \\ M_{l_A,m_A}^{T_A;II} = k_{l_A}^A A_{II;l_A,m_A} R_E^{2l_A+1}. \end{cases} \tag{7.57}$$

The response of the body A, which is described by the Love numbers, is the adiabatic one. However, it is well known that an elastic as well as a fluid body reacts to the tidal perturbation with a damping and a time delay which are due to the internal friction and diffusivities (in other words to the viscosity, ν, and the thermal diffusivity, K, in a non-magnetic body). That allows us to transform the mechanical energy into thermal one which leads us to the dynamical evolution of the studied system (cf. Fig. 7.4). Therefore, we introduce a complex impedance, $Z_{T_A}(\nu, K; \Psi_L)$, with its associated argument, $\delta_{T_A}(\nu, K; \Psi_L)$

$$Z_{T_A}(\nu, K; \Psi_L) = |Z_{T_A}(\nu, K; \Psi_L)| \exp[i\delta_{T_A}(\nu, K; \Psi_L)] \tag{7.58}$$

which describes this damping. We thus substitute the complex Love number

$$\widetilde{k}_{l_A}^A = k_{l_A}^A |Z_{T_A}| \exp[i\delta_{T_A}] \tag{7.59}$$

to $k_{l_A}^A$ in the Eq. (7.57) (see [45]); L corresponds to the indices of the considered tidal Fourier's mode.[3] The different modellings that can be adopted for Z_{T_A} and δ_{T_A} will be extensively discussed in Sect 7.3.

Using Eqs. (7.44) and (7.45), the expression of $M_{l_A,m_A}^{T_A;I}$ and $M_{l_A,m_A}^{T_A;II}$ are obtained

$$\begin{aligned} M_{l_A,m_A}^{T_A;I} = & (-1)^{l_A} \frac{4\pi}{2l_A+1} k_{l_A}^A R_A^{2l_A+1} (1-\delta_{l_A,0})(1-\delta_{l_A,1}) \\ & \times \sum_{l_B,m_B,j,p,q} |Z_{T_A;l_A,m_A,L_I}(\nu, K; \Psi_{l_A+l_B,m_A+m_B,j,p,q})| |M_{l_B,m_B}^B| \gamma_{l_B,m_B}^{l_A,m_A} \\ & \times \frac{1}{a_B^{l_A+l_B+1}} \kappa_{l_A+l_B,j} d_{j,m_A+m_B}^{l_A+l_B}(\varepsilon_A) F_{l_A+l_B,j,p}(I_B) G_{l_A+l_B,p,q}(e_B) \\ & \times \exp[i\,\Delta_{T_A;l_A,m_A,L_I}], \end{aligned} \tag{7.60}$$

where

$$\begin{aligned} \Delta_{T_A;l_A,m_A,L_I} = & \delta_{T_A;l_A,m_A,L_I}(\nu, K; \Psi_{l_A+l_B,m_A+m_B,j,p,q}) - \Psi_{l_A+l_B,m_A+m_B,j,p,q} \\ & - \delta M_{l_B,m_B}^B \end{aligned} \tag{7.61}$$

with $L_I = \{l_B, m_B, j, p, q\}$,

---

[3]Note that each tidal Fourier's mode have its own dissipation rate as it as been shown by Zahn (1966–1977).

$$M^{T_A;II}_{l_A,m_A} = -\frac{1}{M_A}\frac{4\pi}{3}k^A_{l_A}R^{2l_A+1}_A\delta_{l_A,1}\sum_{l'_A,m'_A,l_B,m_B,r,s,u}(-1)^{l'_A+1}(2l'_A+2l_B+1)$$

$$\times\left|Z_{T_A;l_A,m_A,L_{II}}(\nu,K;\Psi_{l'_A+l_B+1,m'_A+m_B+m_A,r,s,u})\right|^{l'_A,m'_A}_{l_B,m_B}\left|M^A_{l'_A,m'_A}\right|$$

$$\times\left|M^B_{l_B,m_B}\right|\gamma^{1,m_A}_{l'_A+l_B,m'_A+m_B}\frac{1}{a^{l'_A+l_B+2}_B}\kappa^{l'_A+l_B+1}_{l'_A+l_B+1,r}d^{l'_A+l_B+1}_{r,m'_A+m_B+m_A}(\varepsilon_A)$$

$$\times F_{l'_A+l_B+1,r,s}(I_B)G_{l'_A+l_B+1,s,u}(e_B)\Delta_{T_A;l_A,m_A,L_{II}}\exp[i\,\Delta_{T_A;l_A,m_A,L_{II}}],\tag{7.62}$$

where

$$\Delta_{T_A;l_A,m_A,L_{II}}=\delta_{T_A;l_A,m_A,L_{II}}(\nu,K;\Psi_{l'_A+l_B+1,m'_A+m_B+m_A,r,s,u})$$

$$-\Psi_{l'_A+l_B+1,m'_A+m_B+m_A,r,s,u}-\delta M^A_{l'_A,m'_A}-\delta M^B_{l_B,m_B}\tag{7.63}$$

with $L_{II}=\{l'_A,m'_A,l_B,m_B,r,s,u\}$. Moreover, $M^B_{l_B,m_B}=M^{S_B}_{l_B,m_B}+M^{T_B}_{l_B,m_B}$[4] and $M^A_{l_A,m_A}=M^{S_A}_{l_A,m_A}+M^{T_A}_{l_A,m_A}$.

As for V^A_T, $\widetilde{V}^A_T$ can be splitted into two components. The first one $\widetilde{V}^A_{T;1}(\mathbf{r},\mathbf{r}_{AB})$ is stationary. It corresponds to the permanent component $V^A_{T;1}$ for which the tidal frequency (σ) vanishes. The second component $\widetilde{V}^A_{T;2}(t,\mathbf{r},\mathbf{r}_{AB})$ is the time-dependent one that corresponds to $V^A_{T;2}$ for which $\sigma\neq0$.

Finally the external potential of A is thus written in its more compact and general form for $|r|\geq R_A$

$$V^A_{ext}(t,\mathbf{r},\mathbf{r}_{AB})=G\sum_{l_A=0}^\infty\sum_{m_A=-l_A}^{l_A}M^A_{l_A,m_A}(t,\mathbf{r}_{AB})\frac{Y_{l_A,m_A}(\theta,\varphi)}{r^{l_A+1}},\tag{7.64}$$

where

$$M^A_{l_A,m_A}=M^{S_A}_{l_A,m_A}+M^{T_A}_{l_A,m_A}=M^{S_A}_{l_A,m_A}+M^{T_A;I}_{l_A,m_A}+M^{T_A;II}_{l_A,m_A}.\tag{7.65}$$

7.2.2.2 Disturbing Function

The goal of this section is to derive the disturbing function, $\mathscr{R}_{A-C}$, due to a tidally perturbed body A, acting on a body C of which dynamics is studied and which can be different from the perturber body B (see Fig. 7.4).

First, the disturbing function is related to the mutual gravitational interaction potential (cf. [21, 114, 115]) through:

$$\mathscr{R}_{A-C}(t,\mathbf{r}_{AC})=-\frac{1}{M_C}V_{A-C};\tag{7.66}$$

the sign being due to the potentials convention adopted here.

---

[4]The tidal multipole moments of B due to A can be derived using the same methodology and substituting A to B for the perturber and vice-versa.

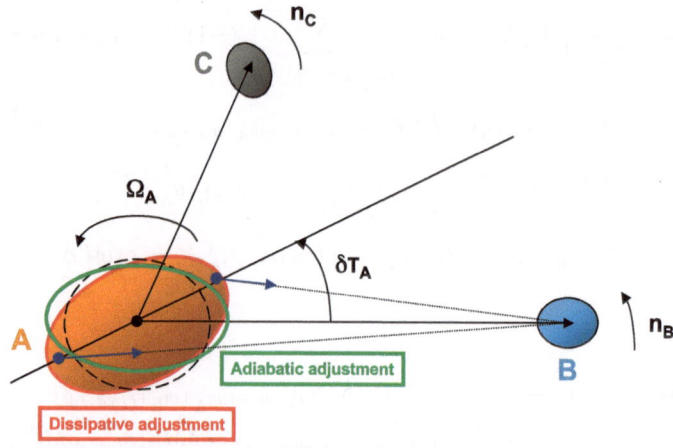

Fig. 7.4 Classical tidal dynamical system. The extended body B is tidally disturbing the extended body A which adjusts itself with a phase lag δ_{T_A} due to its internal friction processes. The dynamics of a third body C (different from B or not) is then studied. Ω_A, n_B, n_C are respectively the spin frequency of A, and the respective mean motions of B and C. (Taken from Mathis and Le Poncin-Lafitte [77]; courtesy Astronomy & Astrophysics)

Using the definition of V_{A-C} given in Eq. (7.47), we deduce the explicit spectral expansion of $\mathscr{R}_{A-C}$ in spherical harmonics

$$
\mathscr{R}_{A-C} = -\frac{G}{Mc} \sum_{l_A=0}^{\infty} \sum_{m_A=-l_A}^{l_A} \sum_{l_C=0}^{\infty} \sum_{m_C=-l_C}^{l_C} M_{l_A,m_A}^{A} M_{l_C,m_C}^{C} (-1)^{l_A} \gamma_{l_C,m_C}^{l_A,m_A}
$$
$$
\times \frac{Y_{l_A+l_C,m_A+m_C}(\theta_{AC},\varphi_{AC})}{r_{AC}^{l_A+l_C+1}}. \tag{7.67}
$$

The M_{l_A,m_A}^{A}, M_{l_C,m_C}^{C} are respectively the mass multipole moments of the body A and of the body C, while r_{AC}, θ_{AC} and φ_{AC} are the spherical coordinates of the center of mass of the body C in the A-frame (cf. Fig. 7.2). Then, using the Kaula's transformation given in Eqs. (7.31), (7.32) and (7.39), $\mathscr{R}_{A-C}$ is expressed as a function of the obliquity (ε_A) and of the Keplerian orbital elements of C (a_C, e_C and I_C)

$$
\mathscr{R}_{A-C} = -\frac{G}{Mc} \sum_{l_A=0}^{\infty} \sum_{m_A=-l_A}^{l_A} \sum_{l_C=0}^{\infty} \sum_{m_C=-l_C}^{l_C} M_{l_A,m_A}^{A} M_{l_C,m_C}^{C} (-1)^{l_A} \gamma_{l_C,m_C}^{l_A,m_A}
$$
$$
\times \frac{1}{a_C^{l_A+l_C+1}} \sum_{v=-(l_A+l_C)}^{(l_A+l_C)} \sum_{w=0}^{l_A+l_C} \sum_{b} \kappa_{l_A+l_C,v} d_{v,m_A+m_C}^{l_A+l_C}(\varepsilon_A) F_{l_A+l_C,v,w}(I_C)
$$
$$
\times G_{l_A+l_C,w,b}(e_C) \exp[i \Psi_{l_A+l_C,m_A+m_C,v,w,b}]. \tag{7.68}
$$

Here, three types of gravitational interaction are treated in our formalism (see also Eq. (7.65)). To describe them, one has first to consider the two causes of the multipolar behaviour of the gravitational potential of a body. The first is due to its internal structure and dynamics. In the case of a solid body, it is due to its proper asymmetry while in the case of a fluid mass, the internal dynamical processes such as rotation or magnetic field will break the ideal spherical hydrostatic symmetry of the body. The second is the deformation of the body due to its response to the tidal perturbation exerted by the perturber(s). In the case studied here, it is the response of the body A to the perturbation exerted by B computed in the previous section. Therefore, we split here the k-indexed mass multipole moments of each body as in Eq. (7.13):

$$M_{l_k,m_k}^k = M_{l_k,m_k}^{S_k} + M_{l_k,m_k}^{T_k}, \qquad (7.69)$$

where $M_{l_k,m_k}^{S_k}$ is the self-structural contribution of the body while $M_{l_k,m_k}^{T_k}$ is the tidal one.

The three type of gravitational interaction are thus identified. The first is the interaction between the structural mass multipole moments of each body, $M_{l_k,m_k}^{S_k} M_{l_{k'},m_{k'}}^{S_{k'}}$ with $k \neq k'$; one should note that $\{l_A = 0, m_A = 0\} - \{l_C = 0, m_C = 0\}$ is the classical interaction between M_A and M_C, M_C being the mass of C. The second corresponds to the mixed interaction between the structural and the tidal mass multipole moments, $M_{l_k,m_k}^{S_k} M_{l_{k'},m_{k'}}^{T_{k'}}$. The third is the interaction between the tidal mass multipole moments of each body, $M_{l_k,m_k}^{T_k} M_{l_{k'},m_{k'}}^{T_{k'}}$. Therefore, the disturbing function could be splitted into three terms

$$\mathcal{R}_{A-C} = \mathcal{R}_{A-C;S-S}(t, \mathbf{r}_{AC}) + \mathcal{R}_{A-C;S-T}(t, \mathbf{r}_{AC}) + \mathcal{R}_{A-C;T-T}(t, \mathbf{r}_{AC}), \qquad (7.70)$$

where $\mathcal{R}_{A-C;S-S}$ is the disturbing function associated to the structure-structure interaction, $\mathcal{R}_{A-C;T-S}$ is associated to the tide-structure interaction and $\mathcal{R}_{A-C;T-T}$ is associated to the tide-tide interaction.

Inserting Eq. (7.69) into Eq. (7.68), the respective Fourier expansions of $\mathcal{R}_{A-C;S-S}$, $\mathcal{R}_{A-C;S-T}$ and $\mathcal{R}_{A-C;T-T}$ are obtained

$$\mathcal{R}_{A-C;S-S} = -\frac{G}{M_C} \sum_{l_A=0}^{\infty} \sum_{m_A=-l_A}^{l_A} \sum_{l_C=0}^{\infty} \sum_{m_C=-l_C}^{l_C} M_{l_A,m_A}^{S_A} M_{l_C,m_C}^{S_C} (-1)^{l_A} \gamma_{l_C,m_C}^{l_A,m_A}$$

$$\times \frac{1}{a_C^{l_A+l_C+1}} \sum_{v,w,b} \kappa_{l_A+l_C,v} d_{v,m_A+m_C}^{l_A+l_C}(\varepsilon_A) F_{l_A+l_C,v,w}(I_C) G_{l_A+l_C,w,b}(e_C)$$

$$\times \exp[i \Psi_{l_A+l_C,m_A+m_C,v,w,b}], \qquad (7.71)$$

$$\mathcal{R}_{A-C;T-S} = -\frac{G}{M_C} \sum_{l_A=0}^{\infty} \sum_{m_A=-l_A}^{l_A} \sum_{l_C=0}^{\infty} \sum_{m_C=-l_C}^{l_C} (M_{l_A,m_A}^{S_A} M_{l_C,m_C}^{T_C} + M_{l_A,m_A}^{T_A} M_{l_C,m_C}^{S_C})$$

$$\times (-1)^{l_A} \gamma_{l_C,m_C}^{l_A,m_A} \frac{1}{a_C^{l_A+l_C+1}} \sum_{v,w,b} \kappa_{l_A+l_C,v} d_{v,m_A+m_C}^{l_A+l_C}(\varepsilon_A) F_{l_A+l_C,v,w}(I_C)$$

$$\times G_{l_A+l_C,w,b}(e_C) \exp[i \Psi_{l_A+l_C,m_A+m_C,v,w,b}] \qquad (7.72)$$

and

$$\mathcal{R}_{\mathrm{A-C;T-T}} = -\frac{G}{M_{\mathrm{C}}} \sum_{l_{\mathrm{A}}=0}^{\infty} \sum_{m_{\mathrm{A}}=-l_{\mathrm{A}}}^{l_{\mathrm{A}}} \sum_{l_{\mathrm{C}}=0}^{\infty} \sum_{m_{\mathrm{C}}=-l_{\mathrm{C}}}^{l_{\mathrm{C}}} M_{l_{\mathrm{A}},m_{\mathrm{A}}}^{\mathrm{T_A}} M_{l_{\mathrm{C}},m_{\mathrm{C}}}^{\mathrm{T_C}} (-1)^{l_{\mathrm{A}}} \gamma_{l_{\mathrm{C}},m_{\mathrm{C}}}^{l_{\mathrm{A}},m_{\mathrm{A}}}$$

$$\times \frac{1}{a_{\mathrm{C}}^{l_{\mathrm{A}}+l_{\mathrm{C}}+1}} \sum_{v,w,b} \kappa_{l_{\mathrm{A}}+l_{\mathrm{C}},v} d_{v,m_{\mathrm{A}}+m_{\mathrm{C}}}^{l_{\mathrm{A}}+l_{\mathrm{C}}}(\varepsilon_{\mathrm{A}}) F_{l_{\mathrm{A}}+l_{\mathrm{C}},v,w}(I_{\mathrm{C}}) G_{l_{\mathrm{A}}+l_{\mathrm{C}},w,b}(e_{\mathrm{C}})$$

$$\times \exp[i \Psi_{l_{\mathrm{A}}+l_{\mathrm{C}},m_{\mathrm{A}}+m_{\mathrm{C}},v,w,b}]. \tag{7.73}$$

This classification of the three different types of interaction allows to explicitly generalise the classical case where the only considered extended body is the tidally perturbed one, A, while B and C are considered as ponctual masses. In this case, the interaction are restricted to the classical gravitational interaction between $M_{l_{\mathrm{A}},m_{\mathrm{A}}}^{\mathrm{S_A}}$, $M_{l_{\mathrm{A}},m_{\mathrm{A}}}^{\mathrm{T_A}}$, M_{B}, the mass of B, and M_{C}.

By now, to lighten the equations, the tidal multipole moments of C are ignored. In a practical case, they have to be derived using Eqs. (7.60)–(7.62) and taken into account. The disturbing function $\mathcal{R}_{\mathrm{A-C}}$ is thus reduced to the two first interactions: the respective structural moments of body A and of body C and the structural moments of body C with the tidal moments of body A. The Fourier expansion of the disturbing function is thus given by:

$$\mathcal{R}_{\mathrm{A-C}} = \mathcal{R}_{\mathrm{A-C;S-S}} + \mathcal{R}_{\mathrm{A-C;T-S}} = \sum_{l_{\mathrm{A}},m_{\mathrm{A}},l_{\mathrm{C}},m_{\mathrm{C}},v,w,b} \mathcal{R}_{l_{\mathrm{A}},m_{\mathrm{A}},l_{\mathrm{C}},m_{\mathrm{C}},v,w,b}(t,\mathbf{r}_{\mathrm{AC}}),$$
$$\tag{7.74}$$

where

$$\mathcal{R}_{l_{\mathrm{A}},m_{\mathrm{A}},l_{\mathrm{C}},m_{\mathrm{C}},v,w,b} = \mathcal{R}_{\mathrm{S-S};l_{\mathrm{A}},m_{\mathrm{A}},l_{\mathrm{C}},m_{\mathrm{C}},v,w,b}(t,\mathbf{r}_{\mathrm{AC}})$$
$$+ \mathcal{R}_{\mathrm{T-S};l_{\mathrm{A}},m_{\mathrm{A}},l_{\mathrm{C}},m_{\mathrm{C}},v,w,b}(t,\mathbf{r}_{\mathrm{AC}}) \tag{7.75}$$

with

$$\mathcal{R}_{\mathrm{S-S};l_{\mathrm{A}},m_{\mathrm{A}},l_{\mathrm{C}},m_{\mathrm{C}},v,w,b}$$
$$= -\frac{G}{M_{\mathrm{C}}} M_{l_{\mathrm{A}},m_{\mathrm{A}}}^{\mathrm{S_A}} M_{l_{\mathrm{C}},m_{\mathrm{C}}}^{\mathrm{S_C}} (-1)^{l_{\mathrm{A}}} \gamma_{l_{\mathrm{C}},m_{\mathrm{C}}}^{l_{\mathrm{A}},m_{\mathrm{A}}}$$
$$\times \frac{1}{a_{\mathrm{C}}^{l_{\mathrm{A}}+l_{\mathrm{C}}+1}} \sum_{v,w,b} \kappa_{l_{\mathrm{A}}+l_{\mathrm{C}},v} d_{v,m_{\mathrm{A}}+m_{\mathrm{C}}}^{l_{\mathrm{A}}+l_{\mathrm{C}}}(\varepsilon_{\mathrm{A}}) F_{l_{\mathrm{A}}+l_{\mathrm{C}},v,w}(I_{\mathrm{C}}) G_{l_{\mathrm{A}}+l_{\mathrm{C}},w,b}(e_{\mathrm{C}})$$
$$\times \exp[i \Psi_{l_{\mathrm{A}}+l_{\mathrm{C}},m_{\mathrm{A}}+m_{\mathrm{C}},v,w,b}] \tag{7.76}$$

and

$$\mathcal{R}_{\mathrm{T-S};l_{\mathrm{A}},m_{\mathrm{A}},l_{\mathrm{C}},m_{\mathrm{C}},v,w,b}$$
$$= -\frac{G}{M_{\mathrm{C}}} M_{l_{\mathrm{A}},m_{\mathrm{A}}}^{\mathrm{T_A}} M_{l_{\mathrm{C}},m_{\mathrm{C}}}^{\mathrm{S_C}} (-1)^{l_{\mathrm{A}}} \gamma_{l_{\mathrm{C}},m_{\mathrm{C}}}^{l_{\mathrm{A}},m_{\mathrm{A}}}$$
$$\times \frac{1}{a_{\mathrm{C}}^{l_{\mathrm{A}}+l_{\mathrm{C}}+1}} \sum_{v,w,b} \kappa_{l_{\mathrm{A}}+l_{\mathrm{C}},v} d_{v,m_{\mathrm{A}}+m_{\mathrm{C}}}^{l_{\mathrm{A}}+l_{\mathrm{C}}}(\varepsilon_{\mathrm{A}}) F_{l_{\mathrm{A}}+l_{\mathrm{C}},v,w}(I_{\mathrm{C}}) G_{l_{\mathrm{A}}+l_{\mathrm{C}},w,b}(e_{\mathrm{C}})$$
$$\times \exp[i \Psi_{l_{\mathrm{A}}+l_{\mathrm{C}},m_{\mathrm{A}}+m_{\mathrm{C}},v,w,b}]. \tag{7.77}$$

Using Eq. (7.56), the Tide-Structure interaction disturbing function is expanded as

$$\mathcal{R}_{T-S;l_A,m_A,l_C,m_C,v,w,b}$$
$$= \sum_{l_B,m_B,j,p,q} \mathcal{R}_{I;L_I}(t,\mathbf{r}_{AC}) + \sum_{l'_A,m'_A,l_B,m_B,r,s,u} \mathcal{R}_{II;L_{II}}(t,\mathbf{r}_{AC}), \qquad (7.78)$$

where $\mathcal{R}_{I;L_I}$ and $\mathcal{R}_{II;L_{II}}$ respectively correspond to the $M_{l_A,m_A}^{T_A;I}$ and $M_{l_A,m_A}^{T_A;II}$ contribution to the external gravitational potential of the tidally perturbed body A, V_T^A.

7.2.2.3 Dynamical Equations

The external potential of the tidally perturbed body A by a body B being now well understood and known, our purpose here is to derive the dynamical equations for the evolution of the angular velocity (the angular momentum in term of Andoyer's variables) and the obliquity of A (ε_A) under the action of the gravitational interaction with a body C which could be different from the perturber B and of the Keplerian orbital elements of this third body (a_C, e_C and I_C). To achieve this aim, we follow the method adopted by [22–24] who used the mutual interaction potential for the variation of the Andoyer's variables and the disturbing function of the orbital elements. Here, gravitational interactions between B and C are not taken into account.

Beginning with the Andoyer's variables (cf. [2]), we respectively get the evolution of the total angular momentum, $L_A = I_A \Omega_A$, I_A being the inertia momentum of A

$$\frac{dL_A}{dt} = \partial_{\Theta_A} V_{A-C}, \qquad (7.79)$$

and of the obliquity (ε_A)

$$L_A \frac{d}{dt} \cos\varepsilon_A = -\partial_{\phi_A} V_{A-C} - \cos\varepsilon_A \partial_{\Theta_A} V_{A-C}. \qquad (7.80)$$

Next, the classical equations of orbital evolution are given by the Lagrange's planetary equations (cf. [12]):

$$\frac{da_C}{dt} = \frac{2}{nc a_C} \partial_{M_C} \mathcal{R}_{A-C}, \qquad (7.81)$$

$$\frac{de_C}{dt} = -\frac{\sqrt{1-e_C^2}}{nc a_C^2 e_C} \partial_{\omega_C} \mathcal{R}_{A-C} + \frac{1-e_C^2}{nc a_C^2 e_C} \partial_{M_C} \mathcal{R}_{A-C}, \qquad (7.82)$$

$$\frac{dI_C}{dt} = -\frac{1}{nc a_C^2 \sqrt{1-e_C^2} \sin I_C} \partial_{\Omega_C^*} \mathcal{R}_{A-C} + \frac{\cos I_C}{nc a_C^2 \sqrt{1-e_C^2} \sin I_C} \partial_{\omega_C} \mathcal{R}_{A-C}. \qquad (7.83)$$

7.2.2.4 When Is it Necessary to Go Beyond the Punctual Mass Approximation for the Tidal Potential?

Here, our goal is to quantify the term(s) of the disturbing function due to the non-ponctual behaviour of the perturber B and to compare it to the one in the ponctual mass case.

To achieve this aim, some assumptions are assumed. First, we adopt the quadrupolar approximation for the response of A to the tidal excitation by B. Thus, we assume that $l_A = 2$ so that $\mathcal{R}_{II;L_{II}} = 0$. Then, we consider the simplified situation where the body of which dynamics is studied is the tidal perturber; therefore $B = C$ and we get

$$
\mathcal{R}_{I;L_I} = -\frac{G}{M_B}\frac{4\pi}{5}k_2^A R_A^5 \left| Z_{T_A;2,m_A,L_I}(\nu, K; \Psi_{2+l_B,m_A+m_B,j,p,q}) \right| \left[\gamma_{l_B,m_B}^{2,m_A}\right]^2 \left| M_{l_B,m_B}^B \right|^2
$$

$$
\times \frac{1}{a_B^{2(2+l_B+1)}}[\kappa_{2+l_B,j}]^2\left[d_{j,m_A+m_B}^{2+l_B}(\varepsilon_A)\right]^2\left[F_{2+l_B,j,p}(I_B)\right]^2\left[G_{2+l_B,p,q}(e_B)\right]^2
$$

$$
\times \exp\left[i\delta_{T_A;2,m_A,L_I}(\nu, K; \Psi_{2+l_B,m_A+m_B,j,p,q})\right]. \tag{7.84}
$$

On the other hand, since we are interested in the amplitude of $\mathcal{R}_{I;L_I}$, we focus on its norm ($|\mathcal{R}_{I;L_I}|$). Finally, as we know that the dissipative part of the tide is very small compared to the adiabatic one (see [126]), we can assume that $|Z_{T_A}| \approx 1$ in this first step.

Let us first derive the term of $|\mathcal{R}_{I;L_I}|$ due to the non-ponctual term of the gravitational potential of B, which has a non-zero average in time over an orbital period of B, $\langle V_{N-P}^B \rangle_{T_B}(\mathbf{r}) = 1/T_B \int_0^{T_B} V_{N-P}^B(t, \mathbf{r}) dt$ that corresponds to the axisymmetric rotational and permanent tidal deformations (see [130]) (the same procedure can of course be applied to the non-stationary and non-axisymmetric deformations, but we choose here to focus only on $\langle V_{N-P}^B \rangle_{T_B}$ to illustrate our purpose). Then, as the considered deformations of B are axisymmetric, we can expand them using the usual gravitational moments of B (J_{l_B}) as

$$
V^B(\mathbf{r}) = \frac{GM_B}{r} + \langle V_{N-P}^B \rangle_{T_B}, \tag{7.85}
$$

where

$$
\langle V_{N-P}^B \rangle_{T_B} = G\sum_{l_B>0}(M_{l_B,0}^{S_B} + M_{l_B,0}^{T_B})\frac{Y_{l_B,0}(\theta, \varphi)}{r^{l_B+1}} \tag{7.86}
$$

with

$$
M_{l_B,0}^{S_B} + M_{l_B,0}^{T_B} = -\frac{J_{l_B}M_B R_B^{l_B}}{\mathcal{N}_{l_B,0}}. \tag{7.87}
$$

Then, we obtain

$$
\left|\mathscr{R}_{I;L_I}^{J_{l_B}}(a_B, e_B, I_B, \varepsilon_A)\right| = \frac{G}{M_B}\frac{4\pi}{5}k_2^A R_A^5 \left[\gamma_{l_B,0}^{2,m_A}\right]^2 \left|M_{l_B,0}^{S_B} + M_{l_B,0}^{T_B}\right|^2
$$

$$
\times \frac{1}{a_B^{2(2+l_B+1)}}\left[\kappa_{2+l_B,j}\right]^2 \left[d_{j,m_A}^{2+l_B}(\varepsilon_A)\right]^2 \left[F_{2+l_B,j,p}(I_B)\right]^2
$$

$$
\times \left[G_{2+l_B,p,q}(e_B)\right]^2. \tag{7.88}
$$

On the other hand, the term of $|\mathscr{R}_{I;L_I}|$ associated to M_B, namely the disturbing function in the case where B is assumed to be a ponctual mass, is given by

$$
\left|\mathscr{R}_{I;L_I}^{M_B}(a_B, e_B, I_B, \varepsilon_A)\right|
$$

$$
= \frac{G}{M_B}\frac{4\pi}{5}k_2^A R_A^5 M_B \frac{1}{a_B^6}\left[\kappa_{2,j}\right]^2\left[d_{j,m_A}^2(\varepsilon_A)\right]^2\left[F_{2,j,p}(I_B)\right]^2\left[G_{2,p,q}(e_B)\right]^2. \tag{7.89}
$$

For this first evaluation of the ratio $|\mathscr{R}_{I;L_I}^{J_{l_B}}|/|\mathscr{R}_{I;L_I}^{M_B}|$, we focus on the configuration of minimum energy. In this state, the spins of A and B are aligned with the orbital one so that $\varepsilon_A = I_B = 0$ (that leads to $j = m_A$ and $p = (2 - m_A + l_B)/2$) and the orbit is circular ($e_B = 0$). Then, we consider:

$$
\mathscr{E}_{m_A,l_B} = \frac{\left|\mathscr{R}_{I;L_I}^{J_{l_B}}(a_B, 0, 0, 0)\right|}{\left|\mathscr{R}_{I;L_I}^{M_B}(a_B, 0, 0, 0)\right|}. \tag{7.90}
$$

Using Eqs. (7.88)–(7.89), we get its expression in function of J_{l_B} and of (R_B/a_B):

$$
\mathscr{E}_{m_A,l_B} = \frac{1}{4\pi}\left[\frac{1}{\mathscr{N}_{l_B}^0}\frac{\gamma_{l_B,0}^{2,m_A}}{\gamma_{0,0}^{2,m_A}}\frac{\kappa_{2+l_B,m_A}}{\kappa_{2,m_A}}\frac{F_{2+l_B,2,\frac{l_B}{2}}(0)}{F_{2,2,0}(0)}\right]^2 J_{l_B}^2\left(\frac{R_B}{a_B}\right)^{2l_B}. \tag{7.91}
$$

As it has been emphasised by [127, 130], the main mode of the dissipative tide ruling the secular evolution of the system is $m_A = 2$. We thus define $\mathscr{E}_{l_B}$ such that

$$
\mathscr{E}_{l_B} = \mathscr{E}_{2,l_B} = \left[\frac{1}{3}F_{2+l_B,2,\frac{l_B}{2}}(0)\right]^2 J_{l_B}^2\left(\frac{R_B}{a_B}\right)^{2l_B}, \tag{7.92}
$$

which can be recast into

$$
\log(\mathscr{E}_{l_B}) = 2\left[\log\left[\frac{1}{3}F_{2+l_B,2,\frac{l_B}{2}}(0)\right] + \log J_{l_B} - l_B \log\left(\frac{a_B}{R_B}\right)\right]. \tag{7.93}
$$

Finally, keeping only into account the quadrupolar deformation of B (J_2), we get:

$$
\log(\mathscr{E}_2) = 2\left[\log\left(\frac{5}{2}\right) + \log J_2 - 2\log\left(\frac{a_B}{R_B}\right)\right]. \tag{7.94}
$$

This gives us the order of magnitude of the terms due to the non-ponctual behaviour of B compared to the one obtained in the ponctual mass approximation. It is directly proportional to the squared J_2, thus increasing with ε_Ω^2 (where $\varepsilon_\Omega = \Omega_B^2/\Omega_c^2$ with $\Omega_c = \sqrt{\frac{GM_B}{R_B^3}}$) in the case of the rotation-induced deformation and with ε_T^2 (where $\varepsilon_T = q(R_B/a_B)^3$ where $q = M_A/M_B$) in the tidal one, while it increases

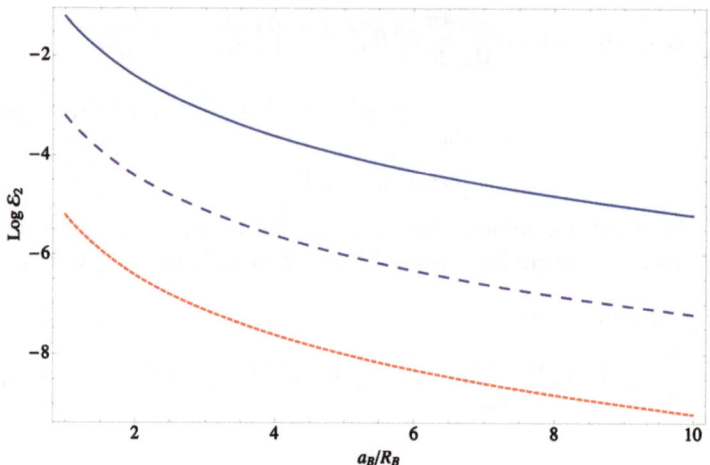

Fig. 7.5 Log $\mathscr{E}_2$ in function of a_B/R_B for $J_2 = 10^{-3}$ (*red dashed line*), 10^{-2} (*purple long-dashed line*), 10^{-1} (*blue solid line*). The non-ponctual terms have to be taken into account for strongly deformed perturbers ($J_2 \geq 10^{-2}$) in very close systems ($a_B/R_B \leq 5$) while they decrease rapidly otherwise. (Taken from Mathis and Le Poncin-Lafitte [77], courtesy Astronomy & Astrophysics)

as $(R_B/a_B)^4$. Therefore, as it is shown in Fig. 7.5, the non-ponctual terms have to be taken into account for strongly deformed perturbers ($J_2 \geq 10^{-2}$) in very close systems ($a_B/R_B \leq 5$) while they decrease rapidly otherwise. This corresponds for example to the case of internal natural satellites of rapidly rotating giant planets as Jupiter and Saturn $J_2 \approx 1.4697 \cdot 10^{-2}$ for Jupiter and $J_2 \approx 1.6332 \cdot 10^{-2}$ for Saturn; (see [41] and references therein). In the case of close Hot-Jupiters which are already synchronised (because of the tidal dissipation the rotation period is close to the orbital one), the rotation period is larger than 2 days (50 hours) that is roughly 5 times slower than Jupiter's rotation (10 hours). In this case, the flattening of Hot-Jupiter is less important and their J_2 should be of the same order of the Earth's value (*i.e.* J_2 runs from 10^{-4} to 10^{-3}). Then, the relative effect of the non-ponctual terms is less important. The situation may be different in the earliest evolutionary stages of those systems.

7.2.2.5 The Classical Two-Body Case with a Punctual Tidal Perturber

We now focus on binary systems (B = C) close enough for the tidal interaction to play a role, but we also consider that the companion is far (or small) enough to be treated as a point mass (*i.e.* $a_B \geq 5\overline{R}_A$, where $\overline{R}_A$ is the mean A radius; cf. Sect. 7.2.2.4). We then are allowed to assume the *quadrupolar approximation*, where we only keep the first mode of the potential, $l_A = 2$:

$$V_T^A(r, \theta, \varphi, t)$$
$$= \mathscr{R}e \left[\sum_{m_A=-2}^{2} \sum_{j=-2}^{2} \sum_{p=0}^{2} \sum_{q \in \mathbb{Z}} V_{m_A,j,p,q}(r) P_2^{m_A}(\cos\theta) e^{i\Phi_{2,m_A,j,p,q}(\varphi,t)} \right], \quad (7.95)$$

where

$$\Phi_{2,m_A,j,p,q}(\varphi,t) = m_A\varphi + \Psi_{2,m_A,j,p,q}(t). \tag{7.96}$$

The functions $V_{m_A,j,p,q}(r,\theta)$ may be expressed in terms of the Keplerian elements (the semi-major axis a_B of the orbit, its eccentricity e_B and its inclination I_B) and the obliquity ε_A of the rotation axis of A, as

$$V_{m_A,j,p,q}(r) = (-1)^{m_A}\sqrt{\frac{(2-m_A)!(2-|j|)!}{(2+m_A)!(2+|j|)!}}$$
$$\times \frac{GM_B}{a_B^3}[d_{j,m}^2(\varepsilon_A)F_{2,j,p}(I_B)G_{2,p,q}(e_B)]r^2. \tag{7.97}$$

If we simplify the expansion of the potential in the case where spins are aligned and perpendicular to the orbital plan, where obliquity ε_A and orbital inclination I_B are zero, Eq. (7.95) reduces to the expression of the potential given by Zahn [130].

Then, the evolution equations of the semi-major axis (a_B), of the eccentricity (e_B), of the inclination (I_B), of the obliquity (ε_A) and of the angular velocity (Ω_A) given in Sect. 2.3.3 become [32, 77, 94]:

$$\frac{dL_A}{dt} = -\frac{8\pi}{5}\frac{GM_B^2R_A^5}{a_B^6}\sum_{m_A,j,p,q}\left\{\frac{k_2^A(\sigma_{2,m_A,p,q})}{Q(\sigma_{2,m_A,p,q})}[\mathcal{H}_{2,m_A,p,q}(e_B,I_B,\varepsilon_A)]^2\right\}, \tag{7.98}$$

$$L_A\frac{d(\cos\varepsilon_A)}{dt}$$
$$= \frac{4\pi}{5}\frac{GM_B^2R_A^5}{a_B^6}$$
$$\times \sum_{m_A,j,p,q}\left\{(j+2\cos\varepsilon_A)\frac{k_2^A(\sigma_{2,m_A,p,q})}{Q(\sigma_{2,m_A,p,q})}[\mathcal{H}_{2,m_A,j,p,q}(e_B,I_B,\varepsilon_A)]^2\right\}, \tag{7.99}$$

$$\frac{1}{a_B}\frac{da_B}{dt} = -\frac{2}{n_B}\frac{4\pi}{5}\frac{GM_B^2R_A^5}{a_B^8}$$
$$\times \sum_{m_A,j,p,q}\left\{(2-2p+q)\frac{k_2^A(\sigma_{2,m_A,p,q})}{Q(\sigma_{2,m_A,p,q})}[\mathcal{H}_{2,m_A,p,q}(e_B,I_B,\varepsilon_A)]^2\right\}, \tag{7.100}$$

$$\frac{1}{e_B}\frac{de_B}{dt} = -\frac{1}{n_B}\frac{1-e_B^2}{e_B^2}\frac{4\pi}{5}\frac{GM_B^2R_A^5}{a_B^8}$$
$$\times \sum_{m_A,j,p,q}\left\{\left[(2-2p)\left(1-\frac{1}{\sqrt{1-e_B^2}}\right)+q\right]\frac{k_2^A(\sigma_{2,m_A,p,q})}{Q(\sigma_{2,m_A,p,q})}\right.$$
$$\times [\mathcal{H}_{2,m_A,p,q}(e_B,I_B,\varepsilon_A)]^2\Bigg\}, \tag{7.101}$$

$$\frac{d(\cos I_B)}{dt} = \frac{1}{n_B} \frac{1}{\sqrt{1-e_B^2}} \frac{4\pi}{5} \frac{GM_B^2 R_A^5}{a_B^8}$$

$$\times \sum_{m_A, j, p, q} \left\{ [j + (2q-2)\cos I_B] \frac{k_2^A(\sigma_{2,m_A,p,q})}{Q(\sigma_{2,m_A,p,q})} \right.$$

$$\left. \times \left[\mathcal{H}_{2,m_A,j,p,q}(e_B, I_B, \varepsilon_A) \right]^2 \right\}, \tag{7.102}$$

where function $\mathcal{H}_{m_A,j,p,q}(e_B, I_B, \varepsilon_A)$ is expressed as

$$\mathcal{H}_{2,m_A,j,p,q}(e_B, I_B, \varepsilon_A) = \sqrt{\frac{5}{4\pi} \frac{(2-|j|)!}{(2+|j|)!}} \, d_{j,m_A}^2(\varepsilon_A) \, F_{2,j,p}(I_B) \, G_{2,p,q}(e_B). \tag{7.103}$$

Note that these expressions are valid for high eccentricities if enough terms are considered in the expansion (see [106]).

The mass redistribution due to the tide is thus at the origin of a tidal torque of non-zero average which induces an exchange of angular momentum between each component and the orbital motion. As shown by previous equations, this tidal torque is proportional to the tidal dissipation k_2^A/Q (see [23, 24]), where

$$k_2^A = |\tilde{k}_2^A| \tag{7.104}$$

(see Eq. (7.59) for the $\tilde{k}_2^A$ definition) is the Love number that describes the adiabatic response of A to the tidal excitation [70] and

$$Q = \frac{|\mathscr{I}m\tilde{k}_2^A|}{|\tilde{k}_2^A|} \tag{7.105}$$

is the tidal quality factor, which scales as the inverse of the ratio of the energy losses during an orbital period to the total energy of the system [39]. For a perfectly adiabatic response to the tidal excitation, A will be elongated in the direction of the line of centers, inducing a torque with periodic variations of zero average, so that no secular exchanges of angular momentum will be possible (see [94, 126]). Next, if dissipative processes are taken into account, the deformation of A presents a time delay Δt with respect to the tidal forcing, which may be measured also by the tidal lag angle 2δ or equivalently by the quality factor Q (see [31, 34]):

$$\sin[\Delta t \times \sigma_{2,m_A,p,q}] = \sin[2\delta_{T_A}(\sigma_{2,m_A,p,q})]$$

$$= \frac{1}{Q(\sigma_{2,m_A,p,q})}. \tag{7.106}$$

Thus, the tidal bulge is no more aligned with the line of centers, as shown in Fig. 7.4. The resulting tidal angle is at the origin of a torque of non-zero average which causes exchange of spin and orbital angular momentum between the components of the system [127].

Then, if the considered two-body system is isolated, two evolutions are possible. In the most common case, provided the system does not loose angular momentum, it tends to a state of minimum energy in which the orbits are circular, the rotation of the components is synchronised with the orbital motion, and the spins are aligned. From previous dynamical equations one may derive the characteristic times of synchronisation, circularisation and spin alignment:

$$\frac{1}{t_{\text{sync}}} = -\frac{1}{\Omega_A - n_B}\frac{d\Omega_A}{dt} = -\frac{1}{I_A(\Omega_A - n_B)}\frac{dL_A}{dt}, \tag{7.107}$$

$$\frac{1}{t_{\text{circ}}} = -\frac{1}{e_B}\frac{de_B}{dt}, \tag{7.108}$$

$$\frac{1}{t_{\text{align}_A}} = -\frac{1}{\varepsilon_A}\frac{d\varepsilon_A}{dt} = \frac{1}{\varepsilon_A \sin \varepsilon_A}\frac{d(\cos \varepsilon_A)}{dt}, \tag{7.109}$$

$$\frac{1}{t_{\text{align}_{\text{Orb}}}} = -\frac{1}{I_B}\frac{dI_B}{dt} = \frac{1}{I_B \sin I_B}\frac{d(\cos I_B)}{dt}. \tag{7.110}$$

However, in very close systems, such final state cannot be achieved: instead, the secondary spirals toward the primary and may be engulfed by it [48, 69].

7.3 Tidal Dissipation Mechanisms in Planetary Systems

7.3.1 The Tidal Kinetic Energy Dissipation: The Driver of Systems Evolution

As it has been emphasised in the previous section, the tidal interaction can be broken down in two steps. First, if we adopt an ideal adiabatic view of the action of the tidal potential exerted by the secondary on the primary, this latter becomes elongated along the line of centers. However, dissipative processes that convert the tidal kinetic energy into heat, such as viscous friction and heat diffusion, have to be taken into account (see Fig. 7.6). Then, the response of the studied body to the tidal excitation presents a delay (the tidal lag), which translates into the tidal angle between the tidal bulge and the line of centers. Those tidal delay and angle are thus directly related to the dissipative mechanisms and their dependence on the tidal frequency. Therefore, to predict the fate of a binary system, one has to identify and to model the dissipative processes achieving the conversion of kinetic energy into thermal energy that take place in fluid and solid layers and at their interfaces, from which one can derive the characteristic times of circularisation, synchronisation and spins' alignment.

7.3.2 Tidal Dissipation in Fluid Bodies

While stars are fluid bodies, planets host gaseous and liquid layers: the deep envelopes of giant planets, as well as the internal core, the atmosphere and the possible ocean of telluric ones (see Figs. 7.7 and 7.8). Therefore, one has to obtain a complete understanding of flows that are tidally excited and dissipated in such regions in planetary systems.

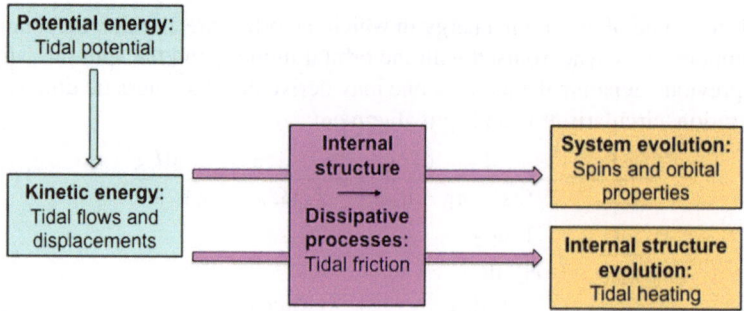

Fig. 7.6 The tidal energy dissipation: first, the gravitational tidal potential energy generates tidal flows with a given kinetic energy. Next, the internal friction, related to the internal structure properties of the studied body, dissipates this kinetic energy into heat. This process leads to the evolution of the system (by modifying its spins and orbital properties) and change the internal structure of its components by tidal heating [68]

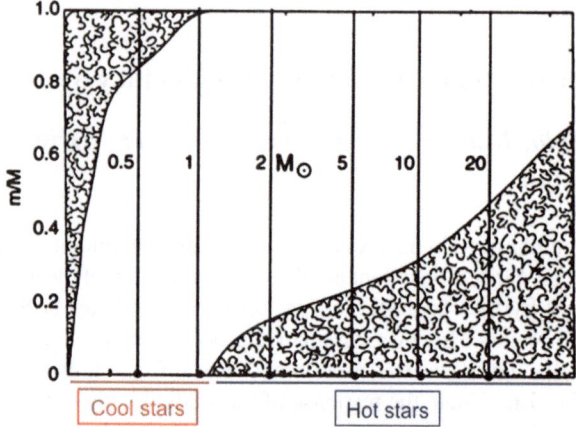

Fig. 7.7 Internal structure of main-sequence stars as a function of their mass represented using Kippenhahn diagram: stellar masses are given in the *horizontal axis* while the fraction in mass and the position of convection zones (*cloudy regions*) and of radiation zones are given along the *vertical axis* (adapted from Kippenhahn and Weigert [56], courtesy Springer). Very low-mass stars are entirely convective. Next, as the stellar mass grows the radiative core becomes more and more important. Finally, a transition occurs because hydrogen is converted into helium through the CNO cycle and a convective core takes place with an external radiative envelope

7.3.2.1 Type of Fluid Tides

Two types of tides operate in stars[5] and in fluid planetary layers: the equilibrium and dynamical tides. On one hand, the equilibrium tide designates the large-scale flow induced by the hydrostatic adjustment of studied fluid layers in response to

---

[5]See also the lecture in this volume on stellar tides by J.-P. Zahn.

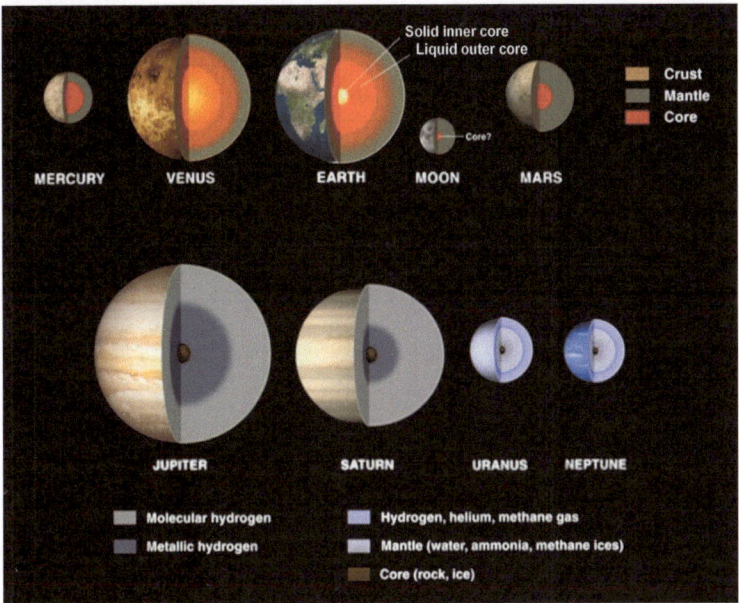

Fig. 7.8 *Top*: internal structure of telluric planets in the Solar system and of the Moon. *Bottom*: internal structure of gaseous giant planets (Jupiter & Saturn) and of icy giant planets (Uranus & Neptune) in the Solar system (see [3, 35, 41, 42]). (Courtesy NASA/JPL-Caltech)

the gravitational force exerted by the companion [126, 127]. On the other hand, the dynamical tide corresponds to the fluid eigenmodes that are excited by the tidal potential. Let us now detail the different types of eigenmodes that should be studied in stellar and fluid planetary regions. First, if Fig. 7.9 is considered, four characteristic frequencies are introduced: the Alfvén frequency ($\omega_{\mathrm{A}} = \frac{B}{\sqrt{\mu\rho}r\sin\theta}$, where B is the field amplitude and μ is the magnetic permeability), the inertial frequency (2Ω), the Brunt-Väisäla frequency (N), and the Lamb frequency (f_{L}). These delimit the frequency domain of corresponding Alfvén waves ($\omega < \omega_{\mathrm{A}}$), inertial waves ($\omega < 2\Omega$), gravity waves (for which $\omega < N$ and that are also called internal waves), and acoustic waves ($\omega > f_{\mathrm{L}}$); these are respectively driven by the magnetic tension force, the Coriolis acceleration, the buoyancy force and the compressibility of the studied layers. If we now focus on low-frequency waves, inertial waves are propagating in convective regions while internal waves are propagating in stably stratified regions. For these latters, if considered frequencies are of the same order of magnitude that the Alfvén and the inertial frequencies, these become gravito-inertial waves if we add the action of the Coriolis acceleration to the one of buoyancy and magneto-gravito-inertial waves if the magnetic field is taken into account. In this picture, tidal excitation is mostly efficient for low-frequency eigenmodes, thus for inertial and internal waves. Moreover, for acoustic waves, which are high-frequency waves, the action of tides is only a perturbation. Therefore, one has to focus on inertial and

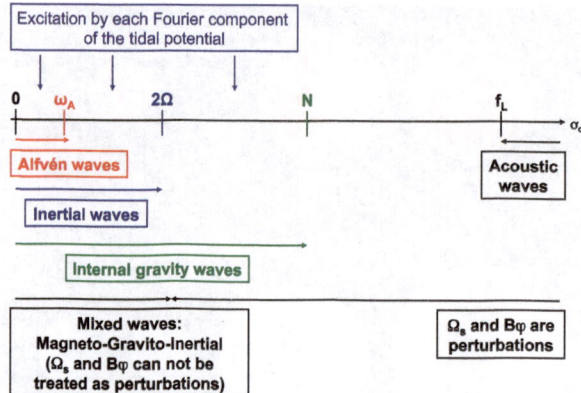

Fig. 7.9 Main wave types in stellar interiors and in fluid planetary layers. Tidal interactions excite low-frequency waves such as inertial waves in convective regions and gravito-inertial waves in stably stratified zones (and associated magneto-inertial and magneto-gravito-inertial waves if magnetic fields are taken into account); high-frequency acoustic waves are only perturbed by the tidal potential. (Adapted from Mathis and de Brye [75], courtesy Astronomy & Astrophysics)

on gravito-inertial waves to study the dynamical tide respectively in convective and in stably stratified regions.

Next, dissipative processes that convert kinetic energy of tidally excited fluid velocities into heat have to be identified.

First, stellar and planetary convective layers host strong turbulent flows because of the high value of the Reynolds number in such celestial bodies. In such regions, the action of turbulence on the tidal flows (the equilibrium tide and the dynamical tide, i.e. the inertial waves excited by the tidal potential) can be modelled as a viscous force with a turbulent viscosity coefficient (see for example [37, 127]). This implicitly assumes that the respective length scales of tidal and convective flows allow to distinguish one from the other. Such rough modeling has now been confirmed using direct numerical simulations of the interaction between an highly turbulent convection and a tidal velocity (see for example [91, 92]; note that in these works, the prescription given by Zahn [127] where the turbulent viscosity scales linearly with the tidal period (and thus the inverse of the tidal frequency) has been confirmed). So, in convective regions, the kinetic energy of tidal flows is dissipated into heat because of the turbulent viscous friction.

Next, in stably stratified stellar and planetary regions, the dynamical tide (i.e. the gravito-inertial waves) is dissipated through viscous and thermal diffusions (see for example [38, 41]). Then the ratio between the viscous and the radiative damping is govern by the Prandtl number ($Pr = \nu/K$, where ν and K are respectively the viscosity and the thermal diffusivity). Moreover, [128] has demonstrated that the dissipation of the equilibrium tide in stellar radiation zones can be neglected.

Finally, planetary interiors host solid/fluid interfaces. There, viscous friction occurs that contributes to the dissipation of tides kinetic energy.

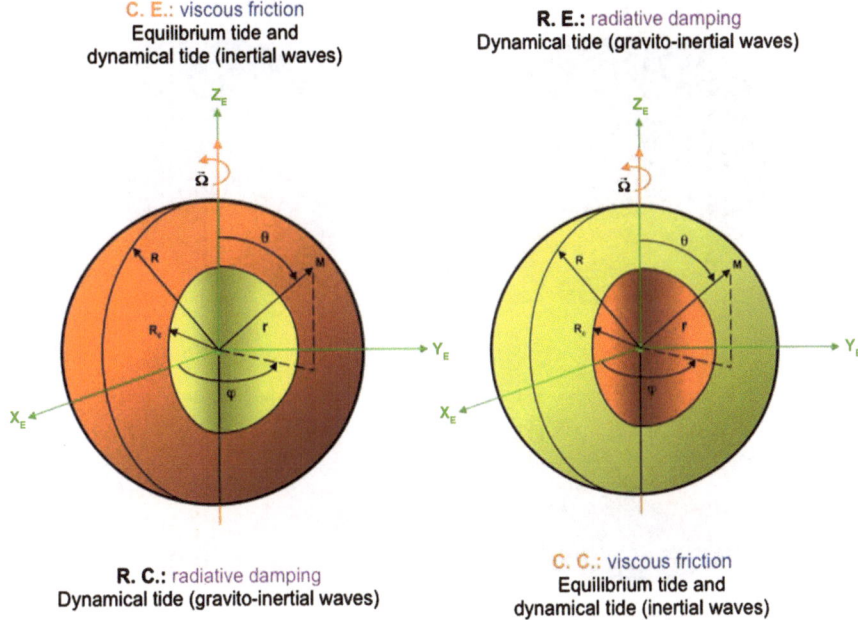

Fig. 7.10 Dissipative processes acting in solar-type stars (*left*) and in massive stars and in Jupiter (and Saturn)-type giant planets (*right*). CE, RE, CC, RC are respectively convective and radiative envelopes and cores. Note that for giant planets, a rocky/icy core could exist [3, 35, 41, 42]

All of this underlines the importance of the internal structure of the studied bodies, because each stellar or planetary layer dissipates the kinetic energy in function of its fluid or solid nature and of its stability with respect to the convective instability for fluid regions. A summary of the dissipation mechanisms that contribute to the tidal friction in liquid and gaseous regions is given in Fig. 7.10.

7.3.2.2 The Fluid Equilibrium Tide

Let us first consider the equilibrium tide. As it has been explained in previous sections, the studied component adjusts in a quasi-hydrostatic way when this is submitted to the tidal potential exerted by the companion. Then, a large-scale flow in phase with the tidal potential is excited as a response to such structural adjustment with an amplitude scaling with the tidal frequency; this is the equilibrium tide (see for example [50, 94, 126, 130]). Since the tidal force is derived from a potential, the density is constant on an isobar, which is also an equipotential of the total potential (the sum of the self-gravitation potential and of the tidal one). Then, by definition of the equilibrium tide, the tidal deformation and the related structural variables (the total gravitational potential and the density) and the equilibrium tide velocity field are time-independent in a frame rotating with the studied Fourier component of the tidal potential; because of that property, the equilibrium tide velocity field is

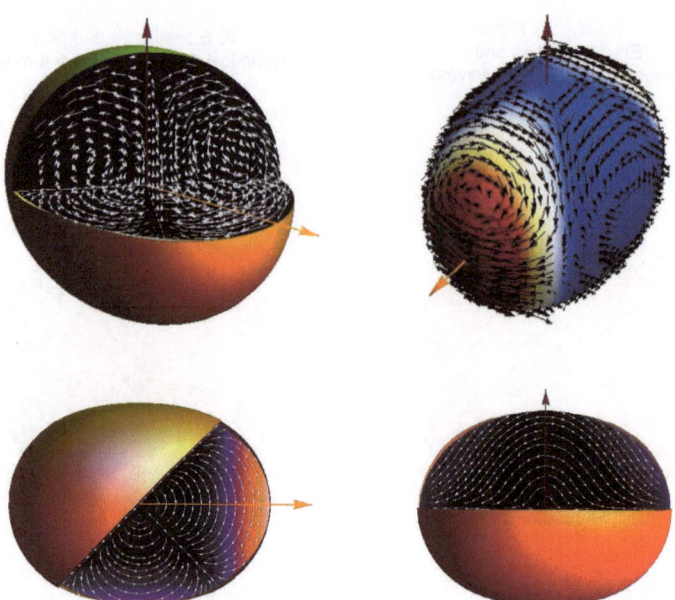

Fig. 7.11 *Top-left*: 3-D view of the total (poloidal and toroidal) adiabatic equilibrium tide velocity field (*white arrows*). The *red* and *orange arrows* indicate the direction of the primary's rotation axis and the line of centers respectively. *Top-right*: Representation of this velocity field at the surface of the primary (*black arrows*); the color-scaled background represents the normalised tidal potential intensity (*blue* and *red* for the minimum and maximum values respectively). *Bottom-left*: View of the velocity field (*white arrows*) in its equatorial plane of symmetry; the color-scaled background represents the velocity value (*black* and *orange* for the minimum and maximum values respectively). *Bottom-right*: View of the velocity field (*white arrows*) in its meridional plane of symmetry; the color-scaled background represents the velocity value (*black* and *purple*) for the minimum and maximum values respectively. (Taken from Remus et al. [94], courtesy Astronomy & Astrophysics)

divergence-free (see [94]) contrary to the claim of [33, 110]. It is also important to point that since this velocity field represented in Fig. 7.11 verifies the momentum equation in this rotating frame, the Coriolis acceleration must be taken into account in its derivation (the equilibrium tide velocity field has thus both a poloidal and a toroidal components). Next, the view of the equilibrium tide we gave above is an adiabatic view, and we must now introduce the associated friction mechanism. For the equilibrium tide this is the turbulent friction in convective regions that will act to convert its kinetic energy into heat (note that [128] has demonstrated that its dissipation in stably stratified regions is negligible). Then, the interaction between turbulent convective flows and the equilibrium tide velocity field has to be examined. From now on, two main assumptions are made: first, we consider that the respective scales of the turbulent convection and the equilibrium tide are different enough to be separated; next, we assume that the action of the turbulent convection on the tidal flow can be modelled as a viscous force, where the used viscosity is a "turbulent viscosity" which is enhanced compared to the molecular viscosity of the plasma. Since the adiabatic equilibrium tide has both poloidal and toroidal components, so does the related viscous force that sustains a secondary toroidal flow, called the "convective" or dissipative equilibrium tide, in quadrature with the tidal potential. By redistributing the density, this velocity field leads to a "dissipative" perturbation of the gravitational potential that drives the secular evolution of the orbit and the spins of the system components.

To describe the dissipation of the kinetic energy of the adiabatic equilibrium tide, namely its amplitude and its dependence on the tidal frequency, the key physical ingredient is the assumed prescription for the turbulent viscosity coefficient. Then, two different regimes can be drawn: the "slow tide" and "fast tide" regimes. In the first one, the orbital period of the perturber is longer that the characteristic convective turn-over time. Then, the turbulent friction can be efficient to dissipate the kinetic energy into heat. Conversely, in the fast tide regime, the orbital period is shorter that the characteristic convective turn-over time and the turbulent friction losses part of its efficiency to convert the kinetic energy into heat, leading to a saturation of the associated energy dissipation. The way in which the dissipation becomes less efficient when the tidal period becomes shorter remains one of the unsolved question in the treatment of the equilibrium tide. Two main prescriptions have been given today in the literature: those by Zahn [127, 131] and by Goldreich and Keeley [37]. In the first one the turbulent viscosity scales linearly with the tidal period while in the second one this scales as the squared tidal period. The most efficient way to probe such prescriptions on the action of turbulence is then to used three-dimensional numerical simulations of highly turbulent convective flows submitted to a periodic forcing, which is often modelled as a shear that oscillates with time. The most recent numerical simulations have been achieved with such set-up in Cartesian coordinates [90–92] that tends to confirm the prescription by Zahn [127]. In a near future, more simulations have to be computed in order to reach flows that are more turbulent and to take into account the spherical (ellipsoidal) geometry of the problem. For the moment, one can at least conclude that the viscous dissipation of the equilibrium tide varies as the product of the tidal frequency (on which depends the velocity field) with the frequency dependence of the turbulent viscosity. In the case of the linear prescription given by Zahn [127], we shall note that in the fast tide case the turbulent viscosity scales as the inverse of the tidal frequency, leading to a constant tidal dissipation, while in the slow tide regime the turbulent viscosity is constant, so that the dissipation scales here with the tidal frequency (see in Fig. 7.12).

To conclude this part on the equilibrium tide, we shall note that in a near future both the differential rotation and the magnetic field have to be taken into account since they modify at the same time the equilibrium tide velocity field, the convective flows and the associated turbulence properties.

7.3.2.3 The Fluid Dynamical Tide: Inertial and Gravito-Inertial Waves

• Inertial waves

Once the equilibrium tide has been studied, it is then necessary to focus on the dynamical tide, *i.e.* the eigenmodes of the studied body, which are excited by the

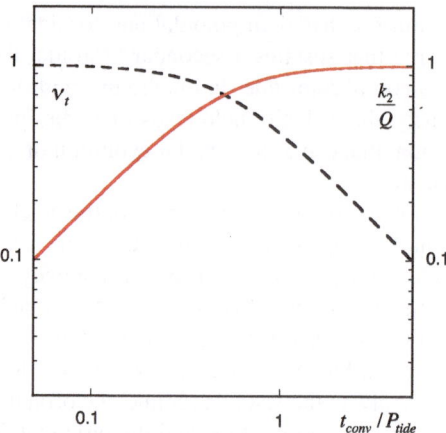

Fig. 7.12 The two regimes of the turbulent dissipation of the equilibrium tide according to the prescription by Zahn [127]. As long as the local convective turn-over time remains shorter than the tidal period ($t_{\mathrm{conv}} < P_{\mathrm{tide}}$), the turbulent viscosity ν_t (in *black dashed line*) is independent of the tidal frequency, and the inverse quality factor k_2/Q (in *red continuous line*) varies proportionally to the tidal frequency (σ_t) (so does also the tidal lag angle). When $t_{\mathrm{conv}} > P_{\mathrm{tide}}$, ν_t varies proportionally to the tidal period, whereas k_2/Q have no longer depend on the tidal frequency. ν_t and k_2/Q have been scaled by the value they take respectively for $t_{\mathrm{conv}}/P_{\mathrm{tide}} \to 0$ and $\to \infty$. (Taken from Remus et al. [94], courtesy Astronomy & Astrophysics)

tidal potential. To achieve this aim, let us first consider convective zones in stellar and planetary interiors. In those regions, if we neglect magnetic field, two type of waves are propagating: the acoustic and the inertial waves. However, as it has been explained above acoustic waves are high-frequency waves and are thus only weakly perturbed by the tidal potential. This is not the case of inertial waves for which the Coriolis acceleration is the restoring force. Then, inertial waves can be efficiently coupled with the tidal potential (see [51, 83–87, 97, 124, 125]). Such coupling can then lead to an important tidal dissipation in the convective envelopes of low-mass stars and giant planets and in the convective cores of stars and telluric planets. Because of inertial waves properties, the necessary condition to get such dissipation is that the tidal frequency is such that $|\sigma_T| \in [0, 2\Omega]$, where from now on σ_T is the tidal frequency.

Let us now examine the properties of tidally excited inertial waves. First, two configurations can occur in stellar and planetary interiors; first, the tidal potential can be coupled with inertial waves that propagate in a full sphere (case of an entirely convective star or fluid planet or of a stellar convective core); next, inertial waves can propagate between concentric spheres with stress-free or no-slip boundary conditions depending on if we are studying stellar or planetary interiors (case of the convective envelopes of low-mass stars and of giant planets if those have an heavy element rocky or icy core). Because of the cylindrical geometry related to the Coriolis acceleration, boundary conditions then strongly influence the excited flows and the related dissipation.

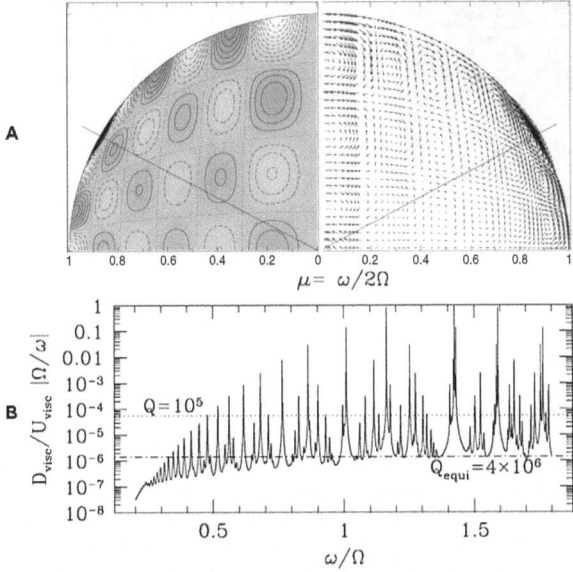

Fig. 7.13 **A**: Density perturbation (*left*) and radial and latitudinal components (*right*) of a tidally excited retrograde inertial mode with $|m| = 2$. The mode amplitude and the wavevector remain relatively uniform over much of the planet and rise sharply toward the surface; this rise is most striking near the critical latitude $\theta_c = \arccos(\sigma_T/2\Omega)$ marked by *straight lines* (Ω is the planet rotation). (Taken from Wu [124], courtesy The Astrophysical Journal.) **B**: Viscous dissipation of such inertial waves excited by the tidal potential in a fully convective planet and the associated value of the tidal quality factor Q as a function of the tidal frequency for an Eckman number $E = \nu/\Omega R^2 = 10^{-7}$, where ν is the viscosity and R the radius of the studied planet. (Taken from Wu [125], courtesy The Astrophysical Journal)

The "Full Sphere" Configuration   The first case of entirely convective sphere has been studied for example by Wu [124, 125] (Fig. 7.13). In this work, the original solution for inertial waves in a full sphere with a constant density derived by Bryan [15] has been generalised to the stratified case. This leads to global modes which then couple to the tidal potential as in the academic case of a forced oscillator. To study the dissipation of their kinetic energy, the same assumptions that in the case of the equilibrium tide (*i.e.* the spatial scales separation between the convective flows and the tidal inertial waves and the modeling of the friction using a viscous force with a turbulent viscosity) are assumed. Numerous resonances between eigenmodes and the tidal potential are obtained and identified. Then, the main important properties to study is the related amplitude of the dissipation and its dependence on the tidal frequency. First, the tidal dissipation can be increased by several order of magnitudes when a resonance is encountered compared to the case of the equilibrium tide. Next, the main difference with the equilibrium tide is the strong dependence on the tidal frequency of the tidal dissipation because of the resonances.

The Cored Configuration   The second case of inertial waves propagating in spherical shells has been studied in [84] for the planetary case and in [85] for the

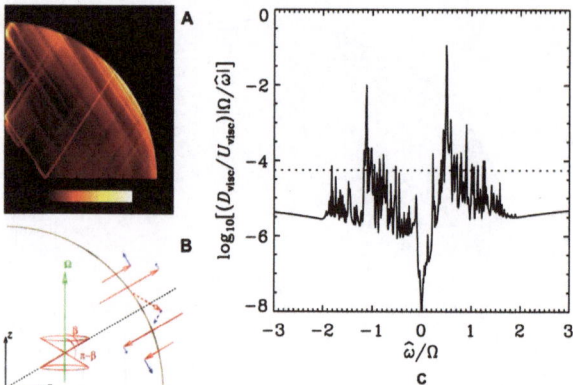

Fig. 7.14 **A**: High-resolution calculation ($E = 10^{-9}$) of the tidal response of a uniformly rotating planet with an internal core. The r.m.s. velocity of the total tide (equilibrium and dynamical) is plotted in a meridional slice through the convective region (the velocity scale is linear, *black* representing zero). Attractors with associated inertial wave beams can be identified. The forcing tidal frequency is chosen to be near the peak of an inertial mode resonance. (Taken from Ogilvie and Lin [84], courtesy The Astrophysical Journal.) **B**: Production of short inertial waves by scattering of the equilibrium tide off the core at critical latitudes to explain results obtained by Ogilvie and Lin [84]. (Taken from Goodman and Lackner [40], courtesy The Astrophysical Journal.) **C**: Viscous dissipation of such inertial waves excited by the tidal potential as a function of the tidal frequency for an Eckman number $E = 10^{-7}$; the *dotted line* corresponds to $Q = 10^5$. (Taken from Ogilvie and Lin [84], courtesy The Astrophysical Journal)

stellar case. Then, because of the conflict between the spherical geometry of the problem and the one related to the Coriolis acceleration, the propagation of inertial waves becomes more complex (Fig. 7.14). First, inertial waves propagate along characteristic rays that are inclined to the rotation axis at a certain angle, which depends on the wave frequency. This angle is necessary preserved in reflections of the waves from boundaries so that a beam is focused of defocused in such a reflection. When propagating in such spherical annulus (in opposite to the first full sphere case), inertial waves are therefore focused onto attractors where an intense dissipation occurs [83, 96, 99, 100]. Then, [82] demonstrated mathematically that the mean dissipation rate associated with waves attractors (in a simplified wave equation) becomes independent of the viscosity in the limit of very small Eckman number (*i.e.* for very small viscosity). This constitutes a remarkable behaviour compared for example to the case of the equilibrium tide dissipation, which is directly proportional to the viscosity. In the complete global modeling by Ogilvie and Lin [84], a similar behaviour is observed in the limit of small viscosity and obtained solutions indicate that the dissipation is typically concentrated along the rays that emanate from the critical latitude on the inner boundary (see also [97]). Goodman and Lackner [40] have proposed a physical interpretation of such phenomena: using WKB methods, they demonstrate the production of short inertial waves by scattering of the equilibrium tide off the core at critical latitudes. The tidal dissipation rate associated with these waves scales as the fifth power of the core radius. They also find that even if the core of rock or ice is unlikely to be rigid, Ogilvie and Lin's mechanism should still operate if the core is substantially denser than its immediate surroundings.

As a partial conclusion, we must point that the viscous dissipation of inertial waves are one of the most important processes to take into account in stellar and in planetary interiors. Let us now draw some perspectives on what should be done to improve the modeling of such processes. First, as in the case of the equilibrium tide, the differential rotation and the magnetic field have to be taken into account. First, convective flows are those that establish the differential rotation that depends both on radius and on latitude in convective regions [13]. Next such regions host dynamo generated magnetic fields [14] that are generated because of the simultaneous action of differential rotation and of convective turbulence that are themselves modified by the magnetic field because of the Lorentz force feed-back in the momentum equation. Then, inertial waves propagation and dissipation will be modified both by the differential rotation and the magnetic fields. Moreover, the dissipation may be modified both by corotation resonances between inertial waves and the sheared rotation (see also the case of gravito-inertial waves) and by the Ohmic heating that constitute supplementary dissipation sources.

Inertial Waves Instabilities Finally, it is important to point the possibility of tidal inertial waves' related instabilities. In the case of inertial waves, main instabilities can come for the interaction with differential rotation (see also the case of gravito-inertial waves) and from the tidal elliptic instability. This latter corresponds to the astrophysical version of the generic elliptical instability, which affects all rotating fluids with elliptically-deformed streamlines [55, 57, 58, 66]. In the astrophysical case, the origin of such elliptic instability is a resonance between inertial waves in rotating stars and planets and the tidal wave, *i.e.* the underlying strain field responsible for the elliptic deformation [118]. This instability is able to generate and sustain large-scale flows (for example the so-called spin-over mode) that superpose to basic flows such that differential rotation or convection in planetary and in stellar interiors (see Fig. 7.15). Then, as in the case of inertial waves, viscous forces can act to dissipate the generated kinetic energy that leads to potential important evolution of the considered system. Recent studies of the tidal elliptic instability have been recently achieved in the context of binary stars [67], planetary cores [17, 18] and extra-solar planetary systems [19]. Then, interesting behaviour of this instability have been isolated. First, this can develop both in convective or stably stratified regions (where inertial waves become gravito-inertial because of the supplementary stabilising buoyancy force) [18, 43, 64, 65]. In the case of convective regions, the elliptic instability can thus develop with a growth-rate that diminishes with the intensity of the convection; thus, the flow generated with the tidal instability can superpose to the convective one. Next, such tidal flow may play an important role in the induction of a magnetic field leading to a "tidal dynamo" [46, 59] in planetary interiors and may be in stellar ones. This last point constitutes one of the must important question to examine in a near future to see a possible impact of tidal interactions on celestial bodies magnetic activity. Finally, we must point that the dissipation related to the elliptic instability depends on boundary conditions that are applied (see

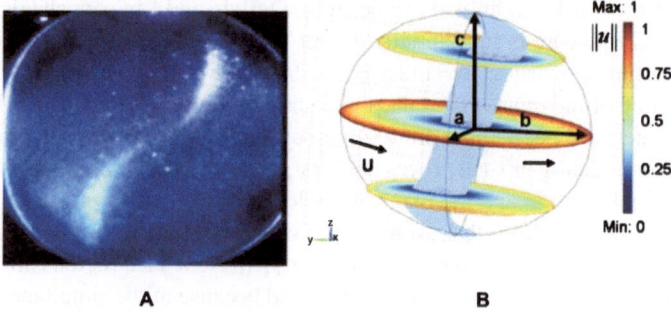

Fig. 7.15 Spin-over mode of the tidal elliptic instability observed in the laboratory (**A**) and computed in numerical simulations (**B**). (Adapted from Cébron [16, 17], courtesy D. Cébron and Physics of the Earth and Planetary Interiors)

Sect. 7.3.4), that can lead to important differences between no-slip boundary conditions (in planetary cores of telluric planets and at the interface between a central core and a surrounding fluid envelope in giant planets) and stress-free conditions (as at giant planets' and stars' surfaces).

- **Internal waves**

Let us now consider the case of stably stratified zones in stellar and planetary interiors. As it has been described in Sect. 7.2.2.1, gravity (and gravito-inertial waves) are propagating in such regions. These are excited at the border with adjacent convective regions both by turbulent movements, and in the case where there is a close companion, by pressure fluctuations induced by tidally excited inertial waves (for example those of inertial waves attractors in the case of external convective envelopes). Then, in the case of binary or multiple systems, gravito-inertial waves will be forced in stellar radiation zones and in stably stratified planetary layers (for example in non-convective layers just below the surface of giant planets in our solar system or of giant extra-solar planets, where those layers can be created because of the heating of the surface by the close star, and in stably stratified regions in telluric planets). Then, the displacement as in the case of convective regions is the sum of the equilibrium tide and of the dynamical tide, which are here gravito-inertial waves. Let us now consider the properties of the tidal dissipation related to gravito-inertial waves. We must here recall that the main dissipative mechanism acting on such waves is the thermal diffusion. Then, as in the case of inertial waves, the tidal dissipation can be increased by several orders of magnitude compared to the one of the equilibrium tide, in particular in resonances that occur for gravity waves has it has been shown by [84, 85, 88, 101–103, 107, 109, 129]. Moreover, because of such increase of the tidal dissipation during resonances, its behaviour is highly dependant on the tidal frequency (see Fig. 7.16).

Let us now discuss the modification of gravito-inertial waves propagation by rotation and magnetic field. First, for the rotation, gravito-inertial waves propagation will strongly depends on the value of the tidal frequency compared to the inertial

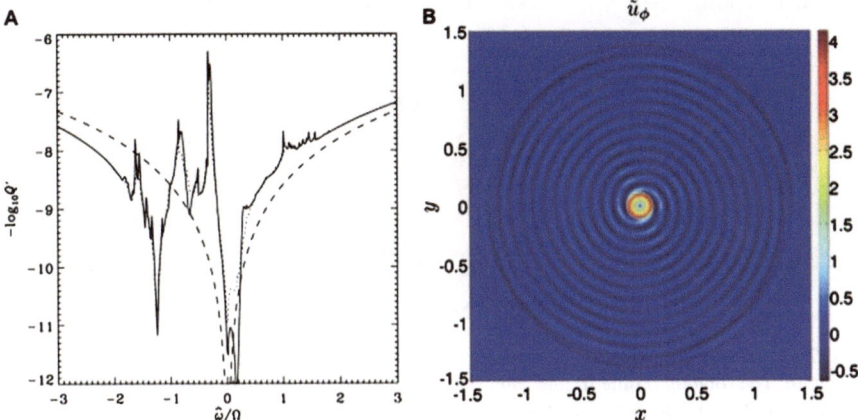

Fig. 7.16 **A**: Tidal dissipation by gravito-inertial waves in a solar-type star and the corresponding quality factor as a function of the tidal frequency computed by Ogilvie and Lin [85]. (Courtesy the Astrophysical Journal.) **B**: Breaking of large-amplitude tidally excited internal waves at the center of a solar-type star computed by Barker and Ogilvie [4]. There, waves deposit their angular momentum that can accelerates the center of the star and makes evolve the orbit of the planetary companion. (Courtesy Monthly Notices of the Royal Astronomical Society)

frequency (2Ω) where we recall that Ω is the rotation of the studied body. First, in the super-inertial regime ($\sigma_T > 2\Omega$), gravito-inertial waves are propagating in the whole sphere. However, in the sub-inertial one ($\sigma_T < 2\Omega$), waves become trapped in an equatorial region, propagating only above a given so-called critical colatitude [74, 78] (see Fig. 7.17). Such kind modification of gravito-inertial propagation is very important for their coupling with tidal-induced displacements in adjacent convective regions. Moreover, has it has been discovered by [108, 120–123], the Coriolis acceleration can lead to the so-called "tapping in resonnance" where retrograde and prograde waves exert respectively a negative and an positive torques that act to block the studied system in a resonant state where the tidal dissipation is very efficient. Then, the tidal dissipation is also dependent on the rotation rate and the associated Coriolis acceleration. Next, for example in the case of solar-type stars, gravito-inertial waves propagation is modified by the presence of magnetic field, for example at the bottom of the convective envelope. Then, waves becomes magneto-gravito-inertial waves where the Lorenz force has to be taken into account. Such waves have been studied for example by Mathis and Brye [75, 76]. First, waves excited with frequencies close to the Alfvén frequency will be vertically trapped. Then, as in the gravito-inertial case, an equatorial trapping can occur depending both on ω_A and 2Ω.

As in the case of inertial waves, we can conclude that the tidal dissipation will be strongly dependent on the tidal frequency, on the rotation and of potential impact of magnetic fields.

Interactions with Shear Flows and Instabilities Finally, let us discuss the interaction of the dynamical tide with differential rotation with the dynamical tide in

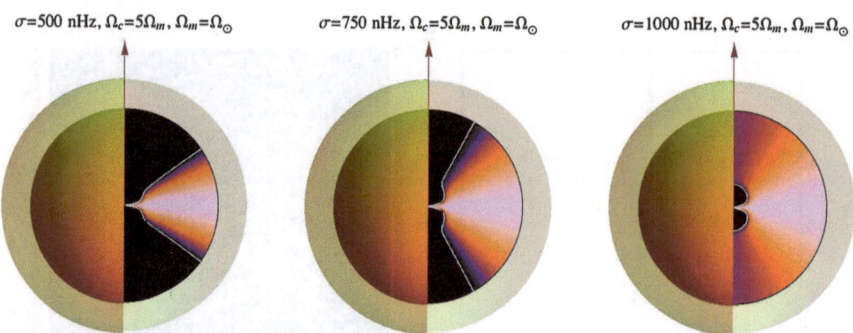

σ=500 nHz, Ω_c=5Ω_m, Ω_m=$\Omega_\odot$ σ=750 nHz, Ω_c=5Ω_m, Ω_m=$\Omega_\odot$ σ=1000 nHz, Ω_c=5Ω_m, Ω_m=$\Omega_\odot$

Fig. 7.17 We consider here a solar-twin star with an external convective envelope and a radiative core. The rotation profile in this latter is flat (i.e. $\Omega = \Omega_m = 430$ nHz) for $r \in [0.2R, 0.7R]$, where R is the stellar radius. In the central region, Ω increases until $\Omega = 5\Omega_m$ at the center. *Coloured regions* correspond to those where gravito-inertial waves are propagative while *black regions* correspond to "dead" zones for waves propagation. We choose three frequencies ($\sigma = \{500, 750, 1000\}$ nHz) that shows that for frequency below $2\Omega_m$ equatorial trapping phenomena appear. The *white line* corresponds to the critical surface (the critical latitude in the case of uniform rotation) at the level of which the wave propagation regime changes. The central region is always a non-propagative region because of the central rapid rotation

stably stratified layers. First, as it has been explained by Goldreich and Nicholson [38, 41], gravity waves can transfer angular momentum only if these are damped by a dissipative mechanism (which is mostly the thermal diffusion) or if these meet corotation resonances during their propagation (*i.e.* if we consider a "shellular" rotation that depends only on r, radii where $\Omega(r_c)$ is proportional to the tidal frequency). Let us first examine the thermal diffusion effect: an important point is that the thermal damping depends on the prograde or retrograde behaviour of the wave because of the Doppler shift. Then, in the case of a differentially rotating body, the synchronisation of each layer will progress from the surface to deeper regions [41]. Let us now focus on corotation region, which are also called the critical layers. There, these are strong interactions between internal waves and the shear of the differential rotation. We can summarise such type of exchanges as follows: first, if the studied layer is stable with respect to the shear instabilities, waves deposit their angular momentum, the damping rate being dependant on the so-called Richardson number, which compares the strength of the stabilisation by the stratification and the destabilisation by the shear gradient; then, if the layer is already turbulent, internal waves can be reflected and transmitted by such layers with an amplitude greater than their initial one because waves take energy for the turbulent flows. In this context, it is important to study the possible instabilities that could affect internal waves dynamics. First, if waves are excited with a large amplitude, waves will break and then, these could overturn the stable stratification (see for example [4, 5] and Fig. 7.16 for dynamical tide dynamics at the center of solar-type star). Then, even for weak amplitude, internal waves can undergo parametric instabilities where a "parent" wave give birth to "daughter" waves that could be then also dissipated [119]. Thus, as in the case of inertial waves, the interaction with the differential rotation as well as their own instabilities could strongly modify the value of the tidal dissipation.

7.3.3 Tidal Dissipation in Rocky or Icy Planetary Regions

As it has been shown in previous sections, the tidal potential is able to excite several types of velocity fields in fluid stellar and planetary layers leading to possible high values of the quality factor Q, which is function of the tidal frequency both for the equilibrium and the dynamical tides. However, planets (and their associated natural satellites) are composed of both fluid and solid layers, and tidal dissipation in these latters and at the fluid-solid interfaces should also be treated.

In this sense, the treatment of what is often called "the bodily tide", in other word the solid tide, has been one of the first studies of tidal dissipation using continuum mechanics (see for example [70]). These studies were of course motivated by the Earth case where tidal interactions with both the Moon and the Sun have to be taken into account [81]. In solid layers, tidal physical mechanisms are similar to those occurring in fluids. First, a solid equilibrium tide is generated that consists on a permanent large-scale displacement (with a zero velocity field that constitutes a difference with the fluid case). In an adiabatic modeling, this displacement is allowed by the elasticity of the material and directed along the line of centers (see the right panel of Fig. 7.18). However, as in the fluid case, solid layers host dissipation because of their anelasticity and the tidal energy is dissipated into heat, leading at the same time to internal heating, to a net applied torque, and to a small delay between the tidal bulge and the line of centers. This anelasticity, which is often modelled as a viscous behaviour that adds to the elasticity as in the Maxwell's body model, depends on the intrinsic properties of the considered material (for example silicates or ices), which are described by its rheology [79]. This latter is given by the constitutive equation that links the strain tensor ($\overline{\overline{\varepsilon}}$) to the stress tensor ($\overline{\overline{\sigma}}$). Under small deformations (as tidal perturbations), it is customary to assume that the strain-stress relationship is linear, and materials that obey this law are called Hookean materials. If we also assume that they are isotropic and incompressible, the Hooke's law states then that: $\overline{\overline{\sigma}} = \tilde{\mu}(\sigma_T)\overline{\overline{\varepsilon}}$, where $\tilde{\mu}$ is the complex shear modulus (also called rigidity, which measures the stiffness of the material) which depends on the tidal frequency (σ_T). Its real and imaginary parts represent respectively the energy storage and the energy losses of the system. Moreover, thanks to the correspondence principle (see [6, 79]), one can calculate the tidal dissipation and the associated quality factor Q for any linear rheology.

Such type of computation has already been performed for the Moon [89], for rocky core of giant planets [29], for icy natural satellites (see for example [116, 117]), and for telluric extrasolar planets as Earth-like planets or Super-Earths [45].

Such type of solid tide occurs for example in a two-layer planet with an internal rocky part and an external fluid envelope as studied by [29, 93] (see Fig. 7.18).

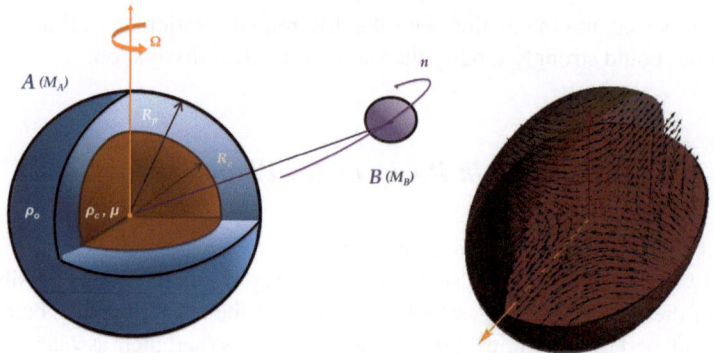

Fig. 7.18 *Left*: Two-layer model of planetary internal structure (body A of mass M_A rotating at angular velocity Ω) formed by an internal rocky core surrounded by a fluid envelope perturbed by a companion (body B of mass M_B orbiting around A with mean motion n). R_c and R_p are respectively the core and the planetary radii, ρ_o and ρ_c the densities of the fluid envelope and of the core, and μ is the shear modulus of the core. *Right*: purely elastic tidal displacement in the solid core free of fluid envelope; the *red* and *orange arrows* indicate respectively the symmetry axis of the planetary core and the direction of the companion. (Taken from Remus et al. [93]; courtesy Astronomy & Astrophysics)

This corresponds to the cases of a telluric planet with an external ocean or atmosphere or a gaseous or icy giant planet with a potential rocky/icy core born during the planetary formation [41, 42]. According to such models, the resulting solid tidal dissipation can reach values greater by several orders of magnitude than those due to fluid tidal velocities described in previous sections, for realistic values of viscoelastic parameters (see Figs. 7.19 and 7.20 for the solid cores of Jupiter- and Saturn-like planets). Observational measurements of tidal dissipation in our Solar system, from astrometry for example (see [60] for Jupiter, and [61] for Saturn confirmed by realistic scenario of natural satellites formation [20]), provide precious constraints to discriminate one tidal process from an other. Indeed, each process present proper characteristics: for example, the study of the tidal frequency-dependence of solid dissipation shows a smooth behaviour (Fig. 7.20) compared to the case of inertial and gravito-inertial waves where the dissipation varies of several orders of magnitude (cf. Sect. 7.3.2). Some other behaviours can be shared by different mechanisms as the sensitivity to the size of the solid cores that is shared by both the core's tidal dissipation and the inertial waves fluid one. This shows how this becomes crucial to get constraints on the size of the rocky core of giant planets from observations and theoretical predictions. Moreover, this large values, as in the fluid case, shows the strong need to go beyond phenomenological prescriptions [30, 31].

7.3.4 Boundary Conditions

As it has been shown previously, tidal interactions excite flows (*i.e.* the equilibrium tide, the dynamical tide, and fluid movements that result from their instabilities).

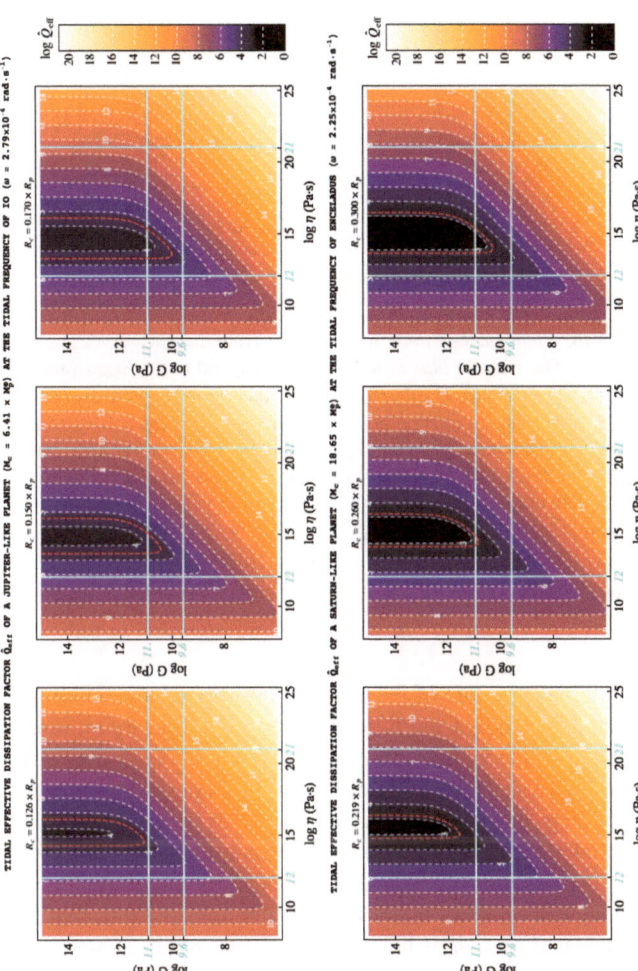

Fig. 7.19 Dissipation quality factor $\widehat{Q}_{\text{eff}}$ of the Maxwell model in function of the viscoelastic parameters G (the stiffness) and η (the viscosity). *Top:* for a Jupiter-like two-layer planet tidally perturbed at the Io's frequency $\sigma_T = 2.79 \times 10^{-4}$ rad $\cdot$ s^{-1}. *Bottom:* for a Saturn-like two-layer planet tidally perturbed at the Enceladus' frequency $\sigma_T = 2.25 \times 10^{-4}$ rad $\cdot$ s^{-1}. The *red dashed lines* correspond to the value of $\widehat{Q}_{\text{eff}}$ obtained by [60] for Jupiter, and by [61] for Saturn also needed by [20] to form its mid-sized satellites. We recall the values of $R_p = \{10.97, 9.14\}$ (in units of $R_p^\oplus$, the Earth radius), $M_p = \{317.8, 95.16\}$ (in units of $M_p^\oplus$, the Earth mass), and $M_c = \{6.41, 18.65\} \times M_p^\oplus$ (see respectively [41, 47]). Moreover, we have followed [36] for the values of the adiabatic Love numbers of Jupiter and Saturn (respectively $k_2 = \{0.379, 0.341\}$). The *blue lines* correspond to the lower and upper limits of the more realistic values taken by the viscoelastic parameters G and η at very high pressure for an unknown mix of ice and silicates. (Taken from Remus et al. [93]; courtesy Astronomy & Astrophysics)

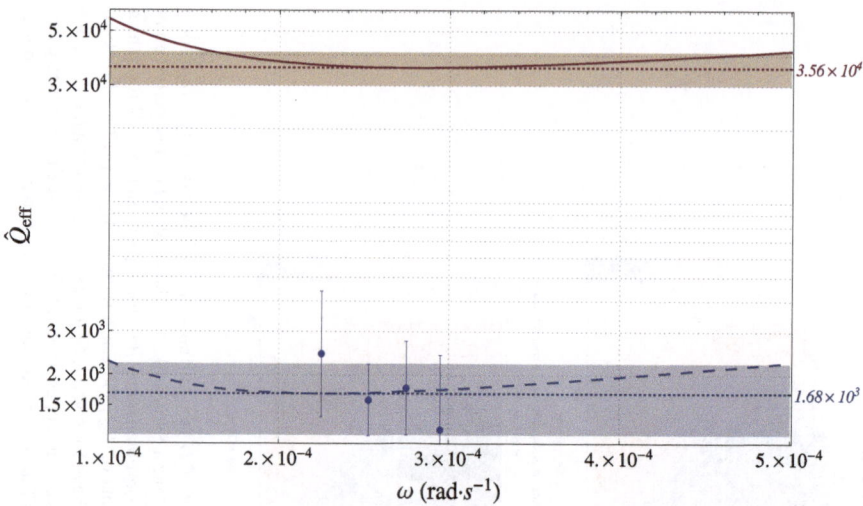

Fig. 7.20 Dependence to the perturbative strain pulsation ω of the tidal quality factor $\widehat{Q}_{\mathrm{eff}}$ Jupiter-like (*red solid line*) and Saturn-like (*blue dashed line*) giant planets. Note that the curves are represented with a logarithmic scale. The *red* and *blue dotted lines* correspond to the mean value of $\widehat{Q}_{\mathrm{eff}} = \{(3.56 \pm 0.56) \times 10^4, (1.682 \pm 0.540) \times 10^3\}$ (for Jupiter and Saturn respectively) determined by Lainey et al. [60, 61]. Their zone of uncertainty is also represented in the corresponding color. The *blue dots* correspond to the values (obtained by Lainey et al. [61]) of the dissipation induced in Saturn by Enceladus, Thetys, Dione and Rhea with their respective error bars. We recall the values of $R_p = \{10.97, 9.14\}$ (in units of $R_p^{\oplus}$, the Earth radius), $M_p = \{317.8, 95.16\}$ (in units of $M_p^{\oplus}$, the Earth mass), $M_c = \{6.41, 18.65\} \times M_p^{\oplus}$, and $R_c = \{0.15, 0.26\} R_p$. We take for the viscoelastic parameters $G = \{2.73, 6.51\} \times 10^{10}$ (Pa), and $\eta = \{8.65 \times 10^{13}, 2.50 \times 10^{14}\}$ (Pa · s) for Jupiter and Saturn respectively. (Adapted from Remus et al. [93]; courtesy Astronomy & Astrophysics)

Then, depending on the internal structure of the studied body, boundary conditions, and particularly at the surface and near solid/fluid interfaces (in planetary interiors) should be examined carefully. Indeed, boundary layer strongly sheared flows can develop there, that may lead to a strong dissipation. Let us first consider the case of the surface of stars or of planets. Tassoul and Tassoul [111–113] have proposed that Eckman boundary layers take place and lead to an important viscous dissipation. However, following [95, 98], we must point that such boundary constitutes a free surface with stress-free boundary conditions for the velocity field and thus that the dissipation related to associated boundary layers will be very weak. However, if we now consider the solid/fluid interfaces, we are in the case of no-slip boundary conditions that correspond to the classical Eckman boundary layers (see Fig. 7.21) where a strong viscous dissipation may occur, particularly if the studied region is turbulent. As a partial conclusion on boundary flows, one has thus to remember that solid/fluid regions (at the top of rocky/icy cores of giant planets and at boundaries of liquid cores in telluric planets) may host strong viscous dissipation while this will not be the case below the surface of stars or planets with a fluid envelope. At solid-fluid boundaries, we must also point that couplings between the fluid dynamical tide

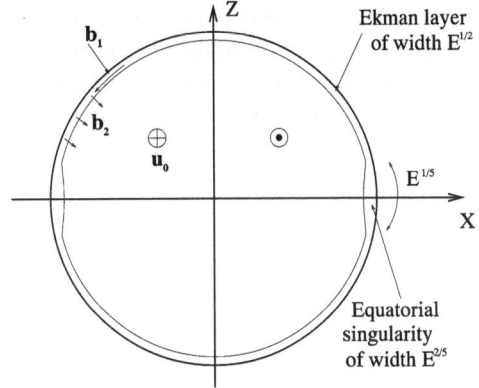

Fig. 7.21 Schematic view of Eckman boundary layers that can be important at boundaries of planetary cores. (Taken from Rieutord and Zahn [98] in which details concerning boundary velocity fields b_1 and b_2 are given, courtesy The Astrophysical Journal)

(inertial or gravito-inertial) and the (an)-elastic tide in solid regions must be studied in a near future since these may modify the related tidal dissipation.

7.3.5 Hierarchy Between Dissipative Physical Processes and the Associated Obtained States

In this review, we have tried to give a complete review of dissipative processes acting on tidally excited velocity fields in stellar and in planetary interiors. It is thus interesting at that point to draw a hierarchy between the intensity of the related dissipations. First, for fluid regions, dynamical tides (inertial and gravito-inertial waves) as well as related instabilities (the elliptic instability for inertial waves and the convective and parametric instabilities for internal waves) lead to a stronger dissipation that can dominate by several order of magnitude the one associated to the equilibrium tide. Next, if we study planetary interiors, we have isolated that tidal dissipation in rocky/icy regions can dominate the one in fluid regions depending on their respective size. Remember also, that each type of tidal dissipation have a different dependence on the tidal frequency that can be constrained by observations to unravel the action of the different physical mechanisms. Finally, it is important to recall that once one have identified all the dissipation processes and derived the associated torques as shown in Sect. 7.2.2, equilibrium states for orbital and spin properties can be obtained (see for example [22–25] for telluric planets) and compared to observational constraints obtained on our Solar system and on exoplanetary systems.

7.4 Conclusion

In this lecture, we have tried to give the must complete picture of tidal interactions in planetary systems. In the first section, we have summarised the most general formalism to model tidal dynamics and we have shown the crucial dependence of systems' evolution on dissipative mechanisms that convert the kinetic energy of tidal flows into heat. Next, we have described the important diversity of mechanisms that take place in stellar and planetary interiors. We have given their main properties, namely their relative amplitude and their dependence on the internal structure and on the tidal frequency that can be very complex. This shows how it is now necessary to study tidal interactions with a good description of the bodies' internal structure and to go beyond the rough approximations that are often adopted to describe systems' evolution due to tidal interactions.

Acknowledgements   We thank the CNRS/INSU for the support given to this Thematic School. This work was supported in part by the Programme National de Planétologie (CNRS/INSU), the Programme National de Physique Stellaire (CNRS/INSU), the EMERGENCE-UPMC project EME0911, and the CNRS *Physique théorique et ses interfaces* program.

References

1. Alexander, M.E.: Astrophys. Space Sci. **23**, 459 (1973)
2. Andoyer, H.: Mécanique Céleste. Gauthier-Villars, Paris (1926)
3. Baraffe, I.: Space Sci. Rev. **116**, 67 (2005)
4. Barker, A.J., Ogilvie, G.I.: Mon. Not. R. Astron. Soc. **404**, 1849 (2010)
5. Barker, A.J., Ogilvie, G.I.: Mon. Not. R. Astron. Soc. **417**, 745 (2011)
6. Biot, M.A.: J. Appl. Geophys. **25**, 1385 (1954)
7. Bodenheimer, P., Lin, D.N.C., Mardling, R.A.: Astrophys. J. **548**, 466 (2001)
8. Borderies, N.: Celest. Mech. **18**, 295 (1978)
9. Borderies, N.: Astron. Astrophys. **82**, 129 (1980)
10. Borderies, N., Yoder, C.F.: Astron. Astrophys. **233**, 235 (1990)
11. Boué, G., Laskar, J.: Icarus **701**, 250 (2009)
12. Brouwer, D., Clemence, G.M.: Methods of Celestial Mechanics. Academic Press, New York (1961)
13. Brun, A.-S., Toomre, J.: Astrophys. J. **570**, 865 (2002)
14. Brun, A.-S., Miesch, M.S., Toomre, J.: Astrophys. J. **614**, 1073 (2004)
15. Bryan, G.H.: Philos. Trans. R. Soc. Lond. A **180**, 187 (1889)
16. Cébron, D.: Ph.D. thesis. Université de Provence, Aix Marseille I (2011)
17. Cébron, D., et al.: Phys. Earth Planet. Inter. **182**, 119 (2010)
18. Cébron, D., Maubert, P., Le Bars, M.: Geophys. J. Int. **182**, 1311 (2010)
19. Cébron, D., et al.: In: EPJ Web of Conferences, vol. 11, id. 03003 (2011)
20. Charnoz, S., et al.: Icarus **216**, 535 (2011)
21. Correia, A.C.M.: Ph.D. thesis. Université Paris VII (2001)
22. Correia, A.C.M., Laskar, J.: J. Geophys. Res. **108**(E11), 9-1 (2003)
23. Correia, A.C.M., Laskar, J.: Icarus **163**, 24 (2003)
24. Correia, A.C.M., Laskar, J., de Surgy, O.N.: Icarus **163**, 1 (2003)
25. Correia, A.C.M., Levrard, B., Laskar, J.: Astron. Astrophys. **488**, L63 (2008)
26. Damour, T., Soffel, M.H., Xu, C.: Phys. Rev. D **45**, 1017 (1992)
27. Darwin, G.H.: Philos. Trans. R. Soc. Lond. **171**, 713 (1880)
28. Darwin, G.H.: Philos. Trans. R. Soc. Lond. **172**, 491 (1881)
29. Dermott, S.F.: Icarus **37**, 310 (1979)
30. Efroimsky, M.: Astrophys. J. **746**, 150 (2012)
31. Efroimsky, M., Lainey, V.: J. Geophys. Res. **112**, E12003 (2007)
32. Efroimsky, M., Williams, J.G.: Celest. Mech. Dyn. Astron. **104**, 257 (2009)

33. Eggleton, P.P., Kiseleva, L.G., Hut, P.: Astrophys. J. **499**, 853 (1998)
34. Ferraz-Mello, S., Rodrìguez, A., Hussmann, H.: Celest. Mech. Dyn. Astron. **101**, 171 (2008)
35. Fortney, J.J., Nettelmann, N.: Space Sci. Rev. **152**, 423 (2009)
36. Gavrilov, S.V., Zharkov, V.N.: Icarus **32**, 443 (1977)
37. Goldreich, P., Keeley, D.A.: Astrophys. J. **211**, 934 (1977)
38. Goldreich, P., Nicholson, P.D.: Astrophys. J. **342**, 1075 (1989)
39. Goldreich, P., Soter, S.: Icarus **5**, 375 (1966)
40. Goodman, J., Lackner, C.: Astrophys. J. **696**, 2054 (2009)
41. Guillot, T.: Planet. Space Sci. **47**, 1183 (1999)
42. Guillot, T.: Annu. Rev. Earth Planet. Sci. **33**, 493 (2005)
43. Guimbard, D., et al.: J. Fluid Mech. **660**, 240 (2010)
44. Hartmann, T., Soffel, M.H., Kioustelidis, T.: Celest. Mech. Dyn. Astron. **60**, 139 (1994)
45. Henning, W.G., O'Connell, R.J., Sasselov, D.D.: Astrophys. J. **707**, 1000 (2009)
46. Herreman, W., Le Bars, M., Le Gal, P.: Phys. Fluids **21**, 046602 (2009) (9 pp.)
47. Hubbard, W.B., Dougherty, M.K., Gautier, D., Jacobson, R.: In: Dougherty, M.K., Esposito, L.W., Krimigis, S.M. (eds.) Saturn from Cassini-Huygens, ISBN 978-1-4020-9216-9, p. 75. Springer, Berlin (2009).
48. Hut, P.: Astron. Astrophys. **99**, 126 (1981)
49. Ilk, K.H.: Ph.D. thesis. Technischen Universitaet, Munich Bayerische Akademie der Wissenschaften (1983)
50. Ivanov, P.B., Papaloizou, J.C.B.: Mon. Not. R. Astron. Soc. **353**, 1161 (2004)
51. Ivanov, P.B., Papaloizou, J.C.B.: Mon. Not. R. Astron. Soc. **407**, 1609 (2010)
52. Kaula, W.M.: Astron. J. **67**, 300 (1962)
53. Kaula, W.M.: Rev. Geophys. Space Phys. **2**, 661 (1964)
54. Kelvin, Lord: The Tides Evening Lecture to the British Association at the Southampton Meeting, Friday, August 25th, 1882. Scientific Papers. The Harvard Classics, New York (1882)
55. Kerswell, R.: Annu. Rev. Fluid Mech. **34**, 83 (2002)
56. Kippenhahn, R., Weigert, A.: Stellar Structure and Evolution. Springer, Berlin–Heidelberg–New York (1990)
57. Lacaze, L., Le Gal, P., le Dizès, S.: J. Fluid Mech. **505**, 22 (2004)
58. Lacaze, L., Le Gal, P., le Dizès, S.: Phys. Earth Planet. Inter. **151**, 194 (2005)
59. Lacaze, L., et al.: Geophys. Astrophys. Fluid Dyn. **100**, 299 (2006)
60. Lainey, V., Arlot, J.-E., Karatekin, Ö., van Hoolst, T.: Nature **459**, 957 (2009)
61. Lainey, V., et al.: Nature **14**, 752 (2012)
62. Lambeck, K.: The Earth's Variable Rotation: Geophysical Causes and Consequences. Cambridge University Press, Cambridge (1980)
63. Laskar, J.: Celest. Mech. Dyn. Astron. **91**, 351 (2005)
64. Lavorel, G., Le Bars, M.: Phys. Fluids **22**, 114101 (2010) (8 pp.)
65. Le Bars, M., Le Dizès, S.: J. Fluid Mech. **563**, 189 (2006)
66. Le Bars, M., Le Dizès, S., Le Gal, P.: J. Fluid Mech. **585**, 323 (2007)
67. Le Bars, M., et al.: Phys. Earth Planet. Inter. **178**, 48 (2010)
68. Leconte, J., Chabrier, G., Baraffe, I., Levrard, B.: Astron. Astrophys. **516**, A64 (2010).
69. Levrard, B., Winisdoerffer, C., Chabrier, G.: Astrophys. J. **692**, L9 (2009)
70. Love, A.E.H.: Some Problems of Geodynamics. Cambridge University Press, Cambridge (1911)
71. MacDonald, G.J.F.: Rev. Geophys. Space Phys. **2**, 467 (1964)
72. Maciejewski, A.J.: Celest. Mech. Dyn. Astron. **63**, 1 (1995)
73. Mardling, R.A., Lin, D.N.C.: Astrophys. J. **573**, 829 (2002)
74. Mathis, S.: Astron. Astrophys. **506**, 811 (2009)
75. Mathis, S., de Brye, N.: Astron. Astrophys. **526**, A65 (2011).
76. Mathis, S., de Brye, N.: Astron. Astrophys. **540**, A37 (2012)
77. Mathis, S., Le Poncin-Lafitte, C.: Astron. Astrophys. **497**, 889 (2009)
78. Mathis, S., Talon, S., Pantillon, F.-P., Zahn, J.-P.: Sol. Phys. **251**, 101 (2008)

79. Melchior, P.: The Earth Tides. Pergamon, New York (1966)
80. Melchior, P.: Physique et Dynamique Planétaire. Vander, Bruxelles (1971)
81. Neron de Surgy, O., Laskar, J.: Astron. Astrophys. **318**, 975 (1997)
82. Ogilvie, G.I.: J. Fluid Mech. **543**, 19 (2005)
83. Ogilvie, G.I.: Mon. Not. R. Astron. Soc. **396**, 794 (2009)
84. Ogilvie, G.I., Lin, D.N.C.: Astrophys. J. **610**, 477 (2004)
85. Ogilvie, G.I., Lin, D.N.C.: Astrophys. J. **661**, 1180 (2007)
86. Papaloizou, J.C.B., Ivanov, P.B.: Mon. Not. R. Astron. Soc. **364**, L66 (2005)
87. Papaloizou, J.C.B., Ivanov, P.B.: Mon. Not. R. Astron. Soc. **407**, 1631 (2010)
88. Papaloizou, J.C.B., Savonije, G.J.: Mon. Not. R. Astron. Soc. **291**, 651 (1997)
89. Peale, S.J., Cassen, P.: Icarus **36**, 245 (1978)
90. Penev, K., Sasselov, D.: Astrophys. J. **731**, 67 (2011)
91. Penev, K., Barranco, J., Sasselov, D.: Astrophys. J. **705**, 285 (2009)
92. Penev, K., Sasselov, D., Robinson, F., Demarque, P.: Astrophys. J. **704**, 230 (2009)
93. Remus, F., Mathis, S., Zahn, J.-P., Lainey, V.: Astron. Astrophys. **541**, A165 (2012).
94. Remus, F., Mathis, S., Zahn, J.-P.: Astron. Astrophys. **544**, A132 (2012)
95. Rieutord, M.: Astron. Astrophys. **259**, 581 (1992)
96. Rieutord, M., Valdettaro, L.: J. Fluid Mech. **341**, 77 (1997)
97. Rieutord, M., Valdettaro, L.: J. Fluid Mech. **643**, 363 (2010)
98. Rieutord, M., Zahn, J.-P.: Astrophys. J. **474**, 760 (1997)
99. Rieutord, M., Valdettaro, L., Georgeot, B.: J. Fluid Mech. **435**, 103 (2001)
100. Rieutord, M., Valdettaro, L., Georgeot, B.: J. Fluid Mech. **463**, 345 (2002)
101. Rocca, A.: Astron. Astrophys. **111**, 252 (1982)
102. Rocca, A.: Astron. Astrophys. **175**, 81 (1987)
103. Rocca, A.: Astron. Astrophys. **213**, 114 (1989)
104. Roxburgh, I.W.: Astron. Astrophys. **377**, 688 (2001)
105. Santos, N.C., et al.: Our non-stable Universe. In: JENAM-2007 (2007)
106. Savonije, G.-J.: EAS Publ. Ser. **29**, 91 (2008)
107. Savonije, G.J., Papaloizou, J.C.B.: Mon. Not. R. Astron. Soc. **291**, 633 (1997)
108. Savonije, G.J., Witte, M.G.: Astron. Astrophys. **386**, 211 (2002)
109. Savonije, G.J., Papaloizou, J.C.B., Alberts, F.: Mon. Not. R. Astron. Soc. **277**, 471 (1995)
110. Scharlemann, E.T.: Astrophys. J. **246**, 292 (1981)
111. Tassoul, M., Tassoul, J.-L.: Astrophys. J. **395**, 259 (1992)
112. Tassoul, M., Tassoul, J.-L.: Astrophys. J. **395**, 604 (1992)
113. Tassoul, M., Tassoul, J.-L.: Astrophys. J. **481**, 363 (1997)
114. Tisserand, F.F.: Traité de Mécanique Céleste, Tome I. Gauthier-Villars, Paris (1889)
115. Tisserand, F.F.: Traité de Mécanique Céleste, Tome II. Gauthier-Villars, Paris (1891)
116. Tobie, G.: Impact du chauffage de marée sur l'évolution géodynamique d'Europe et de Titan. Ph.D. thesis. Université Paris 7 - Denis Diderot (2003)
117. Tobie, G., Mocquet, A., Sotin, C.: Icarus **177**, 534 (2005)
118. Waleffe, F.A.: Phys. Fluids **2**, 76 (1990)
119. Weinberg, N.N., Arras, P., Quataert, E., Burkart, J.: Astrophys. J. **751**, article id. 136 (2012)
120. Witte, M.G., Savonije, G.J.: Astron. Astrophys. **341**, 842 (1999)
121. Witte, M.G., Savonije, G.J.: Astron. Astrophys. **350**, 129 (1999)
122. Witte, M.G., Savonije, G.J.: Astron. Astrophys. **366**, 840 (2001)
123. Witte, M.G., Savonije, G.J.: Astron. Astrophys. **386**, 222 (2002)
124. Wu, Y.: Astrophys. J. **635**, 674 (2005)
125. Wu, Y.: Astrophys. J. **635**, 688 (2005)
126. Zahn, J.-P.: Ann. Astrophys. **29**, 313 (1966)
127. Zahn, J.-P.: Ann. Astrophys. **29**, 489 (1966)
128. Zahn, J.-P.: Ann. Astrophys. **29**, 565 (1966)
129. Zahn, J.-P.: Astron. Astrophys. **41**, 329 (1975)
130. Zahn, J.-P.: Astron. Astrophys. **57**, 383 (1977)
131. Zahn, J.-P.: Astron. Astrophys. **220**, 112 (1989)

Chapter 8
Stellar Tides

Jean-Paul Zahn

Abstract To a first approximation, a binary star behaves as a closed system; therefore it conserves its angular momentum while evolving to its state of minimum kinetic energy, where the orbits are circular, all spins are aligned, and the components rotate in synchronism with the orbital motion. The pace at which this final state is reached depends on the physical processes responsible for the dissipation of the tidal kinetic energy. For stars with an outer convection zone, the dominant mechanism is presumably the turbulent dissipation acting on the equilibrium tide. For stars with an outer radiation zone, the major dissipative process is radiative damping operating on the dynamical tide.

I shall review these physical processes, discuss uncertainties in their present treatment, describe the latest developments, and compare the theoretical predictions with the observed properties concerning the orbital circularization of close binaries.

8.1 Introduction

A fundamental property of isolated mechanical systems is that they conserve their total angular momentum while they evolve. This is true in particular for binary stars, and star-planet(s) systems, as long as one can ignore the angular momentum that is lost by the winds or by gravitational waves. Through tidal interaction, kinetic energy and angular momentum are exchanged between the rotation of the components and their orbital motion. In general, as we shall see, the system evolves toward an equilibrium state of minimum kinetic energy, in which the orbit is circular, the rotation of both stars is synchronized with the orbital motion, and their spin axes are perpendicular to the orbital plane. How rapidly the system tends to that state is determined chiefly by the strength of the tidal interaction, and therefore by the separation of the two components: the closer the system, the faster its dynamical evolution. But it also depends strongly on the efficiency of the physical processes that are responsible for the dissipation of kinetic energy into heat.

J.-P. Zahn (✉)
LUTH, Observatoire de Paris, CNRS, Université Paris-Diderot, 5 place Jules Janssen,
92195 Meudon, France
e-mail: Jean-Paul.Zahn@obspm.fr

J. Souchay et al. (eds.), *Tides in Astronomy and Astrophysics*,
Lecture Notes in Physics 861, DOI 10.1007/978-3-642-32961-6_8,
© Springer-Verlag Berlin Heidelberg 2013

Provided these dissipation processes are understood well enough, the observed properties of a binary system can deliver important information on its evolutionary state, on its past history, and even on the conditions of its formation. The first step is thus to identify these physical processes, and one may wonder why this has not been seriously undertaken until the 1990s, while the tidal theory as such had already reached a high degree of sophistication, starting with the pioneering work of Darwin [2]. The reason can be found in Kopal's classical treatise, where he declares from start that he is interested only in 'dynamical phenomena which are likely to manifest observable consequences in time intervals of the order of 10 or 100 years, and if so, tidal friction can be safely ignored' [14].

But stars live much longer than us human beings, and this is why we shall consider here changes in the properties of binary systems that span their evolutionary time scale; we shall discuss in particular the circularization of their orbits, which is both easy to observe and easy to interpret. We shall deal mainly with binary stars, although much of what follows may be applied also to star-planet systems. In the latter case, however, owing to the stark contrast between the mass of the star and that of the planet, the system may not reach the equilibrium state mentioned above, as we shall see in the next section.

8.2 Equilibrium States

To seek such equilibrium states, we follow here the method introduced by Hut [11]. Consider a binary system whose components (star or planet) are characterized by their mass (M_1, M_2), moment of inertia (I_1, I_2), and rotation vector $(\boldsymbol{\Omega}_1, \boldsymbol{\Omega}_2)$. Their orbits around the center of mass have an eccentricity e and the sum of their semi-major axes is a. The total angular momentum vector of the system is given by

$$L = h + I_1\boldsymbol{\Omega}_1 + I_2\boldsymbol{\Omega}_2, \tag{8.1}$$

where h designates the orbital momentum, with

$$h^2 = G\frac{(M_1 M_2)^2}{M_1 + M_2}a(1 - e^2). \tag{8.2}$$

If one ignores the loss of angular momentum through winds or gravitational waves, L remains constant and it defines an inertial frame perpendicular to it; with respect to that plane, the orbital plane is inclined by an angle i. The Cartesian projections of h on the inertial frame are chosen such that

$$h = (h\sin i, 0, h\cos i). \tag{8.3}$$

The total mechanical energy of the system (kinetic + gravitational) amounts to

$$E = -G\frac{M_1 M_2}{2a} + \frac{1}{2}I_1|\boldsymbol{\Omega}_1|^2 + \frac{1}{2}I_2|\boldsymbol{\Omega}_2|^2. \tag{8.4}$$

A state of equilibrium is achieved when this E reaches a minimum under the constraint of fixed angular momentum, say $L = L_0$. Such a state satisfies the variational equations

$$\frac{\partial}{\partial x_i} E + \lambda \cdot \frac{\partial}{\partial x_i} L = 0, \tag{8.5}$$

where $\lambda = (\lambda_x, \lambda_y, \lambda_z)$ is the Lagrangian multiplier, and where the x_i represent the nine parameters a, e, i, and $\Omega_{1,k}, \Omega_{2,k}$ $(k = x, y, z)$.

The next step is to derive these nine variational equations:

$$G\frac{M_1 M_2}{a} + (\lambda_x \sin i + \lambda_z \cos i)h = 0, \tag{8.6}$$

$$(\lambda_x \sin i + \lambda_z \cos i)\frac{e}{(1 - e^2)}h = 0, \tag{8.7}$$

$$(\lambda_x \cos i - \lambda_z \sin i)h = 0, \tag{8.8}$$

$$\Omega_{1,k} + \lambda_k = \Omega_{2,k} + \lambda_k = 0, \quad k = x, y, z. \tag{8.9}$$

It is easy to check that this system has a unique solution, where the orbits are circular $(e = 0)$, the rotation axes are perpendicular to the orbital plane $(i = 0)$, and where the rotation of the two components is synchronized with the orbital motion: $\Omega_1 = \Omega_2 = \omega$, with the orbital angular velocity ω obeying Kepler's third law $\omega^2 = G(M_1 + M_2)/a^3$.

The angular momentum of these equilibrium states may be expressed as a function of the orbital frequency:

$$\mathcal{L} = \left[\frac{G^2(M_1 M_2)^3}{(M_1 + M_2)}\right]^{1/3} \omega^{-1/3} + (I_1 + I_2)\omega, \tag{8.10}$$

which has a minimum for

$$\omega^2 = \omega_{cr}^2 = \left[\left(\frac{1}{3(I_1 + I_2)}\right)^3 \frac{G^2(M_1 M_2)^3}{(M_1 + M_2)}\right]^{1/2} \tag{8.11}$$

where

$$\mathcal{L} = \mathcal{L}_{cr} = 4\left[\frac{(I_1 + I_2)}{27}\frac{G^2(M_1 M_2)^3}{(M_1 + M_2)}\right]^{1/4}. \tag{8.12}$$

No equilibrium state can exist below the critical value $\mathcal{L}_{cr}$: for $\mathcal{L} < \mathcal{L}_{cr}$ the system evolves with ever increasing orbital frequency (see Fig. 8.1), and this may eventually lead to its coalescence [11]. This occurs when the orbital angular momentum is less than 3 times the rotational angular momentum of the two components. In practice, this can only occur in very close systems, when the mass ratio is small enough:

$$\frac{M_2}{M_1} < \frac{3I_1}{M_1 R_1^2}\left(\frac{R_1}{a}\right)^2. \tag{8.13}$$

This the case for transiting planets, as was shown by Levrard et al. [20]. But here we shall deal mainly with close binary stars, for which $\mathcal{L} > \mathcal{L}_{cr}$, and these will evolve towards a stable equilibrium state (located on the continuous line in Fig. 8.1).

Fig. 8.1 Angular momentum
of the equilibrium states of
binary systems: stable
equilibria are drawn in
continuous line, unstable
equilibria in *dotted line*. No
equilibrium state can be
achieved below the critical
value L_{cr}

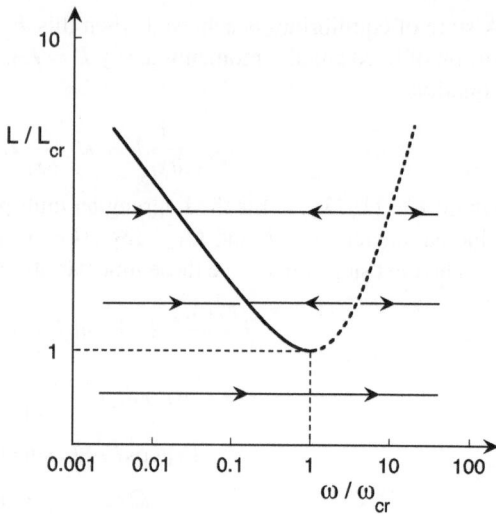

8.3 The Equilibrium Tide

We begin with the most simple concept: that of the equilibrium tide, where one
assumes that the star under consideration is in hydrostatic equilibrium, and that, in
the absence of dissipation mechanisms, it adjusts instantaneously to the perturbing
force exerted by its companion (star or planet).

8.3.1 A Crude Estimate of the Tidal Torque

For simplicity, let us assume that the orbit is circular. When the rotation of the star is
synchronized with the orbital motion, the tidal bulges are perfectly aligned with the
companion star; their elongation δR_1 and mass δM_1 are easily estimated, neglecting
numerical factors of order unity:

$$\frac{\delta M_1}{M_1} \approx \frac{\delta R_1}{R_1} \approx \frac{(f_2 - f_1)}{GM_1/R_1^2} \approx \frac{M_2}{M_1}\left(\frac{R_1}{d}\right)^3, \tag{8.14}$$

where d is the distance between the two components, and f_1 and f_2 the forces that
are exerted on the tidal bulges, as shown in Fig. 8.2. However, when the rotation
is not synchronized, any type of dissipation causes a lag α of the tidal bulges, with
respect to the line of centers, and the star then experiences a torque Γ which tends
to drag it into synchronism:

$$\Gamma \approx (f_2 - f_1)R_1 \sin\alpha \approx -\delta M_1\left[\frac{GM_2R_1}{d^3}\right]R_1 \sin\alpha = -\frac{GM_2^2}{R_1}\left(\frac{R_1}{d}\right)^6 \sin\alpha. \tag{8.15}$$

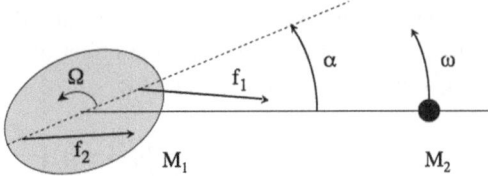

Fig. 8.2 Tidal torque. When the star under consideration rotates faster than the orbital motion ($\Omega > \omega$), its mass distribution is shifted by an angle α from the line joining the centers of the two components, due to the dissipation of kinetic energy. Since the forces applied to the tidal bulges are not equal ($f_1 > f_2$), a torque is exerted on the star, which slows it down and therefore tends to synchronize its rotation with the orbital motion ($\Omega \rightarrow \omega$)

The tidal angle α is a function of the lack of synchronism, since it vanishes for $\Omega \rightarrow \omega$, Ω being the rotation rate and ω the orbital angular velocity. In the simplest case, called the *weak friction approximation*, α is a linear function of the lack of synchronism: $\alpha = (\Omega - \omega)\delta t$, where δt is the time lag of the tidal bulge, and is thus constant in this approximation. That angle depends also on the strength of the physical process that is responsible for the dissipation of kinetic energy, which may be measured by its characteristic time t_{diss}, with α inversely proportional to that time. This leads us to

$$\alpha = \frac{(\Omega - \omega)}{t_{\text{diss}}} \frac{R_1^3}{GM_1}, \tag{8.16}$$

where we have rendered α non-dimensional by introducing the most 'natural' time, namely the dynamical (or free-fall) time $(GM_1/R_1^3)^{-1/2}$.

Inserting this expression of α in (8.15) we obtain the tidal torque

$$\Gamma = -\frac{(\Omega - \omega)}{t_{\text{diss}}} q^2 M R^2 \left(\frac{R}{d}\right)^6, \tag{8.17}$$

where $q = M_2/M_1$ is the mass ratio between secondary and primary components. From here on, when there is no ambiguity, we shall drop the index 1 from R_1 and M_1.

The weak friction law (8.16) is applicable to fluid bodies, such as stars and giant planets, assuming that the dissipation is of viscous nature, and that the viscosity does not depend on the tidal frequency, namely on ($\Omega - \omega$). (As we shall see later on, this condition is not necessarily fulfilled.) In that case the correct expression for the tidal torque, which one derives from the full equations governing the problem, is precisely of the form given above (Eq. (8.17)). From it, we may draw the synchronization time t_{sync}:

$$\frac{1}{t_{\text{sync}}} = -\frac{\Gamma}{I(\Omega - \omega)} = \frac{1}{t_{\text{diss}}} q^2 \frac{M R^2}{I} \left(\frac{R}{a}\right)^6, \tag{8.18}$$

where I is the moment of inertia of the star; here the torque has been averaged over the orbit, whose semi-major axis is a. We shall see later on how the dissipation time t_{diss} may be evaluated.

Since the instantaneous orbital velocity varies along an elliptic orbit, so does also the torque applied to the primary. This has the effect of changing the orbital eccentricity, at a rate given by the circularization time

$$\frac{1}{t_{\text{circ}}} = -\frac{d\ln e}{dt} = \frac{1}{t_{\text{diss}}}\left(9 - \frac{11}{2}\frac{\Omega}{\omega}\right)q(1+q)\left(\frac{R}{a}\right)^8,$$ (8.19)

again in the weak friction approximation; the companion star contributes a similar amount. Note that in binary stars, synchronization proceeds much faster that circularization, because the angular momentum of the orbit is in general much larger than that stored in the stars ($I\Omega \ll Ma^2\omega$); this is not necessarily true in star-planet systems, as we have seen in Sect. 8.2. One verifies that the eccentricity decreases near synchronization, but not for fast rotation: it was Darwin [2] who first pointed out that the eccentricity actually increases when $\Omega/\omega > 18/11$.

8.3.2 Turbulent Convection: The Most Powerful Mechanism for Tidal Dissipation

The dissipation time t_{diss}, which determines the tidal torque and hence the dynamical evolution of the binary system, is often treated as a free parameter, to be adjusted by the observations. We prefer to derive it from the physical processes that convert the mechanical energy of the tide into heat. The first of such processes that comes into mind is viscosity. But in stellar interiors, the viscosity due to microscopic processes is very low: it amounts typically to $\nu \approx 10\text{--}10^3 \text{ cm}^2\text{ s}^{-1}$. Therefore the (global) viscous timescale R^2/ν is much longer than the age of the Universe.

Radiative damping is more efficient: the dissipation time is then of the order of the Kelvin-Helmholtz time: $t_{\text{KH}} = GM/RL$, where L is the luminosity of the star. But (R/a) is raised to such a high power in (8.18) and (8.19) that t_{sync}—and t_{circ} even more so—easily exceed the life-time of the star.

However viscosity still plays a key role in those regions of stars and planets that are the seat of turbulent convection. There the kinetic energy of the large scale flow that is induced by the tide cascades down to smaller and smaller scales, until it is dissipated into heat by viscous friction. The force which acts on the tidal flow may then be ascribed to a 'turbulent viscosity' of order $\nu_t \approx v_t\ell$, where v_t is the r.m.s. vertical velocity of the turbulent eddies, and ℓ their vertical mean free path (or mixing-length). The tidal dissipation time introduced in (8.18) scales as the global convective time:

$$\frac{1}{t_{\text{diss}}} = \frac{6\lambda_2}{t_{\text{conv}}} \quad \text{where } t_{\text{conv}} = \left[\frac{MR^2}{L}\right]^{1/3};$$ (8.20)

the quantity λ_2 is determined by a summation of ν_t over the whole star

$$\frac{\lambda_2}{t_{\text{conv}}} = \frac{4176}{35}\pi\frac{R}{M}\int x^8\rho\nu_t\,dx,$$ (8.21)

where ρ is the density and $x = r/R$ the normalized radial coordinate. This expression is approximate: it applies to a star with a thick convection zone, and it was established assuming that the whole luminosity is carried by convection [32, 44].

The convective dissipation time is very short: $t_{conv} = 0.435$ yr in the present Sun, and for this reason turbulent convection is the most powerful dissipation mechanism acting on the equilibrium tide [44]. It works particularly well in stars possessing an outer convection zone, such as solar-type stars. Assuming that the whole heat flux is carried by convection and that the star is fully convective, $\lambda_2 = 0.019\alpha^{4/3}$, with α (not to be confused with the tidal lag introduced above) being the classical mixing-length parameter [47].

In stars with a convective core, tidal dissipation due to turbulent convection is considerably reduced, since it scales as $(r_c/R)^7$ with the radius r_c of that core [44]. Furthermore, in such cores the convective turnover time easily exceeds the tidal period, and therefore the straightforward definition of the turbulent viscosity taken above, i.e. $\nu_t \approx \nu_t \ell$, can no longer be applied, as we shall see next.

8.3.3 Which Prescription for Fast Tides?

When the local convective turnover time $t_{over} = \ell/\nu_t$ exceeds the tidal period P_{tide}, it seems appropriate to replace the mean free path by the distance that turbulent eddies are traveling during, say, half a tidal period. The turbulent viscosity is then given by

$$\nu_t = \nu_t \ell \min[1, P_{tide}/2t_{over}], \tag{8.22}$$

ignoring numerical coefficients of the order of unity [44]. This reduction occurs mainly in the deepest layers of a convection zone, since the convective turn-over time increases roughly as the 3/2 power of depth.

The same problem was addressed somewhat later by Goldreich and Nicholson [5], when they estimated the tidal damping in Jupiter. They remarked that 'though the largest convective eddies move across distances of order $\ell P_{tide}/t_{over}$ in a tidal period, they do not exchange momentum with the mean flow on this time scale.' Assuming that the Kolmogorov spectrum applies to convective turbulence, they retained in that spectrum only the eddies whose turnover time (or life time) is less than a tidal period; in that case, the turbulent viscosity scales as

$$\nu_t = \nu_t \ell \min\left[1, (P_{tide}/t_{over})^2\right]. \tag{8.23}$$

They concluded that 'tidal interactions between Jupiter and its satellites have played a negligible role in the evolution of the latters' orbits.'

The question of which of these prescriptions should be applied has long been considered as Achilles' heel of tidal theory. One could even question the validity of the very concept of turbulent viscosity, since we know that stratified convection is hardly a diffusive process: the transport of heat and momentum is partly achieved

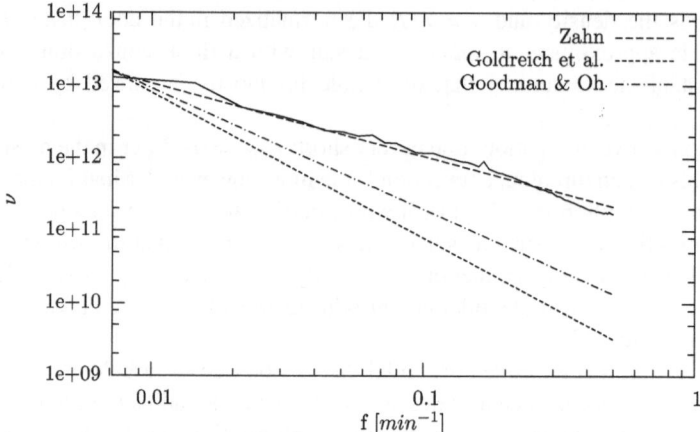

Fig. 8.3 Turbulent viscosity acting on a tidal flow in a stellar convection zone. The vertical component of that viscosity was determined by Penev et al. [30] by applying an oscillating large-scale shear on a numerical simulation of turbulent convection; it decreases with the forcing frequency f. The result (in *solid line*) is compared here with several prescriptions that have been proposed for the loss of efficiency of turbulent friction when the tidal period becomes shorter than the convective turn-over time

by long-lived plumes, and it is not easy to predict how these will interact with the large scale tidal flow.

I believe that the question will be settled through high resolution numerical simulations of turbulent convection. A first step has been taken by Penev et al. [30, 31], who studied the dissipation of a large-scale shear flow, varying periodically in time, when it is imposed on a 3-D convection simulation. They followed the method outlined by Goodman and Oh [9] to derive the viscous stress tensor. They confirmed that convection acts indeed as a turbulent viscosity on such a flow, since the off-diagonal components of the viscous tensor are one order of magnitude smaller than the diagonal components. They also observed that the vertical component of that tensor is about twice that of the horizontal components, due to the anisotropy of turbulent convection. Moreover, as can be seen in Fig. 8.3 borrowed from their article, they found that this turbulent viscosity decreases as f^{-1}, where f is the forcing frequency, which is here lower than the convective frequency. Hence they validated the first recipe (8.22) quoted above, although it remains to be seen whether their result holds in more realistic, hence more turbulent regimes.

It thus appears that turbulent dissipation operates in two regimes, depending on how the tidal period compares with the local convective turn-over time, which in a convection zone varies with depth by several orders of magnitude. To ensure a smooth transition between these two regimes, one may take

$$\nu_t = \nu_t \ell \left[1 + \left(\frac{2t_{\text{conv}}}{P_{\text{tide}}} \right)^2 \right]^{-1/2} , \tag{8.24}$$

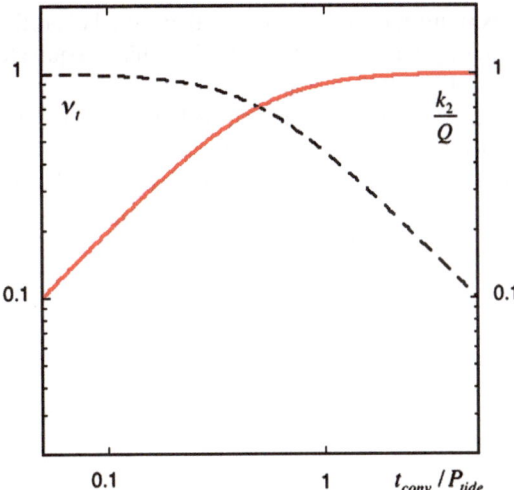

Fig. 8.4 The two regimes of turbulent dissipation (Eq. (8.24)). As long as the local convective turn-over time remains shorter than the tidal period ($t_{conv} < P_{tide}$), the turbulent viscosity ν_t (in *black dashed line*) is independent of the tidal frequency, and the inverse quality factor k_2/Q (in *red continuous line*) varies proportionally to the tidal frequency (σ_l) (so does also the tidal lag angle). When $t_{conv} > P_{tide}$, ν_t varies proportionally to the tidal period, whereas k_2/Q does no longer depend on the tidal frequency. ν_t and k_2/Q have been scaled by the value they take respectively for $t_{conv}/P_{tide} \to 0$ and $\to \infty$ (from Remus et al. [32], courtesy A&A)

as illustrated in Fig. 8.4. In the upper part of a convective envelope, where the convective turnover time is shorter than the tidal period, neither ν_t nor t_{diss} depend on the tidal period; the tidal dissipation varies proportionally to the tidal frequency (cf. Eq. (8.21)), and the tidal bulge has a constant time lag: this is what has been called the weak friction approximation [12]. But in the opposite case, when the life span of the convective eddies exceeds the tidal period, which is likely to occur at the base of convection zones, the tidal torque is independent of the tidal frequency; so are also the tidal lag angle and the quality factor Q which will be discussed next. Note that these two regimes still persist once the summation of ν_t over depth has been performed in (8.21).

8.3.4 The Quality Factor

In planetary sciences one often prefers to characterize the tidal dissipation by a dimensionless quality factor Q defined as

$$Q^{-1} = \frac{1}{2\pi E_0} \oint \left(-\frac{dE}{dt}\right) dt, \qquad (8.25)$$

where E_0 is the maximum energy associated with the tidal distortion and the integral is the energy lost during one complete cycle [7]. This is equivalent to specify the tidal angle, since $\alpha = 1/(2Q)$.

This quality factor Q is always combined with the Love number k_2, which measures the mass concentration in the star; in a homogeneous body $k_2 = 3/2$. In fluid bodies, such as stars with convection zones or giant planets, the tidal torque is given by a summation over the star of the turbulent viscosity, as we have seen above in Sect. 8.3.2, and Q is related to the coefficient λ_2 we have introduced there (Eq. (8.21)):

$$\frac{k_2}{Q} = 4\frac{\lambda_2}{t_{\text{conv}}}\left(\frac{R^3}{GM}\right)(\omega - \Omega). \tag{8.26}$$

Usually Q is treated as a positive quantity, and the sign of the tidal torque is imposed according to that of $(\omega - \Omega)$.

We see that the quality factor Q depends both on intrinsic properties of the star (or the planet) and on the degree of synchronism, and this fact is often overlooked when comparing the Q of different planets or satellites in the solar system. If, as it has been suggested (cf. [29]), the circularization period of late-type binary stars is roughly consistent with $Q = 10^6$, it means that λ_2 is inversely proportional to the tidal frequency $(\omega - \Omega)$, hence that the turbulent viscosity is reduced according to the first prescription (8.22). If one chooses instead the quadratic reduction (8.23), as done in the paper quoted above, Q scales as the tidal frequency.

8.3.5 Beyond the Weak Friction Approximation

When the turbulent viscosity depends on the tidal period, the weak friction approximation no longer applies, and Hut's elegant method can no longer be applied to determine the tidal torque. It is then necessary to break the tidal potential in its multiple Fourier components, of frequencies $\sigma = (j\omega - m\Omega)$, and to sum up the torques exerted by each of these. Keeping only the second order spherical harmonics of the potential, and up to second order terms in eccentricity e, which is sufficient for many purposes, one has

$$U = \frac{GM_2}{a}\left(\frac{r}{a}\right)^2\left\{-\frac{1}{2}P_2(\cos\theta)\left[1 - \frac{3}{2}e^2 + 3e\cos\omega t + \frac{9}{2}e^2\cos 2\omega t\right]\right.$$
$$+ \frac{1}{4}P_2^2(\cos\theta)\left[-\frac{e}{2}\cos(\omega - 2\Omega)t + \left(1 - \frac{5}{2}e^2\right)\cos(2\omega - 2\Omega)t\right.$$
$$\left.\left. + \frac{7e}{2}\cos(3\omega - 2\Omega)t + 17\frac{e^2}{2}\cos(4\omega - 2\Omega)t\right]\right\}. \tag{8.27}$$

Each component of the tidal potential produces a tidal flow of frequency $\sigma = [j\omega - m\Omega]$, which experiences a different turbulent viscosity ν_t, since it depends on the tidal frequency (cf. Sect. 8.3.3). This is reflected in the coefficient λ_2 introduced above in (8.21), which takes a different value $\lambda^{m,l}$ for each tidal frequency. In a star with a deep outer convection zone, such as a late-type main-sequence star or a red giant, this parameter varies approximately as

$$\lambda^{m,l} = 0.019\alpha^{4/3} \left(\frac{3160}{3160 + \eta^2} \right)^{1/2} \quad \text{with } \eta = [j\omega - m\Omega]t_{\text{conv}}, \tag{8.28}$$

where t_{conv} is given in (8.21) and α is the familiar mixing-length parameter.

The equations governing the orbital evolution of the binary system then take the following form, to second order in e and assuming for simplicity that all spins are aligned [47]:

$$\frac{d\ln a}{dt} = -\frac{12}{t_{\text{conv}}} q(1+q) \left(\frac{R}{a} \right)^8 \left(\lambda^{2,2} \left[1 - \frac{\Omega}{\omega} \right] \right.$$
$$+ e^2 \left\{ \frac{3}{8} \lambda^{0,1} + \frac{1}{16} \lambda^{2,1} \left[1 - 2\frac{\Omega}{\omega} \right] \right.$$
$$\left. \left. - 5\lambda^{2,2} \left[1 - \frac{\Omega}{\omega} \right] + \frac{147}{16} \lambda^{3,2} \left[3 - 2\frac{\Omega}{\omega} \right] \right\} \right), \tag{8.29}$$

$$\frac{d\ln e}{dt} = -\frac{3}{t_{\text{conv}}} q(1+q) \left(\frac{R}{a} \right)^8$$
$$\times \left(\frac{3}{4} \lambda^{0,1} - \frac{1}{8} \lambda^{2,1} \left[1 - 2\frac{\Omega}{\omega} \right] - \lambda^{2,2} \left[1 - \frac{\Omega}{\omega} \right] + \frac{49}{8} \lambda^{2,3} \left[3 - 2\frac{\Omega}{\omega} \right] \right), \tag{8.30}$$

plus similar contributions of the secondary star (we recall that $q = M_2/M_1$). Note that we have added here the contribution of the axisymmetric part of the perturbing potential (which varies also in time when the orbit is eccentric, and yields the term in $\lambda^{0,1}$). The angular velocity of the primary star obeys

$$\frac{d}{dt}(I\Omega) = \frac{6}{t_{\text{conv}}} q^2 M R^2 \left(\frac{R}{a} \right)^6 \left(\lambda^{2,2}[\omega - \Omega] \right.$$
$$\left. + e^2 \left\{ \frac{1}{8} \lambda^{2,1}[\omega - 2\Omega] - 5\lambda^{2,2}[\omega - \Omega] + \frac{49}{8} \lambda^{2,3}[3\omega - 2\Omega] \right\} \right), \tag{8.31}$$

and likewise for the secondary star. One verifies that the total angular momentum is conserved, i.e. that

$$\frac{d}{dt} \left[\frac{GM_1 M_2}{(M_1 + M_2)^{1/2}} a^{1/2} (1 - e^2)^{1/2} + I_1 \Omega_1 + I_2 \Omega_2 \right] = 0. \tag{8.32}$$

Equation (8.32) reduces to (8.18) and (8.31) to (8.19) when all $\lambda^{m,j} \to \lambda_2$, in the weak friction approximation.

8.4 Confronting the Theory of the Equilibrium Tide with the Observations

Having identified the most efficient dissipation mechanism, namely turbulent convection acting on the equilibrium tide, we shall now examine how well it accounts for the observed properties in binary stars involving at least one component possessing an outer convection zone. We shall treat in turn the case of solar-type binaries on the main-sequence, that of such binaries during their pre-main sequence phase, and finally that of binaries in which one component has evolved to the giant stage.

8.4.1 Solar-Type Binaries on the Main Sequence

Applying Eq. (8.19) to a binary of equal components of solar mass and age t_{age}, one finds that its orbit should be circular if its period is less than about

$$P_{circ} = 6 \left(\frac{t_{age}}{5 \, \text{Gyrs}} \right)^{3/16} \text{days.} \tag{8.33}$$

To obtain this result we assume that the rotation is synchronized with the orbital motion, and that the eccentricity decreased from $e = 0.30$, a typical value for non-circularized binaries, to $e = 0.02$, taken as detection threshold for the eccentric orbits.

Koch and Hrivnak [13] were the first to compare this theoretical prediction with the distribution $e(P)$ of field binaries drawn from Batten's catalogue of spectroscopic binaries, and they found them to be compatible, although the transition period P_{circ} between circular and elliptic orbits was rather poorly defined, as one may expect with such a sample mixing stars of different mass and age.

But the fact that the transition period is a slowly increasing function of age should be observable, by measuring the eccentricity of coeval cluster binaries. Such a trend was found indeed by comparing the results of several surveys [3, 17, 26]. This incited Mathieu and Mazeh [21] to suggest that the determination of P_{circ} could serve to evaluate the age of a cluster. However for M67, a cluster of about solar age, they found that the transition period was between 10.3 and 11 days, well above the predicted 6 days, suggesting that tidal dissipation was about 20 times more efficient than inferred from the mixing-length theory.

Recently Mathieu et al. [22] gave a summary of the beautiful work accomplished over more than a decade by several dedicated teams (Fig. 8.5). They found that the transition period for circularization increases with age beyond 1 Gyr, but that it is more or less constant for younger stars, around $P_{circ} \approx 7–8$ days. It thus appears that two different mechanisms are at work, one operating on old binaries, and another that circularizes the young binaries even on the PMS.

Fig. 8.5 Transition periods
for circularization, below
which the binary orbits are
circularized, are displayed vs.
age for six coeval stellar
samples: PMS [25], Pleiades
[26], Hyades [3], M67 [17],
NGC 188 [22] and Galactic
halo stars [18]. Note the near
constancy of this period
below 1 Gyr, at about
$P_{circ} \approx 8$ days, and its
increase with age beyond.
(From Mathieu et al. [22];
courtesy ApJ)

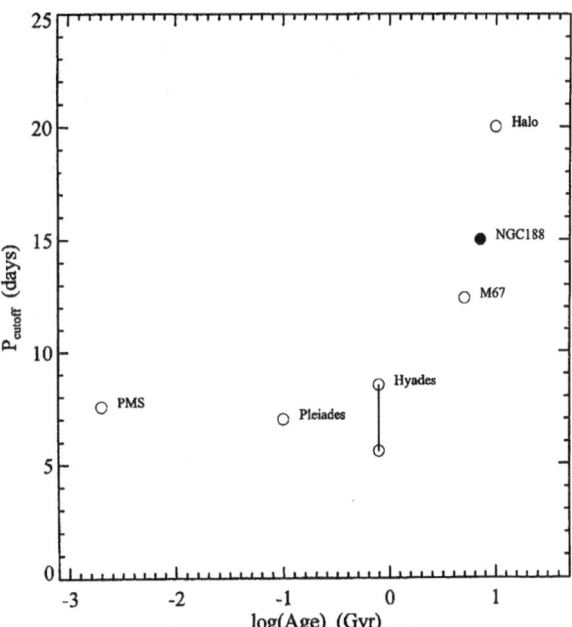

8.4.2 Orbital Circularization During the Pre-Main-Sequence
Phase

The rate at which orbits are circularized depends strongly on the radius of the star:
according to (8.19) $-d \ln e/dt \propto R^8$. Therefore one expects that most of this circularization should occur on the PMS, where the stellar radius is much larger than
later on the main-sequence. This suggestion was first made by Mayor and Mermilliod [23], and I verified it with L. Bouchet by integrating Eqs. (8.30)–(8.32) which
describe the tidal evolution of solar-type binaries, starting at the birthline defined
by Stahler [36, 37]. Since on the PMS the convective turnover time can exceed the
orbital period, it will also exceed the period of most Fourier components present
in the tidal perturbation (cf. (8.28)), and therefore one must take into account the
reduction of the turbulent viscosity, as was discussed in Sect. 8.3.5.

The result is displayed in Fig. 8.6, for a binary consisting of two solar-mass stars.
The initial conditions were taken as $R = 4.79 R_\odot$, $e = 0.3$, $(\Omega/\omega) = 3$, and the orbital period P was chosen such that the eccentricity would drop to 0.005 when the
binary reaches the zero age main-sequence (ZAMS). The rotation quickly synchro-
nizes with the orbital motion (in less than 10^5 yrs), but thereafter the tidal torque
weakens because the convection zone retreats, while the star keeps contracting;
therefore the rotation speeds up again to about $(\Omega/\omega) = 2$ at the ZAMS, with our
choice of initial conditions. Once the star has settled on the MS, synchronization
proceeds unhindered, and is achieved by an age of 1 Gyr. The eccentricity first increases, as long as $(\Omega/\omega) > 18/11$ (cf. Eq. (8.19)), and then it steadily decreases

Fig. 8.6 Evolution in time of the eccentricity e, the orbital period P and of the ratio between rotational and orbital frequencies (Ω/ω), for a system with two components of 1 $M_\odot$. The initial period has been chosen such that the eccentricity would decrease from 0.300 to 0.005 when the binary reaches the zero age main-sequence (indicated by the *arrow*). (From Zahn and Bouchet [49]; courtesy A&A)

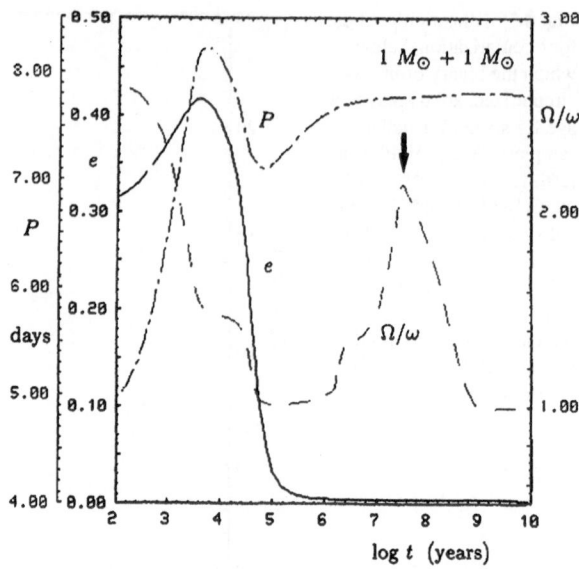

to reach its final value $e = 0.005$ at the ZAMS. Little circularization occurs thereafter on the MS. Angular momentum is transferred from the rotation to the orbit, which explains why the orbital period increases from 5 to 7.8 days. This final period depends rather weakly on the mass of the components, and it represents thus the transition period for circularization, in the absence of other tidal braking mechanisms.

This transition period agrees remarkably well with the properties of late type binaries younger than 1 Gyr, including the PMS stars, and thus there is little doubt that the circularization in these stars is due to the action of the equilibrium tide early on the PMS. The main uncertainties in the theoretical prediction are the initial radius R_i (P_{circ} scales as R_i to the power 15/16) and the recipe used to reduce the turbulent viscosity when the tidal period becomes shorter than the convective turnover time. We took here the linear prescription (8.22); with the other, quadratic prescription (8.23) the predicted transition period would be substantially shorter, contrary to what is observed.

It is important to note that binaries in their early MS stage may be circularized while still not synchronized, which may seem paradoxical since the synchronization time (8.18) is much shorter than the circularization time (8.19). It stresses the necessity of following the whole tidal evolution of a given binary, starting from 'reasonable' initial conditions.

8.4.3 Circularization of Binaries Evolving off the Main-Sequence

Another clever test for the tidal theory was performed by Verbunt and Phinney [40], who chose for that a sample of wide binaries containing a giant star, because they

Fig. 8.7 Observed
eccentricities of binaries
including a giant component
vs. the change in eccentricity
predicted by the tidal theory,
invoking the equilibrium tide
with turbulent dissipation in
the convection zone
(Verbunt [40], courtesy
A&A). In the *upper panel* the
giant components are
assumed to be on the
asymptotic giant branch;
some corrections have been
applied to obtain the result of
the *lower panel* (see text)

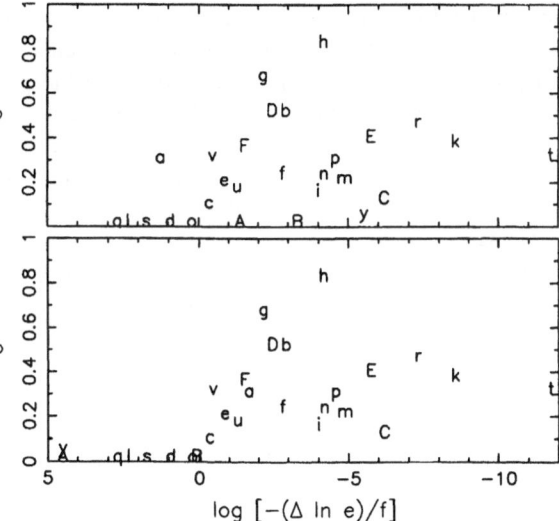

wanted to avoid what they call the 'troublesome problem of pre-main sequence circularization' that we just discussed. Moreover, in such binaries the tidal period exceeds the convective turnover time, so that there is no need to worry about reducing the turbulent viscosity. They considered 29 binaries with giant components in several galactic clusters, whose age and distance are well established. They integrated the circularization equation (8.19) for these binaries from the MS to their present location in the HR diagram, and presented the result in the form $-\Delta \ln e/f$, where $\Delta \ln e$ is the change in eccentricity, and f a factor that depends on the convection theory used to calculate the turbulent dissipation. For the classical mixing-length treatment that was employed in Sect. 8.3.2, f is of order unity.

Figure 8.7 displays the *observed* eccentricity of these binaries (each individually labeled by a letter) as a function of the *predicted* drop in eccentricity $-\Delta \ln e$ (or rather $\log[-\Delta \ln e/f]$ to accommodate the wide range of results). For $\log[-\Delta \ln e/f] > 0$), the orbit should be circularized, whereas it should remain elliptic for $\log[-\Delta \ln e/f] < 0$. Phinney and Verbunt first assumed that all their binaries are presently on the asymptotic giant branch (core helium burning), because they stay there 10 times longer than previously on the red giant branch (shell hydrogen burning).

The result is shown in the upper panel: the great majority of binaries complies with the theoretical prediction, displaying circular orbits for $\log[-\Delta \ln e/f] > 0$ and eccentric obits for $\log[-\Delta \ln e/f] < 0$. However there are 4 notable exceptions: binary 'a' has kept an eccentricity of 0.30, while its orbit should still be circular, and binaries 'A', 'B', 'y' have circular orbits, where these should be elliptic. Phinney and Verbunt concluded that binary 'a' must still be ascending the red giant branch, thus avoiding circularization, and that the other 3 binaries may have undergone an exchange of matter, which very efficiently circularizes the orbit, and therefore that they should have an evolved companion, such as a white dwarf. After these adjustments, the 4 binaries are no longer exceptions, as can be seen in the lower panel; moreover, the fact that the transition from circular to elliptic orbits occurs in the vicinity of $\log[-\Delta \ln e/f] \approx 0$ confirms that the parameter f is indeed of order unity, thus validating the theory of the equilibrium tide with turbulent dissipation.

Two years later Landsman et al. [16] announced that the secondary of S1040 in M67, the binary labeled 'A', is indeed a white dwarf, confirming the brilliant conjecture of Verbunt and Phinney that it must have experienced an episode of mass exchange.

We may thus conclude that turbulent viscosity acting on the equilibrium tide explains most observations, with the important exception of the circularization of main-sequence binaries older than about 1 Gyr, for which it seems that we have to seek another dissipation mechanism. A plausible candidate for that is the dynamical tide, which we shall examine next.

8.5 The Dynamical Tide

Due to its elastic properties, a star can oscillate in various modes: acoustic modes, internal gravity modes, inertial modes, where the restoring force is respectively the compressibility of the gas, the buoyancy force in stably stratified regions, and the Coriolis force in the rotating star. If their frequency is low enough, these modes can be excited by the periodic tidal potential; the response is called the *dynamical tide*.

8.5.1 Gravity Modes Excited by a Close Companion

The modes that have received most attention so far are the tidally excited gravity modes; associated with radiative damping, they have first been invoked for the tidal evolution of massive main-sequence binaries [45]. For these modes, the restoring force is provided by the buoyancy, whose strength is measured by the buoyancy frequency N, given by

$$N^2 = \frac{g\delta}{H_P}\left[\left(\frac{\partial \ln T}{\partial \ln P}\right)_{ad} - \frac{d \ln T}{d \ln P} + \frac{\varphi}{\delta}\frac{d \ln \mu}{d \ln P}\right], \qquad (8.34)$$

using classical notations, and μ being the molecular weight ($\delta = -(\partial \ln \rho/\partial \ln P)_{T,\mu}$ and $\varphi = (\partial \ln \rho/\partial \ln \mu)_{T,P}$ are unity for perfect gas).

The modes that are most excited are those whose frequency is close to the tidal frequency, and these are of high radial order: typically they have more than 10 or 20 radial nodes in the radiation zone, because their wavelength scales as $\lambda_r \propto r\sigma/N$, and because the tidal frequency σ, of the order of days^{-1}, is much lower than the buoyancy frequency N, of the order of 1 hour^{-1}. See Fig. 8.8 for a typical example of such modes, in a 4 $M_\odot$ star of 94 Myr. Dissipation has been neglected, and therefore the mode is an adiabatic standing wave; note that it is evanescent in the convective core, where $N^2 \approx 0$.

These gravity modes couple with the periodic tidal potential in the vicinity of the convective core, whereas their damping occurs mainly near the surface, because the

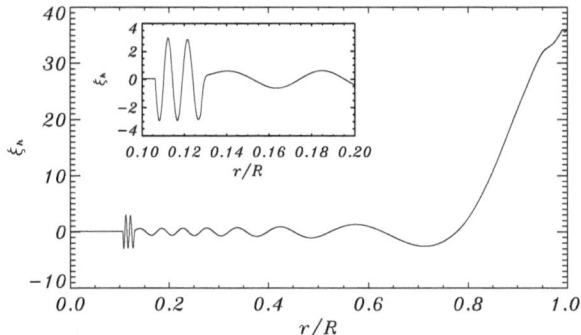

Fig. 8.8 Gravity mode of period 4 days in a 4 $M_\odot$ star of 94 Myr. Only the horizontal displacement ξ_h is shown, scaled such that the radial component $\xi_r = 1$ at the surface; it has 20 radial nodes. The *insert* focuses on the region close to the convective core, where the eigenfunction displays strong oscillations due the steep composition gradient. The effect of rotation is neglected (from Hasan et al. [10], courtesy A&A)

thermal damping rate, which scales roughly as the cube of the temperature, is much higher there than in the deep interior. The angular momentum drawn from the orbit is deposited near the surface, and hence it is the surface layers that are synchronized first with the orbital motion. As was emphasized by Goldreich and Nicholson [6], this synchronization is further sped up because the local tidal frequency experienced by the fluid entrained in the differential rotation, $\sigma = 2\Omega(r) - 2\omega$, tends to zero, and so does also the radial wavelength λ_r, as we have seen above, thus enhancing the damping.

At low enough tidal frequency, the tidal wave is completely damped (meaning that is has become a pure propagating wave), and one can use the WKB treatment to evaluate the total torque applied on the star [45]. For the synchronization time (assuming uniform rotation) one finds

$$\frac{1}{t_{\text{sync}}} = -\frac{d}{dt}\left|\frac{2(\Omega - \omega)}{\omega}\right|^{-5/3} = 5\left(\frac{GM}{R^3}\right)^{1/2}q^2(1+q)^{5/6}\frac{MR^2}{I}E_2\left(\frac{R}{a}\right)^{17/2},$$
(8.35)

and likewise for the circularization time, assuming that synchronization has already been achieved:

$$\frac{1}{t_{\text{circ}}} = -\frac{d\ln e}{dt} = \frac{21}{2}\left(\frac{GM}{R^3}\right)^{1/2}q(1+q)^{11/6}E_2\left(\frac{R}{a}\right)^{21/2};$$
(8.36)

the companion star contributes a similar amount. E_2 is a parameter measuring the coupling between the tidal potential and the gravity mode: it depends sensitively on the size of the convective core, and thus on the mass of the star. Its expression is given in Zahn [45]; it has been tabulated by Claret and Cunha [1] for various stellar models, as shown in Fig. 8.9; for a 10 $M_\odot$ ZAMS star, it is $E_2 \approx 10^{-6}$.

This theory was initially developed for pure gravity modes, and as such it was strictly applicable only to non-rotating stars. It was later extended by Rocca [34]

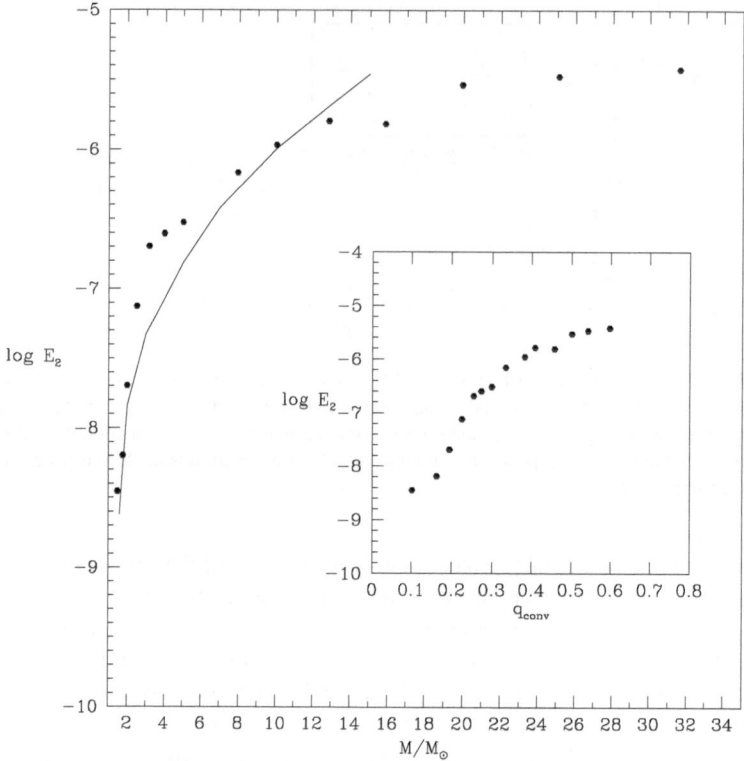

Fig. 8.9 Tidal parameter E_2 characterizing the strength of the dynamical tide, cf. (8.35) and (8.36). It is displayed here on a logarithmic scale, as a function of mass (in solar units), near the ZAMS. The *solid line* reports the results of earlier computations by Zahn [45]. The *insert* shows the dependence of E_2 on the relative size of the convective core (from Claret and Cunha [1]; courtesy A&A)

to (uniformly) rotating stars; she showed that taking the Coriolis force into account modifies only slightly the results presented above.

8.5.2 Circularization of Massive Binaries

Giuricin et al. [4] were the first to compare the predictions of the tidal theory with the properties of early-type binaries, thus possessing an outer radiation zone. Applied to binaries with two identical components of mass between 2 and 15 $M_\odot$, Eq. (8.36) predicts a transition value of $R/a \approx 0.25$ for the normalized radius, i.e. the radius expressed in units of semi-major axis.[1] This value is in good agreement with the

---

[1]This value depends little on mass [46]; if it were translated into tidal periods, the transition periods would spread between 1 to 2 days, depending on mass, which explains why it is preferable to use R/a for the observational test.

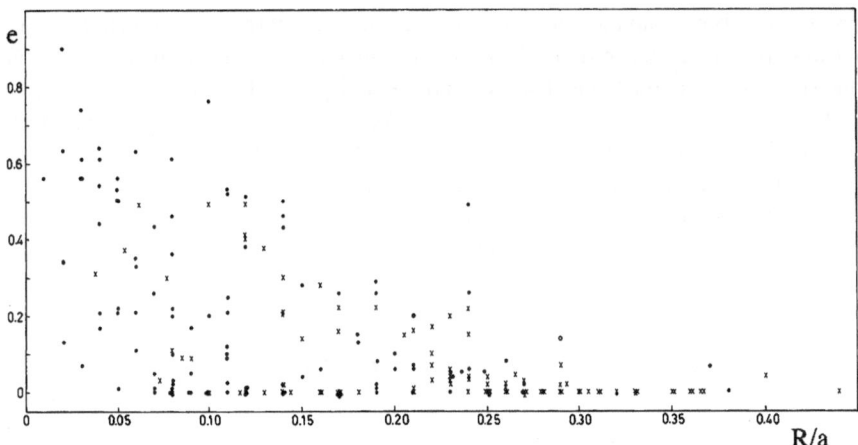

Fig. 8.10 Eccentricity *e* vs. normalized radius R/a for early-type binaries (spectral types O, B, F) listed in Batten's catalogue (from Giuricin et al. [4]; courtesy A&A)

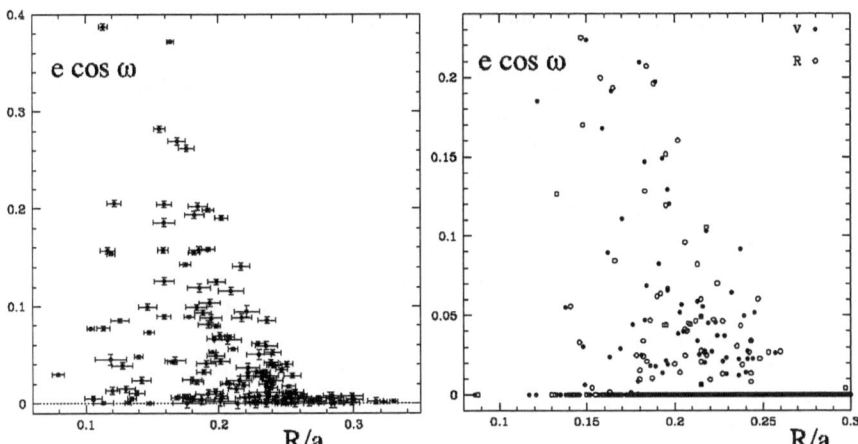

Fig. 8.11 *Left panel*: $e \cos \omega$ vs. relative radius R/a for detached eclipsing binaries in the SMC. *Right panel*: same for the LMC, *full dots* based on *V lightcurves*, *open dots* based on *R lightcurves*. Data from MACHO and OGLE surveys. (From North and Zahn [27]; courtesy A&A)

observed distribution of eccentricities vs. fractional radius displayed in Fig. 8.10, although many binaries are circular for $R/a < 0.25$.

A similar investigation was recently carried out on eclipsing binaries which had been detected in the Magellanic Clouds during the MACHO and OGLE campaigns [27]; the results are shown in Fig. 8.11. Here again the *e* vs. R/a distribution strongly suggests a transition value of $R/a = 0.255$, in excellent agreement with theory. However an important fraction of binaries are circular at lower fractional radius: it is as if there were two populations of binaries, one complying with the predictions above, and the other experiencing another, more efficient tidal damping. Histograms of the eccentricity distribution at given R/a confirm that impression, and so does also a much wider survey carried out by Mazeh et al. [24].

On may wonder why the binaries in the Magellanic Clouds behave so similarly to those in our Galaxy: they have lower metallicities, and therefore somewhat larger convective cores, and one would expect that these differences be reflected in the coefficient E_2. However the radii differ too, and the two effects compensate each other such that the predicted transition periods are very nearly the same.

8.5.3 Resonance Locking in Early-Type Binaries

A decade ago, Witte and Savonije [41, 42] revisited the theory of the dynamical tide, by making full account of the Coriolis force. Instead of projecting the forced oscillations on spherical functions, they solved the governing equations directly in two dimensions (r, θ), for various values of the angular velocity Ω and of the tidal frequency $\sigma = j\omega - 2\Omega$ in the rotating frame. When the orbit is circular and the star rotates in the same sense as the orbital motion, only one retrograde mode can be excited at $\sigma = 2\omega - 2\Omega$. But when the orbit is elliptic, many other tidal frequencies appear: $\sigma_j = j\omega - 2\Omega$ with $|j| = 1, 3$, etc. (see Sect. 8.3.5), and both retrograde and prograde modes can be excited. Therefore it is very likely that a binary undergoes some resonances during its evolution, both because the tidal frequency shifts in the course of synchronization, and because the eigenfrequencies are affected by the structural changes of the stars.

In earlier works [6, 34, 45], the effect of resonances on tidal evolution was largely ignored on the belief that stars would move quickly through such resonances, since their width $\Delta\sigma$ is inversely proportional to their amplitude. But Witte and Savonije [42] pointed out that this is not necessarily true, and that a binary can be trapped into a resonance, for elliptic orbits. Retrograde and prograde modes exert torques of opposite sign, and when they balance each other, they may lock the star into such resonances. Moreover, structural changes also can conspire to favor such locking. The consequence is that circularization is sped up by such resonances, as demonstrated by several specific cases they have studied. The results are rather sensitive to the initial conditions, which may explain the observations mentioned above concerning the Magellanic Clouds binaries, namely that for the same orbital period (or fractional radius), some binaries are circular while the others are not, as if there were two tidal damping mechanisms.

8.5.4 Resonance Locking in Late-Type Binaries

Let us come back to the late-type main-sequence binaries. We have seen that turbulent dissipation of the equilibrium tide, at least in its present state, cannot explain the circularization observed in binaries older than 1 Gyr. This incited Terquem et al. [39] and Goodman and Dickson [8], to examine whether the dynamical tide could not be responsible for the observed circularization. Both teams invoked radiative damping as dissipation mechanism, as had been done previously for early-type stars. But here such damping is rather weak, because the oscillation modes are evanescent in the convection zone, where thermal dissipation would be strongest. Therefore, contrary to what has been found in early type stars, oscillations modes can enter in resonance at very low tidal frequency, i.e. very close to synchronization. This means that one has to deal with modes which have up to thousand radial nodes, which puts a serious burden on the numerical work, as experienced by Terquem et al.; they restricted their exploration to the vicinity of 3 orbital periods, but included turbulent dissipation in the convection zone, where the modes are evanescent. On the contrary, Goodman and Dickson chose a semi-analytical WKB approach, much as in Zahn [46].

Though their quantitative results differ somewhat, the conclusions of the two teams agree, namely that the dynamical tide cannot account for the circularization of the oldest late-type binaries; comparing the predicted transition periods, one sees that it is less efficient than the equilibrium tide.

The problem was re-examined shortly after by Witte and Savonije [43], who anticipated that here also resonance locking could play an important role. Instead of performing the direct 2D calculations as for the early-type binaries, given the high order of the modes, they used the so-called 'traditional approximation', which retains only the radial component of the rotation vector. The r and θ variables then separate again, as in the non-rotating case, the horizontal functions being the so-called Hough functions [35], which contrary to the spherical harmonics depend also on the rotation rate. The tidal torque is displayed in Fig. 8.12, as a function of the forcing frequency.

Today this process of resonance locking in the dynamical tide thus appears as the most efficient process, *on the main-sequence*, among all that have been explored. When starting with quasi-synchronous or super-synchronous stars, the predicted transition period is a slowly increasing function of age; for 5×10^9 yrs, this period is about 7 days, thus higher than that predicted by the equilibrium tide (6 days). But even so, the theoretical predictions are well below the observed ones, unless one allows for very slow, and rather unrealistic initial rotation (such as a period of 100 days). Let us recall that below 1 Gyr the observations agree very well with the transition period derived for the PMS circularization through the equilibrium tide, as we have seen in Sect. 8.4.2

8.6 Tidal Damping Through Inertial Modes

While gravity modes propagate only in stably stratified regions, there is another type of modes, the inertial modes, that are able to propagate also in neutrally stratified convection zones. They owe their existence to the Coriolis force, and hence their frequency, in the frame of the rotating star, is bound by the inertial frequency 2Ω. They may thus be excited by the tidal potential, much as the gravity modes, provided

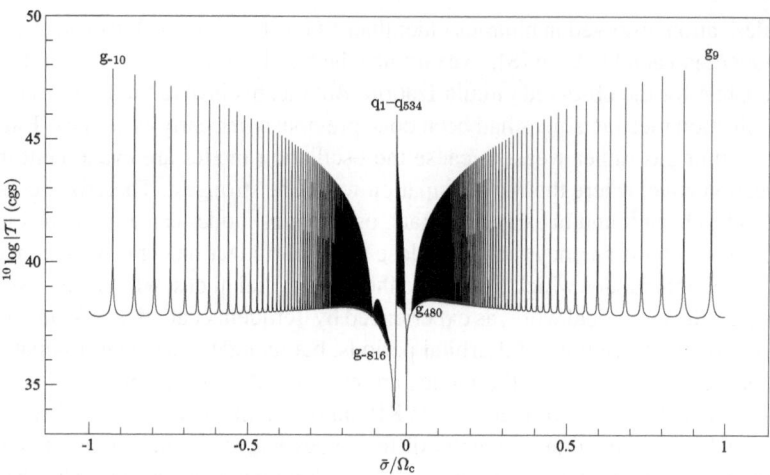

Fig. 8.12 Tidal torque versus forcing frequency $\bar{\sigma}$, scaled by the break-up frequency Ω_c, in a binary of two $1\,M_\odot$ stars, showing the resonances with prograde g-modes ($\bar{\sigma} > 0$), retrograde g-modes ($\bar{\sigma} < 0$), and inertial modes ($-2\Omega < \bar{\sigma} < 0$). (From Savonije and Witte [35]; courtesy A&A)

the tidal frequency is less than the inertial frequency 2Ω. These modes have received little attention so far, until very recently.

Recently Ogilvie and Lin [29] have studied numerically the rôle of these inertial modes in damping the tides, in a solar-type star. The results are depicted in Fig. 8.13. One sees that their contribution (left panel), through their viscous dissipation in the convection zone, can be as large as that of the gravito-inertial modes in the radiation zone (right panel). The dashed lines show the effect of switching off the Coriolis force, and the dotted line, in the left panel, that of increasing the turbulent viscosity by a factor 10. Note that Ogilvie and Lin opted for the quadratic reduction of that turbulent viscosity (Eq. (8.22)), which probably underestimates the contribution of the equilibrium tide.

A remarkable property of these inertial modes is that their peak amplitude, at resonance, does not depend on the strength of the viscosity, as can be seen in the left panel of Fig. 8.13. This is because these modes are described in the inviscid limit by an equation that is spatially hyperbolic, and hence their characteristic rays are focused on wave attractors, where most of viscous dissipation occurs, and whose thickness scales in such a way as to render the dissipation independent of viscosity, as explained in detail by Ogilvie and Lin.

8.7 Conclusion and Perspectives

The reader may wonder why I made no attempt here to reconcile the theoretical predictions for the synchronization of the binary components with their observed

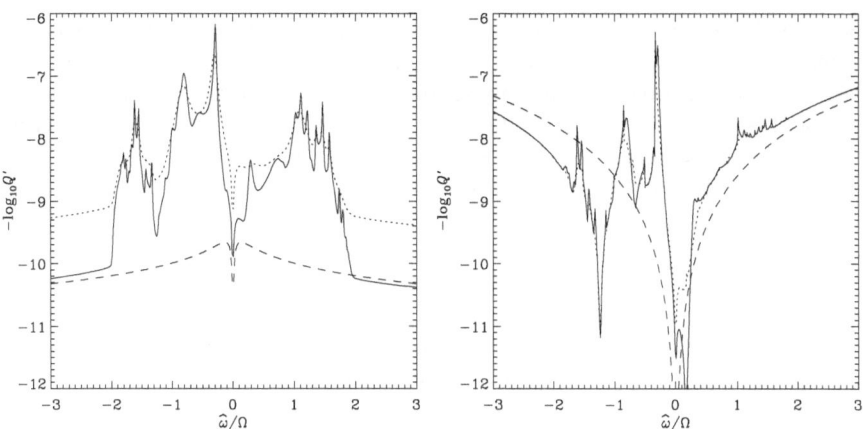

Fig. 8.13 Dissipation rate Q', defined in (8.25) and (8.26), as a function of the tidal frequency $\hat{\omega}$ normalized by the rotation frequency Ω. The solar model has a spin period of 10 d. *Left*: Q' from the viscous dissipation of inertial modes in the convection zone. *Right*: Q' from the excitation of Hough modes in the radiative zone. The *dashed lines* show the effect of omitting the Coriolis force, hence reducing the dissipation to that of the equilibrium tide. The turbulent viscosity has been reduced according to prescription (8.23). The *dotted lines* show the result of increasing that turbulent viscosity by a factor of 10. (From Ogilvie and Lin [29]; courtesy ApJ)

surface rotation. The reason is that in most cases the tidal torque acts mainly on the outermost part of the star, which is thus synchronized much more rapidly than the interior; therefore the interpretation of the surface rotation requires modeling the transport of angular momentum within the star, in particular where it proceeds at the slowest rate, i.e. in the radiation zones. This is a difficult task that only now begins to be undertaken seriously (cf. [38, 48]), but I am confident that we will see much progress in solving this problem in a not too distant future.

To summarize this review, the two tidal dissipation processes that have received most attention so far are turbulent friction acting on the equilibrium tide, which was first described in the 60s [44], and radiative damping on the dynamical tide, which was identified in the 70s [45]. These processes operate respectively in convection zones and in radiation zones, and they have been quite successful in explaining the observed orbital circularization of binary stars. This is particularly true for the early-type MS binaries, for which we have now at our disposal very large samples gathered during the OGLE and MACHO campaigns: their transition period is precisely defined and it agrees extremely well with that predicted by the theory of the dynamical tide, which is thus validated. However many of these binaries are circularized well above this transition period, as if they had experienced another, more efficient tidal dissipation mechanism. A plausible explanation for this behavior is that these binaries have undergone several episodes of resonance locking, as was described by Witte and Savonije [41, 42].

On the other hand, the equilibrium tide damped by turbulent dissipation accounts very well for the properties of binaries containing a red giant, as was demonstrated by Verbunt and Phinney [40]. It also explains the transition period of about 8 days observed in late-type binaries that are younger than about 1 Gyr: the explanation is that these have been circularized during the PMS phase, when they were much larger and fully convective. The only serious discrepancy today seems to be the behavior of late-type main-sequence binaries older than 1 Gyr, whose transition period increases with age and is higher than that predicted when applying straightforward the theory of the equilibrium tide. Here again one may invoke the dynamical tide with resonance locking in the radiative core of these stars, as was shown by Witte and Savonije [43].

Their mechanism appears thus highly promising, and it ought to be further explored. For instance, one should take into account that the tidal torque is applied primarily to specific regions: the outer convection zone in late-type MS stars and the outermost part of the radiation zone in early-type stars. These regions are synchronized more quickly than the rest of the star, and therefore differential rotation develops in their radiation zone. This has the effect of increasing the thermal damping, since the local tidal frequency tends then to zero as the tidal wave approaches the synchronized region, as I explained in Sect. 8.5.1.

For late-type binaries, a highly interesting alternative is offered by the damping of inertial waves in their convective envelope, which is being explored by Ogilvie and Lin [29]. This process is likely to play an important role also in giant planets [28]. The difficulty in studying these waves is that they require highly resolved 2D numerical calculations, since the so-called traditional approximation is no longer applicable to render the problem separable.

Work is in progress on several other points, and I shall quote only a few. Kumar and Goodman [15] have studied the enhanced damping of the oscillations triggered in tidal-capture binaries, due to non-linear coupling between the eigenmodes, which is extremely strong in such highly eccentric orbits. Rieutord [33] is examining the possibility that the so-called elliptic instability may occur in binary stars; this instability is observed in the laboratory when the fluid is forced to rotate between boundaries that have a slight ellipticity, and it leads to turbulence [19]. Even the equilibrium tide in late-type binaries is being revisited [32], solving at last the irritating problem of the 'pseudo-resonances' encountered in Zahn [44].

To conclude, I am very pleased to witness this revival of the theory of stellar tides; it owes much to the discovery of extrasolar planets and to the wide surveys mentioned above, which I didn't anticipate forty-five years ago...

References

1. Claret, A., Cunha, N.C.S.: Astron. Astrophys. **318**, 187 (1997)
2. Darwin, G.H.: Philos. Trans. R. Soc. Lond. **170**, 1 (1879)
3. Duquennoy, A., Mayor, M., Mermilliod, J.-C.: In: Duquennoy, A. Mayor, M. (eds.) Binaries as Tracers of Star Formation, p. 52. Cambridge University Press, Cambridge (1992)
4. Giuricin, G., Mardirossian, F., Mezzetti, M.: Astron. Astrophys. **134**, 365 (1984)
5. Goldreich, P., Nicholson, P.D.: Icarus **30**, 301 (1977)
6. Goldreich, P., Nicholson, P.D.: Astrophys. J. **342**, 1079 (1989)
7. Goldreich, P., Soter, S.: Icarus **5**, 375 (1966)

8. Goodman, J., Dickson, E.S.: Astrophys. J. **507**, 938 (1998)
9. Goodman, J., Oh, S.P.: Astrophys. J. **486**, 403 (1997)
10. Hasan, S.S., Zahn, J.-P., Christensen-Dalsgaard, J.: Astron. Astrophys. **444**, L29 (2005)
11. Hut, P.: Astron. Astrophys. **92**, 167 (1980)
12. Hut, P.: Astron. Astrophys. **99**, 126 (1981)
13. Koch, R.H., Hrivnak, B.J.: Astron. J. **86**, 438 (1981)
14. Kopal, Z.: Close Binary Systems. Chapman & Hall, London (1959)
15. Kumar, P., Goodman, J.: Astrophys. J. **466**, 946 (1996)
16. Landsman, W., Aparaicio, J., Bergeron, P., Di Stefano, R., Stecher, T.P.: Astrophys. J. **481**, L93 (1997)
17. Latham, D.W., Mathieu, R.D., Milone, A.E., Davis, R.J.: In Kondo, Y., Sistero, R.F., Polidan, R.S. (eds.) Evolutionary Processes in Interacting Binary Stars. IAU Symp., vol. 151, p. 471. Kluwer Academic, Dordrecht (1992)
18. Latham, D.W., Stefanik, R.P., Torres, G., Davis, R.J., Mazeh, T., Carney, B.W., Laird, J.P., Morse, J.A.: Astron. J. **124**, 1144 (2002)
19. Le Bars, M., Lacaze, L., Le Dizès, S., Le Gal, P., Rieutord, M.: Phys. Earth Planet. Inter. **178**, 48 (2009)
20. Levrard, B., Winidoerfer, C., Chabrier, G.: Astrophys. J. **692**, 9L (2009)
21. Mathieu, R.D., Mazeh, T.: Astrophys. J. **326**, 256 (1988)
22. Mathieu, R.D., Meibom, S., Dolan, C.: Astrophys. J. **602**, 121 (2004)
23. Mayor, M., Mermilliod, J.-C.: In: Maeder, A., Renzini, A. (eds.) Observational Tests of the Stellar Evolution Theory. IAU Symp., vol. 105, p. 411 (1984)
24. Mazeh, T., Tamuz, O., North, P.: Mon. Not. R. Astron. Soc. **367**, 1531 (2006)
25. Melo, C.H.F., Covino, E., Alcalá, J.M., Torres, G.: Astron. Astrophys. **378**, 898 (2001)
26. Mermilliod, J.-C., Rosvick, J.M., Duquennoy, A., Mayor, M.: Astron. Astrophys. **265**, 513 (1992)
27. North, P., Zahn, J.-P.: Astron. Astrophys. **405**, 677 (2003)
28. Ogilvie, G.I., Lin, D.N.C.: Astrophys. J. **610**, 477 (2004)
29. Ogilvie, G.I., Lin, D.N.C.: Astrophys. J. **661**, 1180 (2007)
30. Penev, K., Sasselov, D., Robinson, F., Demarque, P.: Astrophys. J. **655**, 1166 (2007)
31. Penev, K., Sasselov, D., Robinson, F., Demarque, P.: Astrophys. J. **704**, 930 (2009)
32. Remus, F., Mathis, S., Zahn, J.-P.: Astron. Astrophys. **544**, 132 (2012)
33. Rieutord, M.: In: Eennens, Ph., Maeder, A. (eds.) Stellar Rotation. IAU Symp., vol. 215, p. 394 (2004)
34. Rocca, A.: Astron. Astrophys. **213**, 114 (1989)
35. Savonije, G.J., Witte, M.G.: Astron. Astrophys. **386**, 111 (2002)
36. Stahler, S.W.: Astrophys. J. **274**, 822 (1983)
37. Stahler, S.W.: Astrophys. J. **332**, 804 (1988)
38. Talon, S.: EAS Publ. Ser. **32**, 81 (2008)
39. Terquem, C., Papaloizou, J.C.B., Nelson, R.P., Lin, D.N.C.: Astrophys. J. **502**, 788 (1998)
40. Verbunt, F., Phinney, E.S.: Astron. Astrophys. **296**, 709 (1995)
41. Witte, M.G., Savonije, G.J.: Astron. Astrophys. **341**, 842 (1999)
42. Witte, M.G., Savonije, G.J.: Astron. Astrophys. **350**, 129 (1999)
43. Witte, M.G., Savonije, G.J.: Astron. Astrophys. **386**, 222 (2002)
44. Zahn, J.-P.: Ann. Astrophys. **29**, 489 (1966)
45. Zahn, J.-P.: Astron. Astrophys. **41**, 329 (1975)
46. Zahn, J.-P.: Astron. Astrophys. **57**, 383 (1977)
47. Zahn, J.-P.: Astron. Astrophys. **220**, 112 (1989)
48. Zahn, J.-P.: EAS Publ. Ser. **26**, 49 (2007)
49. Zahn, J.-P., Bouchet, L.: Astron. Astrophys. **223**, 112 (1989)

Chapter 9
Tides in Colliding Galaxies

Pierre-Alain Duc and Florent Renaud

Abstract Long tails and streams of stars are the most noticeable traces of galaxy collisions. However, their tidal origin was recognized only less than 50 years ago and more than 10 years after their first observations. This review describes how the idea of galactic tides has emerged thanks to advances in numerical simulations, from the first simulations that included tens of particles to the most sophisticated ones with tens of millions of them and state-of-the-art hydrodynamical prescriptions. Theoretical aspects pertaining to the formation of tidal tails are then presented. The third part turns to observations and underlines the need for collecting deep multi-wavelength data to tackle the variety of physical processes exhibited by collisional debris. Tidal tails are not just stellar structures, but turn out to contain all the components usually found in galactic disks, in particular atomic/molecular gas and dust. They host star-forming complexes and are able to form star-clusters or even second-generation dwarf galaxies. The final part of the review discusses what tidal tails can tell us (or not) about the structure and the content of present-day galaxies, including their dark components, and explains how they may be used to probe the past evolution of galaxies and the history of their mass assembly. On-going deep wide-field surveys disclose many new low-surface brightness structures in the nearby Universe, offering great opportunities for attempting galactic archeology with tidal tails.

9.1 Preliminary Remarks

The importance of tides on bodies in the Solar System has been understood and quantified for many decades. The various contributions in this Volume reflect the

P.-A. Duc (✉)
AIM Paris-Saclay, CNRS/INSU, CEA/Irfu, Service d'Astrophysique, Université Paris-Diderot, Orme des Merisiers, 91191 Gif sur Yvette cedex, France
e-mail: paduc@cea.fr

F. Renaud
Observatoire de Strasbourg, CNRS UMR 7550 and AIM Paris-Saclay, CEA/Irfu, CNRS/INSU, Service d'astrophysique, Université Paris-Diderot, Orme des Merisiers, 91191 Gif sur Yvette cedex, France
e-mail: florent.renaud@cea.fr

J. Souchay et al. (eds.), *Tides in Astronomy and Astrophysics*, Lecture Notes in Physics 861, DOI 10.1007/978-3-642-32961-6_9, © Springer-Verlag Berlin Heidelberg 2013

maturity of this field of research. Advances in the appreciation of the role of tidal effects on planet/stellar evolution are also remarkable. As far as the extragalactic world is concerned, the situation is paradoxical. Whereas the effects of tidal forces are spectacular—they alter the morphology of the most massive galaxies and may lead to the total destruction of dwarf satellite galaxies—it is only in the 1970s that tides were recognized as actors of galactic evolution. Observations of jet-like structures, antennas, bridges, and plumes occurred well before they were interpreted as "tidal tails". Only the first numerical simulations of galaxy mergers convinced the community about the real nature of these stellar structures, although the straightforward consideration that galaxies are flaccid bodies could have lead to this conclusion much earlier. It is however true that the bulges generated by the Moon and the Sun on the Earth's oceans, which were interpreted as the result of tides soon after the laws of gravity were established, do not resemble the gigantic appendices that emanate from some galaxies, despite the similarity of their origins. What seems obvious now was not 50 years ago.

Having said that, it would be misleading to claim that tidal forces are the only actors of galactic morphological transformations. In fact, the fraction of mass involved in material that is tidally affected is relatively small. Other physical processes such as violent relaxation are more important in shaping galaxies. The nuclear starbursts often associated with galaxy mergers are not directly induced by tidal forces. Furthermore, not all the collisional debris found around mergers are, strictly speaking, of tidal origin. With these preliminary remarks, we wish to make it clear that this review specifically focuses on tides in colliding galaxies and is not an overview of interacting galaxies and associated phenomena. For a more general insight on galaxy-galaxy collisions and mergers, the reader is referred to the somewhat old but comprehensive reviews of Sanders and Mirabel [193] and Schweizer [196], dealing with observations, and Struck [215], more focused on simulations.

We will first present the historical context of the discovery of tails around galaxies, and detail how the role of tides became evident. The tremendous progress in the numerical modeling of tidal tails is detailed before a more theoretical and analytical approach to the formation of tidal tails is presented. In the following sections, we investigate the physical properties of tidal tails, emphasizing what deep multiwavelength observations bring to their study. We then make a close-up on the tails, looking at their sub-structures: from young stars and star clusters to tidal dwarf galaxies. Finally, we examine what tidal features may tell us about galaxies: what they are made of, and how and when they were formed. We hope to convince the reader that tails are not only aesthetic add-ons in images of colliding galaxies but may be used to address fundamental questions of astrophysics.

9.2 Historical Context

In the late 1920s, the observational power of 100-inch (2.5 m) class telescopes allowed Hubble to determine the existence of apparently isolated nebulae outside of our Milky-Way [118]. These so-called "island universes" became of prime importance in the discovery of the expansion of the Universe, thanks to redshift measurements. Rapidly, many more extra-galactic objects have been classified as galaxies and sorted according to their morphology, following the famous Hubble pitch fork diagram.

9.2.1 Discovery of Peculiarities

In the preface of his Atlas of Peculiar Galaxies, Arp [5] noted that "when looked at closely enough, every galaxy is peculiar". While most of the luminous galaxies could be classified as either elliptical, spiral or barred-spiral, it appeared that more and more peculiar morphologies would not fit into these three families. Number of photographic plates of individual systems have been published and revealed twisted shapes and/or faint extensions outside of the central regions of the galaxies (e.g. [76, 128, 241, 252]). These features have been detected in many other objects gathered in atlases and catalogues [5, 233, 234, 255, 256]. This contradicted the persistent idea that the intergalactic space was entirely empty ([254]; see also the discussion in [95] and references therein).

It rapidly appeared that many of these peculiar galaxies were actually double or multiple galaxies, i.e. pairs or small groups, observed close to each other. Really interacting galaxies have been told apart from optical pairs, for which apparent closeness is due to projection effects ([111]; see also [252]). The major signatures of interaction were the detection of long ($\sim 10^{1-2}$ kpc) and thin (~ 1 kpc) filaments either connecting two galaxies or pointing away from them. The former have been named bridges and the latter, tails. This clearly distinguished them from the spiral arms which are located in the more central regions of disk galaxies. However, a confusion persisted because it was noted that tails are sometimes (but not always) in the continuation of spiral arms [180]. Although being faint and thus often difficult to observe, these filaments appear bluer than the disks themselves, suggesting that they host ongoing star formation [2, 254]. But the exact reasons for such morphological features remained opened to debate over the entire 1960 decade.

9.2.2 A Controversial Scenario

Zwicky [253] proposed that collisions of galaxies would enhance the supernovae activity, by increasing the probability of chain explosions. These blasts could then sweep out or eject the galactic material away from the nuclei. With a favorable geometry, such events could even act as "launchers of galaxies", and thus account for the intergalactic filamentary structures. However, this scenario failed to explain the thinness of the filaments and the connection to other galaxies, so that it has rapidly been ruled out [180].

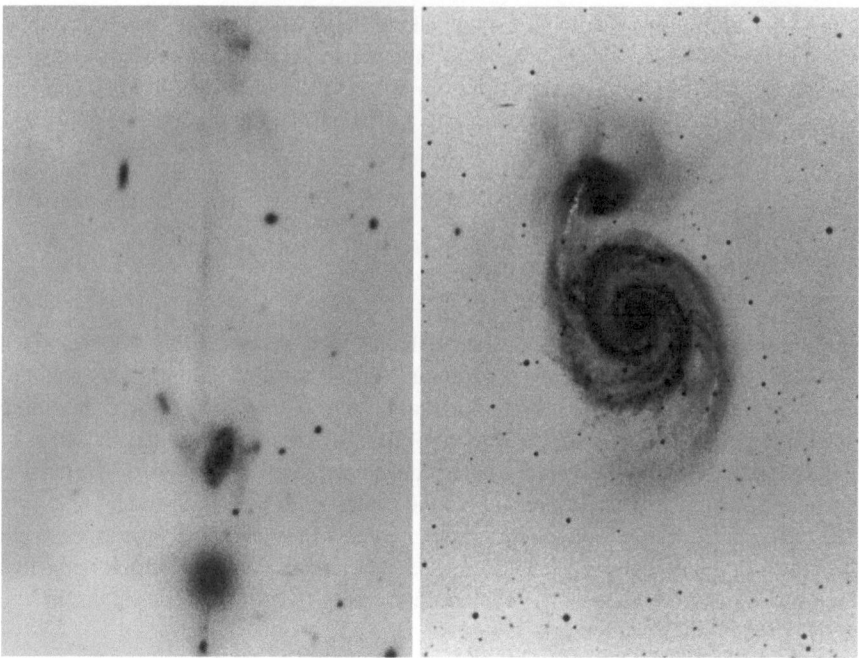

Fig. 9.1 NGC 3561 (*left*) has been seen by Arp [9] as a spiral galaxy having ejected two luminous jets of matter. An high surface brightness object, called "Ambartsumian's knot", can be seen at the tip of the southern jet at the *bottom-edge* of this image. In the case of M 51 (*right*), the companion is situated at the tip of a spiral arm of the main galaxy. Images from the Atlas of Peculiar Galaxies by H. Arp, available in the NASA/IPAC Extragalactic Database, Level 5

Another explanation for the formation of bridges took jets into account [2, 6, 7, 9]. When a massive galaxy ejects a fraction of its matter (gaseous, stellar or both) from its nucleus, a symmetrical pair of jets is formed but rapidly slowed down by the high densities encountered along its path.[1] This would create an overdensity at the tip of the jets that could condense and form a small companion galaxy [202]. All together, the main galaxy, its companion and one of the jets would constitute the interacting pair and the bridge. The absence of galaxy at the end of the second jet (i.e. the tail) was explained by either the escape of the companion to the intergalactic medium, its rapid dissolution, or a delayed formation that has not taken place yet [8]. Illustrative examples of this scenario are NGC 3561 ("the guitar") and M 51 ("Whirlpool galaxy"), as shown in Fig. 9.1. However, Holmberg [113] noted that the condensation of the gravitationally bound companion galaxy would be very unlikely when the jets reach a velocity higher than the escape velocity, which seems to be true in most of the cases. Such an activity from the nuclei of massive galaxies led some authors to classify galaxies with connecting "jets" as radio-galaxies (see e.g. [3]).

---

[1]According to Arp [8], the same mechanism would account for the creation of spiral arms in rotating galaxies.

Meanwhile, tides have been considered as a possible cause for the filaments: a close passage of one galaxy next to another would lead to different gravitational forces over the spatially extended galaxies: the side of the first galaxy facing the second is more attracted than the opposite side. These differential forces would then significantly deform the shape of the galaxies and could even trigger an exchange of some of their stars [112, 139, 217, 250–252]. The pioneer numerical works that addressed this question concluded that, under precise circumstances, tidal structures looking like bridges and tails could form during the close encounter of two galaxies [140, 244].

However, the tidal origin of the tails has been intensively discussed. Vorontsov-Vel'Yaminov [231] argued that the elongation of the tails (sometimes up to a few $\times 100$ kpc, see [161]) was too large to be produced by tides. He added that close pairs of galaxies were not systematically linked to the existence of filaments, and concluded that tails and bridges shared the same origin as the more classical spiral arms. Others followed the same line of arguments and evoked magnetic (or magnetic-like, see [232]) fields to explain the narrow shape of the tails (see e.g. [47, 250]). Tubes of magnetic lines forming at the same time as the galaxy itself would propagate a wave that would trigger the condensation of gas along them. Such an hypothesis would explain the presence of knots of high surface brightness along the tails, as already detected by e.g. Burbidge and Burbidge [46]. Furthermore, Gershberg [91] noted that a collision between two galaxies would heat up the gas too much ($\sim 10^7$ K) to form a thin structure and ruled out this scenario as a possible cause of creation of filaments. Arp [5] summarized the debate by suggesting that forces other than pure gravitation should be at stake in the shaping of peculiar galaxies and their intergalactic structures.

9.2.3 Tidal Origin

The major breakthrough came in the early 1970s, in the newly-born era of computers. Thanks to a series of numerical experiments, Toomre and Toomre [226] showed that the brief but intense tidal forces arising during the encounter of two disk galaxies would be sufficient to create structures as long and thin as the tails referenced in the catalogues. They extended the works of Pfleiderer [179] and Tashpulatov [217, 218] by considering a bound companion galaxy on an very eccentric orbit, as well as disks inclined with respect to the orbital plane. In their study, a single galaxy is represented by a point-mass surrounded by rings of test particles whose masses are zero. When two of such galaxies are set on a given orbit, the central point-mass follows Kepler's law of motion. The test particles feel the net gravitational potential and thus, their motion is affected by both point-masses. However, in this method called restricted simulation, the mass-less test particles themselves do not affect the gravitational field of the galaxy.

Toomre and Toomre [226] noted that close passages could induce a deformation of the disk(s), possibly leading to the creation of bridges and/or tails. By varying

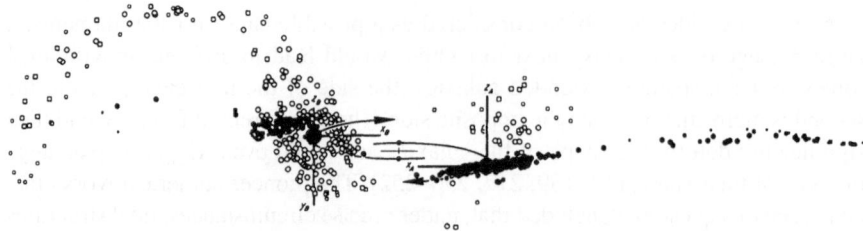

Fig. 9.2 Restricted simulation of the Mice galaxies (NGC 4676) from Toomre and Toomre [226]. The two tails that exhibit very different shapes and thickness have been successfully reproduced numerically by considering tidal interaction only. Images of the real galaxies are shown in Fig. 9.3, second panel, and Fig. 9.7, panel 4

several parameters of the problem such as the inclination of the disks or the eccentricity of the orbit, they have shown that gravitation only was enough to reproduce the structures observed in interacting systems (see Fig. 9.2). This showed the way to many other numerical experiments [80, 129, 137] and allowed to conclude on the tidal origin of several observed features [58, 214, 245].

Since then, gravitational tides have been considered as the major cause of the creation of filaments in interacting galaxies. That is why such features are often referred to as tidal structures.

An examination of the peculiar galaxies with the new light shed by numerical experiments on tides revealed that most of these galaxies would fit into an evolutionary sequence (see Fig. 9.3), called Toomre's sequence [225]. Each step represents a dynamical stage in the evolution of interacting galaxies toward the final coalescence of the merger.[2] With time going, the tidal features created by the first encounters slowly vanish into the intergalactic medium or are captured back by their galaxy. Note however that relics of the tails remain visible for several 10^9 yr [212].

9.2.4 Forty Years of Numerical Simulations

In order to retrieve the steps of the Toomre's sequence and to better understand the role of each parameter involved in interacting galaxies, an important amount of work has been conducted by many authors since the very first (non-numerical) computations in the early 40 s. At that time, Holmberg [112] used the light and the property of the decay of its intensity as r^{-2} as a proxy for gravitation. He set a pair of two "nebulae", each made of 37 light-bulbs, and computed the equivalent gravitational acceleration by measuring the intensity of the light thanks to galvanometers at several positions. This ingenious method allowed him to spot the creation of "spirals"

---

[2]Note that the position of some of the galaxies in the sequence has been recently discussed thanks to new numerical models (see e.g. [126]).

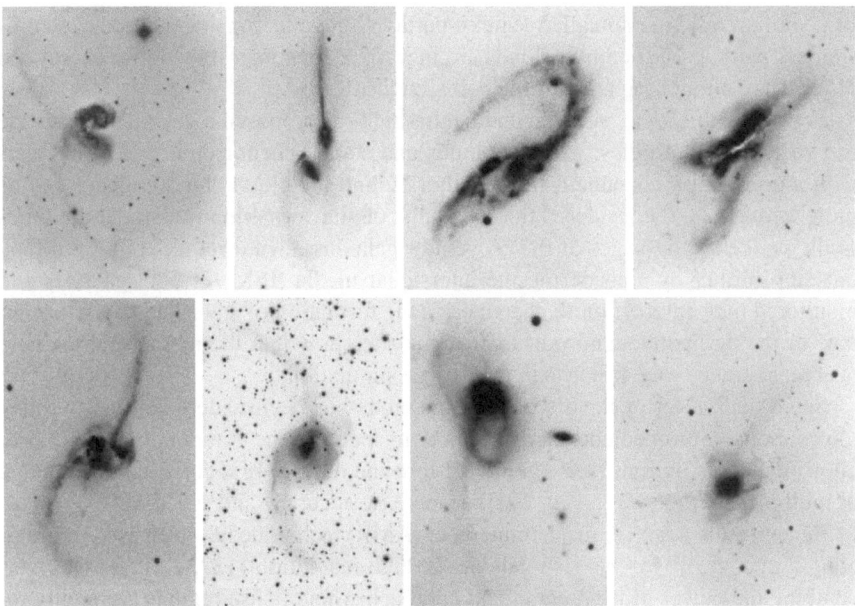

Fig. 9.3 The Tommre's sequence represents the supposed evolution of interacting galaxies. It starts with the early phases, when progenitors have just begun to interact, shows intermediate stages and finishes with the coalescence phase. From *left* to *right*, *top*: NGC 4038/39 (the Antennae), NGC 4676 (the Mice), NGC 3509, NGC 520; *bottom*: NGC 2623, NGC 3256, NGC 3921, NGC 7252 (the Atoms for Peace). Images from the Atlas of Peculiar Galaxies by H. Arp, available in the NASA/IPAC Extragalactic Database, Level 5

during a close encounter. But it is in the numerical era that most of the progresses have been done.

Despite their success in reproducing observed systems, the restricted simulations of Toomre and Toomre [226] lacked the orbital decay due to dynamical friction. The problem was solved when considering self-consistent ("live") galaxies, i.e. models where all the particles interact with each other (e.g. [90, 239]). However, the cost of such computations was very high at that time. That is why tree-codes [20, 99] and multipole expansions techniques [228, 240] have been introduced to decrease the computation time, or equivalently to increase the reachable resolution.

Barnes [12] presented the first simulation of self-consistent multi-components galaxies. He showed that the presence of a dark-matter halo increases significantly the dynamical friction, thus favoring the merger of the galaxies.

In the same time, Hernquist and Katz [102] gathered the tree-code method and the smooth particle hydrodynamics (SPH) technique [92, 146] to treat both the gravitation and the hydrodynamics within a particle-based code. In SPH simulations, the physical properties of the particles are smoothed over a kernel of finite size, centered on the particle itself. Thanks to this Lagrangian approach, SPH does not suffer from the limitations of grid codes [110], i.e. mainly the waste of computational power in areas of nearly vacuum, an omnipresent situation in the case of galaxy mergers.

In a similar way, the so-called "sticky-particle" method considers clouds as collisionless particles. When two clouds are in a close encounter, they loose energy via dissipation, mimicking an inelastic collision [169].

Following this idea, Noguchi and Ishibashi [173] proposed a galaxy model made of two types of particles: gaseous clouds and stars. When such a galaxy interacts with a point-mass encounter, these authors found the cloud-cloud collisions to be more frequent, and considered this as a burst of star formation (mostly at the times of the pericenter passages of the progenitor galaxies). Mihos et al. [154–156] took one step further by considering the interstellar media (ISM) of both galaxies and monitored their interaction to characterize the formation of stars. They took advantage of the dissipative nature of their models to show that the merger phase could take place up to twice faster than in gas-free simulations.

Since then, a lot of flavors of these methods has been widely applied to many topics. Some improvements also appeared, to speed up the computation and thus to allow higher resolutions (see e.g. [60]). More and more hybrid codes take advantage of multiples methods (e.g. [26, 201]) to increase accuracy and speed-up.

Recently, the adaptive mesh refinement (AMR) technique has been used for modeling a merger of two gas-rich galaxies at high resolution [132, 220]. AMR codes combine the power of the Lagrangian approach where dense regions are highly resolved, and the continuous description of the ISM on grids (e.g. [87, 175, 219]). The computational domain is meshed on a (usually Cartesian) grid, which is refined at the regions of interest, typically those of highest densities. Two different snapshots of a numerical model using the AMR technique are shown in Fig. 9.4 and Fig. 9.11 (left panel).

As seen in the literature since Toomre and Toomre [226], simulations of interacting galaxies can follow two approaches:

- the systematic exploration of a large volume of the parameter space, with the goal of understanding the influence of certain parameters on the evolution of the merger and its stellar population (see e.g. [96, 125, 165, 174, 210, 235]). Among them, the GalMer project [63] gathers ~ 1000 SPH simulations of mergers and the associated star formation histories, and makes them publicly available online.[3] With such databases, the simulations can be interpreted statistically, thus strengthening the physical conclusions.
- the simulation of specific, observed galaxies in order to bring new lights when interpreting the observations (see e.g. [12, 14, 21, 64, 71, 105, 126, 156, 185, 220]). Several pairs of interacting galaxies have been numerically reproduced (see an example in Fig. 9.4) by putting the effort on finding a set of parameters that best describe the pair, generally by trial-and-error. Intuition and experience play an important role in such a study. However, this process has recently been automatized thanks to new numerical tools: these codes make a series of restricted, fast simulations ("à la Toomre and Toomre") and slightly modify one parameter of the initial conditions at each iteration, to improve the match with observational data

---

[3]http://galmer.obspm.fr/.

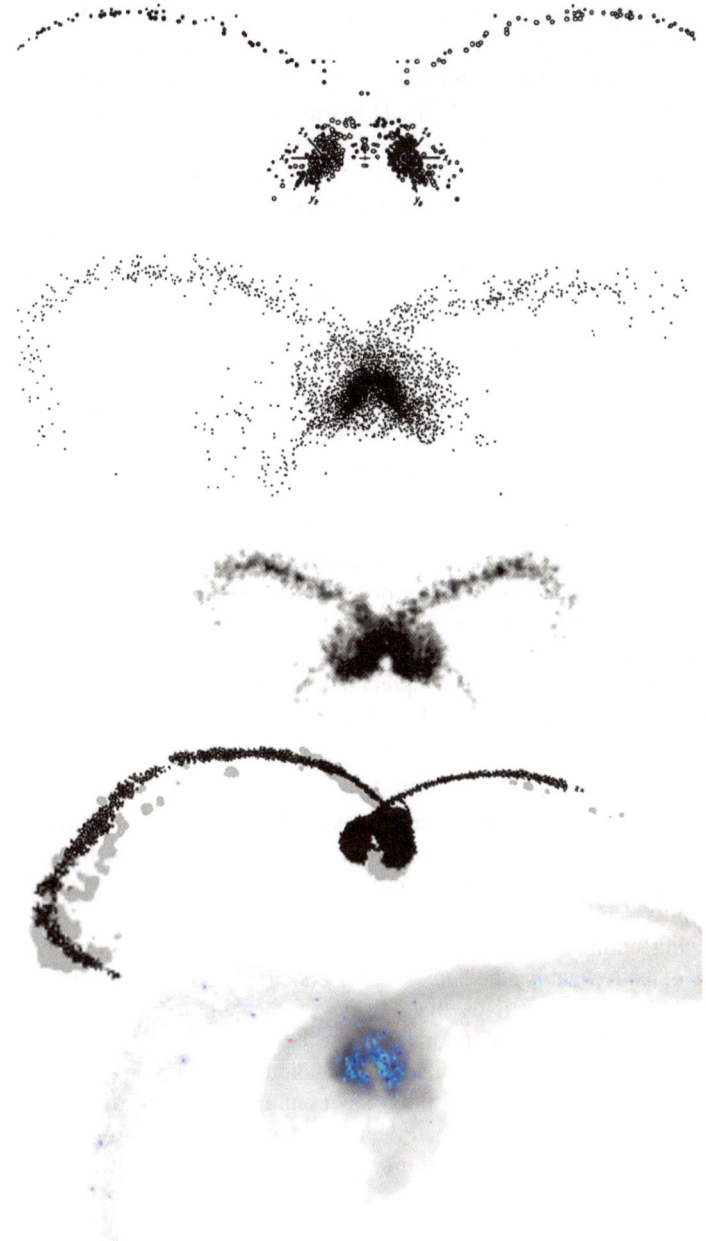

Fig. 9.4 Numerical simulations of the Antennae galaxies (NGC 4038/39) within four decades. From *top* to *bottom*: restricted simulation of Toomre and Toomre [226]; first self-consistent simulation of the Antennae by Barnes [12]; hydrodynamic run of Mihos et al. [156]; recent models with SPH by Karl et al. [126] and with AMR by Teyssier et al. [220]. Improvements in both the techniques and the set of parameters allowed the models to get closer and closer to the observational data (see Fig. 9.7, panel 6)

(see for example Identikit by Barnes and Hibbard [19], Barnes [15], and AGC by Smith et al. [207]). Genetic algorithms have also been implemented to optimized the search of the parameters (see the MINGA code by Theis and Kohle [221], Theis and Spinneker [222]). This way, a large range of parameters can be covered very quickly to find which set best matches the observational data. However because the simulations are restricted, they do not account for the orbital decay of the galaxies due to dynamical friction, which represents an important limitation for such methods, in particular when multiple collisions occur. Starting from the set of parameters suggested by such fast codes and fine-tuning them in self-consistent re-runs could be a good compromise.

The simulation of interacting galaxies is not limited to pairs. However, numerical models of observed (compact) groups of galaxies are still very rare, due to the difficulty to set a consistent scenario for an entire group. Each galaxy-galaxy interaction has to take place in a system already perturbed by the previous interactions, such that the mass and the orbit of the progenitor are to be re-evaluated constantly during the evolution of the group. (Some attempts have been made in the case of Stephan's Quintet, see [119, 187]).

While they face an increasing need of resolution and accuracy, these state-of-the-art numerical methods can efficiently provide a solution to the questions raised by observations at higher and higher resolution. Simulations of interacting galaxies still represent an important part of the numerical work done in astrophysics. The models of individual galaxies are regularly updated to fit the most recent theories on galaxy formation and evolution and include better descriptions of the physical processes. Nowadays, the research on interacting galaxies is mainly threefold:

- The cosmological approach, mostly based on the Λ Cold Dark Matter (CDM) theory, focuses on galaxy formation through repeated accretion of satellites (e.g. [50, 170]). In particular, this hierarchical scenario describes the formation of elliptical galaxies as remnants of a merger but also provides clues on the dynamical status and evolution of groups of galaxies.
- A growing number of works focus on the central region of mergers. The formation of active galaxy nucleus (AGN) and the associated feedback is intensively discussed, as well as the pairing of black holes in mergers (see e.g. [28, 59, 135, 168, 204], among many others).
- The stellar populations of the interacting galaxies and the properties of the star clusters and the dwarf galaxies they may contain is also a widely covered topic (e.g. [36, 64, 238]). In this respect, the tides play an important role on the physics of these subsystems. This last point will be further developed in the following sections.

9.3 Theory of the Tidal Tail Formation in Interacting Galaxies

After having reproduced numerically some of the extragalactic tidal structures observed in the Universe, several physical and mathematical descriptions of the phenomenon have been proposed to better understand the tides at galactic scale. The complexity of the task comes from the diversity of possible configurations, which translates into a large number of parameters. In this section, we review the role of the first order parameters and illustrate their respective effects thanks to numerical simulations of interacting galaxies. A mathematical description of the tidal field is also presented.

9.3.1 Gravitational Potential and Tidal Tensor

By definition, the tides are a differential effect of the gravitation. Let's consider a galaxy, immersed in a given gravitational field. At the position of a point within the galaxy, the net acceleration can be split into the effect from the rest of the galaxy $\mathbf{a}_{\text{int}}$, and the acceleration due to external sources $\mathbf{a}_{\text{ext}}$. The latter can itself be seen as a part common to the entire galaxy (usually the acceleration of the center of mass), and the differential acceleration, that differs from point to point within the galaxy. In other terms, the net acceleration at the position r_P, in the reference frame of the center of mass of the galaxy (which lies at the position r_g), is given by

$$\mathbf{a}(\mathbf{r_P}) = \mathbf{a}_{\text{int}}(\mathbf{r_P}) + \left[\mathbf{a}_{\text{ext}}(\mathbf{r_P}) - \mathbf{a}_{\text{ext}}(\mathbf{r_g})\right]. \tag{9.1}$$

For small $\delta = \mathbf{r_P} - \mathbf{r_g}$ with respect to $\mathbf{r_g}$, one can develop at first order and get

$$\mathbf{a}(\mathbf{r_P}) = \mathbf{a}_{\text{int}}(\mathbf{r_P}) + \delta \, d\mathbf{a}_{\text{ext}}, \tag{9.2}$$

which also reads

$$\begin{pmatrix} a_x(\mathbf{r_P}) \\ a_y(\mathbf{r_P}) \\ a_z(\mathbf{r_P}) \end{pmatrix} = \begin{pmatrix} a_{\text{int},x}(\mathbf{r_P}) \\ a_{\text{int},y}(\mathbf{r_P}) \\ a_{\text{int},z}(\mathbf{r_P}) \end{pmatrix} + \begin{pmatrix} [\delta_x \frac{d}{dx} + \delta_y \frac{d}{dy} + \delta_z \frac{d}{dz}] a_{\text{ext},x} \\ [\delta_x \frac{d}{dx} + \delta_y \frac{d}{dy} + \delta_z \frac{d}{dz}] a_{\text{ext},y} \\ [\delta_x \frac{d}{dx} + \delta_y \frac{d}{dy} + \delta_z \frac{d}{dz}] a_{\text{ext},z} \end{pmatrix}, \tag{9.3}$$

or simpler

$$a_P^i = a_{\text{int}}^i(\mathbf{r_P}) + \delta_j \partial^j a_{\text{ext}}^i, \tag{9.4}$$

when using Einstein's summation convention. The effect of the external sources on the galaxy are described by the term

$$T^{ji} \equiv \partial^j a_{\text{ext}}^i, \tag{9.5}$$

which is the j, i term of the 3×3 tensor T called tidal tensor [185]. Such a tensor encloses all the information about the differential acceleration within the galaxy. Therefore, the (linearized) tidal field at a given point in space is described by the tensor evaluated at this point.

Note that the tidal tensor is a static representation of the tidal field: the net effect on the galaxy also depends on its orbit in the external potential, or in other words, on the variations of intensity and orientation of the tidal field. This can be accounted for by writing to pseudo-accelerations (centrifugal, Coriolis and Euler) in the co-rotating (i.e. non-inertial) reference frame, or by the means of a time-dependent effective tidal tensor in the inertial reference frame. For simplicity, in the following we focus on static, purely gravitational tides and refer the reader to Renaud et al. [188] for more details.

Because the acceleration $\mathbf{a}_{\text{ext}}$ derives from a gravitational potential φ_{ext}, one can write

$$T^{ji} = -\partial^j \partial^i \varphi_{\text{ext}} = -\partial^i \partial^j \varphi_{\text{ext}} = T^{ij}. \tag{9.6}$$

(Several examples of tidal tensors of analytical density profiles are given in [186], see also the Appendix B of [184].) It is important to note that these considerations are scale-free and applies to any spatially extended object, such as galaxy clusters, galaxies, star clusters, stars, planets, etc.

For example, let's consider the Earth-Moon system and compute the tidal field with the Moon as source of gravitation. It can been seen from the Earth as a point-mass, and yields a potential of the form

$$\varphi_{\text{ext}} = -\frac{GM}{r}, \tag{9.7}$$

with $r = \sqrt{x_i^2 + x_j^2 + x_k^2}$. The components of the tidal tensor are

$$T^{ij} = \frac{GM}{r^5} \left(3 x_i x_j - \delta^{ij} r^2 \right) \tag{9.8}$$

where $\delta^{ij} = 1$ if $i = j$ and 0 otherwise. When computed at the distance d along the i-axis (i.e. for $r = d$ and $x_j = x_k = 0$), the tidal tensor becomes

$$\mathsf{T}(d, 0, 0) = \frac{GM}{d^3} \begin{pmatrix} 2 & 0 & 0 \\ 0 & -1 & 0 \\ 0 & 0 & -1 \end{pmatrix}. \tag{9.9}$$

The signs of the diagonal terms (which are, in this case, the eigenvalues because the tensor is written in its proper base) denotes differential forces pointing inward along the i-axis, and outward along the other two axes. A rapid study of the differential forces around the Earth (see Fig. 9.5) shows indeed, that they point toward the Earth along the axes perpendicular to the direction of Moon. One speaks of a compressive effect. Along the Earth-Moon axis however, the differential forces point away from the planet: the effect is extensive.

9.3.2 Compressive Tides

Back to the general case, it follows from Eq. (9.6) that any tidal tensor is symmetric. Because it is also real-valued, it can be set in diagonal form, by switching to its proper base. In this case, three eigenvalues $\{\lambda_i\}$ denote the strength of the tides along the associated eigenvectors. The trace of the tensor (which is base-invariant) reads

$$\mathrm{Tr}(\mathsf{T}) = \sum_i \lambda_i = -\partial^i \partial^i \varphi_{\text{ext}} = -\nabla^2 \varphi_{\text{ext}}, \tag{9.10}$$

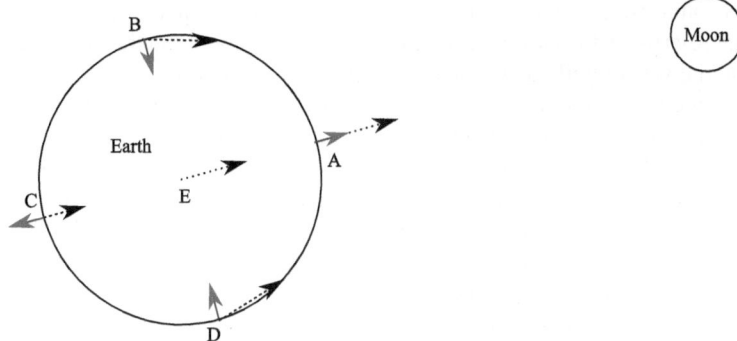

Fig. 9.5 Gravitational attraction (*black, dotted line*) of the Moon on the Earth, and the differential forces (*grey*). The tidal effect appears to be extensive in A and C, while it is compressive in B and D

which can be connected to the local density ρ thanks to Poisson's equation:

$$\mathrm{Tr}(\mathsf{T}) = -4\pi G\rho \leq 0. \tag{9.11}$$

The condition on the sign of the trace implies that it is impossible to compute simultaneously three strictly positive eigenvalues. Remains the cases of two, one or no positive eigenvalues, as mentioned by Dekel et al. [61]. For two or one positive λ's, the tidal field is called (partially) extensive, like e.g. in our Earth-Moon example. When all three eigenvalues are negative, the tides are (fully) compressive. By noticing that T is minus the Hessian matrix of the potential, one can show that a change of curvature of the potential implies a change of sign for T. Therefore, compressive tides are located in the cored regions of potentials only, and never in cusps.

Note that a compressive mode (three negative λ's) implies that the local density due to the source of gravitation is non-zero. Although such a situation does not exist with point-masses, it can occurs when considering extended mass distributions, like e.g. for galaxies embedded in a dark matter halo.

The duality of compressive/extensive tidal modes plays a role in the formation, early evolution and dissolution rates of star clusters. It has been noted that observed young clusters were preferentially found in the regions of compressive tides (see [185], in the case of the Antennae galaxies), and a compressive mode would slow down the dissolution of young globulars [188].

9.3.3 Formation of Tidal Tails and Bridges

In isolation, a galaxy keeps its material, which is made of dark matter, stars, gas and dust, bound thanks to the gravitation. However, when it moves in an external potential, created for instance by neighbor galaxies, it can experience gravitational forces which are different from one side of the galaxy to the other. In other words, the galaxy is plunged in a tidal field. As a result, its material undergoes deforming effects that re-arrange the individual components of the galaxy. On the one hand, when this material was initially distributed in an (almost) random way in phase-space (as opposed to e.g. sharing a common velocity pattern), the net tidal effect does not translate into a clear global change for an entire region of the galaxy. Therefore, such tides are difficult to detect. On the other hand, when large scale, regular patterns exists in the distribution of the galactic material in phase-space (e.g. a disk), the tides have a similar impact on stars that already lied in the same region of phase-space. All these stars are affected the same way and thus, the effect is much more visible. In the end, a given tidal field is easier to detect when it affects a regular, organized distribution of matter, than when it applies to isotropic structures. This is the reason why tidal features like tails and bridges are well visible around disk galaxies where the motion is well-organized, and merely inexistent in ellipticals, which yield much more isotropic distributions of positions and velocities. This last point can be extended to all structures with a high degree of symmetry (halos, bulges, and so on), as opposed to axisymmetric components like disks.

As a consequence, the tidal structures gather the matter that occupy a well-defined region in phase-space. Figure 9.6 (top row) shows the N-body toy-simulation of an encounter between a composite galaxy (disk + bulge + dark matter halo) and a point mass. Particles being part of one of the tails are tagged so that it is possible to track them back in time to their initial position in the disk. As mentioned above, these particles are distributed in a more or less confined region of phase-space at the time of the pericenter passage of the intruder, so that their individual motions are re-organized in a similar way. It is interesting to note that they cover a wide range of radii in the disk and thus, because of the differential rotation, the zone they occupy before the interaction is far from being symmetrical.

When the same experiment is repeated with an elliptical galaxy (Fig. 9.6, bottom row), the velocities are distributed almost isotropically and thus, no structure is created by the tidal field. As a conclusion, strong galactic tidal bridges and tails are formed from the material of disks galaxies. Note that the experiment we conducted above applies to any mass element, and thus can be, in principle, extended to both the gaseous and stellar components of a galaxy.

In the case of a flyby, the galaxies do not penetrate in the densest regions of their counterpart, do no loose enough orbital energy to become bound to each other, and thus they escape without merging. However, when the exchange of orbital angular momentum (through dynamical friction) is too high, the mean distance between the progenitors rapidly decreases (as a damped oscillation) before they finally merge, forming a unique massive galaxy. On the external regions of the merger, the tidal tails (if they exist) expand in the intergalactic medium and slowly dissolve. Because the tails are generally long-lived, they can indicate past interactions, as discussed in Struck [215]. As a result, tidal features can point to interacting events, even when what has caused their creation (i.e. a counterpart progenitor) has disappeared in a merger or has flown away.

Note that D'Onghia et al. [65] followed an approach based on the quasi-resonant theory to describe the response of disk to a tidal perturbation. Their analytical formalism gives a good match with numerical experiments.

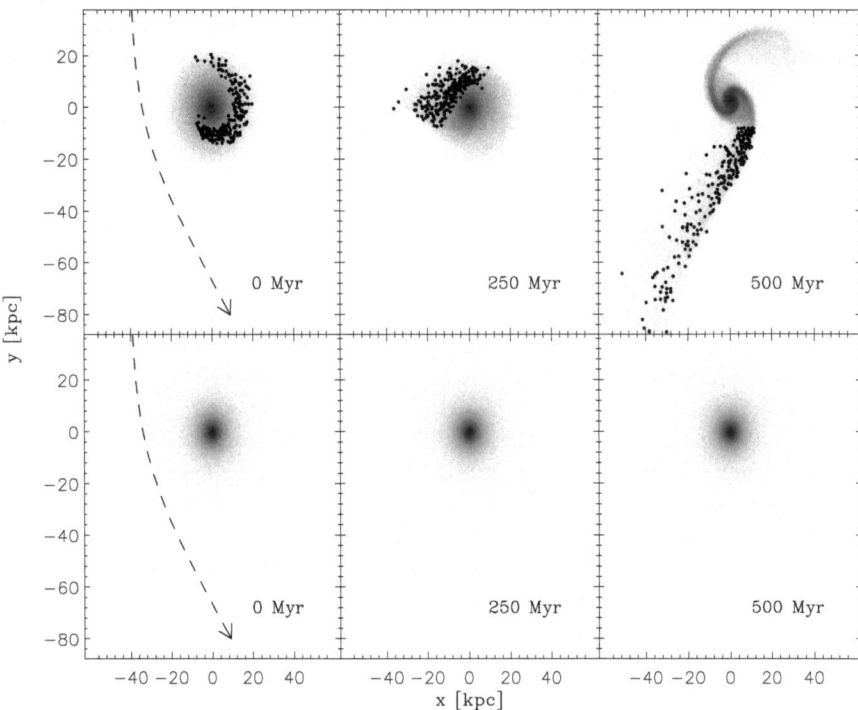

Fig. 9.6 *Top*: morphology of a disk galaxy, seen face-on, during its coplanar interaction with a point-mass (mass ratio = 1), before the interaction (*left*), at pericenter (*middle*) and after (*right*). The *dashed line* indicates the trajectory of the point-mass (from *top* to *bottom*). The *black dots* tag a subset of particles that are situated in one of the tidal tails at $t = 500$ Myr. *Bottom*: same but for an elliptical galaxy. No tidal structures are visible

9.3.4 Gas Dynamics

The response of the gas to a galactic interaction can be seen as either an outflow or an inflow. For distant, non-violent encounters, a large fraction of the hot gas $(T > 10^3$ K$)$ can be tidally ejected into the intergalactic medium, thus forming broad gaseous tails and/or halos around galaxies (see e.g. [132]). It has been noted that while the least bound material would expand widely, more bound structures could easily fall back into the central region of the galaxies within less than ~ 1 Gyr [105, 106].

During a first, distant passage, some galactic material is stripped off thanks to the transformation of the orbital energy of the progenitor galaxies. As a result, and because of dynamical friction, the interacting pair becomes more and more concentrated and can, under precise conditions (see e.g. [239]), experience other passage(s) and finally end as a merger [17]. During such a second, closer interaction, tidal forces can induce shocks covering a large fraction of the galactic disk, which gives the gas a significantly different behavior than that of the stars [169]. In particular, when stellar and gaseous bars form, the symmetry of the galaxy is broken: gravitational torques remove the angular momentum of this gaseous structure [51] and make it fall onto the nucleus of the merger (< 1 kpc, see e.g. [16, 103, 153, 172]). Such an inflow fuels the central region of the merger and participates in the nuclear starburst [13, 166, 210] often observed as an excess of infrared light or a strong nuclear activity [89, 136, 144, 209, 248]. At large radii in a disk, one gets the opposite effect: the gravitational torques push the material to the outer regions. This outflow enhances the formation of the tails already formed by the tidal field itself [32].

Note that star formation in mergers is also considered to be triggered by energy dissipation through shocks [14]. This is, however, quite sensible to the orbital parameters of the galaxies. Details about merger-induced starbursts are out of the scope of the present document. The reader can find a mine of information on this topic in Hopkins et al. [114], Robertson et al. [191], Di Matteo et al. [63], Cox et al. [56], Hopkins et al. [115], Teyssier et al. [220] and references therein.

Interestingly, Springel and Hernquist [211] showed that the collision between two gas-dominated disks could form a spiral-like galaxy instead of an elliptical one, as one could expect. In this case, a significant fraction of the gas is not consumed by the burst of star formation induced by the merger. Through conservation of the angular momentum, dissipation transforms the gaseous structure into a star-forming disk [115]. Owning that the gas fraction in galaxy increases with redshift (as suggested by Faber [81], Lotz et al. [145]), this last point sheds light on the formation history of low-redshift spiral galaxies.

9.3.5 Influence of the Internal and Orbital Parameters

The details on the formation of tidal structures are adjusted by several parameters that mainly concern the orbit of the galaxies, i.e. the way one sees the gravitational potential of the other. Because an analytical study of the influence of these parameters is very involved, many authors conducted numerical surveys to highlight the trends obtained from several morphologies.

9.3.5.1 Spin-Orbit Coupling

In their pioneer study, Toomre and Toomre [226] already mentioned the influence of the spin-orbit coupling of the progenitors. For simplicity, let's consider two galaxies A and B separated by a distance r_{AB}, and whose disks lie in the orbital plane. The norm of the velocity of an element of mass of the galaxy A situated at a radius r, relative to the galaxy B is $r_{AB}\Omega \mp r\omega$, where Ω denotes orbital rotational velocity and ω the (internal) rotation speed of the galaxy A (i.e. the spin). The sign of the second term depends on the alignment of ω with Ω. For a prograde encounter, the spin (ω) and the orbital motion (Ω) are coupled (i.e. aligned). Therefore, the relative velocity is lower ($r_{AB}\Omega - r\omega$) than for a retrograde encounter ($r_{AB}\Omega + r\omega$)

and the net effect of the tides is seen for a longer period of time. As a result, the structures formed during prograde encounters are much more extended than those of retrograde passages.

Although this conclusion can be exported to inclined orbits, the strongest responses of the disks are seen for planar orbits, i.e. with a zero-inclination. The highly inclined configurations, called polar orbits, give generally birth to a single tail, as opposed to the bridge/tail pairs [116]. In short, because an observed tidal effect does not only depend on the strength of the differential forces, but also on the duration of their existence, long tails are associated with prograde configurations.

9.3.5.2 Mass Ratio

Another key parameter is the mass ratio of the progenitors. In the hierarchical scenario, the galaxies form through the repeated accretion of small satellites (see e.g. [213], and references therein), and interactions between a main galaxy and number of smaller progenitors would occur more or less continuously. It is usual to distinguish the major mergers where the mass ratio is smaller than 3:1 (i.e. almost equal-mass galaxies), from the minor mergers involving a larger ratio (e.g. 10:1). In the last case, tidal tails are generally thin and small, while the same features are more extended and survive for a longer time in major mergers [167].

The dependence of the structure of the remnant of the interaction (disky or boxy elliptical, as opposed to more symmetric galaxies) on the mass ratio of the progenitors has been extensively debated but is not directly connected to the tidal activity, and thus is out of the scope of this review (see [16, 17, 39, 40, 100, 101, 165, 195], for much more details).

9.3.5.3 Impact Parameter

During the interaction, the impact parameter plays an indirect role: a close, penetrating encounter will drive one galaxy deep inside the high density regions of the other, which implies a strong dynamical friction (see e.g. [27]). In this case, the separation of the progenitors after such a passage would be much smaller than for a more distant encounter.

Furthermore, a close passage generally corresponds to a significant tidal stripping. This situation occurs repeatedly for satellites orbiting within the halo of major galaxies [183]. Only the densest satellites can survive such a disruption [200], while more fragile object would be converted into stellar streams [123, 150, 178], as observed in the local Universe [120].

However, the mass captured by a more massive companion (mass ratio close to 1:1) seems to be higher for short pericenter distances, as noted by Wallin and Stuart [235]. The lost of material into the intergalactic medium is also higher under these circumstances.

9.3.5.4 Dark Matter Halo

In addition to the effect of orbital parameters, several authors noted the role played by the dark matter halo of the progenitor on the morphology of the merger, mainly the length of the tails. E.g. Dubinski et al. [66] showed that long, massive tidal tails are associated with light halos, while the deep potential created by more massive ones would prevent the creation of extended structures. Note that, for a given mass, a dense halo appears to be more efficient in retaining the stellar component bound [157]. An important conclusion of this work was that galaxies exhibiting striking tails are likely to have relatively light halo (i.e. a dark to baryonic mass ratio smaller than $\sim 10 : 1$).

However, Springel and White [212] qualified this by stating that the important parameter is in fact the ratio of escape velocity to circular velocity of the disk, at about solar radius (see also [67]). Therefore, even massive halos (e.g. mass ratio $40 : 1$) can allow the growth of tails, provided the kinetic energy of the disk material is high enough to balance the depth of the gravitational potential of the massive dark matter halo. See Sect. 9.6.3 for more details.

9.3.6 Rings, Ripples, Shells, and Warps

Although they are the most visible structures formed during galactic interactions, the tidal tails and bridges are not the only signatures of encounters. Other mechanisms (not directly of tidal origin) lead to disrupted morphology. We briefly mentioned them here, for the sake of completeness.

- Shells or "ripples" describe the arcs and loops showing sharp edges in the envelope of galaxies. They originate from the collision between a massive galaxy and a small companion, 10 to 100 times lighter [182]. The material of the satellite is spread by an extensive tidal field in the potential well of the primary, along a given orbit of low-angular momentum (see [10], and references therein). A sharp ridge forms near the turnaround points of the orbit. The multiplicity of the shells corresponds to an initial spread in energy, leading to several possible radii for the ridges.
- A ring galaxy forms from the head-on collision between a large disk and a compact, small perturber [86, 223]. The density wave created by the collision empties the central region of the disk and forms a ring in radial expansion ([147], see also [4] for an observational and theoretical review).
- Warped disks can be created by gravitational torques due to an infalling satellite galaxy (mass ratio $\sim 10 : 1$, see [117, 189]). Note that warps and bending instabilities can also form through the torques exerted by a misaligned dark halo, or via accretion of matter [11, 122, 190].

9.3.7 Differences with Tides at Other Scales

The galactic tides are a purely gravitational effect, which means that they rely on scale-free quantities like the relative mass of the galaxies, the inclination of the orbits, their relative velocities and so on. Therefore, the conclusions presented above can be applied to any scales, from planetary to cosmological. If true in principle, this statement must be qualified because the requirements of the galactic-type tides themselves do not exist at all scales.

In the case of planetary tides, for example in the Earth-Moon system, the source of gravity does not penetrate in the object experiencing tides, and is generally situated at a distance large enough that it can be approximated by a point-mass. Furthermore, the binding energy of a solid and/or dense body like a planet is much higher than those of the galaxy on its stars. That is, the planetary tidal effects are weaker than the galactic ones. Note however that both the planetary and the galactic tides can destroy an object, like the comet Shoemaker-Levy 9 pulled apart by Jupiter's tidal field, or dwarf galaxies that dissolves in the halo of a larger galaxy, generally forming streams.

Another major difference arises from the periodicity of the motion. While a binary star or a planet is orbiting in a regular, periodic way, the galaxies show more complex trajectories, highly asymmetric, and rarely closed (because of high velocity dispersion and/or orbital decay). As a consequence, the tides at stellar or planetary scales can be seen as a continuous, or at least periodic effect, while they are rather well-defined in time and never occur twice the same way at galactic scales.

Therefore, the tidal effects seen at planetary or stellar scales, like the deformation of the oceans, atmospheres or external stellar envelops strongly differ from their equivalent phenomena in galaxies. At intermediate scale, the star clusters share properties of both tidal regimes. When orbiting an isolated galaxy, they undergo rather regular tidal effects and can, by filling their Roche lobe, evacuate stars through the Lagrange points. As a result, some globular clusters exhibit tidal tails, as seen in observations and reproduced by simulations (see e.g. [24, 83, 134], and references therein).

9.4 Multi-wavelength Observations of Tidal Tails

Tidal tails have originally been discovered on deep photographic plates (see Sect. 9.2.1) revealing the optical light emitted by stars. This monochromatic, black and white, view hides the variety of components and physical processes hosted by collisional debris. Their multi-wavelength observation and analysis were boosted in the 90s [194]. We present here an overview of the recent colorful view of tidal tails.

The average *optical color* of tidal tails is consistent with the bulk of their stellar population being older than the interaction, and originally born in the disk of the parent galaxies. Tidal tails however host bluer regions, whose light is dominated by OB-type stars. Given the life time of OB stars (less than 10 Myr) and the typical dynamical age of tidal tails (100 Myr), the young stellar component has been formed in-situ. These giant star-forming complexes are usually compact and appear detached from the rest of the tails, explaining why they were once believed to be ejected galaxies (see Sect. 9.2.2). Knots of star-formation are responsible for the bulk of the *ultraviolet emission* also emitted by tidal tails. Star formation is partially hidden by dust cocoons. Heated dust causes the *infrared emission* of tidal tails. The formation of stars requires the presence of gas. The main reservoir is atomic hydrogen, detected through the emission of the *radio 21 cm* hyperfine line; it hosts pockets of molecular gas in which stars are born and that are detectable in the *millimetric domain*, using emission lines of molecules such as carbon monoxide.

This section emphasizes the importance of multi-wavelength observations for studies of the physical properties of tidal tails, especially those formed during major mergers.

9.4.1 Where the Mass Is: Atomic Hydrogen in Tidal Tails

One of the first galaxies to have been fully mapped at 21 cm is the Antennae galaxies ([229], see also our Fig. 9.7, panel 6). These early observations obtained with the Westerbork Synthesis Radio Telescope (WSRT) revealed that about 70 % of the total amount of hydrogen in the galaxy pair is distributed along the optical tidal tails. For comparison, tidal tails account for only a few percent of the stellar component of colliding galaxies. The H I gas appears as the most massive ingredient of tidal tails and is thus one of its best tracer.

Furthermore, since the H I component is almost always more extended than the stellar disk (by a factor 2–5, depending on the morphological type of the parent galaxy), it is less gravitationally bound than the stellar disk. As a consequence, gaseous tails are more easily produced than stellar ones. Hibbard [104] used the Very Large Array (VLA) to carry out one of the first systematic study of H I in pairs of galaxies. Observing systems of the so-called Toomre sequence (see Fig. 9.3), he was able to reconstruct the evolution of the gaseous component during a merger. Together with numerical simulations, these data show how part of the gas is stripped, forming the tidal tails, while a fraction of it sinks into the central regions, sometimes via a bar and fuels there a nuclear starburst or an active galactic nucleus. Finally, observations of the 21 cm H I line have the additional advantage of providing the radial velocity over large scales. As emphasized in Sect. 9.3.5, a large variety of orbital parameters and corresponding models should be explored to reproduce the morphology of interacting systems. Having the complete radial velocity field restricts the parameter space. Tidal tails are too diffuse to allow spectroscopic measurements in the optical regime except in compact HII regions. Emission line regions are however not numerous enough in tidal tails to allow a correct sampling of the velocity field, contrary to the H I probe.

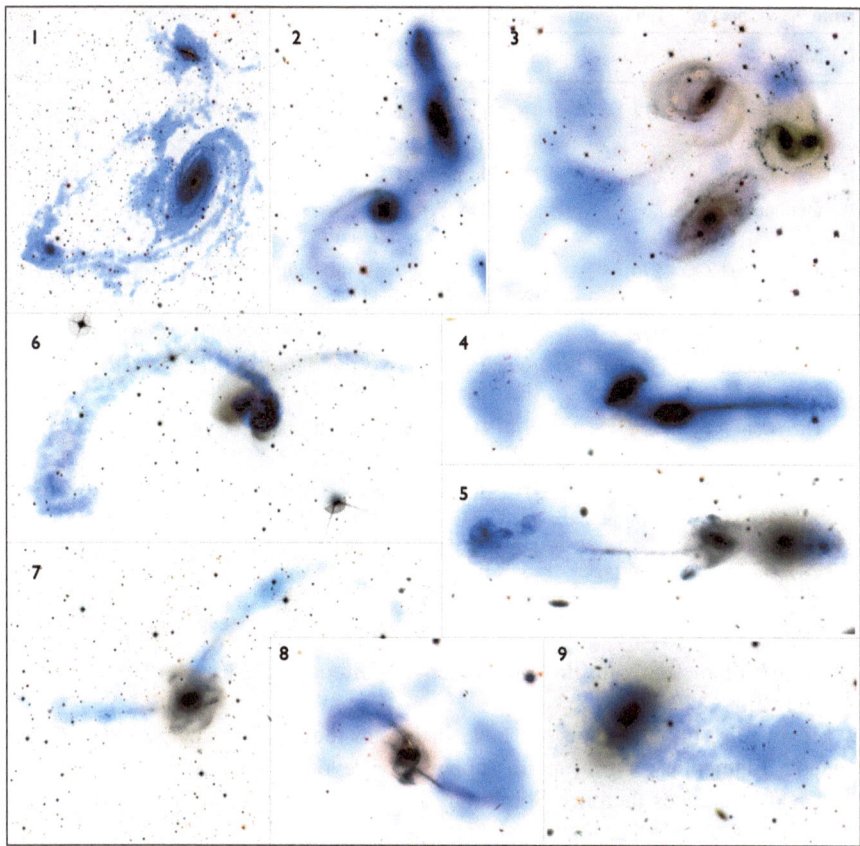

Fig. 9.7 A sample of interacting systems covering the various stages of major mergers, from the initial phases of the encounter (*top left*), to the last ones and formation of a relaxed object (*bottom right*). The gaseous component (atomic hydrogen) is superimposed on true color optical images of the galaxies, showing the distribution of the young and old stars. See Table 9.1 for details on the data

9.4.2 When Components Are Missing: HI Without Optical Counterparts; Stellar Tails Without Gas

Since the pioneer HI observations mentioned above, numerous colliding systems have been mapped with the WSRT, the VLA, the Australia Telescope Compact Array (ATCA) or the Giant Metrewave Radio telescope (GMRT). In a vast majority of cases, there is a very good match between the HI and the stellar components. The old stellar components and gas perfectly overlap, whereas young stars are formed at the HI peaks. In a few rare cases, an offset is observed between the gas and the stars (see e.g. System 8 in Fig. 9.7). The origin of the star/gas offset is debated: it may be due to different initial distributions of both components (e.g. [107]) or additional processes that act on one component and not the other. For instance, ram pressure

Table 9.1 Observed systems and sources of data

ID	Name	Sources
1	M81/M82/NGC 3077	HI: VLA Yun et al. [249]; optical: DSS; H_α: KPNO/36 in
2	NGC 2992/93	HI: VLA Duc et al. [72]; optical: CTIO/SSRO (Courtesy D. Goldman); H_α: ESO/NTT
3	Stephan's Quintet	HI: VLA Williams et al. [242]; optical: HST (NASA); H_α: Calar Alto/2.2 m (Courtesy J. Iglesias)
4	The Mice (NGC 4676)	HI: VLA Hibbard and van Gorkom [106]; optical: HST (NASA); H_α: CFHT
5	The Guitar (Arp 105)	HI: VLA Duc et al. [70]; optical: CFHT; H_α: CFHT
6	The Antennae (NGC 4038/39)	HI: VLA Hibbard et al. [108]; optical: NOAO/AURA/NSF (B. Twardy); H_α: CFHT + Palomar/1.5 m
7	The Atom for Peace	HI: VLA (Belles et al. [23]); optical: ESO/WFC; H_α: KPNO/2.1 m
8	NGC 2623 (Arp 243)	HI: VLA (Courtesy J. Hibbard); optical: HST/NASA/ESA (A. Evans); H_α: CFHT
9	NGC 4694	HI: VLA Duc et al. [74] (Courtesy VIVA Collaboration); optical: ESO/NTT; H_α: KPNO/0.9 m

due to the interaction with the intergalactic medium might strip the HI gas further away [152].

Meanwhile, blind HI surveys have disclosed the presence of numerous intergalactic, filamentary, HI structures apparently devoid of stars (see the review by [45], and references therein). Some may be of tidal origin. Spectacular examples are visible in the M81 group of galaxies (see System 1 in Fig. 9.7). Looking at the optical image of the M81 field, it might be difficult to infer that the three visible main galaxies are involved in a tidal interaction. The HI map of the same region provides a different picture and reveals a complex network of tails and bridges connecting the three galaxies. The HI Rogues gallery compiled by J. Hibbard[4] exhibits similar cases, emphasizing the role of HI as the most sensitive tracer of on-going tidal interactions.

As a matter of fact, this may be a too simple picture. Recently the optical regime had its revenge: with the availability of sensitive, large field of view CCD cameras, the surface brightness limit reached in the optical has gained several magnitudes. Diffuse light up to 29 mag arcsec^{-2} can be probed [85, 158]. In the nearest systems for which individual stellar counts are possible (with current technologies, in the Local Group), limits of 32 mag arcsec^{-2} are reachable [151]. At these limits, the most massive HI tails do exhibit a stellar counterpart. This is likely the case for the M81 group [164], and many other interacting galaxies with available ultra-deep optical images [75]. An example of a newly discovered optical tidal tail, discovered as part of the Atlas^{3D} survey [48] is shown in Fig. 9.8 (top panel).

---

[4]http://www.nrao.edu/astrores/HIrogues/.

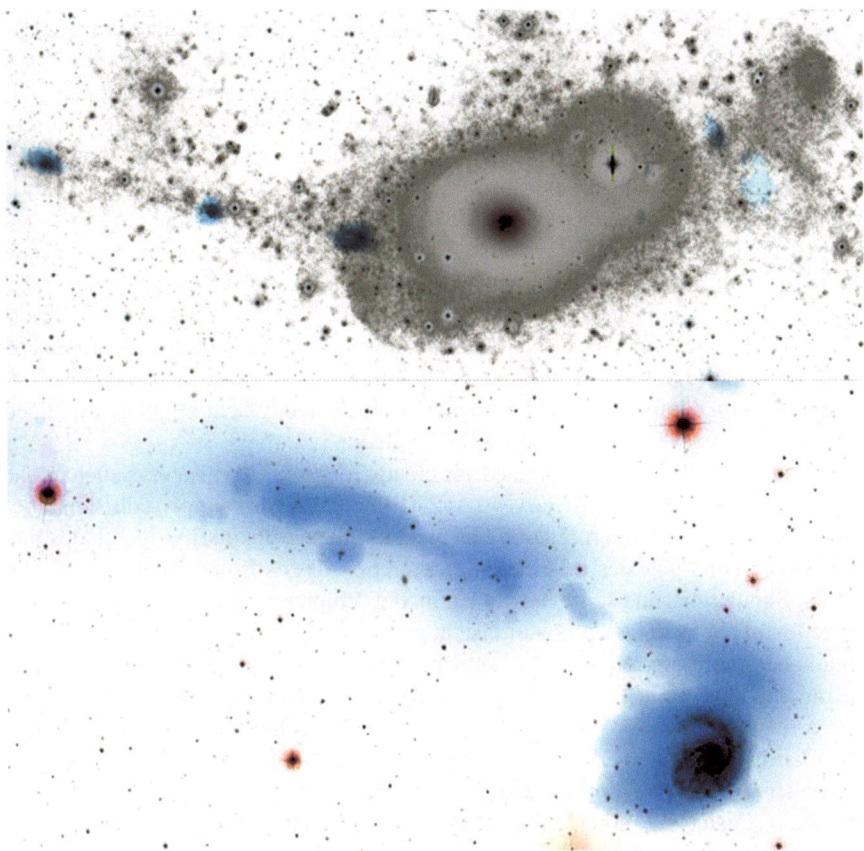

Fig. 9.8 Examples of a gas-poor and gas-rich tidal tails: NGC 5557 (*top*, [75]) and NGC 4254 (*bottom*, [69]), The HI component is superimposed in *blue* on optical images. For both systems, the tails result from an encounter with a massive galaxy, which has merged in the case of NGC 5557 or just flied by for NGC 4254. The tidal tails of NGC 5557 are best visible in the optical as extremely low surface brightness structures, whereas the HI emission is patchy and concentrated towards a few optical condensations. The tail of NGC 4254 has no optical counterpart and was once believed to be part of a dark galaxy, known as VirgoHI21

HI intergalactic structures without any stars are thus much less frequent than once believed. A few of them however have escaped an optical detection. The Magellanic Stream in the local group is the most famous of them. This HI structure is the largest tidal tail detected in the Local Group [171]. It has long been attributed to a tidal interaction between the Magellanic Clouds and possibly our Milky Way (e.g. [52]). However the absence of stars along the stream[5] was used to claim that this structure might in fact result from ram-pressure [163]. Indeed, ram-pressure only acts on the gas. The HI is stripped along filaments that may be mistaken with tidal tails.

---

[5]Though, a stellar component has been found associated to the Magellanic Bridge.

Instances of long H<small>I</small> tails, likely of ram pressure origin, may be found in the Virgo cluster [49]. However, flybys, i.e. collisions at high velocity which do not result in a merger, might as well produce tails without any stars, provided that the companion is massive enough to grasp gas from the target galaxy but resulting gravitational forces too weak to drag the stars [69]. Such tails are of tidal origin but could themselves be mistaken with filaments created by ram-pressure. According to a more exotic scenario, star-less intergalactic H<small>I</small> clouds might reveal the presence of so-called "dark galaxies", i.e. galaxies embedded in a massive dark matter halo that would contain very few baryons, only in the form of gas (see the proceedings of IAU symposium 244 dedicated to dark galaxies and [160]). An example of such objects is VirgoHI21, near the spiral NGC 4254 shown in Fig. 9.8 (bottom panel). The elongated cloud exhibits a strong velocity gradient, as if it was rotated and moved by an unseen dark component. But here again, such a velocity field might be explained by streaming motions generated by a tidal collision [22, 69].

It is not unlikely that the isolated H<small>I</small> clouds found in deep surveys such as the Arecibo Legacy Fast ALFA Survey (ALFALFA, [130]) are simply collisional debris. Finally, another interpretation has recently gained popularity: the clouds and filaments around galaxies might divulge accretion of gas from so-called cold filaments. Simulations and some theoretical models emphasize the key role of external accretion of gas in the evolution of distant galaxies [62]. Primordial accreted clouds should be devoid of stars and have a low metallicity, whereas the metallicity of tidal debris should be high. This characteristic provides a method to disentangle tidal and cosmological origins for starless gas clouds. In practice the measurement of element abundances is extremely difficult for objects with no optical counterpart.

Stellar tails without any gas are rather common around massive galaxies. Usually such streams are rather narrow and associated with tidally disrupted satellites. The gas of the progenitors might have been stripped, evaporated or consumed well before the satellites were destroyed by their giant hosts. Stellar streams are regularly discovered in our own Milky Way: the Sagittarius and Monoceros streams are among the most famous ones [25, 121, 247]. Numerical simulations show how a satellite might be stripped of its stars, wrap around the main host galaxy before eventually falling in (e.g. [149]). A spectacular example of a disrupted dwarf in the halo of a spiral galaxy is shown in Fig. 9.9.

9.4.3 Sparse Components: Molecular Clouds, Dust, and Heavy Elements

If old stars and H<small>I</small> gas are the main components of the mass budget of tidal tails, they are far from being the only ones. In fact collisional debris contain all the usual constituents of the interstellar medium of galaxies. A key ingredient is obviously the molecular gas in which stars are formed. Braine et al. [42] reported the first detection of carbon monoxyde at the tip of two tidal tails. Surveys of colliding galaxies with HI-rich tidal tails lead to several other detections [41, 141, 205]. Follow-up CO(1-0)

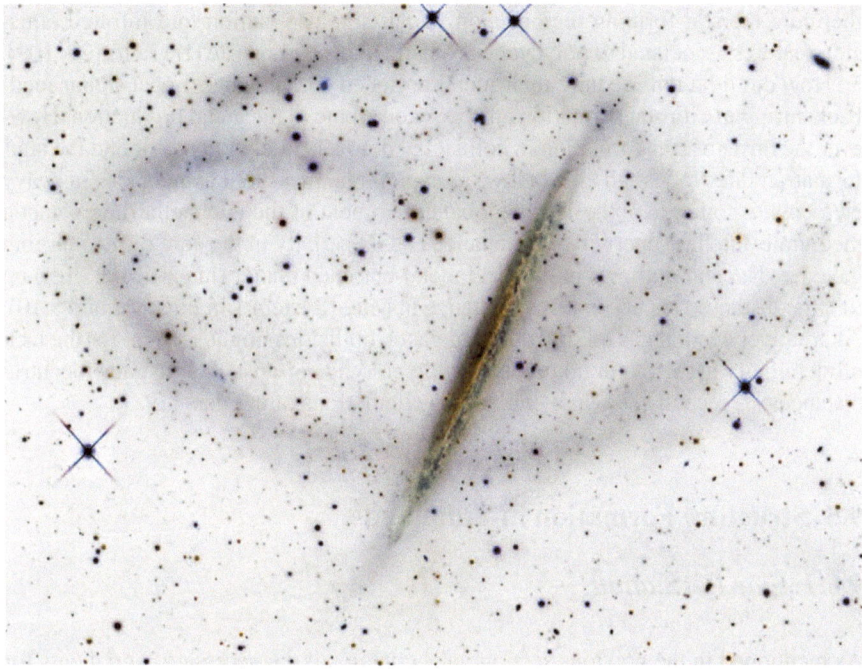

Fig. 9.9 Faint stellar streams wrapping around the spiral galaxy NGC 5907, seen edge-on. They are most probably due to a disrupted small dwarf spheroidal satellite. Such minor collisions are quite common around galaxies, including around our own Milky Way. Courtesy of R.J. Gabany in collaboration with Martínez-Delgado et al. [148]

mapping with interferometers has been achieved in a few systems [142, 236]. CO clouds were detected towards local H$_I$ peaks (and HII regions), with observed H$_2$/H$_I$ mass ratio ranging from 0.02 to 0.5. This supports the hypothesis that the molecular gas has been formed locally out of collapsing H$_I$ clouds [42]. However the later-on CO mapping of entire tidal tails revealed the presence of molecular clouds outside the H$_I$ peaks, leaving open the possibility that the molecular component (or part of it) might have been directly stripped form the colliding galaxies at the same time as the H$_I$ and stellar components [74, 143].

The detection of CO at the tip of tidal tails indirectly reveals that heavy elements are present in that environment. The oxygen abundance could be determined in HII regions located along the tails (e.g. [237]). Typical values are between one third and half solar, even at distances of 100 kpc from the parent galaxies. For comparison, abundances in the very outskirts of isolated spiral galaxies range between one tenth to one third solar [43, 84]. The disk of spiral galaxies usually exhibit a strong metallicity gradient, with a possible flattening in the outmost regions [43]; no such gradient has yet been measured in tidal tails [131].

The presence of cold dust in tidal tails has been first disclosed on far-IR images obtained with the ISO satellite [243]. Dust continuum emission in collisional debris has later-on been mapped by Spitzer [29, 206] and more recently by Herschel. Furthermore the star-forming regions along tidal tails also exhibit mid-infrared emission features associated to polycyclic aromatic hydrocarbon (PAH) grains [29, 109].

How did tidal tails acquire their metal-enriched components? Local stellar feedback during in-situ star formation episodes contribute to the metal production. However the onset of star-formation in collisional debris is likely too recent and the star-formation rates too small (see below) to explain the measured abundances in heavy elements. Another hypothesis is a global enrichment of the interstellar/intergalactic medium by stellar superwinds or enhanced AGN activity in the core of the merging galaxies. Nuclear outflows might eject metal-enriched matter (in particular dust) up to large distances, as observed for instance in M82 (System 1 in Figs. 9.7 and 9.10). Alternatively, radial gas mixing during galactic collisions might account for the lack of metallicity gradients in tidal tails and the presence of dust at large galactocentric distances, as recently shown by numerical simulations of mergers [192].

9.5 Structure Formation in Tidal Tails

9.5.1 Star Formation

As mentioned in the previous section, tails contain all the necessary ingredients for the onset of star-formation, in particular molecular gas and dust, and indeed young stars are often observed in collisional debris.

A census of star-forming regions in tidal tails has been carried out using a variety of tracers, such as the ultraviolet [29, 208], H_α [34, 227] or mid-infrared emission [30, 206]. These tracers may be combined to further constrain the star formation history (see composite image on Fig. 9.10). Star-forming regions in collisional debris may consist of extremely compact and tiny knots with star formation rate (SFR) as low as 0.001 $M_\odot$/yr (see examples at the tip of the tails of systems 3, 6 and 8 in Fig. 9.10) or giant complexes with SFR reaching 0.1 $M_\odot$/yr (see systems 5, 7 in Fig. 9.10).

A few studies have detailed the star-formation process in tidal tails, from the observational and theoretical point of view (e.g. [77]). Tidal objects are a priori a special environment simultaneously characterized by (a) the same local chemical conditions as in spiral galaxies (ISM composition, metallicity), (b) the lack of an underlying massive stellar disk, like dwarf irregular and low surface brightness galaxies, (c) the kinematical conditions typical of mergers, i.e. an enhanced gas turbulence and possibly shocks. Does then star-formation in collisional debris obey the rules that prevail (a) in regular massive disks, (b) in low-metallicity dwarfs, characterized by a low star-formation efficiency (SFE, the ratio between the star-formation rate and molecular gas content), (c) in the central regions of mergers where deviations from the so-called Kennicut-Schmidt relation (a correlation between the star-formation rate per unit area and the gas surface density) have been measured [57]? The SFE estimated in several tidal objects favors the first hypothesis: its value is close to that usually measured in galactic disks [31, 41].

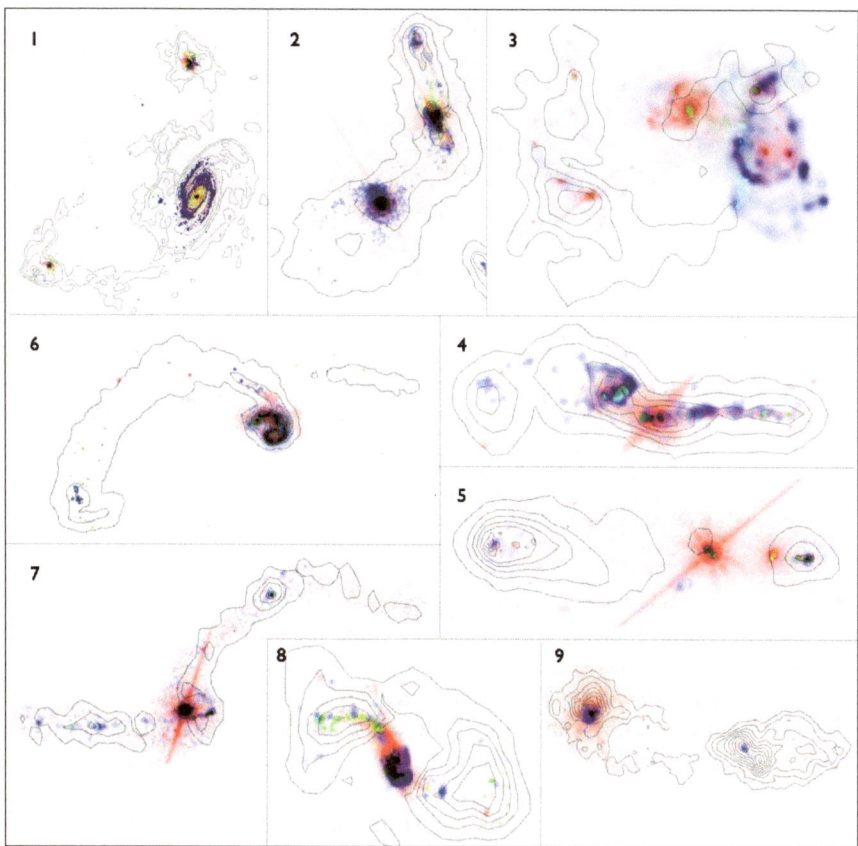

Fig. 9.10 Star formation in a sequence of merging galaxies. The displayed systems are the same as in Fig. 9.7. The contours of the H I 21 cm emission are superimposed on a composite image combining light emission from three tracers of on-going or recent star-formation: the ultraviolet (*blue*), the H_α line emission (*green*) and the mid-infrared (*red*). The most active star-forming regions belong to the so-called tidal dwarf galaxies

Therefore, with respect to star-formation, tidal tails do not appear as exotic objects. The properties of the pre-enriched interstellar medium inherited from their parent galaxies govern their star-formation capabilities rather than the violent episode at their origin or the large-scale (intergalactic) environment in which they now evolve.

9.5.2 Star Cluster Formation

Galaxy mergers do not only enhance star-formation. The increase of the gas pressure during mergers triggers the formation of star clusters as well. The Hubble Space

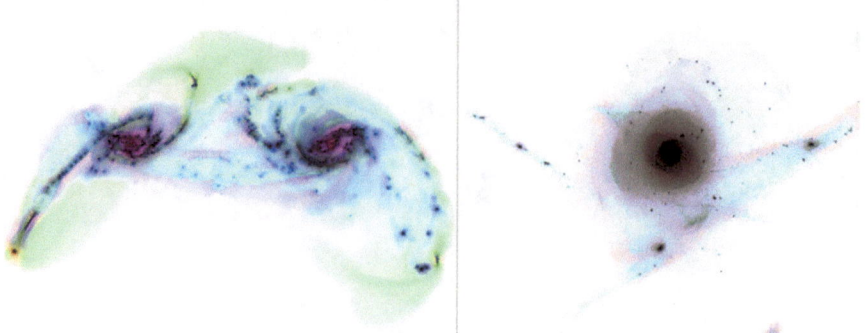

Fig. 9.11 Formation of stellar structures in high-resolution numerical simulations of major mergers. *Left*: after the first pericentric passage, with the hydrodynamical AMR code RAMSES [220]. *Right*: at the merger stage, with a sticky-particle code [37]. On the electronic version of this figure, gas is rendered in *green*, young stars in *blue* and old stars in *magenta/brown*. Both models show the formation of stellar objects (rendered in *yellow/white*): compact knots with properties similar as Super Star Clusters, or more massive, extended structures resembling Tidal Dwarf Galaxies

Telescope has revealed the presence of a large population of young Super Star Clusters (SSCs) in nearby merging systems, including along tidal tails (see [197], for a review). The most massive of them are believed to evolve into globular clusters (GCs), thus making mergers a possible origin of GCs. Numerical simulations at high resolution support this hypothesis [37]. Figure 9.11 presents two different models that were able to form SSCs. Globally, the cluster formation rate follows the star-formation rate. The infant mortality of SSCs less than 10 Myr after their formation appears however to be very high. SSCs in particular suffer from sudden gas loss due to feedback effects that alter their dynamical stability. There are special locations in merging systems, where local compressive tidal modes might contribute to (at least partially) protect them and increase their life-time [186]. Large volumes (up to 10 kpc wide) of compressive modes have been located in the tidal tails of major mergers, with an intensity comparable to that found in the central regions. But the lower gas density and turbulence in such an environment do not seem to particularly favor the formation of SSCs in tails. Note that attempts to connect hydrodynamical simulations of star-cluster formation with a semi-analytical formalism of tidal shocks have recently been made [133].

9.5.3 Formation of Tidal Dwarf Galaxies

Tidal tails host the most massive structures that may be born during galaxy mergers: the Tidal Dwarf Galaxies (TDGs). As indicated by their name, TDGs have the mass of classical dwarf galaxies, i.e. above 10^8 $M_\odot$. They have originally been detected on optical images as prominent and generally blue (thus star-forming) condensations at the end of tidal tails. Follow-up radio observations revealed that they were associated with massive HI clouds (see [68], for a recent review on TDGs). Detailed kinematical studies of the ionized, HI or molecular gas indicate that TDGs are gravitational bound entities that are kinematically decoupled from their parent galaxies. They exhibit velocity curves that are typical of rotating objects. In practice, the kinematical study of tidal tails suffers from strong projection effects: tidal tails are highly curved filaments; when seen edge-on, several components of the tail may be projected along the same line of sight. This creates an artificial velocity gradient that may be mistaken with a genuine rotation curve. Projection effects are especially critical near the end of tidal tails where most TDGs are precisely located [34].

Numerical simulations have provided clues on the formation mechanism of tidal dwarf galaxies (see examples in Fig. 9.11). Several scenarios have been proposed:

- growth of condensations born following local gravitational instabilities in the stellar component [18] or in the gaseous component [238],[6]
- multiple mergers of super star clusters [82],
- formation and survival of massive star clusters thanks to the fully compressive mode of tidal forces ([186], see above),
- formation of massive gas clouds in the outskirts of colliding disks, following the increased gas turbulence, that become Jeans-unstable and collapse once in the intergalactic medium [77],
- accumulation and collapse of massive gaseous condensations at the end of the tidal tails, following a top-down scenario [73].

In the context of this review, we detail here the latter scenario as it grants to the shape of tidal forces a key role in the formation of TDGs. In the potential well of disk galaxies, constrained by extended massive dark matter halo (see Sect. 9.6.3), the tidal field carries away the outer material, while keeping its high column density—the radial excursions are constant, as illustrated in Fig. 9.12. Gas may pile up at the tip of tidal tails before self-gravity takes over and the clouds fragment and collapse. Toy models show that the local shape of the tidal field plays the key role in structuring tidal tails and enabling the formation of TDGs.

The presentation of the long term evolution and survival of TDGs is behind the scope of this review. Details on the predictions of numerical simulations and observations of old TDGs may be found in Duc et al. [68].

9.6 Tidal Structures as Probes of Galaxy Evolution

Tidal tails, and more generally the fine structures that surround galaxies (stellar streams, rings, bridges, shells) are among the least ambiguous signposts of galaxy evolution. Indeed, whereas other galactic properties such as the presence of spiral

---

[6]Wetzstein et al. [238] claimed that the clumps formed in N-body models that do not include gas are numerical artifacts.

Fig. 9.12 The effect of tidal
forces on the potential well
corresponding to an extended
dark matter halo. Amplitude
of the radial excursions of
matter as a function of the
initial radius in a numerical
model made of concentric
annuli. Above a certain
distance, it becomes constant,
enabling an accumulation of
gas in tidal tails, the seed of
tidal dwarf galaxies. Adapted
from Duc et al. [73]

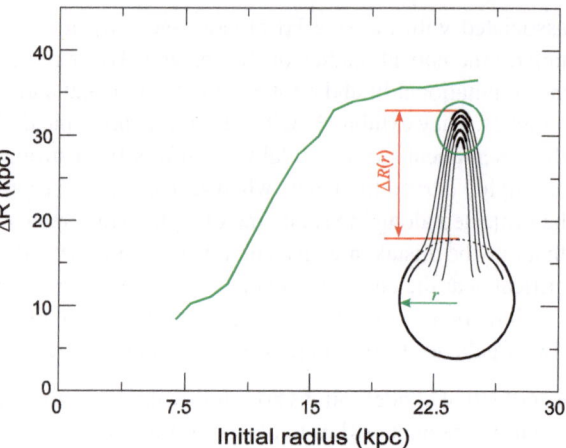

structures, bars, warps or even starbursts, may be accounted for by secular and internal evolution, the formation of stellar filaments can only be explained by a past collision between galaxies. Numerical cosmological simulations predict the formation of many such structures (see among many others [124, 176]). However, their census and interpretation face a number of issues.

- Fine structures are faint, with typically optical surface brightness fainter than 26 mag arcsec^{-2}, HI gas column densities below 10^{19} cm^{-2}, and thus difficult to observe. Nevertheless, the current generation of optical surveys as well as deep blind HI surveys have now the required sensitivity to detect a significant fraction of the large number of fine-structures predicted by numerical models of galaxy evolution.

- The properties of the fine structures depend on the properties of the parent galaxies: a wet merger (a collision involving gas-rich galaxies) will generate gaseous streams; stars from hot stellar systems (early-type galaxies) will not make long tidal tails. Prograde encounters produce more narrow tidal tails. Conversely, by studying the shape and inner characteristics of collisional debris, one may learn about the properties of their ancestors.

- Fine structures may be short lived. It takes a few 100 Myr to form a long tidal tail and a similar time to destroy them: tidal material may be dispersed or fall back at a rate that depends on the distance to the parent galaxy, from a few hundred Myr [53] up to a few Gyr [105]. Conversely the discovery of a tidal tail around an object might provide an age estimate of the last major merger event. With the support of a numerical model of the collision, one may even reconstruct the history of the collision (or at least have a model consistent with it) and predict its future.

- Fine structures are fragile and quickly react to their environment. For instance, in clusters of galaxies, tidal tails appear more diffuse as the interaction with the additional potential well of the cluster will accelerate the evaporation of their stars. As a consequence tidal tails should be less visible in dense environments and larger stellar halos are expected there, which is indeed observed [158].

Such issues might be addressed by combining predictions from numerical simulations and observations. We present below examples of the use of tidal structures as probes of galaxy evolution and the mass assembly of galaxies.

9.6.1 Determining the Merger Rate Evolution with Tidal Tails

Early deep observations with the Hubble Space Telescope revealed that distant galaxies ($z > 1$) seemed to be much more morphologically perturbed than local galaxies [1, 93, 97], supporting the idea that a smaller, denser and younger Universe favored galaxy-galaxy collisions. Since then, many studies based on deep surveys, such as the one illustrated in Fig. 9.13, have tried to quantify the evolution of the merger rate as a function of time, without in fact reaching a consensual value. A variety of methods have been used, based on:

- the census of close galaxy pairs (e.g. [127, 138]). The method assumes that galaxies observed in pairs are physically linked and doomed to merge.
- the identification of perturbed kinematics using Integral Field Spectroscopy. The method has been recently used as part of the IMAGES [246], MASSIV [55], SINS [203] and AMAZE/LSD [94] surveys at redshifts of 0.6, 1.3, 2 and 3 respectively. This method is very time consuming and may only be applied to limited samples.
- the census of morphologically perturbed galaxies showing for instance anisotropies in their stellar distribution (e.g. [54]). This requires a reliable algorithm to automatically measure the degree of perturbation.
- the direct detection of tidal tails (e.g. [44]), which, as argued earlier, is likely the most direct technique.

However a few remarks need to be made at this stage: first, the most massive component of tidal tails formed in major mergers is by far the atomic hydrogen. As mentioned before, HI surveys might disclose collisional debris that are hardly visible in the optical. Unfortunately, the current technology and antennas sensitivity limit the detection of the 21 cm emission line to redshifts less than 0.3. In the more distant Universe, tidal tails may only be observed through the emission of their stars. Intrinsic dimming with redshift as well as band shifting make them less and less visible and bias surveys in favor of UV emitting, star forming structures. Other difficulties arise at high redshift. The gas fraction of galaxies was higher and their gaseous disks more unstable. Prominent star forming condensations formed in the disk may be mistaken with either multiple nuclei of merging galaxies or even condensations within tidal tails [79]. Among these "clumpy" galaxies, only a fraction of them (for instance the so-called "tadpoles" systems) may be genuine interacting systems [78]. One usual hypothesis when counting the number of tidally perturbed systems is that disk-disk collisions at low and high redshift produce similar external structures. However if the colliding progenitors are the gas-rich clumpy disks

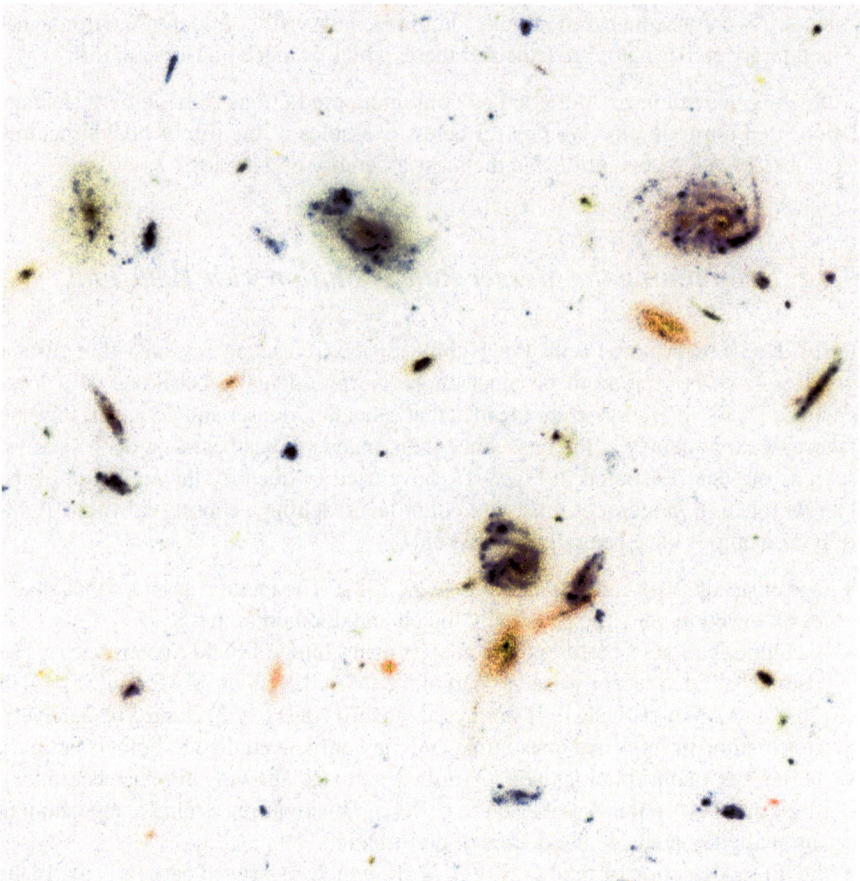

Fig. 9.13 The Hubble Ultra Deep Field, showing several tidally perturbed distant galaxies. The evolution of the fraction of collisions as a function of redshift is the subject of strong debates. Credit: NASA/ESA

mentioned above, the mutual interaction between their clumps (which have masses comparable to that of dwarf galaxies) might prevent the formation of tidal tails [38]. Thus, when trying to measure the evolution of the past merger rate by looking at the level of tidal perturbations, one should keep in mind that distant tidally interacting galaxies might differ from those observed in the Local Universe.

A last word of caution: when comparing the merger rate at low and high redshift, it is assumed that the fraction of galaxies involved in a tidal interaction is well known in the nearby Universe (and considered not to exceed a few percent, see [162]). However even tidal tails from past major mergers might have been missed at $z = 0$ because of their low surface brightness. Indeed, the extremely deep mapping of the Andromeda region, in the Local Group, has revealed an extremely faint stellar bridge between M31 and M33 [151], suggesting that the two spirals are involved in a tidal collision. Prominent tidal tails of very low-surface brightness were also recently discovered around apparently relaxed massive ellipticals [75]. An example is shown on the top panel of Fig. 9.8. Such observations indicate that, in the local Universe, the fraction of tidally interacting galaxies is likely underestimated: serious issues plague the determination of the merger rate even at low redshift.

9.6.2 Determining the Mass Assembly History of Galaxies with Tidal Tails

The existence of a tidal tail unambiguously establishes the occurrence of merger event in the past history of the host galaxy. Therefore, the census of the collisional debris might, in principle, constrain the recent mass assembly of nearby galaxies. Now that surface brightness limits of unprecedented depth may be reached with the current generation of optical cameras with large-field of view, this method of galactic archeology might be very powerful. However, it faces a number of issues:

- The frequency of tidal features, and degree of tidal perturbation (which one would like to link to other properties of the parent galaxies to constrain their mode of for-mation), is difficult to quantify. Tidal indexes proportional to the degree of mor-phological asymmetries have been introduced (e.g. [216]); most often however, more subjective "fine structure" indexes determined by eye are used to classify merging pairs or more evolved systems [198, 199].
- Not all collisions and mass accretion events produce tidal tails. The method is biased against mergers involving hot, pressure supported, galaxies. Indeed tidal forces most efficiently act on (rotating) disks. This indirectly means that tidal tails trace wet mergers rather than dry ones.
- Tidal features fade with time, either because they fall back onto their progenitors or evaporate into the intergalactic medium. Their detectability, and the ability to trace back past merging events, strongly depend on the surface brightness limit achieved by the observations.
- The destruction rate of tidal features depends on the environment. Dense environ-ments such as galaxy clusters contribute to erase collisional debris [216]. Tidal tails may also be destroyed during successive merger events.

As a consequence, it might be difficult to probe collisions older than a few Gyrs.

9.6.3 Constraining the Distribution of Dark Matter with Tidal Tails

Tidal tails might not only tell us about the baryonic content of their parent galaxies and how it reacted to the environment; they are as well insightful to constrain the structure and distribution of the most massive component of galaxies: dark matter (DM). Rotation curves of galaxies reveal how much gravitational matter is located within the radius at which velocities are measured but do not constrain the extent and 3D shape of the dark matter halo. The halo of CDM models is very extended, at least 10 times the optical radius. The rotation curve cannot be easily probed at these large distances. Tidal tails produced during major mergers have however sizes that can exceed 100 kpc, reaching the outskirts of the dark matter halos: tails are thus a priori a convenient tool to probe the structure of cosmological halos. Numerical simulations have been used to study the effect of the size of the DM halo on the shape of tidal tails. Apparently contradictory results have been obtained, claiming or not a dependence with the halo mass, size, concentration or spin [66, 67, 157, 212].

The shape of the DM halo, its triaxiality and presence of sub-halos might be probed by smaller, thiner tidal tails from minor mergers that wrap around galaxies. Those found around the Milky Way, such as the Sagittarius stream, are the target of numerous studies (e.g. [98, 150, 177, 230]).

While no direct correlation between the size of the DM halo and the size of tidal tails has yet been established, the internal structure of tidal tails might be connected to the DM extent. Bournaud et al. [33] argued that the massive condensations at the tip of tidal tails, associated with TDGs, cannot be formed if the halo of the parent galaxy is truncated. Duc et al. [73] provided a toy model showing that in the case of a truncated halo, the tidal material is stretched along the tidal tails, preventing its collapse and the formation of massive sub-structures. When the halo is large enough, this stretching does no longer occur beyond a certain distance, and apparent massive condensations near the tip of the tail might form TDGs (see Fig. 9.12). The observation of TDGs is thus consistent with the extended dark matter halos predicted by the CDM theory.

If large DM halos seem to be required to form TDGs and shape the inner structures of tidal tails, tails should themselves not contain large quantities of dark matter. Indeed the current picture of DM makes them collisionless particles distributed in a hot halo on which tidal forces have little impact. The tidal material originates from the disk, which is predicted to contain almost no DM. In practice, the DM content of tidal tails is difficult to probe. However in some special circumstances, it may be measured using the traditional method of rotation curves. Tidal dwarfs are gravitationally bound systems; their DM content may thus simply be derived determining their dynamical mass and subtracting it from the luminous one (consisting of HI, H_2, stars and dust). This exercise has been carried out for a few systems ([35, 74], Belles et al. [23]). Even if the error bars are large, these measurements yield reliable dynamical to luminous mass ratios of 2–3. Assuming that the CDM theory is correct, one should conclude that TDGs (and thus the galactic disks) contain non-conventional dark matter, likely traditional baryonic matter which has not yet been detected by existing surveys. A possible candidate is molecular gas not accounted for by CO observations. The observations of dust in the far infrared by the Planck satellite supports the hypothesis of an unseen, dark, component in the gaseous disk of galaxies, which might contribute to the global budget of the missing baryons in the Local Universe [181]. Alternatively, CDM might be wrong, as claimed by several groups who push for modified gravity. Modified Newtonian Dynamics (MOND) has retrieved the rotation curves of galaxies, including TDGs, without the need of a dark matter halo [88, 159]. Numerical simulations of galactic collisions in the

MOND framework have been carried out: they also reproduce the long tidal tails made with classical Newtonian dynamics [224]. The main difference is the absence of dynamical friction during the collision, which contributes to extend the time scale of the collision, and decrease the probability of a final coalescence.

9.7 Conclusions

It is an undeniable fact that tidal forces and the formation of tidal tails are overall a second-order process in galaxy evolution. The fraction of stars expelled in the inter-galactic medium is low, at most a few percent in major mergers. The fraction of gas is more important, but the bulk of the gaseous reservoir is funneled into the central regions. Collisional debris may host star-forming regions, but their contribution to the total star formation rate is minimal. Clearly, most of the activity occurs in the more central and nuclear regions where starbursts and/or AGN fueling are triggered. However, one of the aims of the present review is to emphasize the idea that tidal debris can provide insightful information about the properties of galaxies, the same way as garbage in trash cans tells us much about the way of life of their owners.

The presence of tidal features is an unambiguous proof that a major/minor merger occurred in the recent past, and that at least one of the colliding galaxies had a stellar and/or gaseous disk. The converse is not true though, as not all collisions produce prominent tidal features. Determining when the merger took place is less straight-forward. However, numerical simulations done in cosmological context will soon be able to constrain the survival time of collisional debris and thus give predic-tions on their age. Comparisons between observations and simulations should then allow us to reconstruct the mass assembly of galaxies. The current generation of wide field-of-view cameras and the on-going extremely deep surveys of the nearby Universe detect numerous new tidal features of very low surface brightness, offer-ing interesting prospects to galactic archeology. At high redshift, the census of tidal perturbations is much more complex, not only because of dimming and band shift-ing issues, but also because distant galaxies are much more gas-rich and therefore are intrinsically irregular. This makes the separation between secular and external effects rather ambiguous.

Multi-wavelength surveys have revealed the presence in collisional debris of all the constituents of regular galaxies though with different proportions: young and old stars, atomic gas, molecular gas, even possibly dark gas, heavy elements, and dust. Star formation seems to proceed there in a similar way as in isolated spiral disks, despite the very different environment at large scale.

Tidal tails may in principle be used even to probe some fundamental aspects of physics, including, of course, the properties of tidal forces but also the laws of gravitation, as shown by recent experiments with modified gravity. The fact that tidal forces can be compressive and for instance contribute to the stability of star clusters whereas they are usually associated with destruction processes has only recently been understood. The shape of the tidal tensor explains why massive tidal dwarf galaxies may be formed only within an extended dark matter halo. A theoretical study on the nature and the role of tidal forces in galaxies remains largely to be done and might provide further surprises.

Acknowledgements First of all, we express our gratitude to the two main organizers of the school, Jean Souchay and Stéphane Mathis. We not only enjoyed the marvelous premises—Cargèse in Corsica—but also the very stimulating discussions that took place between experts in terrestrial, planetary, stellar and galactic tides. We are very grateful to Frédéric Bournaud for daily discussions on galaxy collisions and numerical simulations. Many thanks as well to Pierre-Emmanuel Belles for his careful reading of the manuscript. We finally wish to thank all our collaborators and colleagues for their crucial contributions to the various works presented here.

References

1. Abraham, R.G., Tanvir, N.R., Santiago, B.X., Ellis, R.S., Glazebrook, K., van den Bergh, S.: Mon. Not. R. Astron. Soc. **279**, L47 (1996)
2. Ambartsumian, V.A.: Astron. J. **66**, 536 (1961)
3. Ambartsumian, V.A.: Highlights Astron. **3**, 51 (1974)
4. Appleton, P.N., Struck-Marcell, C.: Fundam. Cosm. Phys. **16**, 111 (1996)
5. Arp, H.: Atlas of Peculiar Galaxies. California Inst. Technology, Pasadena (1966)
6. Arp, H.: Astrophys. J. **148**, 321 (1967)
7. Arp, H.: Publ. Astron. Soc. Pac. **80**, 129 (1968)
8. Arp, H.: Astron. Astrophys. **3**, 418 (1969)
9. Arp, H.C.: Ejection of small compact galaxies from larger galaxies. In: Evans, D.S., Wills, D., Wills, B.J. (ed.) External Galaxies and Quasi-Stellar Objects. IAU Symposium, vol. 44, p. 380 (1972)
10. Athanassoula, E., Bosma, A.: Annu. Rev. Astron. Astrophys. **23**, 147 (1985)
11. Bailin, J., Steinmetz, M.: Astrophys. Space Sci. **284**, 701 (2003)
12. Barnes, J.E.: Astrophys. J. **331**, 699 (1988)
13. Barnes, J.E.: Mon. Not. R. Astron. Soc. **333**, 481 (2002)
14. Barnes, J.E.: Mon. Not. R. Astron. Soc. **350**, 798 (2004)
15. Barnes, J.E.: In: Galaxy Mergers in an Evolving Universe. ASP Conf. Ser., in press. arXiv:1112.4186
16. Barnes, J.E., Hernquist, L.E.: Astrophys. J. **370**, L65 (1991)
17. Barnes, J.E., Hernquist, L.: Annu. Rev. Astron. Astrophys. **30**, 705 (1992)
18. Barnes, J.E., Hernquist, L.: Nature **360**, 715 (1992)
19. Barnes, J.E., Hibbard, J.E.: Astron. J. **137**, 3071 (2009)
20. Barnes, J., Hut, P.: Nature **324**, 446 (1986)
21. Bekki, K.: Mon. Not. R. Astron. Soc. **390**, L24 (2008)
22. Bekki, K., Koribalski, B.S., Kilborn, V.A.: Mon. Not. R. Astron. Soc. **363**, L21 (2005)
23. Belles, P.-E.: (2012, in prep.)
24. Belokurov, V., Evans, N.W., Irwin, M.J., Hewett, P.C., Wilkinson, M.I.: Astrophys. J. **637**, L29 (2006)
25. Belokurov, V., Zucker, D.B., Evans, N.W., Gilmore, G., Vidrih, S., et al.: Astrophys. J. **642**, L137 (2006)
26. Berczik, P., Hensler, G., Theis, C., Spurzem, R.: Astrophys. Space Sci. **284**, 865 (2003)
27. Bertin, G., Liseikina, T., Pegoraro, F.: Astron. Astrophys. **405**, 73 (2003)
28. Blecha, L., Cox, T.J., Loeb, A., Hernquist, L.: Mon. Not. R. Astron. Soc. **412**, 2154 (2011)
29. Boquien, M., Duc, P., Wu, Y., Charmandaris, V., Lisenfeld, U., Braine, J., Brinks, E., Iglesias-Páramo, J., Xu, C.K.: Astron. J. **137**, 4561 (2009)

30. Boquien, M., Duc, P.-A., Galliano, F., Braine, J., Lisenfeld, U., Charmandaris, V., Appleton, P.N.: Astron. J. **140**, 2124 (2010)
31. Boquien, M., Lisenfeld, U., Duc, P.-A., Braine, J., Bournaud, F., Brinks, E., Charmandaris, V.: Astron. Astrophys. **533**, 19 (2011)
32. Bournaud, F.: Star formation and structure formation in galaxy interactions and mergers. In: Smith, B., Higdon, J., Higdon, S., Bastian, N. (eds.) Galaxy Wars: Stellar Populations and Star Formation in Interacting Galaxies. Astronomical Society of the Pacific Conference Series, vol. 423, p. 177 (2010)
33. Bournaud, F., Duc, P.-A., Masset, F.: Astron. Astrophys. **411**, L469 (2003)
34. Bournaud, F., Duc, P.-A., Amram, P., Combes, F., Gach, J.-L.: Astron. Astrophys. **425**, 813 (2004)
35. Bournaud, F., Duc, P.-A., Brinks, E., Boquien, M., Amram, P., Lisenfeld, U., Koribalski, B.S., Walter, F., Charmandaris, V.: Science **316**, 1166 (2007)
36. Bournaud, F., Duc, P., Emsellem, E.: Mon. Not. R. Astron. Soc. **389**, L8 (2008)
37. Bournaud, F., Duc, P.-A., Emsellem, E.: Mon. Not. R. Astron. Soc. **389**, 8 (2008)
38. Bournaud, F., Chapon, D., Teyssier, R., Powell, L.C., Elmegreen, B.G., Elmegreen, D.M., Duc, P.-A., Contini, T., Epinat, B., Shapiro, K.L.: Astrophys. J. **730**, 4 (2011)
39. Bournaud, F., Jog, C.J., Combes, F.: Astron. Astrophys. **437**, 69 (2005)
40. Bournaud, F., Jog, C.J., Combes, F.: Astron. Astrophys. **476**, 1179 (2007)
41. Braine, J., Duc, P.-A., Lisenfeld, U., Charmandaris, V., Vallejo, O., Leon, S., Brinks, E.: Astron. Astrophys. **378**, 51 (2001)
42. Braine, J., Lisenfeld, U., Duc, P.-A., Leon, S.: Nature **403**, 6772 (2000)
43. Bresolin, F., Ryan-Weber, E., Kennicutt, R.C., Goddard, Q.: Astrophys. J. **695**, 580 (2009)
44. Bridge, C.R., Carlberg, R.G., Sullivan, M.: Astrophys. J. **709**, 1067 (2010)
45. Briggs, F.H.: Intergalactic HI clouds. In: Duc, P.-A., Braine, J., Brinks, E. (eds.) Recycling Intergalactic and Interstellar Matter. IAU Symposium, vol. 217, p. 26 (2004)
46. Burbidge, E.M., Burbidge, G.R.: Astrophys. J. **130**, 23 (1959)
47. Burbidge, E.M., Burbidge, G.R., Hoyle, F.: Astrophys. J. **138**, 873 (1963)
48. Cappellari, M., Emsellem, E., Krajnović, D., McDermid, R.M., Scott, N., et al.: Mon. Not. R. Astron. Soc. **413**, 813 (2011)
49. Chung, A., van Gorkom, J.H., Kenney, J.D.P., Vollmer, B.: Astrophys. J. **659**, L115 (2007)
50. Cole, S., Helly, J., Frenk, C.S., Parkinson, H.: Mon. Not. R. Astron. Soc. **383**, 546 (2008)
51. Combes, F., Gerin, M.: Astron. Astrophys. **150**, 327 (1985)
52. Connors, T.W., Kawata, D., Gibson, B.K.: Mon. Not. R. Astron. Soc. **371**, 108 (2006)
53. Conselice, C.J.: Mon. Not. R. Astron. Soc. **399**, L16 (2009)
54. Conselice, C.J., Bershady, M.A., Dickinson, M., Papovich, C.: Astron. J. **126**, 1183 (2003)
55. Contini, T., Garilli, B., Le Fèvre, O., Kissler-Patig, M., Amram, P., Epinat, B., Moultaka, J., Paioro, L., Queyrel, J., Tasca, L., Tresse, L., Vergani, D., López-Sanjuan, C., Perez-Montero, E.: MASSIV: Mass Assemby Survey with SINFONI in VVDS. I. Survey description and global properties of the $0.9 < z < 1.8$ galaxy sample. Astron. Astrophys. **539**, A91 (2012). doi:10.1051/0004-6361/201117541. arXiv:1111.3631
56. Cox, T.J., Jonsson, P., Somerville, R.S., Primack, J.R., Dekel, A.: Mon. Not. R. Astron. Soc. **384**, 386 (2008)
57. Daddi, E., Elbaz, D., Walter, F., Bournaud, F., Salmi, F., Carilli, C., Dannerbauer, H., Dickinson, M., Monaco, P., Riechers, D.: Astrophys. J. **714**, L118 (2010)
58. Danziger, I.J., Schuster, H.: Astron. Astrophys. **34**, 301 (1974)
59. Debuhr, J., Quataert, E., Ma, C.-P.: Mon. Not. R. Astron. Soc. **412**, 1341 (2011)
60. Dehnen, W.: Astrophys. J. **536**, L39 (2000)
61. Dekel, A., Devor, J., Hetzroni, G.: Mon. Not. R. Astron. Soc. **341**, 326 (2003)
62. Dekel, A., Birnboim, Y., Engel, G., Freundlich, J., Goerdt, T., Mumcuoglu, M., Neistein, E., Pichon, C., Teyssier, R., Zinger, E.: Nature **457**, 451 (2009)
63. Di Matteo, P., Combes, F., Melchior, A., Semelin, B.: Astron. Astrophys. **468**, 61 (2007)
64. Dobbs, C.L., Theis, C., Pringle, J.E., Bate, M.R.: Mon. Not. R. Astron. Soc. **403**, 625 (2010)

65. D'Onghia, E., Vogelsberger, M., Faucher-Giguere, C.-A., Hernquist, L.: Astrophys. J. **725**, 353 (2010)
66. Dubinski, J., Mihos, J.C., Hernquist, L.: Astrophys. J. **462**, 576 (1996)
67. Dubinski, J., Mihos, J.C., Hernquist, L.: Astrophys. J. **526**, 607 (1999)
68. Duc, P.-A.: Birth, Life and Survival of Tidal Dwarf Galaxies. In: Papaderos, P., Recchi, S., Hensler, G. (eds.) Dwarf Galaxies: Keys to Galaxy Formation and Evolution, p. 305 (2012). arXiv:1101.4834
69. Duc, P.-A., Bournaud, F.: Astrophys. J. **673**, 787 (2008)
70. Duc, P.-A., Brinks, E., Wink, J.E., Mirabel, I.F.: Astron. Astrophys. **326**, 537 (1997)
71. Duc, P., Brinks, E., Springel, V., Pichardo, B., Weilbacher, P., Mirabel, I.F.: Astron. J. **120**, 1238 (2000)
72. Duc, P.-A., Brinks, E., Springel, V., Pichardo, B., Weilbacher, P., Mirabel, I.F.: Astron. J. **120**, 1238 (2000)
73. Duc, P.-A., Bournaud, F., Masset, F.: Astron. Astrophys. **427**, 803 (2004)
74. Duc, P.-A., Braine, J., Lisenfeld, U., Brinks, E., Boquien, M.: Astron. Astrophys. **475**, 187 (2007)
75. Duc, P.-A., Cuillandre, J.-C., Serra, P., Michel-Dansac, L., Ferriere, E., et al.: Mon. Not. R. Astron. Soc. **417**, 863 (2011)
76. Duncan, J.C.: Astrophys. J. **57**, 137 (1923)
77. Elmegreen, B.G., Kaufman, M., Thomasson, M.: Astrophys. J. **412**, 90 (1993)
78. Elmegreen, D.M., Elmegreen, B.G., Ferguson, T., Mullan, B.: Astrophys. J. **663**, 734 (2007)
79. Elmegreen, D.M., Elmegreen, B.G., Marcus, M.T., Shahinyan, K., Yau, A., Petersen, M.: Astrophys. J. **701**, 306 (2009)
80. Eneev, T.M., Kozlov, N.N., Sunyaev, R.A.: Astron. Astrophys. **22**, 41 (1973)
81. Faber, S.M.e.a.: Astrophys. J. **665**, 265 (2007)
82. Fellhauer, M., Kroupa, P.: Mon. Not. R. Astron. Soc. **330**, 642 (2002)
83. Fellhauer, M., Evans, N.W., Belokurov, V., Wilkinson, M.I., Gilmore, G.: Mon. Not. R. Astron. Soc. **380**, 749 (2007)
84. Ferguson, A.M.N., Gallagher, J.S., Wyse, R.F.G.: Astron. J. **116**, 673 (1998)
85. Ferrarese, L., Côté, P., Cuillandre, J.-C., Gwyn, S.D.J., Peng, E.W., MacArthur, L.A., Duc, P.-A., Boselli, A., Mei, S., Erben, T., McConnachie, A.W., Durrell, P.R., Mihos, J.C., Jordán, A., Lançon, A., Puzia, T.H., Emsellem, E., Balogh, M.L., Blakeslee, J.P., van Waerbeke, L., Gavazzi, R., Vollmer, B., Kavelaars, J.J., Woods, D., Ball, N.M., Boissier, S., Courteau, S., Ferriere, E., Gavazzi, G., Hildebrandt, H., Hudelot, P., Huertas-Company, M., Liu, C., McLaughlin, D., Mellier, Y., Milkeraitis, M., Schade, D., Balkowski, C., Bournaud, F., Carlberg, R.G., Chapman, S.C., Hoekstra, H., Peng, C., Sawicki, M., Simard, L., Taylor, J.E., Tully, R.B., van Driel, W., Wilson, C.D., Burdullis, T., Mahoney, B., Manset, N.: The Next Generation Virgo Cluster Survey (NGVS). I. Introduction to the survey. Astrophys. J., Suppl. Ser. **200**, 4 (2012). doi:10.1088/0067-0049/200/1/4
86. Freeman, K.C., de Vaucouleurs, G.: Astrophys. J. **194**, 569 (1974)
87. Fryxell, B., Olson, K., Ricker, P., Timmes, F.X., Zingale, M., Lamb, D.Q., MacNeice, P., Rosner, R., Truran, J.W., Tufo, H.: Astrophys. J. Suppl. Ser. **131**, 273 (2000)
88. Gentile, G., Famaey, B., Combes, F., Kroupa, P., Zhao, H.S., Tiret, O.: Astron. Astrophys. **472**, L25 (2007)
89. Genzel, R., Tacconi, L.J., Rigopoulou, D., Lutz, D., Tecza, M.: Astrophys. J. **563**, 527 (2001)
90. Gerhard, O.E.: Mon. Not. R. Astron. Soc. **197**, 179 (1981)
91. Gershberg, R.E.: Sov. Astron. **9**, 259 (1965)
92. Gingold, R.A., Monaghan, J.J.: Mon. Not. R. Astron. Soc. **181**, 375 (1977)
93. Glazebrook, K., Ellis, R., Santiago, B., Griffiths, R.: Mon. Not. R. Astron. Soc. **275**, L19 (1995)
94. Gnerucci, A., Marconi, A., Cresci, G., Maiolino, R., Mannucci, F., Calura, F., et al.: Astron. Astrophys. **528**, 88 (2011)
95. Gold, T.: Nature **164**, 1006 (1949)
96. González-García, A.C., van Albada, T.S.: Mon. Not. R. Astron. Soc. **361**, 1030 (2005)

97. Griffiths, R.E., Casertano, S., Ratnatunga, K.U., Neuschaefer, L.W., Ellis, R.S., et al.: Astrophys. J. **435**, L19 (1994)
98. Helmi, A.: Mon. Not. R. Astron. Soc. **351**, 643 (2004)
99. Hernquist, L.: Astrophys. J. Suppl. Ser. **64**, 715 (1987)
100. Hernquist, L.: Astrophys. J. **400**, 460 (1992)
101. Hernquist, L.: Astrophys. J. Suppl. Ser. **86**, 389 (1993)
102. Hernquist, L., Katz, N.: Astrophys. J. Suppl. Ser. **70**, 419 (1989)
103. Hernquist, L., Mihos, J.C.: Astrophys. J. **448**, 41 (1995)
104. Hibbard, J.E.: PhD thesis. Columbia University (1995)
105. Hibbard, J.E., Mihos, J.C.: Astron. J. **110**, 140 (1995)
106. Hibbard, J.E., van Gorkom, J.H.: Astron. J. **111**, 655 (1996)
107. Hibbard, J.E., Yun, M.S.: Astron. J. **118**, 162 (1999)
108. Hibbard, J.E., van der Hulst, J.M., Barnes, J.E., Rich, R.M.: Astron. J. **122**, 2969 (2001)
109. Higdon, S.J., Higdon, J.L., Marshall, J.: Astrophys. J. **640**, 768 (2006)
110. Hockney, R.W., Eastwood, J.W.: Computer Simulation Using Particles. Hilger, Bristol (1988)
111. Holmberg, E.: Ann. Obs. Lund **6**, 1 (1937)
112. Holmberg, E.: Astrophys. J. **94**, 385 (1941)
113. Holmberg, E.: Ark. Astron. **5**, 305 (1969)
114. Hopkins, P.F., Somerville, R.S., Hernquist, L., Cox, T.J., Robertson, B., Li, Y.: Astrophys. J. **652**, 864 (2006)
115. Hopkins, P.F., Cox, T.J., Younger, J.D., Hernquist, L.: Astrophys. J. **691**, 1168 (2009)
116. Howard, S., Keel, W.C., Byrd, G., Burkey, J.: Astrophys. J. **417**, 502 (1993)
117. Huang, S., Carlberg, R.G.: Astrophys. J. **480**, 503 (1997)
118. Hubble, E.P.: Astrophys. J. **69**, 103 (1929)
119. Hwang, J.-S., Struck, C., Renaud, F., Appleton, P.N.: Models of Stephan's quintet: hydrodynamic constraints on the group's evolution. Mon. Not. R. Astron. Soc. **419**, 1780–1794 (2012). doi:10.1111/j.1365-2966.2011.19847.x
120. Ibata, R., Irwin, M., Lewis, G., Ferguson, A.M.N., Tanvir, N.: Nature **412**, 49 (2001)
121. Ibata, R., Lewis, G.F., Irwin, M., Totten, E., Quinn, T.: Astrophys. J. **551**, 294 (2001)
122. Jiang, I., Binney, J.: Mon. Not. R. Astron. Soc. **303**, L7 (1999)
123. Johnston, K.V., Zhao, H., Spergel, D.N., Hernquist, L.: Astrophys. J. **512**, L109 (1999)
124. Johnston, K.V., Bullock, J.S., Sharma, S., Font, A., Robertson, B.E., Leitner, S.N.: Astrophys. J. **689**, 936 (2008)
125. Kapferer, W., Knapp, A., Schindler, S., Kimeswenger, S., van Kampen, E.: Astron. Astrophys. **438**, 87 (2005)
126. Karl, S.J., Naab, T., Johansson, P.H., Kotarba, H., Boily, C.M., Renaud, F., Theis, C.: Astrophys. J. **715**, L88 (2010)
127. Kartaltepe, J.S., Sanders, D.B., Scoville, N.Z., Calzetti, D., Capak, P., Koekemoer, A., Mobasher, B., Murayama, T., Salvato, M., Sasaki, S.S., Taniguchi, Y.: Astrophys. J. Suppl. Ser. **172**, 320 (2007)
128. Keenan, P.C.: Astrophys. J. **81**, 355 (1935)
129. Keenan, D.W., Innanen, K.A.: Astron. J. **80**, 290 (1975)
130. Kent, B.R., Giovanelli, R., Haynes, M.P., Saintonge, A., Stierwalt, S., Balonek, T., Brosch, N., Catinella, B., Koopmann, R.A., Momjian, E., Spekkens, K.: Astrophys. J. **665**, L15 (2007)
131. Kewley, L.J., Rupke, D., Zahid, H.J., Geller, M.J., Barton, E.J.: Astrophys. J. **721**, L48 (2010)
132. Kim, J., Wise, J.H., Abel, T.: Astrophys. J. **694**, L123 (2009)
133. Kruijssen, J.M.D., Pelupessy, F.I., Lamers, H.J.G.L.M., Portegies Zwart, S.F., Bastian, N., Icke, V.: Formation versus destruction: the evolution of the star cluster population in galaxy mergers. Mon. Not. R. Astron. Soc. **421**, 1927–1941 (2012). doi:10.1111/j.1365-2966.2012.20322.x

134. Küpper, A.H.W., Kroupa, P., Baumgardt, H., Heggie, D.C.: Mon. Not. R. Astron. Soc. **401**, 105 (2010)
135. Lagos, C.D.P., Cora, S.A., Padilla, N.D.: Mon. Not. R. Astron. Soc. **388**, 587 (2008)
136. Larson, R.B., Tinsley, B.M.: Astrophys. J. **219**, 46 (1978)
137. Lauberts, A.: Astron. Astrophys. **33**, 231 (1974)
138. Le Fèvre, O., Abraham, R., Lilly, S.J., Ellis, R.S., Brinchmann, J., Schade, D., Tresse, L., Colless, M., Crampton, D., Glazebrook, K., Hammer, F., Broadhurst, T.: Mon. Not. R. Astron. Soc. **311**, 565 (2000)
139. Lindblad, P.O.: Stockh. Obs. Ann. **21**, 4 (1960)
140. Lindblad, P.O.: Sov. Astron. **5**, 376 (1961)
141. Lisenfeld, U., Braine, J., Duc, P.-A., Leon, S., Charmandaris, V., Brinks, E.: Astron. Astrophys. **394**, 823 (2002)
142. Lisenfeld, U., Braine, J., Duc, P.-A., Brinks, E., Charmandaris, V., Leon, S.: Astron. Astrophys. **426**, 471 (2004)
143. Lisenfeld, U., Mundell, C.G., Schinnerer, E., Appleton, P.N., Allsopp, J.: Astrophys. J. **685**, 181 (2008)
144. Lonsdale, C.J., Persson, S.E., Matthews, K.: Astrophys. J. **287**, 95 (1984)
145. Lotz, J.M., Jonsson, P., Cox, T.J., Primack, J.R.: Mon. Not. R. Astron. Soc. **404**, 590 (2010)
146. Lucy, L.B.: Astron. J. **82**, 1013 (1977)
147. Lynds, R., Toomre, A.: Astrophys. J. **209**, 382 (1976)
148. Martínez-Delgado, D., Gabany, R.J., Crawford, K., Zibetti, S., Majewski, S.R.: Astron. J. **140**, 962 (2010)
149. Mayer, L., Governato, F., Colpi, M., Moore, B., Quinn, T., Wadsley, J., Stadel, J., Lake, G.: Astrophys. J. **559**, 754 (2001)
150. Mayer, L., Moore, B., Quinn, T., Governato, F., Stadel, J.: Mon. Not. R. Astron. Soc. **336**, 119 (2002)
151. McConnachie, A.W., Irwin, M.J., Ibata, R.A., Dubinski, J., Widrow, L.M., Martin, N.F., Côté, P.: Nature **461**, 66 (2009)
152. Mihos, J.C.: Astrophys. J. **550**, 94 (2001)
153. Mihos, J.C., Hernquist, L.: Astrophys. J. **464**, 641 (1996)
154. Mihos, J.C., Richstone, D.O., Bothun, G.D.: Astrophys. J. **377**, 72 (1991)
155. Mihos, J.C., Richstone, D.O., Bothun, G.D.: Astrophys. J. **400**, 153 (1992)
156. Mihos, J.C., Bothun, G.D., Richstone, D.O.: Astrophys. J. **418**, 82 (1993)
157. Mihos, J.C., Dubinski, J., Hernquist, L.: Astrophys. J. **494**, 183 (1998)
158. Mihos, J.C., Harding, P., Feldmeier, J., Morrison, H.: Astrophys. J. **631**, 41 (2005)
159. Milgrom, M.: Astrophys. J. **667**, L45 (2007)
160. Minchin, R., Davies, J., Disney, M., Grossi, M., Sabatini, S., Boyce, P., Garcia, D., Impey, C., Jordan, C., Lang, R., Marble, A., Roberts, S., van Driel, W.: Astrophys. J. **670**, 1056 (2007)
161. Mirabel, I.F., Lutz, D., Maza, J.: Astron. Astrophys. **243**, 367 (1991)
162. Miskolczi, A., Bomans, D.J., Dettmar, R.-J.: Astron. Astrophys. **536**, 66 (2011)
163. Moore, B., Davis, M.: Mon. Not. R. Astron. Soc. **270**, 209 (1994)
164. Mouhcine, M., Ibata, R.: Mon. Not. R. Astron. Soc. **399**, 737 (2009)
165. Naab, T., Burkert, A.: Astrophys. J. **597**, 893 (2003)
166. Naab, T., Jesseit, R., Burkert, A.: Mon. Not. R. Astron. Soc. **372**, 839 (2006)
167. Namboodiri, P.M.S., Kochhar, R.K.: Bull. Astron. Soc. India **13**, 363 (1985)
168. Narayanan, D., Cox, T.J., Kelly, B., Davé, R., Hernquist, L., Di Matteo, T., Hopkins, P.F., Kulesa, C., Robertson, B., Walker, C.K.: Astrophys. J. Suppl. Ser. **176**, 331 (2008)
169. Negroponte, J., White, S.D.M.: Mon. Not. R. Astron. Soc. **205**, 1009 (1983)
170. Neistein, E., Dekel, A.: Mon. Not. R. Astron. Soc. **383**, 615 (2008)
171. Nidever, D.L., Majewski, S.R., Butler Burton, W., Nigra, L.: Astrophys. J. **723**, 1618 (2010)
172. Noguchi, M.: Astron. Astrophys. **203**, 259 (1988)
173. Noguchi, M., Ishibashi, S.: Mon. Not. R. Astron. Soc. **219**, 305 (1986)
174. Olson, K.M., Kwan, J.: Astrophys. J. **361**, 426 (1990)

175. O'Shea, B.W., Bryan, G., Bordner, J., Norman, M.L., Abel, T., Harkness, R., Kritsuk, A.: Introducing Enzo, an AMR cosmology application. In: Plewa, T., Linde, T., Weirs, V.G. (eds.) Adaptive Mesh Refinement—Theory and Applications. Lecture Notes in Computational Science and Engineering. Springer, Berlin (2004)
176. Peirani, S., Crockett, R.M., Geen, S., Khochfar, S., Kaviraj, S., Silk, J.: Mon. Not. R. Astron. Soc. **405**, 2327 (2010)
177. Peñarrubia, J., Benson, A.J., Martínez-Delgado, D., Rix, H.W.: Astrophys. J. **645**, 240 (2006)
178. Peñarrubia, J., Navarro, J.F., McConnachie, A.W., Martin, N.F.: Astrophys. J. **698**, 222 (2009)
179. Pfleiderer, J.: Z. Astrophys. **58**, 12 (1963)
180. Pikel'Ner, S.B.: Sov. Astron. **9**, 408 (1965)
181. Planck Collaboration, Ade, P.A.R., Aghanim, N., Arnaud, M., Ashdown, M., Aumont, J., Baccigalupi, C., Balbi, A., Banday, A.J., Barreiro, R.B., et al.: Planck early results. XIX. All-sky temperature and dust optical depth from Planck and IRAS. Constraints on the "dark gas" in our Galaxy. Astron. Astrophys. **536**, A19 (2011). doi:10.1051/0004-6361/201116479. arXiv:1101.2029
182. Quinn, P.J.: Astrophys. J. **279**, 596 (1984)
183. Read, J.I., Wilkinson, M.I., Evans, N.W., Gilmore, G., Kleyna, J.T.: Mon. Not. R. Astron. Soc. **366**, 429 (2006)
184. Renaud, F.: PhD thesis. Universities of Vienna and Strasbourg (2010)
185. Renaud, F., Boily, C.M., Fleck, J., Naab, T., Theis, C.: Mon. Not. R. Astron. Soc. **391**, L98 (2008)
186. Renaud, F., Boily, C.M., Naab, T., Theis, C.: Astrophys. J. **706**, 67 (2009)
187. Renaud, F., Appleton, P.N., Xu, C.K.: Astrophys. J. **724**, 80 (2010)
188. Renaud, F., Gieles, M., Boily, C.M.: Mon. Not. R. Astron. Soc. **418**, 759 (2011)
189. Revaz, Y., Pfenniger, D.: N-body simulations of warped galaxies. In: Hibbard, J.E., Rupen, M., van Gorkom, J.H. (eds.) Gas and Galaxy Evolution. Astronomical Society of the Pacific Conference Series, vol. 240, p. 278 (2001)
190. Revaz, Y., Pfenniger, D.: Astron. Astrophys. **425**, 67 (2004)
191. Robertson, B., Bullock, J.S., Cox, T.J., Di Matteo, T., Hernquist, L., Springel, V., Yoshida, N.: Astrophys. J. **645**, 986 (2006)
192. Rupke, D.S.N., Kewley, L.J., Barnes, J.E.: Astrophys. J. **710**, L156 (2010)
193. Sanders, D.B., Mirabel, I.F.: Annu. Rev. Astron. Astrophys. **34**, 749 (1996)
194. Schombert, J.M., Wallin, J.F., Struck-Marcell, C.: Astron. J. **99**, 497 (1990)
195. Schweizer, F.: Astrophys. J. **252**, 455 (1982)
196. Schweizer, F.: Observational evidence for interactions and mergers. In: Kennicutt Jr., R.C., Schweizer, F., Barnes, J.E., Friedli, D., Martinet, L., Pfenniger, D. (eds.) Galaxies: Interactions and Induced Star Formation, Saas-Fee Advanced Course 26, p. 105 (1998)
197. Schweizer, F.: Globular Cluster Formation in Mergers. In: Richtler, T., Larsen, S. (eds.) Globular Clusters—Guides to Galaxies. Springer, New York (2006)
198. Schweizer, F., Seitzer, P.: Astron. J. **104**, 1039 (1992)
199. Schweizer, F., Seitzer, P., Faber, S.M., Burstein, D., Dalle Ore, C.M., Gonzalez, J.J.: Astrophys. J. **364**, L33 (1990)
200. Seguin, P., Dupraz, C.: Astron. Astrophys. **310**, 757 (1996)
201. Semelin, B., Combes, F.: Astron. Astrophys. **388**, 826 (2002)
202. Sérsic, J.L.: Bull. Astron. Inst. Czechoslov. **19**, 105 (1968)
203. Shapiro, K.L., Genzel, R., Förster Schreiber, N.M., Tacconi, L.J., Bouché, N., Cresci, G., et al.: Astrophys. J. **682**, 231 (2008)
204. Sijacki, D., Springel, V., Haehnelt, M.G.: Mon. Not. R. Astron. Soc. **414**, 3656 (2011)
205. Smith, B.J., Struck, C.: Astron. J. **121**, 710 (2001)
206. Smith, B.J., Struck, C., Hancock, M., Appleton, P.N., Charmandaris, V., Reach, W.T.: Astron. J. **133**, 791 (2007)
207. Smith, B.J., Carver, D.C., Pfeiffer, P., Perkins, S., Barkanic, J., Fritts, S., Southerland, D., Manchikalapudi, D., Baker, M., Luckey, J., Franklin, C., Moffett, A., Struck, C.: The au-

tomatic galaxy collision software. In: Smith, B., Higdon, J., Higdon, S., Bastian, N. (eds.) Galaxy Wars: Stellar Populations and Star Formation in Interacting Galaxies. Astronomical Society of the Pacific Conference Series, vol. 423, p. 227 (2010)

208. Smith, B.J., Giroux, M.L., Struck, C., Hancock, M.: Astron. J. **139**, 1212 (2010)
209. Soifer, B.T., Rowan-Robinson, M., Houck, J.R., de Jong, T., Neugebauer, G., Aumann, H.H., Beichman, C.A., Boggess, N., Clegg, P.E., Emerson, J.P., Gillett, F.C., Habing, H.J., Hauser, M.G., Low, F.J., Miley, G., Young, E.: Astrophys. J. **278**, L71 (1984)
210. Springel, V.: Mon. Not. R. Astron. Soc. **312**, 859 (2000)
211. Springel, V., Hernquist, L.: Astrophys. J. **622**, L9 (2005)
212. Springel, V., White, S.D.M.: Mon. Not. R. Astron. Soc. **307**, 162 (1999)
213. Stewart, K.R., Bullock, J.S., Wechsler, R.H., Maller, A.H., Zentner, A.R.: Astrophys. J. **683**, 597 (2008)
214. Stockton, A.: Astrophys. J. **190**, L47 (1974)
215. Struck, C.: Phys. Rep. **321**, 1 (1999)
216. Tal, T., van Dokkum, P.G., Nelan, J., Bezanson, R.: Astron. J. **138**, 1417 (2009)
217. Tashpulatov, N.: Sov. Astron. **13**, 968 (1970)
218. Tashpulatov, N.: Sov. Astron. **14**, 227 (1970)
219. Teyssier, R.: Astron. Astrophys. **385**, 337 (2002)
220. Teyssier, R., Chapon, D., Bournaud, F.: Astrophys. J. **720**, L149 (2010)
221. Theis, C., Kohle, S.: Astron. Astrophys. **370**, 365 (2001)
222. Theis, C., Spinneker, C.: Astrophys. Space Sci. **284**, 495 (2003)
223. Theys, J.C., Spiegel, E.A.: Astrophys. J. **212**, 616 (1977)
224. Tiret, O., Combes, F.: Interacting galaxies with MOND. In: Funes, J.G., Corsini, E.M. (eds.) Formation and Evolution of Galaxy Disks (2007)
225. Toomre, A.: Mergers and some consequences. In: Tinsley, B.M., Larson, R.B. (eds.) Evolution of Galaxies and Stellar Populations, p. 401 (1977)
226. Toomre, A., Toomre, J.: Astrophys. J. **178**, 623 (1972)
227. Torres-Flores, S., Mendes de Oliveira, C., de Mello, D.F., Amram, P., Plana, H., Epinat, B., Iglesias-Páramo, J.: Astron. Astrophys. **507**, 723 (2009)
228. van Albada, T.S.: Mon. Not. R. Astron. Soc. **201**, 939 (1982)
229. van der Hulst, J.M.: Astron. Astrophys. **71**, 131 (1979)
230. Varghese, A., Ibata, R., Lewis, G.F.: Mon. Not. R. Astron. Soc. **417**, 198 (2011)
231. Vorontsov-Vel'Yaminov, B.: Interaction of multiple systems. In: McVittie, G.C. (ed.) Problems of Extra-Galactic Research. IAU Symposium, vol. 15, p. 194 (1962)
232. Vorontsov-Vel'Yaminov, B.A.: Sov. Astron. **8**, 649 (1965)
233. Vorontsov-Vel'Yaminov, B.A., Arkhipova, V.P.: Morphological catalogue of galaxies. Part 1. In: Morphological Catalogue of Galaxies, vol. 1 (1962)
234. Vorontsov-Vel'Yaminov, B.A., Arkhipova, V.P.: Morphological catalogue of galaxies. Part 2. In: Morphological Catalogue of Galaxies, vol. 2 (1964)
235. Wallin, J.F., Stuart, B.V.: Astrophys. J. **399**, 29 (1992)
236. Walter, F., Martin, C.L., Ott, J.: Astron. J. **132**, 2289 (2006)
237. Weilbacher, P.M., Duc, P.-A., Fritze-v. Alvensleben, U.: Astron. Astrophys. **397**, 545 (2003)
238. Wetzstein, M., Naab, T., Burkert, A.: Mon. Not. R. Astron. Soc. **375**, 805 (2007)
239. White, S.D.M.: Mon. Not. R. Astron. Soc. **184**, 185 (1978)
240. White, S.D.M.: Astrophys. J. **274**, 53 (1983)
241. Wild, P.: Publ. Astron. Soc. Pac. **65**, 202 (1953)
242. Williams, B.A., Yun, M.S., Verdes-Montenegro, L.: Astron. J. **123**, 2417 (2002)
243. Xu, C.K., Lu, N., Condon, J.J., Dopita, M., Tuffs, R.J.: Astrophys. J. **595**, 665 (2003)
244. Yabushita, S.: Mon. Not. R. Astron. Soc. **153**, 97 (1971)
245. Yabushita, S.: Mon. Not. R. Astron. Soc. **178**, 289 (1977)
246. Yang, Y., Flores, H., Hammer, F., Neichel, B., Puech, M., Nesvadba, N., Rawat, A., et al.: Astron. Astrophys. **477**, 789 (2008)
247. Yanny, B., Newberg, H.J., Grebel, E.K., Kent, S., Odenkirchen, M., Rockosi, C.M., Schlegel, D., Subbarao, M., Brinkmann, J., Fukugita, M., Ivezic, Ž., Lamb, D.Q., Schneider, D.P.,

York, D.G.: Astrophys. J. **588**, 824 (2003)
248. Younger, J.D., Fazio, G.G., Ashby, M.L.N., Civano, F., Gurwell, M.A., Huang, J., Iono, D., Peck, A.B., Petitpas, G.R., Scott, K.S., Wilner, D.J., Wilson, G.W., Yun, M.S.: Mon. Not. R. Astron. Soc. **407**, 1268 (2010)
249. Yun, M.S., Ho, P.T.P., Lo, K.Y.: Nature **372**, 530 (1994)
250. Zasov, A.V.: Sov. Astron. **11**, 785 (1968)
251. Zwicky, F.: Phys. Today **6**, 7 (1953)
252. Zwicky, F.: Ergeb. Exakten Naturwiss. **29**, 344 (1956)
253. Zwicky, F.: Publ. Astron. Soc. Pac. **74**, 70 (1962)
254. Zwicky, F.: Leafl., Astron. Soc. Pac. **9**, 17 (1963)
255. Zwicky, F., Herzog, E., Wild, P.: Catalogue of Galaxies and of Clusters of Galaxies, vol. I. California Institute of Technology, Pasadena (1961)
256. Zwicky, F., Herzog, E., Wild, P.: Catalogue of Galaxies and of Clusters of Galaxies, vol. 2. California Institute of Technology, Pasadena (1963)

Index

A

Action variables, 139
Adaptive mesh refinement, 334
Adiabatic, 219, 222
Airy's canal theory, 20
Amphidromic points, 104, 105
Amphidromic system, 104
Amplitudes of the tides, 49, 56
Andoyer variables, 138, 142
Anelasticity, 293
Angle variables, 139
Angular momentum, 146, 205, 210, 301–304,
    306, 311, 314, 317, 323
Angular velocity, 303, 305, 311, 320
Annual retrograde nutation, 156
Antennae galaxies, 346
Apparent angular diameter, 3
Atmospheric tides, 79
Axis of angular momentum, 142
Axis of figure, 119, 159
Axis of rotation, 117, 119

B

Bay of Fundy, 16
BCRS (Barycentric Celestial Reference
    System), 132, 162
Black hole, 30
Bradley, 115, 116
Bridges, 329

C

Cassini, 168, 188–190, 195
Chandler Wobble (CW), 155, 157, 159
CIO (Celestial Intermediate Origin), 132, 138
CIP (Celestial Intermediate Pole), 132, 137,
    163

Circular nutations, 153
Circularization, 302, 306, 310, 312–318, 320,
    321, 323
Clumpy disks, 357
Co-tidal line, 104
Compressive, 338, 339
Compressive tidal modes, 354
Conservation of mass, 68, 70
Core-accretion, 242
Core-mantle boundary (CMB), 160
Coriolis acceleration, 71
Coriolis force, 26
Corotation torque, 202, 205, 210, 220,
    223–225, 247
Crossed nutation effects, 151
Cryovolcanism, 192, 195

D

Daily amplitude of the tide, 16
D'Alembert, 116
Dark galaxies, 350
Dark matter, 359
Dark-matter halo, 333, 344
Darwin's expansions, 93
De Sitter precession, 149
Departure point, 140
Differential forces, 331, 338, 339, 343
Direct planetary effects, 148, 150
Dissipation, 302, 304–309, 312, 315, 316,
    320–323
Dissipation factor, 175, 179, 183, 184, 195
Disturbing potential, 143
Diurnal component, 103
Diurnal inequality, 49, 103, 107
Diurnal (or tesseral) waves, 98
Diurnal tide, 108
Doodson expansion, 92

J. Souchay et al. (eds.), *Tides in Astronomy and Astrophysics*,
Lecture Notes in Physics 861, DOI 10.1007/978-3-642-32961-6,
© Springer-Verlag Berlin Heidelberg 2013